General Principles of Biochemistry of the Elements

BIOCHEMISTRY OF THE ELEMENTS

Series Editor: Earl Frieden
Florida State University
Tallahassee, Florida

A Continuation Order Plan is available for this series. A continuation order will bring delivery of each new volume immediately upon publication. Volumes are billed only upon actual shipment. For further information please contact the publisher.

General Principles of Biochemistry of the Elements

Ei-Ichiro Ochiai

Juniata College
Huntingdon, Pennsylvania

PLENUM PRESS • NEW YORK AND LONDON

Library of Congress Cataloging in Publication Data

Ochiai, El-ichiro, date
  General principles of biochemistry of the elements.

  (Biochemistry of the elements; v. 7)
  Includes bibliographies and index.
  1. Chemical elements—Physiological effect. 2. Bioinorganic chemistry. I. Title. II.
Series. [DNLM: 1. Biochemistry. 2. Elements. QU 130 B6144 1980 v. 7]
QP531.O24 1987   574.19′24   87-20236
  ISBN 978-1-4684-5373-7   ISBN 978-1-4684-5371-3 (eBook)
  DOI 10.1007/978-1-4684-5371-3

© 1987 Plenum Press, New York
Softcover reprint of the hardcover 1st edition 1987
A Division of Plenum Publishing Corporation
233 Spring Street, New York, N.Y. 10013

To Dr. Shoji Makishima,
Professor Emeritus of the University of Tokyo

Preface

The present book might be regarded as a sequel to my previous work, *Bioinorganic Chemistry: An Introduction* (Allyn and Bacon, 1977). The latter is essentially a collection of chemical and physical data pertinent to an understanding of the biological functions of the various elements and the proteins dependent on them. The ten years since its publication have seen an enormous increase in research activity in this area, hence of research papers. A number of monographs and review series on specific topics have also appeared, including the volumes in the series of which the present volume is a part.

Nevertheless, a gap has developed between the flood of information available at a detailed level (papers and reviews) and a general description of the underlying principles of biofunctions of the elements as presently conceived. It is hoped that this book will help bridge this gap and at the same time provide an overview of the entire Biochemistry of the Elements series. Specifically, the work attempts to focus on "why" questions, especially, "Why has an element been chosen by organisms for a specific biofunction?" and "Why does an element behave the way it does in biological systems?" It therefore complements my 1977 book and, together with *Laboratory Introduction to Bio-Inorganic Chemistry* (E.-I. Ochiai and D. R. Williams, Macmillan, 1979), completes a trilogy on the topic of bioinorganic chemistry.

This book consists of five parts. Two chapters constitute Part I. Chapter 1 provides a global view of the biochemistry of the elements by examining their macroscopic spatial and temporal movements. Chapter 2 attempts to delineate the most basic principles pertaining to the questions posed above, a goal which necessitates a discussion of the fundamental properties of the elements. Part II (Chapters 3–7) deals with the chemical bases for the biological functions of specific elements in terms of several important reaction categories, again with emphasis on the basic questions. Part III (Chapters 8–10) treats the chemistry underlying spatial movement of the elements with respect to biological systems, i.e., their uptake and transfer. The final two parts deal with the biological aspects of elemental behavior: metabolism (Part IV; Chapters 11 and 12) and toxicity (Part V; Chapters 13–15). Here also the emphasis is on delineating underlying general principles. Regrettably, space constraints dictated the omission of a number of interesting topics, including electron transfer in proteins, carcinogenicity, and pharmacological effects.

This book was conceived and work on it begun shortly after the 1977 publication of its predecessor. The intervening years have brought a number of changes in the author's life, in addition to producing a mountain of new research papers and data. Most of the new data represent extensions or minor revisions of information collected in the 1977 book, but there have also been a number of significant new discoveries that necessitate a rethinking of certain basic concepts; examples include the [3Fe-xS] proteins and Ni-containing redox proteins. To the extent that such developments pertain to basic principles they have been incorporated in the text, but no discussion is provided of specific physical/chemical data for element-containing biological systems. Instead, the focus is restricted to more general principles. Because of the nature of the inquiry (i.e., seeking answers to "why" questions) the discussion tends at times to be rather speculative. The author makes no pretense of being an expert in the details of every single research area. Some may regard the treatment as biased in favor of certain topics or points of view, and the validity of the speculation will be determined only by further inquiry. Much of the merit of the book may lie in its attempt to offer a coherent view of the subject from the perspective of the author. It is hoped that new insight has been provided on some topics; if so, it should stimulate further debate and experimentation.

The language of the book is such that it will be understood by upper-level undergraduates as well as graduate students and experts in the subject area. Since topics like the interpretation of ESR, EXAFS, or Mössbauer spectral data have been avoided the treatment does not presuppose specialized knowledge of chemical physics. Thus it should be suitable as a textbook for both undergraduate and graduate courses, but it is recommended that the instructor supplement the book with relevant experimental data and their interpretation.

This manuscript, like most, underwent a number of revisions, especially linguistic polishing. The author is deeply indebted to his colleague, Dr. William E. Russey of Juniata College, for his efforts to improve the English of the entire manuscript at several early stages. At the final stage, Professor Earl Frieden of Florida State University, the editor of this series, gave the author invaluable suggestions for improvements, as well as comments and corrections applicable to the entire manuscript. I am very grateful to Professor Frieden for his help; any shortcomings that may remain are solely my responsibility.

It is a pleasure to be able to dedicate this book to my mentor, Dr. Shoji Makishima, Professor Emeritus of the University of Tokyo, whose lifelong influence on my work I gratefully acknowledge.

E.-I. Ochiai

Contents

Part II. Chemical Principles of the Biochemistry of the Elements

3. Oxidation–Reduction and Enzymes and Proteins 53

Part III. Chemical Principles of Transport of the Elements

8. Chemistry of Uptake—Thermodynamic and Kinetic Factors in Passive Transport

9. Ionophores, Channels, Transfer Proteins, and Storage Proteins

Part V. Biological Aspects II—Toxicity of and Defense against the Elements

Overview

I

Global Aspects of the
Biochemistry of the Elements

1.1 Introduction

Life is a product of chemical evolution that took place on the earth, and both life and the biosphere that encompasses it have evolved under the constraints this planet imposes. The biosphere could not exist in the absence of its surroundings—the atmosphere, the hydrophere, and the lithosphere. In other words, life—whether taken as an individual living cell or as the biosphere as a whole—is an open system, constantly interacting with its surroundings. The form that this interaction takes is the exchange of material and energy. Inorganic materials are processed and altered by organisms, and then are returned to the surroundings. In view of these interactions, it is not surprising that a large number of elements are not only processed by organisms but that as many as 30 of them are essential to organisms. Figure 1-1 summarizes the known biological functions of elements. Readers are referred to Frieden (1985a,b) for recent reviews of essential trace elements.

The essential elements are obviously processed by organisms, but other elements, nonessential or toxic, can also be processed because those elements do enter organisms, though inadvertently, and the organisms must be able to deal with them.

Autotrophic organisms, including algae, cyanobacteria, and green plants, absorb CO_2 and H_2O to produce carbohydrates and free O_2 with the aid of energy from the sun. The O_2 is returned to the atmosphere and hydrosphere. Heterotrophic organisms, including animals, utilize the carbohydrates produced by the autotrophs to generate energy, in the process producing CO_2 and H_2O, which are returned to the environment. In addition, some CO_2 is fixed as carbonate minerals. This brief sketch using as an example the cycling of carbon illustrates the nature of the interaction between the biosphere and its surroundings. The overall purpose of the

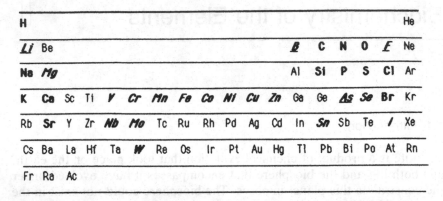

H																	He
Li	Be											*B*	C	N	O	*F*	Ne
Na	*Mg*											Al	Si	P	S	Cl	Ar
K	Ca	Sc	Ti	*V*	*Cr*	*Mn*	*Fe*	*Co*	*Ni*	*Cu*	*Zn*	Ga	Ge	*As*	*Se*	Br	Kr
Rb	Sr	Y	Zr	*Nb*	*Mo*	Tc	Ru	Rh	Pd	Ag	Cd	In	*Sn*	Sb	Te	*I*	Xe
Cs	Ba	La	Hf	Ta	*W*	Re	Os	Ir	Pt	Au	Hg	Tl	Pb	Bi	Po	At	Rn
Fr	Ra	Ac															

Figure 1-1. Biofunctions of elements. The elements in boldface are macronutrients, and used for constructing the body and for maintaining the body fluid (osmotic pressure); the elements in italic boldface are micronutrients (essential trace elements), and the majority of them function as catalysts (in enzymatic reactions); the elements underlined are "suspected" essential trace elements; all the other elements are not yet known to be essential to organisms.

present chapter is to discuss in some depth what is known about the chemical and geochemical aspects of the processing of elements by organisms.

1.2 The Distribution and Bioutilization of the Elements

Figure 1-2 shows the current relative abundance of major elements in the earth's crust and in seawater. As will be discussed shortly, the elemental composition of seawater has changed significantly over the lifetime of the earth. Nevertheless, the overall picture portrayed here may well approximate the situation throughout the earth's history, requiring only small modifications involving a few elements during particular areas.

Figure 1-2a compares terrestrial elemental abundances in units of gram atoms per ton of crust. As a rule of thumb, lighter elements are more abundant, as indicated by the dashed curve. Those elements known to be required in large quantities (macronutrient) are shown in boldface, whereas essential trace elements (micronutrients) are indicated by italic boldface. Underlined elements are suspected of being essential to certain organisms. The universally required elements are lighter ones, those lighter than zinc (atomic number = 30). Two elements that are heavier than zinc—molybdenum and selenium—seem to be fairly widely required. There are a few other essential heavy elements such as strontium, tungsten, and iodine, but these seem essential only to a limited number of organisms.

At any rate, we can safely say that the majority of the essential elements are light; also, they are relatively abundant. Organisms would be expected to use those elements that are most accessible—i.e., abundant—since otherwise they would have to expend extra energy to secure essential elements. If we define more or less arbitrarily the lower abundance limit for an element being essential to organisms as 0.1 gram atom per ton of the earth's crust, then the majority of the essential elements fall within this range, as seen in Fig. 1-2a. There are some peculiar elements,

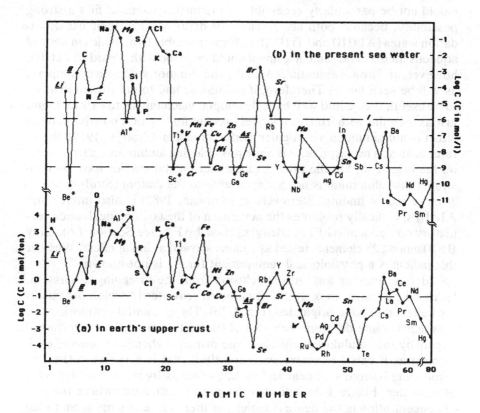

Figure 1-2. Distribution of elements in (a) the earth's crust (moles or gram atoms/ton), and (b) the present seawater (moles/liter). The elements in boldface are macronutrients, those in italic boldface are essential trace elements, and those underlined are "suspected" essential trace elements.

however, ones that are very abundant but are not known to be essential to any organisms. These are indicated by asterisks in the figure; examples include beryllium, aluminum, scandium, titanium, and gallium. Beryllium is relatively rare and any function it might perform could probably be carried out more easily or better by the more abundant (and chemically similar) magnesium. As a matter of fact, beryllium is very toxic.

Aluminum is one of the most abundant elements and yet it is apparently not used by organisms. The same can be said of titanium. At least two explanations can be suggested for these findings. One is that despite their abundance, these elements may not be sufficiently available for use by organisms. A second possibility is that these two elements simply have no useful role in organisms and hence are not utilized. That these elements would not be particularly accessible to organisms seems at first a strong possibility, because both are extensively hydrolyzed in their usual oxidation states [Al (III) and Ti (IV)], to form insoluble hydroxides in neutral aqueous media. The same argument could be made with regard to Fe(III), however, and iron is essential. (Actually the situation with iron is complex, as will be seen later.) Therefore, if organisms had found these elements to be useful, they could well have developed mechanisms to extract them, just as they did with iron. A relatively recent finding of a positive growth effect of aluminum on a particular marine organism (Stoffyn, 1979), therefore, comes as no surprise. It is quite possible that aluminum exerts effects on the organism. On the other hand, experimental studies have demonstrated that aluminum is not indispensable to the diatom (Stoffyn, 1979). A more recent finding (Sternweis and Gilman, 1982) is also interesting: Al(III) specifically promotes the activation of the guanine nucleotide regulatory component (G/F) of adenyl cyclase by F^- (see Section 7.6). Only Be(II) among 28 elements tested as cations showed a similar effect. Whether this indicates a physiological requirement for Al is not known.

Before drawing any firm conclusion from the foregoing discussion, let us take a closer look at the situation in seawater in which organisms are believed to have originated (Fig. 1-2b). The elemental composition of seawater is quite different from that of the earth's crust. It is determined largely by the solubility equilibria of the principal chemical forms of each element as it exists in seawater (currently its pH is 8.1) and in the corresponding bottom sediment and rock, as well as by the redox state (E_h) of seawater. Figure 1-2b indicates the present situation, where the unit of concentration in the figure is moles per liter. The elements seem to fall into well-defined (through perhaps arbitrary) categories. Those elements whose concentrations are higher than 10^{-5} M are referred to as macronutrients; they are required in large quantity by organisms and would thus need to be present at high concentration. Again there are exceptional elements—fluorine and bromine—whose essentiality has not been established. Most of the plausible functions of fluorine and bromine would be satisfied by chlorine, although fluorine may yet turn out to be essential, particularly since some aspects of its chemistry are so distinct from the other halogens.

A second significant concentration range can be defined rather arbitrarily as 10^{-6}–10^{-9} M; in this range fall most of those essential elements that are required in trace amounts. They are called micronutrients or trace essential elements. Aluminum and titanium again appear in this range, indicating that they would be accessible to organisms if there were reasons for them to be used.

Two notable differences occur in the elemental composition of seawater as compared to that of the earth's crust: molybdenum and iodine. Both are relatively rare in the crust, but occur as trace elements in seawater, as do the more common trace elements.

Cobalt, despite its low abundance (being located in the lowest portion of this range), seems to be fairly universally required, probably as a component of vitamin B_{12}. This implies that there is no satisfactory alternative to cobalt for the tasks it performs, but also that trace amounts will suffice.

The elemental composition given in Fig. 1-2 for the earth's crust and for seawater are only averages; some local elemental concentrations could differ significantly from those shown in the figure. The local abundance of a particular element may be one factor that renders that element essential to those organisms that inhabit that locale. In other words, organisms, confronted with an unusually high concentration of a given element, may have sought ways to incorporate and make use of it. This kind of argument could, for example, be applicable to the elements chromium, selenium, tin, and tungsten.

Once an organism had adopted a certain element for its use, then it would have had to develop reliable mechanisms to take up, store, excrete, and control the levels of that element. The control of level is crucial, because even essential elements may disrupt the functioning of an organism if present in excess. Nonessential elements pose a problem: they are potentially dangerous to organisms at any level. Most organisms have developed some mechanisms to enable them to reject and/or tolerate those elements that are abundant and may be absorbed inadvertently. Examples include aluminum and titanium. It is known that an enormous amount of aluminum succinate is deposited in some species of trees, including *Oites excelsa*, as well as some club moss. In the case of elements that are scarce and would hence be rarely encountered by organisms, it is unlikely that defense mechanisms would have developed. As a result, such elements might well be toxic to most organisms; this is true with the typical heavy metals, such as cadmium, mercury, and lead. Some organisms, however, undoubtedly would have come across heavy metals in the course of their evolution, and might thus have developed at least partial defense mechanisms against them. Selenium is a case of an element that is scarce and at the same time extremely toxic to most organisms, but has nonetheless found some use in certain organisms.

The difference in elemental composition between the earth's crust and seawater might be expected to reflect itself in the elemental requirements of terrestrial organisms versus marine organisms. For example, the elements vanadium and chromium, adjacent in the periodic table, differ considerably in their relative abundances in seawater and the earth's crust. Vanadium is considerably more abundant in seawater, compared to other similar transition elements. One would expect vanadium to be required by marine organisms rather than terrestrial organisms. This is indeed the case; vanadium has been found to be essential to some algae and to the Ascidiacea, a group of primitive marine organisms. It should be noted, however, that this argument does not exclude the possibility of use of vanadium by some terrestrial organisms, and it is quite possible that such use will be discovered.

Iodine, too, is comparatively more abundant in seawater than in the earth's crust. This fact is consistent with the pattern of its utilization by marine and terrestrial organisms. Marine organisms, especially brown algae, contain high levels of iodine; an absolute requirement for iodine has been established in brown algae such as *Ectocarpus fasciculatus*, *Petalonia fascia*, and *Polysiphonia urceolata* (Gauch, 1972). On the other hand, terrestrial organisms, including mammals, require only trace amounts of iodine.

The marine concentration of chromium, on the other hand, is unusually low; it is significantly more abundant in the earth's crust. Therefore, it is to be expected that those organisms that utilize chromium are likely to be terrestrial rather than marine. This, too, seems to be the case.

Selenium is known to be concentrated in special restricted locales such as arid terrains. Plants inhabiting such regions, particularly some species of *Astragalus*, have been found to contain high levels of selenium and apparently require it for proper growth. Selenium at very low levels is essential to a number of organisms, both prokaryotes and eukaryotes, though, as noted earlier, it is quite toxic to most organisms even at low levels. These facts are all in accord with its differential distribution and its general low abundance.

1.3 Biogeochemical Cycling of the Elements

1.3.1 Introduction

The supply of each of the elements on the planet is limited, and can be regarded as being confined within a rather narrow region of the earth, at least as far as organisms are concerned. This confine includes the atmosphere, hydrosphere, lithosphere, and biosphere. The atmosphere is a rather thin film around the earth and is presently composed mainly of nitrogen, oxygen, argon, neon, helium, carbon dioxide, and water vapor. The atmospheric composition has changed drastically since the time of the earth's formation. The major contributor to this change has been plants, starting with cyanobacteria around 2.5 billion years ago and followed later by green algae and then by land plants; this will be discussed later on. Oceans, lakes, and rivers constitute the hydrosphere. Water, of course, is its major component, although the hydrosphere also contains a large number of inorganic elements and compounds as minor components. The lithosphere is the solid part of the earth. Only the upper layers are significantly involved in the cycling of the elements, though the inner core plays a part in the process over the long term. The biosphere is the region that the organisms inhabit and includes the organisms themselves.

Each element on earth, viewed temporally, is constantly moving from one of these spheres to another, completing a cycle within a certain reasonably definite period of time. The rate and route of the cycling depend on the nature of the individual element, as well as on a number of other factors. In particular, it is important to recognize at the outset that the cycling of bioelements is greatly influenced by organisms within the biosphere.

It is possible to state a few general principles that apply to the cycling of virtually all the elements:

1. An element usually changes its chemical form as it moves from one medium (sphere) to another.
2. If a change in chemical form does not occur, that particular step of the cycle must be accomplished by purely physical means.
3. The conversion of an element from one chemical form to another must be effected by some processes: either a chemical process or a biological process.
4. There are usually one or two large pools of an element somewhere in the cycling process. Material residing in such a large pool will exchange only slowly with that in smaller and perhaps more readily

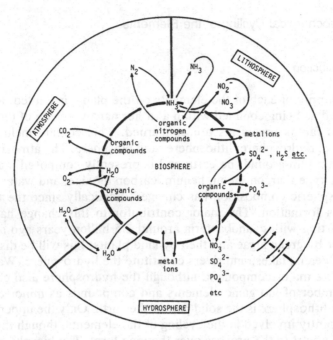

Figure 1-3. A generalized biogeochemical cycling of elements.

accessible pools. The transport of an element in and out of pools is a result of chemical and/or biological processes.

A few examples will illustrate these points. The cycling of water is carried out mostly by means of physical processes. Starting from the hydrosphere, water either moves into the atmosphere by evaporation or is absorbed into the biosphere by organisms; the latter translocation can be regarded as a biological process. Water absorbed by organisms is then either excreted or is transpired by plants. Water in the atmosphere precipitates and returns to the hydrosphere either directly or indirectly through the lithosphere. Calcium enters the hydrosphere from the lithosphere (calcium-containing rock) by erosion. Soluble Ca(II) is absorbed by organisms and becomes converted to such biocompounds as $CaCO_3$ in invertebrate shells, $Ca_3(PO_4)_2$ in vertebrate skeletons, and numerous calcium-containing proteins. The calcareous shells of dead invertebrates and other organisms (including coccolithophores) deposit on the sea bottom and eventually form sedimentary rock consisting largely of $CaCO_3$. This $CaCO_3$ rock serves as the "large pool" in the calcium cycle. It also serves as a large reserve in the carbon cycle, since it slowly releases CO_2, which reenters its cycle as outlined earlier. A generalized outline of cycles of the elements is shown in Fig. 1-3. The cycle of each individual element will be examined in more detail in ensuing sections of this chapter.

Within the biosphere, the elements circulate according to the so-called food chain. A generalized food chain can be described as follows. Photoautotrophs, algae and green plants, are eaten by herbivores. The herbivorous animals are then eaten by primary carnivorous animals, which are in turn consumed by secondary carnivores and so on. The bodies of dead plants and animals are decomposed by saprophytic scavengers such as bacteria, fungi, and some invertebrates (see Fig. 1-4).

In most cases the concentration of a given element in an organism is much higher than that in the environment. The phenomenon that leads to this result is often called "bioaccumulation," "bioconcentration," or "enrichment." Bioaccumulation is most prominent among aquatic photoautotrophs (algae) and is accomplished either by active uptake of particular elements, entailing distinct chemical mechanisms, or by their passive uptake, including simple adsorption. The ratio of the concentration of an element in an organism to that in its surroundings is called an "enrichment factor." Enrichment factors on the order of 10^4–10^5 are not uncommon, especially with regard to trace essential elements. Certain

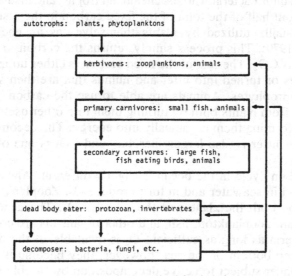

Figure 1-4. A generalized food chain.

elements accumulate at high concentration in those organisms high in the food chain ladder, a phenomenon known as "bioamplification." This is caused by a metabolic mechanism in which the excretion rate is lower than the uptake rate. An example is the very high concentration of DDT found in those sea birds and mammals that turn out to be the highest organisms in their respective food chains. The explanation for this finding is that DDT, being fat-soluble, is difficult for organisms to metabolize and hence it tends to remain unexcreted, and eventually is passed to a predator.

1.3.2 Cycles of Macronutrients

1.3.2.1 Carbon Cycle (Bolin, 1970)

CO_2 gas serves as a convenient starting point. At present, the average CO_2 content in the atmosphere is estimated to be about 320 ppm. Atmospheric CO_2 is in equilibrium with CO_2 in the biosphere. On land it is assimilated by green plants, which convert it into a number of organic compounds. In the hydrosphere, algae assimilate CO_2 and process an amount that is about the same order of magnitude (or even a little more) as that taken up by land plants. A minor proportion of CO_2 fixation is carried out by autotrophic microorganisms in soils; these include sulfur bacteria, nonsulfur bacteria, various chemoautotrophs, and methane-forming bacteria. About half of the total of the fixed carbon in the resulting compounds is usually utilized by plants themselves as an energy source (Woodwell, 1970). This process simply returns the carbon to the atmosphere again as CO_2. The rest of the fixed carbon is either utilized directly by herbivores or turned into litter and humus that are then slowly consumed by saprophytes. Animals are able to use the carbon compounds they acquire from plants both by turning them into other useful materials and by transforming them aerobically into energy. The decomposition of humus and/or litter is also largely aerobic, and CO_2 is one of the major end products.

The carbon cycle in the ocean is slightly different. Algae assimilate CO_2 dissolved in seawater and in turn produce O_2. Zooplankton and fish then consume both the algae and the released O_2. The dead bodies of phytoplankton, zooplankton, fish, and other organisms are consumed by aerobic bacteria as long as sufficient O_2 is available. If the dead bodies reach the deep bottom of the sea, however, they become isolated from O_2 and no longer subject to active decomposition by aerobic bacteria. In this and other sites of deposition that are rather stagnant and devoid of oxygen, anaerobic decomposition takes place. One group of anaerobic bacteria decompose organic compounds by reduction, thereby producing methane; examples are *Methanobacterium* and *Methanococcus*. These organisms also convert CO_2 or simple compounds such as acetic acid to methane, which can become part of the "natural gas" supply (a fossil fuel carbon reservoir). This particular pool is currently being extensively exploited, so that its CH_4 is being converted to an unusually high rate back to CO_2. There are also a few bacteria that can oxidize methane as a source of energy, e.g., *Pseudomonas* and *Methanomonas*. As a result of both natural factors and human intervention, it is estimated that the atmospheric CO_2 content has increased by about 0.7% in recent years.

A significant portion of the aqueous CO_2 in the hydrosphere combines with dissolved calcium to form calcium carbonate. Calcareous shells of organisms also contribute to calcium carbonate rock formation. This rock serves as a slowly acting carbon reservoir. Another possible source of CO_2 lies outside the formal cycle: volcanic activity. Various studies indicate that terrestrial plants annually produce organic compounds equivalent to 20 to 30 billion tons of carbon, and that the production in the sea is comparable (40 billion tons). The rate of carbon turnover has been estimated to be about 33,300 years per cycle. The total carbon reserve in carbonate and other sedimentary rocks (including fossil fuels) is believed to be aboout 2.5×10^{18} tons.

1.3.2.2 Oxygen Cycle (Cloud and Gibor, 1970)

Although oxygen constitutes about 20% of the atmosphere, most of the oxygen on earth exists in combined form as various oxides, silicates, carbonates, and sulfates; water, ferric oxides (rock), aluminosilicate (soil and rock), calcium carbonate (calcite), and calcium sulfate (gypsum) are a few important examples. The major source of molecular oxygen is water, which is decomposed by photoautotrophs and by ultraviolet radiation. The formation of O_2 from water is highly endothermic, which means that the reverse is highly exothermic. This thermodynamic factor is a major reason why O_2 is utilized as the ultimate oxidizing agent in the metabolism of organic compounds in organisms. The major pathway of the oxygen cycle is illustrated in Fig. 1-5. One cycle of this process has been estimated to take about 2000 years. Other potential oxygen sources, such as oxide, carbonate, and sulfate minerals, represent only very slowly exchanging reservoirs.

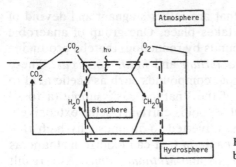

Figure 1-5. The biogeochemical cycle of carbon and oxygen.

1.3.2.3 Nitrogen Cycle (Delwiche, 1970)

The atmosphere is the largest reservoir of nitrogen. N_2, constituting about 80% of the atmosphere, is slowly converted into NH_3 by nitrogrenfixing organisms. These include many cyanobacteria, various anaerobic bacteria, and certain aerobic bacteria, either free-living or symbiotically associated (e.g., with the roots of higher plants). This process represents almost the only means by which gaseous nitrogen enters the active cycle. NH_3 thus formed either is incorporated into organic nitrogenous compounds or is oxidized to NO_2^- and then NO_3^- by microorganisms. The first step, $NH_3 \rightarrow NO_2^-$, is effected by *Nitrosomonas*, *Nitrococcus*, *Nitrosospira*, *Nitrosocystis*, and *Nitrosogloea*. These organisms, the so-called "chemoautotrophs," are present in most soils and use the energy released from the oxidation of NH_3 in order to produce carbohydrates. The important genus that oxidizes NO_2^- to NO_3^- is *Nitrobacter*, and it also uses the energy released by the oxidation.

A limited amount of nitrogen fixation is carried out by the chemical effect of radiation and lightning on the mixture of N_2 and O_2 in the atmosphere. This process yields nitrogen oxides, which are eventually turned into nitrite and nitrate. NO_2^- and NO_3^- are utilized by many organisms, which reduce them to either N_2O, N_2, or NH_3. Most organisms, especially autotrophs, have mechanisms permitting them to reduce NO_2^- and/or NO_3^-, and they assimilate the product NH_3. Other microorganisms, those known as facultative, use NO_2^- and/or NO_3^- as a terminal electron acceptor in place of the usual oxygen. These organisms are heterotrophic and do not require inorganic NH_3 as a source of nitrogen because they feed on ready-made organic nitrogenous compounds. Therefore, they do not need a means of reducing NO_2^- or NO_3^- all the way to NH_3. Moreover, the last step of this reduction, that to NH_3, seems to be highly energy-consuming. Thus, reduction in these organisms usually proceeds only as far as N_2O or N_2. An example of such an organism is *Thiobacillus*

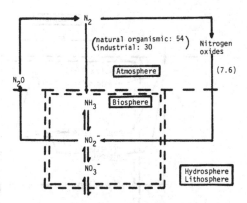

Figure 1-6. The biogeochemical cycle
of nitrogen.

denitrificans. N_2 produced by these species is simply lost to the atmosphere.

The most important organic nitrogen compounds in living organisms are the amino acids, the nucleotides and their derivatives. The metabolites from these compounds include NH_3, urea, and uric acid. NH_3 can be directly utilized by many organisms. Urea, uric acid, and other organic nitrogenous metabolites slowly decompose and eventually turn into simple inorganic compounds such as NO_3^-. The nitrogenous compounds in dead bodies of organisms are likewise decomposed aided by a swarm of microorganisms. Human activities are now resulting in the artificial synthesis of substantial amounts of nitrogenous compounds, especially ammonia and urea. The overall nitrogen cycle is summarized in Fig. 1-6.

1.3.2.4 Sulfur Cycle (Wetzel, 1975)

This cycle involves all of the major oxidation states of sulfur, i.e., $SO_4^{2-}(+VI)$, $SO_3^{2-}(+IV)$, $S_2O_3^{2-}(+III)$, $S(0)$, and $H_2S(-II)$. The oxidation–reduction reactions of key sulfur compounds are reviewed in Section 3.5. Sulfate is the major form of sulfur found in the hydrosphere and in the aerated portion of the lithosphere. SO_4^{2-} is absorbed and assimilated directly by organisms; they reduce it to SO_3^{2-} and then H_2S, the latter being incorporated into the sulfur-containing amino acids (cysteine, methionine) and into other organic sulfhydryl compounds such as glutathione, lipoic acid, and coenzyme A. SO_4^{2-} is also incorporated into organic compounds directly in the form of sulfate esters. Particularly important sulfate esters are those of simple sugars and their polymers (including the so-called mucopolysaccharides).

In addition to ordinary plants and animals, a great variety of minor

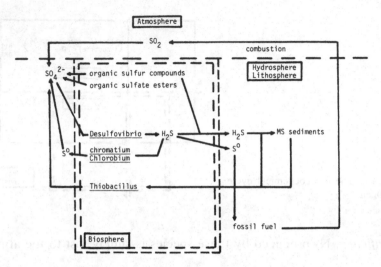

Figure 1-7. The biogeochemical cycle of sulfur.

microorganisms, the so-called sulfur bacteria, are involved in the geo-chemical cycle of sulfur. The reduction of SO_4^{2-}, for example, is carried out by *Desulfovibrio* species, anaerobes that utilize SO_4^{2-} as an electron acceptor. The oxidation of H_2S or $S(0)$ is accomplished by different groups of bacteria.. One group is autotrophic, and utilizes the electrons from H_2S or S ($S_2O_3^{2-}$ in some cases) to reduce CO_2. Primary subgroups are the purple sulfur bacteria (such as *Chromatium*) and the green sulfur bacteria (e.g., *Chlorobium*). The second major group is chemoautotrophic; mem-bers of this group oxidize H_2S or S using oxygen, and use the resulting energy to synthesize carbohydrates. Representative of this type are the *Thiobacillus* species, including *T. denitrificans*, *T. thiooxidans*, and *T. ferroxidans*. The last-mentioned bacterium can oxidize Fe(II) as well as sulfur.

Under anaerobic conditions (e.g., the deep sea bottom), H_2S is the major metabolite, as mentioned previously. As H_2S is released, it com-bines with metallic cations, forming the insoluble sulfides Cu_2S, FeS, and so on. These compounds are then attacked by *Thiobacillus* species and are slowly leached out. Sulfur is also a minor component of fossil fuel; there it arose from biological source material as a result of the action of anaerobic bacteria. Consumption of fossil fuels by man leads to a signif-icant contribution of SO_2 to the atmosphere and the environment. The overall sulfur cycle is outlined in Fig. 1-7.

1.3.2.5 Phosphorus Cycle

Whereas all the elements mentioned so far undergo oxidation–reduction during their biogeochemical cycles, at least the major portion of the cycle of phosphorus operates without oxidation–reduction. The most important natural form of phosphorus is phosphate, PO_4^{3-}. All of the important organic biocompounds containing phosphorus are phosphate esters; examples include ATP, ADP, AMP, other nucleotides (e.g., GTP, NAD) and their derivates, DNA and RNA, phospholipids, and the phosphate esters of carbohydrates, which are intermediates in metabolism. An important inorganic phosphorus compound in biological systems is calcium phosphate, the main constituent of bones and dentin. Microorganisms attack dead plant and animal bodies and at least partially decompose the organic phosphates, thereby releasing orthophosphate and smaller phosphate esters.

Most of the phosphorus in the hydrosphere, and especially in the lithosphere is bound up in organic compounds. One of the major phosphates in humic soil, for example, is phytic acid:

$$
\begin{array}{c}
H_2O_3PO \qquad OPO_3H_2 \\
OPO_3H_2 \\
OPO_3H_2 \\
H_2O_3PO \qquad OPO_3H_2 \\
OPO_3H_2
\end{array}
$$

Soil phytic acid is usually present in the form of its Ca(II) or Mg(II) salts. Orthophosphate forms insoluble compounds with many cations. Thus, orthophosphate released to the ithosphere is likely to be fixed as phosphate salts such as $Ca_3(PO_4)_2$ and $FePO_4$, which are solubilized by the action of bacteria. For example, certain bacteria produce inorganic or organic acids that convert the insoluble $Ca_3(PO_4)_2$ to the more soluble forms $Ca(H_2PO_4)_3$ or $Ca(HPO_4)_2$. Under anaerobic conditions, bacteria release H_2S, which then displaces the PO_4^{3-} group in such compounds as $FePO_4$.

The ultimate source of phosphate is phosphate-containing rocks, from which the phosphate is slowly leached out by mechanisms similar to those described above. Man is now introducing phosphate into the biosphere in the form of phosphate fertilizers obtained by mining phosphate rock and the guano derived from seabird excrement.

Oxidation–reduction of phosphorus is not entirely nonexistent in the biosphere. A number of microorganisms have been shown to use phosphite, PO_3^{3-}, as a sole source of phosphorus (Malacinski and Konetzka, 1967). This indicates that these organisms have some mechanism to oxidize PO_3^{3-} to PO_4^{3-}. There are also reports of microorganisms reducing

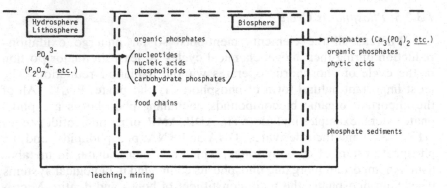

Figure 1-8. The biogeochemical cycle of phosphorus.

PO_3^{3-} to PH_3 (Iverson, 1968). The reduction potentials of these processes are fairly low; $E_0'(\text{pH } 7)$ is about -0.5 V for PO_4^{3-}/PO_3^{3-} and -0.75 V for PO_3^{3-}/PH_3. It appears, therefore, that any PO_4^{3-} reducing system would have to be so strongly reducing that its role would be limited to highly acidic environments where the reduction potentials increase. The biological cycle of phosphorus is summarized in Fig. 1-8, from which, however, the oxidation–reduction of phosphorus has been omitted.

1.3.2.6 Biogeochemical Cycles of Alkali and Alkaline Earth Metals

The common salts of sodium and potassium are quite soluble in water. Therefore, the geocycles of these elements include a facile leaching from ores containing these salts, so that the only complex feature of the cycle involves transport between the hydrosphere and the biosphere. Because of the differential distribution of Na(I) and K(I) in living cells and their surroundings, all organisms require mechanisms for pumping excess Na(I) out and K(I) in.

Mg(II) is similar to K(I) in terms of both the solubility of its salts and its distribution in and out of the cell. Because the magnesium concentration is usually higher inside a cell than outside, Mg(II) has to be actively absorbed by cells. Magnesium is required by all organisms, but the demand is especially high in green plants and algae, which need relatively large amounts of Mg(II) for their chlorophyll.

Magnesium in rocks is largely in the forms of $Mg_3(PO_4)_2$, $MgCO_3$, or magnesium silicates. The first two compounds are found in minerals but are also deposited in the bones or shells of organisms along with $Ca_3(PO_4)_2$ (bones) or $CaCO_3$ (shells). In soil, Mg(II) is present adsorbed on humic substances or clay minerals. An outline of the geochemical cycle of magnesium is as follows. Mg(II) leaches out of rocks and either enters the soil or passes into rivers and then the ocean. The Mg(II) is next picked up by microorganisms and plants in the soil or by algae and other organisms in the hydrosphere, after which it moves from one organism to another through the food chain. Mg(II) is never concentrated under normal conditions because all organisms seem to have mechanisms to control their Mg(II) levels. Finally, Mg(II) is either excreted or returned to the soil or the hydrosphere by the action of the usual organisms that decompose dead matter. It is conceivable that Mg(II) also returns to its large slowly exchanging pool (i.e., rocks), but the extent of this return seems to be rather small compared, for example, to the case of calcium. This is due to the higher solubility of most Mg(II) compounds.

The distribution of calcium in and out of cells is virtually opposite that of magnesium. While Mg(II) is important in intracellular fluid, Ca(II) is, as it were, an extracellular cation. It is not unimportant intracellularly, but its main functions in a quantitative sense are extracellular. As mentioned above, the bulk of the calcium in organisms is either in the form of $CaCO_3$, as in the shells of invertebrates and avian eggs, or in the form of $Ca_3(PO_4)$, as in bones and dentin in vertebrates. Shells, particularly those of marine organisms, pile up at the bottom of the sea and eventually turn into carbonate rocks. The white cliffs along the Dover Strait of the British Isles are made up of $CaCO_3$ deposits from ancient sea organisms. Organisms that do not require large amounts of calcium have been found to develop mechanisms to keep Ca(II) out because of the prevalent concentration gradient in the opposite direction, especially in the sea. In summary, the major pathway of the geochemical cycle of calcium takes the form: the hydrosphere→the biosphere→rocks (the lithosphere)→the hydrosphere.

Strontium is similar to calcium in its chemical properties. Sr(II) can, for example, be incorporated into shells and bones along with Ca(II). Indeed, natural shells and bones always contain Sr(II); its level is usually commensurate with its natural abundance relative to calcium. The radioactive isotope Sr-90, which is one of the fission products of uranium and is contained in the fall-out from nuclear explosions, can thus be incorporated into the bones of mammals and its radioactivity is thought to cause a serious damage to the blood-producing apparatus in the bone marrow. This Sr-90, unlike calcium, comes from the atmosphere in the form of fallout particles, which are readily adsorbed on the surfaces of vegetation. Grazing animals are, therefore, especially liable to be exposed to Sr-90. High Sr-90 concentrations have been found in the bone and flesh of reindeer and caribou on arctic tundra (Kulp et al., 1960). These are animals that feed on the matlike tundral vegetation comprised of sedges, lichens, and grasses, all of which are efficient fallout traps. Significant amounts of Sr-90 have also been found in humans as a result of ingestion of cow's milk.

1.3.3 Biogeochemical Cycles of Some Trace Elements

1.3.3.1 Cycle of Iron (Konetzka, 1977)

Iron is required by virtually all organisms. Iron metabolism in organisms is discussed in Chapters 9, 11, and 12. Several special microorganisms play important roles in the biogeochemical cycle of iron; these organisms are collectively called "iron bacteria." Examples include *Sphaerotilus natans*, *Leptothrix ochracea*, *Gallionella terruginea*, and *Siderocapsa treubii*. These bacteria oxidize Fe(II) to Fe(OH)$_3$, which is deposited in peculiar extracellular substances of structures (sheath or stalk). They apparently do not use the energy released by the oxidation of Fe(II). This energy, however, is used by *Thiobacillus ferroxidans*, which is a true iron-oxidizing bacterium. The mechanism of Fe(II) oxidation in *T. ferroxidans* is not well understood, but it is believed to occur in the cell envelope.

The microbial reduction of Fe(III) seems to be accomplished by practically every facultative or anaerobic heterotroph. Some microorganisms appear to be able to reduce Fe(III) using S^0 as the electron donor. One particularly interesting observation is that *Desulfovibrio* (sulfate-reducing

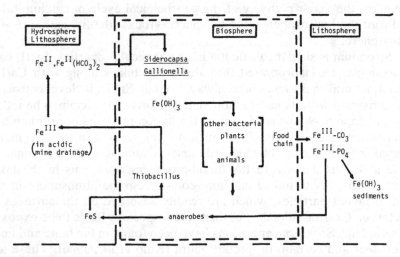

Figure 1-9. The biogeochemical cycle of iron.

bacterium) can, under certain conditions, produce magnetic FeS; these bacteria thus respond to a magnetic field (magnetotacticity). *Desulfovibrio* and *Desulfotomaculum* have also been shown to effectively extract trace amounts of metals from their culture media and, if the iron level is high in the media, they accumulate electron-dense particles (presumably FeS) within their cells. These organisms play an extremely important role in the transformation of macroquantities of iron.

Fe(III) precipitates from the hydrosphere as $Fe(OH)_3$, $FePO_4$, or $Fe_2(CO_3)_3$ and enters into sedimentary rock formation. Under anaerobic conditions, $Fe(OH)_3$ [and other Fe(III) compounds] is converted to FeS (or FeS_2). The FeS is then leached out by *Thiobacillus* species, particularly *T. ferroxidans*, and the iron again enters the hydrosphere. This process is especially active in areas of so-called "acid mine drainage." The biogeochemical cycle of iron is outlined in Fig. 1-9.

1.3.3.2 Cycle of Manganese (Konetzka, 1977)

The cycle of manganese is similar to that of iron. Mn(II) in soil and water is oxidized to Mn(IV) in the form of MnO_2 by a variety of microorganisms, including the usual soil bacteria and fungi. The *Sphaerotilus* and *Leptothrix* species mentioned above in connection with the oxidation of iron are known also to oxidize Mn(II) to MnO_2 and to accumulate MnO_2 in their sheaths or filaments. It has been found that *Leptothrix* secretes a protein that catalyzes the oxidation of Mn(II). The growth of *Sphaerotilus discophorus* has been shown to be markedly enhanced upon addition of $MnSO_4$ to the culture medium. This result suggests that the bacterium can utilize the energy released in the oxidation of Mn(II) to MnO_2. *Hyphomicrobia* (a budding bacterium) has been suggested to be responsible for manganese deposits found in freshwater pipelines. A similar bacterium, *Metallogenium*, is widely distributed in soil and water, and it also produces very thin filaments encrusted with MnO_2. This particular organism can develop only in symbiosis with a fungus.

There seem to be a number of microorganisms that can reduce MnO_2 to Mn(II) under anaerobic conditions. They appear to be able to utilize MnO_2 as the electron acceptor (in place of oxygen). It has been reported that some bacteria capable of reducing Mn(IV) can be isolated from ferromanganese nodules brought up from the Atlantic Ocean bottom. A protein designated as MnO_2 reductase was shown to be inducible using MnO_2 as the inducer.

Manganese is essential to virtually all organisms, especially to plants, which utilize it in photosynthesis. Microorganisms capable of reducing the insoluble MnO_2 to Mn(II) undoubtedly play a very important role in making manganese available to plants.

1.3.3.3 Cycles of Other Essential Trace Elements (Konetzka, 1977)

Little work has been done to elucidate the biogeochemical cycles of copper, zinc, cobalt, or molybdenum. There is, however, little doubt that *Thiobacillus* species, especially *T. ferroxidans*, play important roles in leaching metals out of a variety of metallic sulfide ores. *T. ferroxidans* has been shown to oxidize Cu_2S and Cu_2Se to CuS and Cu(II) + Se^0, respectively. The same bacterium can also oxidize sulfides in other ways as, for example, $ZnS \rightarrow ZnSO_4$, $CuS \rightarrow CuSO_4$, and $As_2S_3 \rightarrow As(III)$ or As(IV) + SO_4^{2-}.

1.3.4 Biogeochemical Cycles of Some Toxic Elements

Organisms have developed mechanisms to detoxify or dispose of some often-encountered toxic elements. Unfortunately, the detoxified product form of an element excreted by one organism is not necessarily safe to other organisms. An important example is methyl mercury. Some bacteria, as shown below, methylate Hg(II) as a detoxifying device; the resulting CH_3Hg^+ circulates through the biosphere and proves to be very toxic to most organisms. It is a general rule that toxic elements released by organisms in their efforts to protect themselves are often taken up inadvertently by other organisms lacking efficient mechanisms to detoxify or excrete the materials. The result is that the poisonous elements are bioamplified through the food chain. Two specific toxic elements have been selected for detailed discussion below: mercury and arsenic.

1.3.4.1 Cycle of Mercury (Konetzka, 1977)

Biological defense mechanisms against Hg(II) can be grouped into three or four classes. The first entails the conversion of Hg(II) to elemental Hg^0, which is volatile and much less toxic than Hg(II). The second mechanism involves the conversion of Hg(II) to the volatile and nontoxic dimethyl mercury, $(CH_3)_2Hg$. This compound decomposes to CH_3Hg^+, which is water-soluble and stable but also highly toxic. The third means employs the conversion of Hg(II) to a very insoluble material HgS or HgSe. The last mechanism requires the production of a metallothionein that sequesters heavy metal cations including Hg(II). The first two devices seem to be utilized principally by microorganisms, which, being single-celled, are able to eliminate mercury by simply allowing it to evaporate in the form of Hg^0 or $(CH_3)_2Hg$. The third and fourth devices are typical of higher organisms, including mammals as well as some microorganisms.

The methylation of mercury has been demonstrated in a variety of microorganisms, including such bacteria as *Methanobacterium* strain M.o.H., *Clostridium cochlearium*, *Escherichia coli*, *Enterobacter aerogenes*, *Pseudomonas fluorescens*, and *Mycobacterium*, as well as the fungi *Aspergillus niger*, *Neurospora crassa*, and *Saccharomyces cerevisiae*. The specific mechanisms used for this methylation have yet to be conclusively established. One hypothesis invokes methylcobalamin to serve as the methylating agent. Methyl group transfer from methylcobalamin to Hg(II) does occur *in vitro* without the aid of an enzyme, but whether it also occurs *in vivo* remains to be investigated. This seems unlikely to be the only methylation path employed, however, since some organisms known to methylate Hg do not contain cobalamin; examples include *Methanobacterium* strain M.o.H. and the fungus (mold) *Neurospora crassa*. Alternative methylating agents are not hard to imagine: *S*-adenosylmethionine and N^5-methyltetrahydrofolate (CH_3—THF) would be possible candidates.

There are numerous organisms that are capable of converting Hg(II) to Hg(0). Examples include the prokaryotes *Pseudomonas aeruginosa*, *E. coli*, and *Staphylococcus aureus*, as well as some eukaryotic algae, such as *Chlamydomonas* and a yeast, *Cryptococcus*. In prokaryotes, the gene for mercury resistance is contained in the plasmid carrying the resistance factors to antibiotics. The plasmid also contains a transmittance factor. Both resistant and sensitive strains are known in at least one species. Resistance to mercury poisoning appears sometimes to be transmitted from one strain to another, athough in some cases resistant strains survive and proliferate in the presence of mercury whereas the sensitive strains are killed.

Certain microorganisms are known to be capable of oxidizing elemental mercury (Hg^0), perhaps to Hg(II). *Bacillus subtilis* and *B. megaterium*, for example, have been shown to oxidize Hg(0), but other organisms oxidize Hg(0) very little, if at all. These include *Pseudomonas aeruginosa*, *Ps. fluorescens*, *E. coli*, and *Citrobacter*. An outline of the biogeochemical cycle of mercury is given in Fig. 1-10.

1.3.4.2 Cycle of Arsenic (Konetzka, 1977)

The main culprits implicated in arsenic toxicity are considered to be arsenite $AsO_2{}^-$ ion and arsonic acid derivatives, $RAsO(OH)_2$. There are a number of microorganisms that participate in the oxidation, reduction, and methylation of arsenic compounds.

Soil microorganisms such as *Pseudomonas*, *Xanthomonas*, and *Ach-*

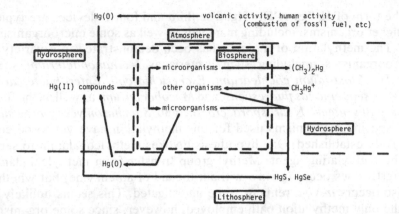

Figure 1-10. The biogeochemical cycle of mercury.

romobacter have been found to oxidize arsenite to arsenate. The enzyme responsible for this oxidation is designated as arsenite dehydrogenase and is known to be inducible with arsenite. *Thiobacillus ferroxidans*, the versatile oxidizing bacterium, was demonstrated to be capable of oxidizing the arsenic present in such minerals as arsenopyrite ($FeS_2 \cdot FeAs_2$), enargite ($3Cu_2S \cdot As_2S_5$), and orpiment (As_2S_3).

The reduction of arsenate to arsenite is rather easy, and can be effected by virtually any reduction system, not only in microbes such as *Micrococcus lactilyticus*, but also in higher plants and animals. The details of the further reduction of arsenite are not clear, however, although a number of fungi and *Methanobacterium* species are known to produce arsine AsH_3 from arsenate.

A more prevalent pathway for the metabolism of arsenate or arsenite seems to be its methylation. Three different methylated products can result: dimethyl arsine $(CH_3)_2AsH$, trimethylarsine $(CH_3)_3As$, and tetramethyl arsonium ion $(CH_3)_4As^+$. It has been shown that anaerobic methane bacteria, including *Methanobacterium* strain M.o.H., produce dimethylarsine (McBride *et al.*, 1978), whereas aerobic fungi such as *Scopulariopsis brevicaulis* and *Candida humicola* synthesize trimethylarsine from both arsenate and its metabolic intermediates (such as methylarsonate and cacodylate). These species also methylate other organic arsenicals such as $(C_6H_5)AsO(OH)_2$ and $(C_2H_5)(C_3H_7)AsO(OH)$ leading to $(C_6H_5)As(CH_3)_2$ and $(C_2H_5)(C_3H_7)As(CH_3)$, respectively (McBride *et al.*, 1978; Challenger, 1945). A possible pathway for the formation of the methyl derivatives is discussed in Section 15.2. The methyl (CH_3^+) donor here is believed to be *S*-adenosylmethionine (McBride *et al.*, 1978), but the reduction system has yet to be elucidated. Trimethylarsine and di-

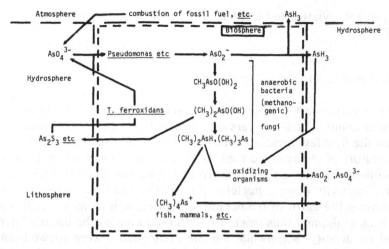

Figure 1-11. The biogeochemical cycle of arsenic.

methylarsine once formed can readily diffuse out of cells or be exhaled. In organisms with urinary excretion systems, trimethylarsine seems to further react with some source of CH_3^+, thereby being converted to tetramethylarsonium salts, which are water-soluble and readily excreted. It appears that soil contains microorganisms capable of oxidizing both the reduced and methylated arsenicals, although these have not been identified. An outline of the biogeochemical cycle of arsenic is given in Fig. 1-11.

1.4 Historical Perspective on the Biochemistry of the Elements

The foregoing sections have discussed the relationship of the biosphere to its surroundings in terms of the present-day earth, emphasizing factors that place constraints on the use of particular elements by organisms. Life, however, is definitely a reflection of a historical process and thus has been subjected to different constraints in the past. Specifically, it is certain that the availability of at least some elements has changed with time, and this variation must have had its consequences with respect to the origin and development of life. I shall briefly review this subject in this section. The problem of the origin of life, when and how it occurred, is still not well understood despite the recent surge of scientific inquiry concerning it (see Schopf, 1983; Ponnamperuma, 1983; and Day, 1984). The description that follows is, inevitably, highly speculative. It is also sharply abbreviated and emphasizes the inorganic aspects (see Ochiai, 1983a).

1.4.1 Outline of a Possible Origin of Life

The earth started to form as a result of accretion of gas and dust particles about 4.6 billion years ago and it is believed to have taken shape within the first few hundred million years (Wetherill, 1981). At first, the temperature of the newly formed planet was low enough to retain volatile constituents within the still uniform rocky structure. Eventually, however, the radioactivity of such nuclei as ^{40}K, ^{232}Th, and ^{235}U heated the rocks comprising the earth to temperatures high enough to melt silicates and other minerals, and thus began the differentiation of the earth's interior based on density. Meanwhile volatile components were spewed out in the form of volcanic eruptions and fumaroles, thereby creating the first atmosphere. This gaseous material consisted primarily of hydrogen, nitrogen, ammonia, water vapor, carbon monoxide, carbone dioxide, and a small amount of the acids of sulfur and chlorine. Oxygen gas (dioxygen) may have been present, but if so, it would have quickly reacted with reduced material, thus ensuring that it would be a very minor component of the atmosphere. There is strong reason to believe that dioxygen was virtually nonexistent in the Hadean (4.6–3.5 billion years ago) and early Archean (3.5–2.5 billion years ago) periods.

The earliest biogenic microfossils known are about 3.4–3.5 billion years old (Schopf and Walter, 1983), which suggests that the creation of life must have occurred sometime between 4.0 and 3.5 billion years ago. How? One possible scenario (see Day, 1984; Ferris, 1984) is as follows.

The abiogenic synthesis of biologically essential compounds, including the amino acids, some purine and pyrimidine bases, and some carbohydrates, has been demonstrated repeatedly in simulation experiments starting with simple compounds such as H_2O, N_2 (and NH_3), CO, and CO_2 (and CH_4) and employing such energy sources as electric discharge, ultraviolet light, and shock wave. Some of the bases appear to be most easily derived from HCN, which can also be produced from the simple compounds listed above. This first stage of the prebiotic chemical evolution seems relatively well delineated, though certain problems with it remain unsolved, particularly the formation from HCHO of carbohydrates with specific structure, as well as the origin of lipids. Especially, the latter problem seems not to have been examined carefully.

The second stage in chemical evolution is the formation of derivatives from the component materials, including polymeric substances. It is still an open question as to how, e.g., ATP (adenosine triphosphate) might have been formed from its components adenine, ribose, and phosphate. A number of ways are known by which amino acids polymerize to form polypeptides, but it is not so clear that they represent a realistic picture of what actually happened on the primitive earth. The same can be said about the formation of polynucleotides. It is this kind of uncertainty which led Cairns–Smith to propose an entirely new hypothesis, the inorganic origin of genetic material (Cairns–Smith, 1982, 1985).

In the third and final stage of chemical evolution, appropriate groups of compounds, presumably including polypeptides and polynucleotides, were compartmentalized by incorporation into lipid-bound droplets. The subsequent interaction between polynucleotides themselves or between polypeptides and polynucleotides is assumed to have somehow led to a self-replicating system (see Eigen and Schuster, 1977, 1978; Eigen et al., 1981).

1.4.2 Outline of the Evolution of Life on the Primitive Earth

The first organisms to appear on the earth are believed to have been heterotrophic prokaryotes that fed on the organic compounds that had been produced abiogenically. Before long, however, and before this supply of organic compounds was depleted, photoautotrophic organisms appeared. These may have been sulfur bacteria, either green or purple, which extracted electrons from sulfur-containing compounds. Alternatively (or concurrently), nonsulfur photoautotrophic bacteria could have emerged and utilized organic compounds as their source of electrons. These would have been the first photoautotrophs. Since the extraction of electrons from these sources is chemically much easier than the extraction of electrons from water, a less sophisticated photosynthetic apparatus would have been utilized than that in cyanobacteria and modern plants which decompose water. Probably the so-called archeabacteria (methanogens, halophiles, and thermophiles) originated about the same time, derived from the original heterotrophs (Woese, 1981; Fox et al, 1980). Sulfur bacteria and nonsulfur bacteria then diverged into other eubacteria. All of these bacteria were still anaerobic, since no significant amount of dioxygen was available in the atmosphere. Perhaps this condition lasted until about 2.0–2.5 billion years ago.

One of the most significant events in the early history of life on earth was the emergence of cyanobacteria equipped with a photosynthetic apparatus capable of decomposing water and simultaneously releasing dioxygen. There is much controversy over when this occurred. A filamentous microfossil found in 3.5-billion-year-old stromatolitic cherts of the Warrawoona group (Australia) is though to be of cyanobacterial type (see Schopf and Walter, 1983). A 3.4-billion-year-old stromatolite found in South Africa (Onverbacht group) is also believed to be derived from the activities of cyanobacteria, as are modern stromatolites (see Walter, 1983). If these interpretations are correct, cyanobacteria might have been among the earliest photoautotrophs. Other indications, however, including studies of the evolution of the atmosphere and certain geological data, place the event in the period 3.0–2.5 billion years ago (i.e., late Archean period). The complexity of the cyanobacterial photosynthetic apparatus is also consistent with this later date.

The proliferation of cyanobacteria, perhaps starting about 2.5 billion years ago, changed the atmosphere from an anoxic state to the oxygenic state [see Holland (1984) for a more recent review of the early atmosphere]. This event, which must have been gradual rather than sudden, seems to have occurred about 2.2 billion years ago. The increase in atmospheric oxygen brought about by cyanobacterial activities permitted aerobic metabolism to evolve in the biosphere, leading ultimately to the emergence of eukaryotes about 1.5 billion years ago (Vidal, 1984; Day, 1984). All eukaryotes are aerobes. Cyanobacteria, too, are aerobes, but some still carry the nitrogen-fixation capability typical of anaerobic organisms. A tentative summary of the evolution of life on primitive earth is given in Fig. 1-12.

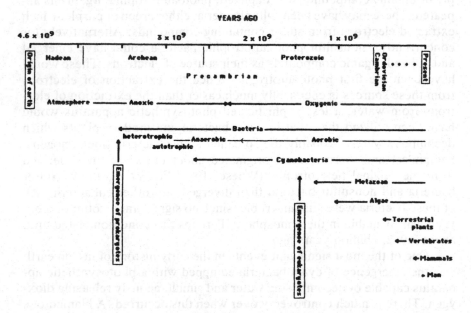

Figure 1-12. An outline of the history of life on earth.

1.4.3 Environmental (Elemental) Constraints on the Origin and Evolution of Life

The emergence and evolution of life as outlined above must have been subjected to a number of serious environmental constraints, including ultraviolet radiation from the sun, high ambient temperatures, and so on. Organisms are obviously made of material, so their continued existence must critically depend on the availability of that material in the form of the appropriate elements. About 30 elements are now recognized as being essential to organisms. Some of them (C, H, N, O, S, P, Na, K, Cl, Mg, Ca, and Zn) are essential to all organisms; three additional elements, Fe, Mo, and Mn, are essential to almost all organisms, while a number of others, including Si, V, Cr, Co, Ni, Cu, Se, Sr, I, and W, are utilized by only limited numbers of organisms.

The availability and distribution of the elements have changed continuously throughout the long history of the earth, largely due to the effect of the geological activities in continental and oceanic crust, change in the pH of the ocean, and increase in the dioxygen content of the atmosphere.

It certainly seems plausible that change in the availability of the elements would have had significant effects on the emergence and evolutionary course of life (Ochiai, 1978c). Simply put, this thesis asserts that at a time in which a given element was not available in a readily accessible form, no organism that required this particular element would have emerged.

By about 4.0 billion years ago, the basic crustal structure of the earth had been established. Effects of the nature of the crust on evolution thus can be inferred by studying its current makeup. Some of the universally essential elements, including Na, K, Mg, Ca, and Zn, are present as cations in the hydrosphere and the lithosphere, and their compounds are fairly soluble. Their abundance in seawater is regulated essentially by their content in the oceanic crust and the appropriate dissolution equilibria. Various factors could still change their distribution, especially pH and temperature. The average pH of present-day seawater is about 8, but many believe that seawater was somewhat more acidic in the Hadean and Archean periods, partly because the gases from early volcanic activity contained more HCl or H_2SO_4 than is presently the case, and partly because the CO_2 concentration was higher as a result of the virtual absence of CO_2-consuming (i.e., photoautotrophic or methane-forming) organisms. This difference in pH between the contemporary and primordial seawater would not have grossly altered the seawater content of the soluble elements, however, and it can be concluded that they were as available in the early stages of evolution as they are now.

Most of the other essential elements including C, H, N, O, S, Fe, Mo, V, Cr, Mn, Cu, and Se are subject to oxidation–reduction reactions; hence, their status and availability would be critically dependent on the oxidative state of the environment, especially that of the hydrosphere. The virtually oxygen-free atmosphere of the Hadean and Archean earth means that these elements would have had to be present in relatively low oxidation states: H_2S (or other sulfides), Fe(II), Mo(IV) (MoS_2?), Cu(0) or Cu_2S, and so forth. NH_3, the most reduced form of nitrogen, is though unlikely to have survived for any length of time after its initial emergence through degassing. Likewise, CH_4 is believed to have disappeared relatively quickly. Thus, the main components of the primordial atmosphere are thought to have been CO, CO_2, H_2O, and N_2. These must have been the starting materials from which the essential organic compounds were derived. The reactions that led to the formation of simple organic compounds and then subsequent reactions by which these were converted into complicated or polymeric compounds certainly took place in the aforementioned environment. Some of these latter compounds could eventually have acted as catalysts or other kinds of reaction mediators (e.g., template) for the creation of still more complicated materials. Possibilities in this regard have not been well studied.

As the oxidative state of the environment is largely determined by the oxygen content of the atmosphere, it is possible to infer what elements might have been available or unavailable at various times, using as data the known historical changes in atmospheric oxygen content. Unfortunately, the data are not regarded as very accurate (Kasting, 1984; Holland, 1984). Estimates of the oxygen content during the Archean and early Proterozoic period (up to 2.2 billion years ago) range from 10^{-11} atm (Kasting, 1984) to 2×10^{-3} atm (Holland, 1984). The oxygen content is known to have risen rapidly beginning about 2.2 billion years ago. The low oxygen content prior to that point is presumably due to a combination of limited free oxygen production counterbalanced by the presence of large oxygen sinks. Any oxidizable material could act as an oxygen sink, but the most significant one seems to have been Fe(II) dissolved in seawater. The oxidation of Fe(II), which may have fluctuated on a seasonal basis, produced the vast quantity of material known as the banded iron formation (BIF). BIF today provides the majority of the world's extractable iron ore. It was formed throughout the Archean and the early Proterozoic periods, stopping abruptly about 1.8 billion years ago (see *Econ. Geol.* Vol. 68, 1973, for BIF).

Manganese also remained mainly as Mn(II) in seawater but until much later than iron. Reduction potential data indicate that molybdenum would have become MoO_4^{2-} as early as the Archean period, so that it would have been available to primitive organisms in the ocean since the Archean period. As the original abiogenic organic compounds became scarce, photosynthesis evolved. Nitrogen fixation must also have had to arise about the same time because of the unavailability of NH_3. The elements in the nitrogen fixation reaction as we know it (Fe, S, and Mo) were, fortunately, available, perhaps in the forms of FeS (or Fe(II) + S^{2-}) and MoO_4^{2-}. Nitrogen fixation is distributed today among anaerobic bacteria, as well as some aerobic bacteria and cyanobacteria.

The emergence of sulfate-reducing bacteria, *Desulfovibrio* species may have been delayed until about 2.0 billion years ago, the time when sulfate became abundant (according to the reduction potential).

One of the most interesting elements from an evolutionary point of view is copper (Ochiai, 1983b). Because of the high reduction potential of Cu(II), copper would have become available to organisms only as recently as 1.8–1.7 billion years ago. Consistent with this late date of its availability is the fact that the majority of known copper enzymes and proteins are restricted to eukaryotes; only three copper proteins (plastocyanin, azurin, and cytochrome oxidase) have been discovered in some aerobic prokaryotes, suggesting that these particular prokaryotes are latecomers, appearing within the last 1.7 billion years.

Although the examples above would have to be regarded as speculative, they are certainly illustrative of the possible role of constraints placed by elemental availability on the origin and evolution of life.

Some Basic Principles of the Biochemistry of the Elements

Essentially two different categories of basic principles are dealt with here. One is a set of principles dealing with the biological selection of elements (Section 2.1), whereas those in the other category are related to physicochemical properties of elements and their compounds (Section 2.2.). Biological systems could be expected to adapt themselves to the use of particular elements for their needs as a result of many factors, important among which should be "usefulness" and "availability". The extent to which an element and its compounds are found to be appropriate will depend heavily on physicochemical properties relevant to enzymatic or other biological functions, such as reduction potential and acid–base character. Some of the basic parameters that determine these properties will be discussed in Section 2.2.

2.1 Basic Principles in Biological Selection of Elements

The majority of the key biological functions are carried out by organic compounds, including genetic transmission by polynucleotides and catalysis by metal-free enzymes (and proteins). However, some crucial functions cannot be effected by organic compounds alone, i.e., they require the added presence of inorganic materials. That this is true is a consequence of certain limitations in the capabilities of organic materials. For example, their mechanical strength (at least that of naturally occurring organic compounds) is never very great. Organic compounds also are not very suitable for catalyzing reactions of the oxidation–reduction type. As a result, some inorganic compounds are essential to organisms; those now considered to be in the essential category are summarized in Fig. 1-1.

In Chapters 3 through 7, we shall address ourselves to our main question: why are certain inorganic elements (compounds) specifically required for specific biological functions and hence essential to organisms? In order to answer this question, the following basic principles (basic rules) will be invoked (Ochiai, 1978a,b): (1) Rule of Basic Fitness (Chemical Suitability), (2) Rule of Abundance, (3) Rule of Efficiency, and (4) Evolutionary Adaptation (Evolutionary Pressure).

The basic fitness rule asserts that a certain element(s) has an inherent capacity (potentiality) for a specific biological function. Any element for which this is the case obviously becomes essential for the biological system in question. There could, of course, be more than one such element for a particular function. If so, then the second rule (rule of abundance) applies and dictates which of the elements is to be selected. Obviously, given a choice, the more abundant or more readily available element would be utilized. If there is no alternative and the essential element is not readily available, the organism would simply be unable to exist as such or would have to develop a means of securing the needed element. When two or more suitable elements are equally available, the more efficient one would be preferred (rule of efficiency). Once essentiality is established for a given element, an organism would be expected to try to make the best possible use of it, thus developing a system employing the element as efficiently as possible. In so doing, the system would be expected to become very specifically adapted to the use of that particular element so that an alternative, one that had the potential for being equally effective (basically capable) but that was not chosen because of its more limited availability, would no longer be an effective substitute for the end product of evolution. This whole process of an organism's adaptation to (or its effort to make the possible best use of) a particular element is what is called the rule of evolutionary adaptation [or the evolution pressure effect (rule 4)].

The role of availability in essentiality (i.e., rule 2) has already been discussed in Chapter 1. The major part (i.e., Part II) of this book is devoted to a discussion of the application of rules 1 and 4.

2.2 Some Relevant Inorganic Principles

In this section I will review some important inorganic chemical principles pertinent to the ensuing discussion in an attempt to ascertain what factors determine the physicochemical characteristics of an element in its compounds. Basic concepts such as ligand field theory and its applications to thermodynamic and kinetic properties will be found in most standard inorganic or coordination chemistry textbooks. Here I will take a wider and somewhat less quantitative view of the subject.

Table 2-1. Some Important Properties of Metal
Coordination Compounds

Continuous property	Discrete property
Reduction potential	Oxidation state
Lewis acidity	Number of electrons
Electron transfer rate	Coordination
Effective charge (of central metal)	number
	Shape of orbitals
	Spin state

Table 2-1 lists some important parameters relevant to assessing the chemical suitability (basic fitness) of elements.

2.2.1 Reduction Potential

2.2.1.1 General Formulations

The reduction potential is to be defined for an equilibrium of the following type:

$$M(+n) + me \rightleftharpoons M(+n-m) \qquad E_h^0, \Delta G^0, K \qquad (2\text{-}1)$$

$$\Delta G^0 = -mFE_h^0 = -RT \ln K \qquad (2\text{-}2)$$

The conventional way of expressing E_h (related to E_h^0) is:

$$\Delta G = \Delta G^0 + RT \ln \{M(+n-m)\}/\{M(+n)\} \qquad (2\text{-}3)$$

$$E_h = E_h^0 - (RT/mF) \ln \{M(+n-m)\}/\{M(+n)\} \qquad (2\text{-}4)$$

where $\{x\}$ represents the activity of species x.

If one expresses the K value in terms of activities,

$$K = \{M(+n-m)\}/\{M(+n)\}/\{e\}^m \qquad (2\text{-}5)$$

Taking the logarithm of both sides gives:

$$\log K = \log \{M(+n-m)\}/\{M(+n)\} - m \log\{e\} \qquad (2\text{-}6)$$

Sillen defines $-\log \{e\}$ as the pE in analogy to the definition of pH (Sillen, 1965, 1966); thus:

$$\log K = \log \{M(+n-m)\}/\{M(+n)\} + m\text{pE} \qquad (2\text{-}7)$$

Multiplication of both sides by $2.303RT$ gives:

$$2.303RT \log K = 2.303RT \log \{M(+n-m)\}/\{M(+n)\} +$$
$$2.303RTmpE \quad (2\text{-}8)$$

The left-hand side is equal to $mFE_h{}^0$; therefore,

$$(2.303RT/F)pE = E_h{}^0 - 2.303RT \log \{M(+n-m)\}/\{M(+n)\} \quad (2\text{-}9)$$

It can readily be seen that Equation 2-9 is equivalent to Equation 2-4 if one equates $E_h{}^0$ to $(2.303RT/F)pE$. These equations are the general expressions required in dealing with the reduction potential.

2.2.1.2 Standard Reduction Potential: Determining Factors

It is hardly possible to make a general statement that will describe the reduction potential values of various elements and their compounds. Equation 2-2 indicates that $E_h{}^0$ depends on the stability of the oxidized state relative to the reduced state. Thus, $E_h{}^0$ would rise as the reduced state becomes more stable relative to the oxidized state. Therefore, the problem of predicting the reduction potential of a substance is transferred into that of estimating the relative stabilities of two different oxidation states under specific conditions.

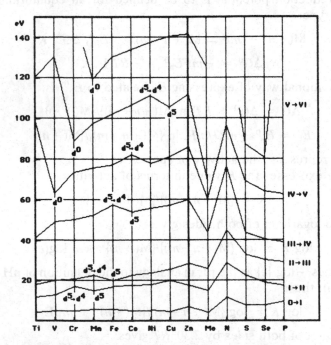

Figure 2-1. Ionization potentials of some important elements.

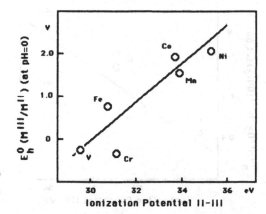

Figure 2-2. Correlation between E_h^0 value (M^{III}/M^{II}) at pH 0 and ionization potential (II–III).

Let us first consider an element in a vacuum. It can be ionized (oxidized) by the addition of an appropriate ionization energy (potential). The reverse process is a reduction so that the reduction potential will be related to the inverse of the ionization energy. Therefore, it seems reasonable that there would be some correlation between the ionization potentials and the corresponding reduction potentials, at least for simpler compounds. Figure 2-1 shows the ionization potentials of some relevant elements. Particularly for the transition metal series a few interesting features emerge. Dips can be seen in the lines at points corresponding to d^0 and d^5. d^0 represents the configuration in which the readily ionizable penultimate d electrons are fully removed, and any subsequent ionization would be expected to be very difficult. d^5 represents a half-filled configuration in which the Coulomb repulsion term is minimized, and hence this state is somewhat stable. Therefore, the subsequent ionization $d^5 \rightarrow d^4$ is notably difficult, appearing as a peak in the lines representing the ionization potentials of the series.

In Fig. 2-2, the reduction potentials M(III)/M(II) in aqueous media at pH 0 are plotted against the ionization potential II→III. Figure 2-3 shows a similar correlation between the reduction potential $MO_2/M(OH)_3$ at pH 7 and the ionization potential III→IV. Obviously it is a gross simplification to try to correlate such reduction potentials solely with the ionization potentials, because there are a number of other factors to be considered, including hydration energy. Nevertheless, it is seen from Figs. 2-2 and 2-3 that there is some degree of correlation between these two sets of values.

At higher oxidation states, a metal ion becomes (in general) more strongly oxidative and more strongly polarizing (because of a larger z/r value, see Section 2.2.3), and hence it can be stabilized only by binding

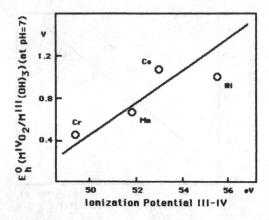

Figure 2-3. Correlation between E_h^0 value (M^{IV}/M^{III}) at pH 7 and ionization potential (III–IV).

with highly electronegative anions such as F^- and O^{2-}. This is the reason why the only usual compounds containing metals (or other elements) in extremely high oxidation states are oxyanions, such as VO_2^+, CrO_4^{2-}, MnO_4^-, NO_3^-, SO_4^{2-}, SeO_4^{2-}, and PO_3^{3-}. In terms of ionization potentials (see Fig. 2-1), it is reasonable to expect that oxidative power would decrease in the order: $MnO_4^- > CrO_4^{2-} > VO_2^+ > MoO_4^{2-}$, and $NO_3^- > SO_4^{2-} > SeO_4^{2-}$. This prediction is in accord with the known facts, with the exception that the reduction potential of SeO_4^{2-} is higher than that of SO_4^{2-}. Again it is an oversimplification to attempt to estimate reduction potential solely on the basis of ionization potentials. The stability of any given oxyanion (and hence its reduction potential) depends also, among other factors, on the binding energy between the central ion (in the high oxidation state) and O^{2-}.

Let us now turn to more subtle effects caused by the presence of ligands. In this case the reduction reaction is expressed by:

$$M(+n)L + me \rightleftharpoons M(+n-m)L' \qquad E_L^0, \Delta G_L^0 \qquad (2\text{-}10)$$

where L and L' may or may not be the same. The reduction potentials cited in Figs. 2-2 and 2-3 are those for complexes with $L,L' = (H_2O)_p$ or $(O^{2-})_q$. For an aquo complex, reaction 2-10 can be written as:

$$M(+n)(HOH)_p + me \rightleftharpoons M(+n-m)(HOH)_q \qquad E_h^0, \Delta G_h^0 \qquad (2\text{-}11)$$

where the subscript h is utilized to indicate an aquo complex.

The more general equilibrium 2-10 can be broken down into:

$$M(+n)(HOH)_p + L \rightleftharpoons M(+n)L + pHOH \qquad \Delta G_n^0 \qquad (2\text{-}12a)$$

$$M(+n)(HOH)_p + me \rightleftharpoons M(+n-m)(HOH)_q \qquad E_h^0, \Delta G_h^0 \quad (2\text{-}12b)$$

$$M(+n-m)(HOH)_q + L' \rightleftharpoons M(+n-m)L' + qHOH \qquad \Delta G_{n-m}^0$$
$$(2\text{-}12c)$$

Equation 2-10 is equivalent to 2-12b + 2-12c − 2-12a; therefore,

$$\Delta G_L^0 = \Delta G_h^0 + (\Delta G_{n-m}^0 - \Delta G_n^0) \qquad (2\text{-}13)$$

The effects of the ligands (L and L') on the redox equilibrium can thus be translated into their effect on the stabilities of the complexes with M($+n$) and M($+n-m$). It is this effect that forms the basis for the alteration of the reduction potential of an element by ligands. Because of limits on the potential choice of ligands in biological systems, the E_h^0 value of a given element may be limited to within a certain range; but nevertheless it can be varied fairly continuously within that range from one extreme to the other by subtle modification by the ligands.

Some of the more general effects on the stability (and hence the reduction potential) are listed below:

1. Ligands of low polarizability (so-called "hard" bases), such as OH^-, O^{2-}, and F^-, tend to stabilize higher oxidation states and hence reduce the reduction potentials. Examples include reduction potentials of $+0.77$ V for Fe(III)aq/Fe(II) $>$ $+0.61$ V for Fe(III)–PO_4^{3-}/Fe(II) $>$ $+0.41$ V for Fe(III)–F^-/Fe(II).

2. Ligands of high polarizability (so-called "soft" bases) and ligands with low-lying vacant π orbitals tend to stabilize lower oxidation states and hence raise reduction potentials; e.g., $+0.77$ V for Fe(III)aq/Fe(II)aq $<$ $+1.14$ V for Fe(III)(phen)$_3$/Fe(II) $<$ $+1.25$ V for Fe(III)(nitro-phen)$_3$/Fe(II) where phen $=$ phenanthroline.

3. The contribution of a ligand field effect to the stability of a complex is called the ligand field stabilization energy (LFSE). The magnitude of the LFSE depends largely on the ligand field strength, which is usually expressed in terms of Dq. The energy gap between the two sets of d orbitals, splitting of which is caused by the octahedral ligand field, is defined as $10Dq$. The LFSE depends also on the structure of the complex and its spin state; the latter also being controlled by the ligand field strength. In the case of iron complexes, the LFSE is $0Dq$ for high-spin Fe(III) versus $4Dq$ for high-spin Fe(II), and $20Dq$ for low-spin Fe(III) versus $24Dq$ for low-spin Fe(II). Dq values are different (usually much larger) in Fe(III) complexes than in Fe(II) complexes. Thus, high-spin iron complexes tend to show higher reduction potentials than low-spin complexes; for example, compare the value of $+0.36$ V for Fe(III)(CN)$_6$/Fe(II) (low-spin) to $+0.77$ V for Fe(III)aq/Fe(II) (high-spin). Those ligands cited in point 2 above, however, give rise to greater Dq values for Fe(II) than for Fe(III). In other words, complexes containing such ligands would have higher reduction potentials, despite their being low-spin species.

4. Structural factors: Different oxidation states of a metal ion can

have different preferred structures. For example, the favored structure of Cu(II) complexes is square-planar, whereas Cu(I) complexes prefer a tetrahedral structure. Therefore, in the couple Cu(II)/Cu(I), the reduced state would be highly stabilized relative to the oxidized state whenever the structure of the complex is tetrahedral. Put another way, the closer a structure to being a tetrahedron (or the farther it is from square-planar), the higher the reduction potential is for Cu(II)/Cu(I).

2.2.1.3 pH Dependence, E_h–pH Diagram

For the reduction of CrO_4^{2-} to Cr^{3+}:

$$CrO_4^{2-} + 8H^+ + 3e \rightleftharpoons Cr^{3+} + 4HOH \qquad (2\text{-}14)$$

the reduction potential can be expressed by:

$$E_h = E_h^0 - (RT/3F)\ln\{Cr(III)\}\{HOH\}^4/\{Cr_4^{2-}\}/\{H^+\}^8$$

$$= E_h^0 - (RT/3F)\ln\{Cr(III)\}\{HOH\}^4/\{Cr_4^{2-}\}$$

$$- (2.303RT/3F)8pH \qquad (2\text{-}15)$$

If the standard state of 25°C and the activities $\{Cro_4^{2-}\} = \{Cr(III)\} = \{H_2O\} = 1$ M, the following expression will be obtained:

$$E_h = E_h^0 - (0.0592)(8/3)pH \qquad (2\text{-}16)$$

which gives the dependence of E_h^0 on pH for system 2-14. In general, one can express E_h by the following formula, assuming that all entities except H^+ are in their standard state of unit activity:

$$E_h = E_h^0 - (0.0592)(n/m)pH \qquad (2\text{-}17)$$

where m is the number of electrons involved in the reduction and n is the number of protons consumed in the net reaction. It is obvious from this result that reduction potential generally decreases with a rise in pH of the medium if the reaction involves H^+. A diagram showing E_h values as a function of pH is called an "E_h–pH" diagram, and such diagrams can be extremely useful in discussing redox behavior in aqueous media (see Garrels and Christ, 1965).

2.2.2 Lewis Acidity

Characterization of an element or a cation with regard to acid–base property would be useful in any discussion of the catalytic activity of the metal in an enzymatic reaction of the acid–base type. Figure 2-4 collects some relevant data on divalent cations.

2.2.2.1 Charge-to-Radius Ratio (or Ionic Potential) and Effective Nuclear Charge

The force (attractive) between an anion of charge z' and a cation of charge z and radius r with center-to-center distance D is given by:

$$F = zz'/D^2 \qquad (2\text{-}18)$$

The energy required to transport such an anion from the surface of its associated cation to infinity is given by

$$E = \int_{r}^{\infty} (zz'/D^2)dD = zz'/r \qquad (2\text{-}19)$$

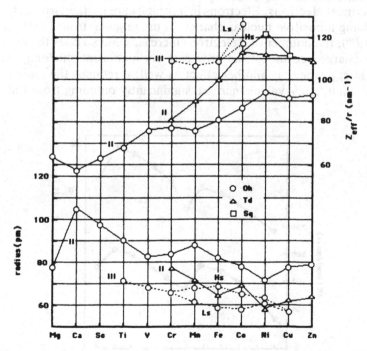

Figure 2-4. Some important parameters of cations. II = +2 oxidation state; III = +3 oxidation state; Hs = high spin; Ls = low spin; Oh = octahedral; Td = tetrahedral; Sq = square-planar. Z_{eff} value is estimated from the Raimondi and Clementi effective nuclear charge value for 4s orbital (in the case of the first transition-series elements; or an appropriate one for Mg and Ca) plus 2 (in the case of II) or plus 3 (in the case of III).

This energy is thus proportional to z/r and the cation may be regarded as creating an electric field of z/r. If a polarizable entity were to be placed in this field, its polarization would be proportional to z/r. Therefore, the charge-to-radius ratio represents a kind of estimate of the polarizing power of a cation; it is, in this sense, related to the acidity of the cation. A cation with a high effective z/r value could effectively polarize a ligand's electron cloud, leading to a rather covalent bonding. In the case of ligands that are not easily polarizable, z/r may simply represent the electrostatic effect produced by the cation.

The "z" value to be employed in these calculations would not be the nominal one, but should rather be an "effective" value. For example, although the nominal electric charges of all the divalent cations of the series Ca(II) through Zn(II) are the same, the effective charge of the ion increases with atomic number. As the atomic number (and hence the nuclear charge) increases in the series, the number of electrons also increases, thus cancelling a portion of the increase in the nuclear charge. However, this shielding effect by electrons is never perfect except with the innermost electrons. Electrons in d orbitals are particularly inefficient in shielding a positive nuclear charge. Thus, in going from Ca(II) (d^0) to Zn(II) (d^{10}), the number of d electrons increases and so does the effective nuclear charge because of the progressively inefficient shielding. An increase in z_{eff} value has another effect as well: it reduces the ionic radius r. As a result, Z_{eff}/r values increase significantly on going from Ca(II) to

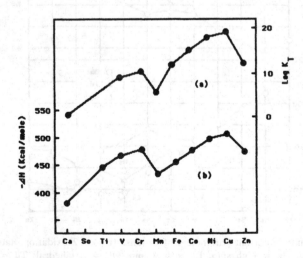

Figure 2-5. Irving–Williams series. (a) Total stability constant (log K_T) for [M^{II}(en)$_3$] and (b) hydration enthalpy.

Zn(II). The Z_{eff} value can be estimated from Clementi and Raimondi values (Clementi and Raimondi, 1963); Z_{eff}/r values estimated in this way are shown in Fig. 2-4. Among the more common bioelements (in their divalent state), Zn(II) and Cu(II) seem to be the most effective acids from the standpoing of Z_{eff}; Zn(II) and Cu(II) are assumed here to take tetrahedral and square-planar structures, respectively.

Transition metal cations with d^n ($n = 1$–9) configurations could obtain extra stabilization energy (LFSE) by forming coordination compounds (see Huheey, 1983). The magnitude of the LFSE for the series of divalent cations, assuming octahedral structures, can readily be calculated (see Table 2-2). The overall effect of superimposing this LFSE further on the "effective charge" (i.e., Z_{eff}/r plus LFSE) leads to the so-called "Irving–Williams" series: Mn(II) < Fe(II) < Co(II) < Ni(II) < Cu(II) > Zn(II). This series summarizes the general trend in thermodynamic stability of complexes containing the listed cations. Two examples of the trend are shown in Fig. 2-5.

2.2.2.2 Acidity of a Complexed Metal Cation

The stability constants K_n for the following equilibria were determined to be $K_1 = 10^{2.1}$, $K_2 = 10^{1.6}$, $K_3 = 10^{1.0}$, $K_4 = 10^{0.8}$, $K_5 = 10^{-0.2}$, and $K_6 = 10^{-0.6}$. A statistical factor is partly

$$[Co^{II}(HOH)_{7-n}(NH_3)_{n-1}] + NH_3 \rightleftharpoons [Co^{II}(HOH)_{6-n}(NH_3)_n] \quad K_n \quad (2\text{-}20)$$

responsible for the decrease noted in the series of stability constant. After correction for the statistical factor, the stability constants can be shown to be $10^{2.1}$, $10^{1.7}$, $10^{1.2}$, $10^{1.1}$, $10^{0.3}$, and $10^{0.2}$, respectively. Since NH_3 is a stronger base than HOH, the replacement of a coordinated water molecule by an ammonia molecule would be expected to reduce the effective residual acidity of the central metal ion [Co(II) in this case]. Therefore, the binding strength of each successive NH_3 molecule is reduced. It follows that binding a stronger base would reduce the acidity of the metal cation even more. An alternative way of stating the matter is to attribute the change to the reduction of the effective positive charge of the metal cation. If the change in the binding constants is regarded as a measure of the change in acidity, the latter can be said to vary over approximately 10^2 in this series. In the similar case of $[Co^{III}(HOH)_{7-n}(NH_3)_{n-1}]$, the K values (and hence the acidity) range from a high of $10^{7.3}$ ($= K_1$) to a low of $10^{4.4}$ ($= K_6$, $10^{5.2}$ after a statistical correction); i.e., the range of variation is again about 10^2. On average, the replacement of one HOH molecule by an NH_3 molecule in complexes of either Co(II) or Co(III) seems to

Table 2-2. LFSE of d^n Configuration in High-Spin Octahedral and Tetrahedral Structure

	d^0	d^1	d^2	d^3	d^4	d^5	d^6	d^7	d^8	d^9	d^{10}
Octahedral (in Dq)	0	4	8	12	6	0	4	8	12	6	0
Tetrahedral (in Dq')	0	6	12	8	4	0	6	12	8	4	0
Tetrahedral (in Dq)[a]	0	2.7	5.3	3.6	1.8	0	2.7	5.3	3.6	1.8	0

[a] Theoretically, $Dq' = (4/9)Dq$; values calculated on this basis.

reduce the corresponding stability constant and, hence, the acidity by a factor of about $10^{0.4}$.

As another example, consider:

$$[Cd^{II}(CN)_{n-1}] + CN^- \rightleftharpoons [Cd^{II}(CN)_n] \qquad K_n \qquad (2\text{-}21)$$

In this series, $K_1 = 10^{5.5}$, $K_2 = 10^{5.1}$, $K_3 = 10^{4.6}$, and $K_4 = 10^{3.6}$; these translate to $10^{5.5}$, $10^{5.2}$, $10^{4.9}$, and $10^{4.2}$ after the statistical correction. The average change in the stability constant upon addition of one CN^- ligand to Cd(II) is thus approximately $10^{0.4}$. Because of the tetrahedral structure of these complexes, repulsive electrostatic effects between CN^- ligands do not seem to affect the relative K_n values significantly. In an octahedral complex on the other hand, such repulsive effects appear to be more significant. For example, in the case of $[Cr^{III}(NCS)_n]$, $K_1 = 10^{2.1}$, $K_2 = 10^{1.7}$ ($10^{1.8}$ after the statistical correction), $K_3 = 10^{1.0}$ ($10^{1.2}$), $K_4 = 10^{0.3}$ ($10^{0.6}$), $K_5 = 10^{-0.7}$ ($10^{-0.2}$), and $K_6 = 10^{-1.6}$ ($10^{-0.8}$). Here the decrement in K upon addition of NCS^- is, on the average, approximately $10^{0.6}$, which is significantly higher than the value from the Cd-tetrahedral complexes discussed above, presumably as a result of the mutual proximity of the charged ligands. For another octahedral series, the $[Fe^{III}(NCS)_n]$ system, the K values range from $K_1 = 10^{2.9}$ to $K_6 = 10^{-1.2}$ ($10^{-0.3}$); the average decrement here, too, is $10^{0.6}$.

A better measure of the acidity of a metal cation in a complex may be the pK_a value derived for an equilibrium of the following type:

$$[M^{+n}L(HOH)] \rightleftharpoons [M^{+n}L(OH)] + H^+ \qquad K_a \qquad (2\text{-}22)$$

For Co(III) complexes, the corresponding pK_a values have been deermined to be 0.7 for $[Co(HOH)_6]$, 4.5 for trans-$[Co(en)_2(HOH)_2]$, 6.1 for cis-$[Co(en)_2(HOH)_2]$, 6.0 for cis-$[Co(NH_3)_4(HOH_2]$, and 6.6 for $[Co(NH_3)_5(HOH)]$. In the Cr(III) series, pK_a's have been shown to be 3.8 for $[Cr(HOH)_6]$, 4.1 for trans-$[Cr(en)_2(HOH)_2]$, 4.8 for cis-$[Cr(en)_2(HOH)_2]$, and 5.3 for $[Cr(NH_3)_5(HOH)]$. These data also indicate that substitution of HOH by NH_3 or en (= ethylenediamine) reduces the acidity or electron-withdrawing power of the central metal ion.

2.2.3 Number of Electrons versus Structures

The most favored structure for a complex of a metal cation is determined largely by two factors: (1) the ionic radius of the metal ion versus the size of the ligands, and (2) the electronic configuration of the metal ion. If the metal cation is relatively small in radius, it has a limited ability to accommodate ligands for simple steric reasons. For example, the usual

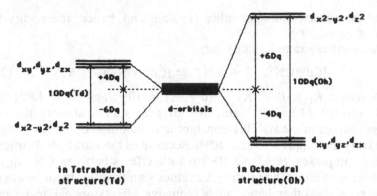

Figure 2-6. *d*-orbital splittings in octahedral and tetrahedral structures.

structure for complexes of the rather small cation Zn(II) ($r = 74$ pm) is tetrahedral, whereas that of the larger Cd(II) ($r = 106$ pm) is octahedral.

Let us take two common structures, tetrahedron and octahedron, to illustrate the effect of the electronic configuration. The expected energy splitting of *d* orbitals is shown for these two cases in Fig. 2-6. Furthermore, for simplicity we will for now restrict our analysis to the complexes in high-spin states. We can evaluate the LFSE for each of the electronic configurations d^0–d^{10} in terms of Dq (for octahedron) or Dq' (for tetrahedron). The results of such calculations are shown in Table 2-2. The LFSE for a tetrahedral structure can be recalculated and expressed in terms of the octahedral parameter Dq, assuming the theoretical relation $Dq' = (4/9)Dq$. The data in Table 2-2 permit the following conclusions to be drawn:

1. There is no difference in LFSE between the two structures in the d^0, d^5, and d^{10} cases. Therefore, metal cations with these electronic configurations can, by this criterion, equally well form octahedral or tetrahedral complexes.
2. In the case of d^3, d^4, d^8, and d^9, the octahedral structure is far more favorable than the tetrahedral one.
3. In the case of d^1, d^2, d^6, and d^7, the octahedral structure is preferred, but the LFSE difference between octahedra and tetrahedra is not very large. This suggests that tetrahedral structures may form nearly as readily as octahedral ones and may even be preferred with specific ligands.

The structures of the d^0, d^5, and d^{10} complexes of Mg(II), Ca(II), Mn(II), and Zn(II) are dictated by the size of the cation, as mentioned above, as well as by the binding energy obtained in the complex formation; the binding energy in these coordination compounds consists mainly of the electrostatic energy as described by Equation 2-19. Many complexes of d^0 with higher oxidation states [such as Ti(IV), V(V), Cr(VI), Mn(VII)] assume tetrahedral structures, both because of the small size of the metal ions and because of the electrostatic repulsions between ligands, usually O^{2-}. Mn(II) and Fe(III) are two typical examples of d^5 configuration. Although some tetrahedral complexes of these metal ions exist (e.g., $MnCl_4^{2-}$, $FeCl_4^-$), their usual coordination structure is octahedral. The main reason for this seems to be the increase in stability associated with coordinating more ligands.

The LFSE for octahedral complexes in low-spin states is $0Dq$ for d^0, $4Dq$ for d^1, $8Dq$ for d^2, $12Dq$ for d^3, $16Dq$ for d^4, $20Dq$ for d^5, $24Dq$ for d^6, $18Dq$ for d^7, $12Dq$ for d^8, $6Dq$ for d^9, and $0Dq$ for d^{10}. These values indicate that there should be a strong tendency toward formation of low-spin octahedral complexes for d^4, d^5, d^6, and d^7 ions. Examples of these configurations are Mn(III) (d^4), Fe(III) (d^5), Co(III) (d^6), and Co(II) (d^7). For d^5 and d^6, however, there is a strong opposing factor due to increased coulombic repulsive force present in low-spin cases. This implies that the tendency toward formation of a low-spin state should be a little less pronounced than the LFSE values suggest, especially if the value of Dq itself is not very large. There is also an increased coulombic repulsion effect in the case of low-spin octahedral complexes of d^4 and d^7, too; it is not as strong as that found in d^5 and d^6. These considerations would suggest that low-spin octahedral complexes should be most common among Co(III) complexes and somewhat rarer among Fe(III) and Fe(II) complexes.

In the discussion above, one factor called the "Jahn–Teller effect" has been ignored. A Jahn–Teller effect destabilizes an orbitally generate ground state such as E_g and T_{2g} so that the ground state splits into an orbitally singlet state and other states. The Jahn–Teller effect is especially significant for the E_g state but also has a slight effect on the T_{2g} state. An E_g ground state in an octahedral field is found with high-spin d^4, low-spin d^7 and d^9 configurations. Jahn–Teller effects would distort these structures, usually resulting in a tetragonal species. This is the reason why most common Cu(II) (d^9) complexes are square-planar.

Another relevant feature to consider is the tendency of a d^8 configuration to assume a tetragonal (or square-planar) structurre. This phenomenon can easily be explained by reference to Fig. 2-7. Such a system would be stabilized by lowering the ligand field effect on the d_{z^2} orbital, i.e., displacing axial ligands farther away from the metal center. The effect of this kind is seen in many Ni(II) complexes.

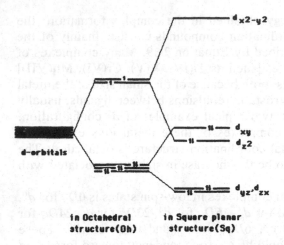

in Octahedral   in Square planar
structure(Oh)   structure(Sq)

Figure 2-7. The tendency to assume square-planar (tetragonal) structure in the case of d^8 configuration.

2.2.4 Constraints Imposed by Biomolecules

The structural principles outlined above are strictly applicable only in ideal cases in which no factors other than coordination bond forces are operative. In reality, however, a number of other constraints must be taken into account. Especially important among these are the spatial/energetic constraints imposed upon metal coordination sites by specific biomolecules, most notably proteins.

A system tends to proceed to a situation in which its overall potential energy is minimized. The observed coordination structure in a biomolecule is dictated by a complex interplay of the potential energy contributions arising from conformational adjustments by the organic ligands available and the direct effects of atomic orbitals presented by the metal atoms. Many metalloproteins undergo little conformational change upon removal of the metal cation. In these cases the coordination structure of the metal binding site is imposed by the protein conformation, irrespective of whether that particular structure is a preferred one for the metal. The conformation potential energy profile of the particular protein in such a case is relatively deep, so that it would require considerable energy for the protein to change its conformation to one coinciding more closely with the preferred structure of the metal ion. This result is not surprising, since the structure of a protein will have been molded through the long history of evolution, so as to function most efficiently as a whole, and there is no reason to expect that it will necessarily hold its active cation in its most favorable conformation (from the cation's viewpoint). The resulting "poised" structure has been described by Vallee and Williams (1968) as the result of an "entatic effect"; see Chapter 5 for a more detailed discussion of the matter. The rigid protein structures of this kind generally contain a larger proportion of β-sheet structure than α helix.

On the other hand, there are a number of proteins that do change their conformations upon binding a metal cation (or for that matter, upon binding an anionic entity such as phosphate): see Chapter 6 for a more detailed discussion of such an effect. Thus, the cation is here imposing its directing and/or electrostatic influence on the local area of a protein, which in turn determines the protein's overall conformation. It appears that most metal cations that produce conformational changes in proteins are acting as allosteric modifiers and not as catalytic elements. Even here it is reasonable to expect that the resulting structure is a compromise of these two opposing forces. However, proteins of this class appear to contain more α-helical segments; they are connected to each other by flexible random coil segments, so that the relative positions of these helical segments can readily be altered upon binding a metal ion.

Chemical Principles of the Biochemistry of the Elements

Oxidation–Reduction and Enzymes and Proteins

3.1 Introduction

Table 3-1 summarizes some of the important biological oxidation–reduction reactions and the elements associated with them. In the present chapter we are concerned with both the electron-carrier proteins and the enzymes in which inorganic elements function as electron mediators (i.e., the first two groups in Table 3-1), and the oxidation–reduction reactions involving sulfur and selenium. Proteins/enzymes involved in dealing with dioxygen and its derivatives will be discussed in the next chapter. Our main interest lies in the question: why would a specific element have been selected for a specific biological function? There are two important types of parameters to consider: thermodynamic and kinetic. These are illustrated in Fig. 3-1. The kinetic parameter expresses how fast the reaction proceeds and is dependent on the activation energy ΔH^* or ΔG^*. This will be considered later.

The thermodynamic factor is the free energy (or enthalpy) change ΔG (ΔH) between the initial and final states. In general,

$$S_1 \quad + \quad M(+n) \rightarrow S_1^+ + M(+n-1) \qquad \Delta G_1$$

$$\text{(substrate 1)} \quad \text{(metal ion)}$$

$$\langle \text{initial state} \rangle \rightarrow \langle \text{final state} \rangle \qquad \qquad (3\text{-}1)$$

The reduced species, $M(+n-1)$, usually is oxidized back to $M(+n)$ by another substrate S_2; that is:

$$S_2 + M(+n-1) \rightarrow S_2^- + M(+n) \qquad \Delta G_2 \qquad (3\text{-}2)$$

Thus, $M(+n)$ is recovered. Reduction potentials are the most readily available data for oxidation–reduction reactions and will be used hereafter instead of ΔG. The two parameters are related in the following way: $\Delta G = -nFE$, where E is the reduction potential, n is the number of electrons

Table 3-1. Bioelements Involved in Oxidation–Reduction Reactions

Type of reaction		Element	Examples
Electron transfer	Fe	Iron–sulfur	Ferredoxin, rubredoxin
		Heme iron	Cytochrome c, cytochrome a
	Cu	Blue-copper protein	Plastocyanin, azurin
Two-electron transfer	Mo		Xanthine oxidase, aldehyde oxidase, nitrogenase
(Hydride transfer			Nicotine adenine dinucleotide)
$2S^- - 2e \rightarrow -S-S-$	S		Lipoic acid, glutathione
	S		$SO_4^{2-} \rightarrow SO_3^{2-} \rightarrow H_2S$
Reversible O_2 binding	Fe		Hemoglobin, hemerythrin
	Cu		Hemocyanin
O_2 activation	Fe	Heme iron	Cytochrome P-450, tryptophan dioxygenase
		Nonheme iron	Dioxygenases, dihydroxylases
	Cu	Nonblue enzyme	Tyrosinase, dopamine-β-hydroxylase
Reduction of O_2 to HOH	Cu	Blue-copper enzyme	Ceruloplasmin, ascorbate oxidase, laccase
	Cu	Nonblue	Tyrosinase
(Reduction of O_2 to HOOH			Flavin adenine dinucleotide)
Peroxidase/catalase	Fe	Heme iron	Catalase, horseradish peroxidase
	Mn		Catalase-like protein
	Se		Glutathione peroxidase
Superoxide dismutase (SOD)	Fe		Fe-SOD
	Mn		Mn-SOD
	Cu		Cu/Zn-SOD
Others	Fe		Ribonucleotide reductase
	Co		Ribonucleotide reductase

involved in the process, and F is the Faraday constant. In order for reaction 3-1 to occur, the reduction potential of $M(+n)$ should be higher than that of S_1. Likewise, the reduction potential of S_2 should be higher than that of $M(+n)$ for reaction 3-2 to proceed to a significant degree. Thus, the reduction potential of $M(+n)/M(+n-1)$ should lie between those of S_1 and S_2, in order for the whole process (3-1 and 3-2) to be favorable. The situation is illustrated in Fig. 3-2. It must be pointed out that a metal M, whose reduction potential lies below but not too far below that of S_1, may function effectively in such a reaction, provided that the forward rate of reaction 3-1 is high despite the unfavorable thermodynamic conditions (energy of the final state relative to the initial state).

Reduction potential data can thus indicate the suitability of an element

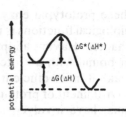

Figure 3-1. The potential energy profile of a reaction.

for involvement in a given reaction. However, before we proceed further, we must examine what sorts of compounds the original enzymes (and proteins) might have been, keeping in mind that reduction potentials of metal complexes are dependent on many factors, including the nature of ligands and the surrounding media. In other words, what we are here interested in is information about the nature of portotypal enzymes and proteins (or their precursors) and how their metallic components might have been picked up or adapted for their jobs by primitive organisms or their precursors. We know virtually nothing about this matter, and we must therefore base our arguments on a plausible hypothesis about how enzymes containing inorganic elements might have evolved. The following will be our working hypothesis (Ochiai, 1978a).

Prototypal enzymes were simple compounds, or to the extent they were complex, they were bound with nonspecific protein precursors such as proteinoids (as contended by Fox and Harada, 1958). Examples would include FeS bound to a proteinoid (Hall *et al.*, 1976), Fe-porphyrins, and simple Fe-complexes and their products of association with proteinoids. Iron–sulfur cores have been shown to form spontaneously from Fe(II), S^{2-}, and RSH (Berg and Holm, 1982). This may well have been the way in which the first primitive ferredoxin-like molecule was formed. These might have had the capacity to perform certain chemical tasks, though they are unlikely to have been as efficient and specific or selective as their progeny, i.e., present-day enzymes and proteins. The hypothesis that

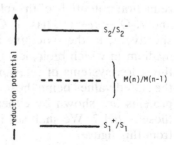

Figure 3-2. The reduction potential of the catalytic element $M(n)/M(n-1)$ versus those of the substrates S_1 and S_2.

these prototypal enzymes had the basic potentialities required for their biological functions is the "rule of basic fitness" mentioned in the previous chapter. Having adopted a specific element or its simple complex (free or bound with a proteinoid), an organism would then have attempted to make it more efficient and more specific by providing a suitably tailored environment of protein surrounding the element or complex. This process must have involved mutation in the genetic material, and is what was termed "evolutionary adaptation (pressure)" in the previous chapter. In the course of this process the thermodynamic potentialities of the element, such as its reduction potential, would have been modified specifically, although it can be assumed that they would not have varied greatly from their original values. It is for this reason that we can justify dealing with the capabilities or potentials of simple metal compounds as we attempt to explore consequences of the "rule of basic fitness." It must be pointed out, however, that a small modulation of a thermodynamic factor of a catalytic metal atom should also be considered to be an important function of the protein, a subject to which I will return later.

3.2 Basic Fitness Rule as Applied to Oxidation–Reduction Reactions

Figure 3-3 shows the reduction potentials of several important inorganic complexes and other redox systems. The points ($\bigcirc$) on the pH 0 and pH 14 lines represent the reduction potentials of the metal ions in aqueous media at pH 0 and pH 14. The two points for an ion have been connected simply by a straight line, although this does not necessarily represent the pH dependency of the reduction potential of that ion. It does, however, suggest a reasonable range of possible reduction potentials for the ion. Thus, Fig. 3-3 is not the same as an ordinary E_h–pH diagram. The other points, those shown by empty circles on the extended lines, are the reduction potentials of other simple compounds. The two dashed lines connecting the points at pH 0 and pH 14 represent the pH-dependent reduction potentials of the following reaction: $(1/2)O_2 + 2H^+ + 2e \rightarrow H_2O$ and $H_2O + e \rightarrow (1/2)H_2 + OH^-$. These two reactions correspond, respectively, to the oxidative and reductive decomposition of water, the medium in which biological reactions occur. The reduction potentials of the redox systems of relevance are shown on the left side of the figure, the stated values being those at pH 7. The potentials for the enzymes and proteins are shown by either $\bullet$ or $\blacktriangle$; these values, too, are nominally those at pH 7. We shall see in the ensuing discussion what we can learn from this figure.

To begin with, any species or redox couple whose reduction potential

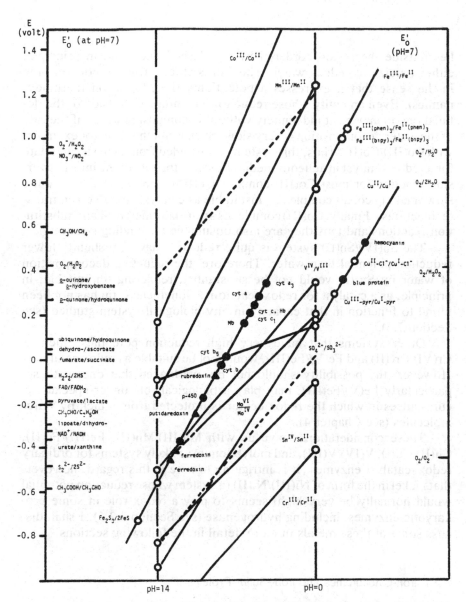

Figure 3-3. Reduction potentials of biologically relevant elements and systems. ○ represent the reduction potential of the aquo (or hydroxo) complex of a metal at pH 0 and pH 14; those two points are connected by a straight line, which does not necessarily represent the pH dependence of the reduction potential. There are a few other reduction potential data (represented by ⊘) of commonly encountered metal complexes on the extension of the line. The reduction potentials (at pH 7) of some representative proteins and enzymes containing iron or copper are shown by ● or ▲ on the respective line. All the other reduction potential values are those at pH 7. The dashed lines represent the oxidative and the reductive decomposition of water and their pH dependence; specifically, the top line is for $(1/2)O_2 + 2H^+ + 2e \rightarrow HOH$ and the bottom line is for $2H^+(HOH) + 2e \rightarrow H_2$.

lies outside the region bordered by the dashed lines can, in principle, either oxidize or reduce water. This holds strictly true, of course, only in the sense that it expresses a potentiality (i.e., based on thermodynamics). Even an entity whose reduction potential lies outside of the defined region may exist indefinitely without decomposing water, if the rate of the water decomposition is very slow. Such is not the case, for example, with Co(III)/Co(II). Thus, this system is precluded from being a candidate for a redox catalyst in aqueous media. Besides, the rate of the interconversion reaction for most Co(III)-complex/Co(II)-complex systems is quite slow, making cobalt complexes unsuitable as catalysts for electron transfer reactions. Finally, Co(III) complexes are usually inert to ligand substitution reactions and thus they are also unsuitable for binding substrates.

The Sn(IV)/Sn(II) system is quite reducing, having a slightly lower reduction potential than water. Therefore, the reductive decomposition of water by Sn(II) would not be substantial, rendering tin suitable, in principle, as a catalyst for redox reactions. Nonetheless, tin has not been found to function in that capacity in any biological system studied (see Section 7.5).

Other systems that have very high reduction potentials, such as Cr(VI)/Cr(III) and Fe(V)/Fe(III), would not be suitable as redox catalysts. However, the possibility should not be overlooked that one of these, particularly Fe(V)/Fe(III), may play a biological role under special circumstances in which the redox center is protected from attack by water molecules (see Chapter 4).

These considerations leave us with Mn(III)/Mn(II), Fe(III)/Fe(II), Cu(II)/Cu(I), V(IV)/V(III), and molybdenum as likely systems for ordinary redox–catalyst enzymes. It is intriguing to note in this regard, however, that nickel in the form of Ni(III)/Ni(II) (a system whose reduction potential would normally be very high) seems to play a redox role in some prokaryotic enzymes, including hydrogenase (see Section 3.2.5). I shall discuss some of these metals in more detail in the following sections.

3.2.1 Iron-Cytochromes and Iron–Sulfur Proteins

The most important oxidation states of iron are $+3$ and $+2$, although it has been suggested on the basis of some evidence that higher oxidation states such as $+4$ and $+5$ are involved in some special cases. The higher oxidation states of iron will be dealt with in Chapter 4. As seen in Fig. 3-3, the reduction potentials for Fe(III)/Fe(II) in hemoglobin, myoglobin, and most cytochromes lie within ± 0.3 V of the reduction potential of Fe(III)aq/Fe(II)aq at pH 7. The reduction potential of one particular cytochrome, P-450, is fairly low, however. One of the axial ligands of the heme in P-450 has been shown to be a cysteinyl sulfide (Hahn *et al.*, 1982; Poulos *et al.*, 1985). Negatively charged S atom(s) or O atom(s) are known to stabilize the Fe(III) state over the Fe(II) state, as is found in the low reduction potential of $Fe_2S_3/2FeS$. Catalase and peroxidase, which also show rather low reduction potentials, have a carboxylate O-anion as a ligand.

Ferredoxins were apparently devised for the purpose of transferring electrons in the course of photosynthesis; the electron acceptor here is $NAD(P)^+$, whose reduction potential is about -0.32 V. In order to be effective, any ferredoxin involved in this crucial step of photosynthesis must have a reduction potential lower than -0.32 V. The $Fe_2S_3/2FeS$ system falls in the proper range, and it is not surprising that organisms (or their precursors) selected this system for the job. It has been pointed out (Hall *et al.*, 1976) that ferredoxins are among the most ancient of biomolecules and that the primary structures of ferredoxins are not very different from those of abiotically produced proteinoids. In one bacterial ferredoxin [of the $2(Fe_4S_4)$ type] from *Clostridium butyricum*, it has been shown that 50 of the 55 amino acid residues consist of the nine most primitive (and abiotically available) amino acids, i.e., glycine, alanine, valine, proline, glutamic acid, aspartic acid, leucine, serine, and threonine (Hall *et al.*, 1976).

It is quite likely that the prototypes of cytochromes and other heme-containing enzymes and proteins consisted of some sort of porphyrin–iron complex (free or bound to proteinoids) rather than of simple compounds. Nevertheless, it remains true that simple iron complexes such as Fe-aquo and Fe-sulfide have similar electrochemical potentials or capacities as the enzymes and proteins utilized today. It seems also to be true that their potentials, though modified by evolutionary changes to suit specific requirements, have not changed very drastically. This relative constancy over time implies that the job performed by today's enzymes and proteins could also have been carried out by the simpler compounds discussed above, lending credence to an argument based on the rule of basic fitness.

It is well known (and clearly evident from Fig. 3-3) that the reduction potentials of cytochromes in the respiratory system are exceptionally well adjusted in relation to those of NADH, FAD, and ubiquinone. This adjustment can be viewed as a result of fine tuning accomplished under evolutinary pressure (see Section 3.4.2).

Both ferredoxin and flavodoxin (Mayhew and Ludwig, 1975) function as electron carriers in ways that are similar, the two being interchangeable in certain cases. Flavodoxin contains flavin mononucleotide (FMN) as its prosthetic group, a moiety whose reduction potential is close to that of flavin adenine dinucleotide (FAD), i.e., -0.22 V. Ferredoxin's functional unit, on the other hand, is comprised of iron–sulfur complexes whose potentials range from -0.45 to -0.24 V. Flavodoxin is, in general, less efficient than ferredoxin, and the synthesis of flavodoxin in a number of microorganisms occurs only if they are forced to grow in iron-poor media. For example, in the case of *Peptococcus elesdenii*, iron-rich cells were found not to contain flavodoxin, whereas iron deficiencies brought about its *de novo* synthesis. Here we have a case in which the rule of efficiency (rule 3) is applicable.

3.2.2 Other Redox Proteins Containing Iron

There are a number of proteins that contain so-called "nonheme" iron. Most are involved in the reactions with dioxygen and its derivatives. Examples include superoxide dismutase, hemerythrin, pyrocatechase, protocatechuate 4,5-dioxygenase, and prolyl hydroxylase. These proteins and enzymes will be discussed in Chapter 4.

Ribonucleotide reductase found in higher organisms contains iron that seems to participate somehow in a redox reaction, along with some free-radical source. The details of the reaction mechanism are unknown, however (Lammers and Follman, 1983).

3.2.3 Similarity between Iron and Manganese

A number of cases have now been discovered in which both iron- and manganese-containing proteins exist, performing the same biological function. The first to be discovered was superoxide dismutase (OSD); both Fe- and Mn-containing SODs are known, and their structural properties (primary and secondary) seem to be similar. In fact, it has recently been discovered (Clare *et al.*, 1984) that Fe-SOD from *E. coli* is actually heterogeneous, in that some of it is a Mn-containing variant, with both forms being functional.

Ribonucleotide reductases from various organisms differ considerably from one another in their requirements for an activator. Enzymes dependent on vitamin B_{12} coenzyme are found in anaerobic bacteria. The corresponding enzymes in most animals and plants, on the other hand, are dependent on nonheme iron. Recently, a new type of ribonucleotide reductase has been isolated from gram-negative bacteria such as *Brevibacterium ammoniagenes* and *Micrococcus luteus* (Schimpff-Weiland *et al.*, 1981). This enzyme has an absolute requirement for Mn(II) for its activation.

Catalase is another enzyme for which both Fe- and Mn-containing varieties are known. Ordinary catalase is a heme-containing protein, and is widely distributed. Several types of organisms are devoid of heme, however, and they seem to contain Mn-dependent catalase, instead. One such protein isolated from *Lactobacillus plantarum* seems to employ the valence change Mn(III)/Mn(II) (Kono and Fridovich, 1983a,b).

Dioxygenases generally require nonheme iron, although a Mn(II)-containing dioxygenase has been isolated from *Bacillus brevis* (Que *et al.*, 1981).

Acid phosphatase also apparently exists in both Fe and Mn forms; here oxidation–reduction reactions are not involved, however.

The above brief survey suggests that there are fundamental similarities between Fe and Mn. Let us examine their chemical properties in an attempt to see what parallels can be found. The range of known reduction potentials for Mn(III)/Mn(II) is shown in Fig. 3-3, from which it can be seen that the values for Fe and Mn are similar in aqueous media. A more detailed diagram (E_h–pH) for Fe and Mn (Fig. 3-4) confirms this general

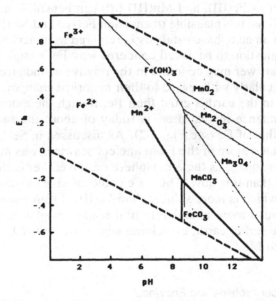

Figure 3-4. E_h–pH diagram for Fe and Mn. The thick lines represent the borderlines of the stability ranges of various states of iron/HOH/CO_3^{2-} system; the thin lines represent those of manganese system; the dashed lines represent oxidative (upper) and reductive (lower) decomposition of water. (After Maynard, J.B., 1983. *Geochemistry of Sedimentary Ore Deposits*, Springer-Verlag, Berlin.)

redox similarity between the two elements; the resemblance is clearly greater than that between Fe and any other common transition element (Cr, Co, Ni, and Cu). The ionic radii (6-coordinate and in high spin state) are also similar: 92 pm Fe(II)/97 pm Mn(II) and 78.5 pm Fe(III)/78.5 pm Mn(III). Again, the similarity is greater than any found between Fe and other transition elements. Since these are the very properties that should be important for biological functions, it is no surprise that Fe and Mn could occasionally replace one another in enzymes catalyzing such reactions.

This is about as far as we can go in establishing unique similarities between the two elements. Obviously there are fundamental differences as well. The major difference is in the electronic configuration. Mn(II) is d^5, whereas Fe(II) is d^6. Mn(II), being half-filled, would not benefit from ligand field stabilization, and hence it forms complexes that are much weaker than the corresponding ones for Fe(II). This would restrict somewhat the usefulness of Mn and may be the reason why, for example, there are no known naturally occurring porphyrins containing Mn(II). Other reasons may also be involved, however, including nonchemical ones, as discussed below. On the other hand, the acidity of Mn(III) is essentially as great as that of Fe(III), and Mn(III) (d^4) can benefit from ligand field stabilization. Thus, it is plausible to envision Fe(III) and Mn(III) behaving similarly when an acid–base catalysis is required as in acid phosphatase.

The last question to be asked concerns why Fe is more widely used than Mn. The answer may be found in the relative abundances of the two elements, particularly with regard to their historical changes. Mn is much less abundant in the earth's crust than Fe, though the more meaningful value, its abundance in seawater, is today of about the same order of magnitude as that of Fe (see Fig. 1-2). As discussed in Section 1.4, it is thought that much more of the Fe in ancient seawater was in the form of Fe(II) than it is today, as the atmosphere on the earlier earth was much more reducing than it is now. Thus, Fe could survive in water of neutral pH as Fe(II), which is more soluble than Fe(III). From this we conclude that Fe was much more abundantly and readily available to organisms than Mn on the earlier earth, explaining why it was utilized by organisms in preference to Mn.

3.2.4 Blue-Copper Proteins and Enzymes

The situation with copper is not so straightforward as that with iron, as the Cu(I)-aquo complex is not stable and disproportionates into Cu(II) and Cu(0). Therefore, a simple Cu(II)aq/Cu(I)aq system could not be regarded as a good candidate for the prototype of copper enzymes and proteins. Other simple possibilities include $CuCl_2$ and Cu(II)-amino acids. $CuCl_2$ may well have been one of the predominant species in the ancient sea. Furthermore, a Cu(II)-histidine complex would be likely to have a reduction potential similar to that of $Cu(pyridine)_2$ (see Fig. 3-3). The range of the reduction potentials of these compounds seems to embrace the reduction potential shown by the blue-copper centers in the so-called blue-copper enzymes and proteins, perhaps implying that simple complexes such as $CuCl_2$ and Cu(II)-histidine could have been the prototypes for those enzymes and proteins.

As mentioned in Chapter 1, copper enzymes and proteins are considered to have emerged fairly late in the course of evolution of life. Thus, by the time the first organisms picked up copper, they had already developed sophisticated protein systems, some of which may have been utilized (with some modification) to bind copper. Therefore, it is quite possible that the prototype of blue-copper enzymes and proteins might not, in fact, have been simple compounds like those mentioned above.

The next question that may be asked is why copper is utilized at all rather than the much more abundant iron. The so-called blue copper enzymes act on the compounds of the *p*- or *o*-benzoquinone type, as well as ascorbic acid. The reduction potentials of these compounds range from +0.05 to +0.4 V, which is rather high in the range of the reduction potentials for iron complexes and iron proteins but closer to (and below) the range of those for the simpler copper complexes proposed above for the prototypal copper enzymes. This implies that such copper complexes might be intrinsically more suitable for the oxidation of *p*- or *o*-quinones and ascorbic acid than would iron complexes. The way in which the rule of basic fitness applies to other types of copper enzymes and proteins will be discussed later.

3.2.5 Nickel as a Possible Redox Center

Increasingly enzymes are being discovered in which nickel is the activator. Some of these are of the acid–base type; examples include ureases found in plants and certain bacteria. Other Ni-containing enzymes are involved in oxidation–reduction reactions.

Methanobacterium thermoautotrophicum produces two hydrogenases containing iron and nickel (Kojima *et al.*, 1983; Lindahl *et al.*, 1984; Tan *et al.*, 1984). In addition, it produces methyl reductase whose chrom-

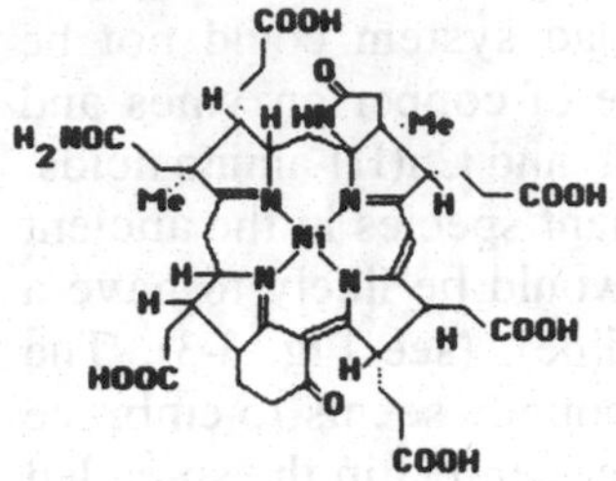

Figure 3-5. F_{430}, a Ni-containing factor found in methane-forming bacteria. (After Pfaltz et al., 1982.)

ophore is F_{430}, a Ni-containing tetrapyrole (Ellefson et al., 1982; Thauer, 1983); the tetrapyrole has been determined to be a corphine (Pfaltz et al., 1982; see Fig. 3-5). Hydrogenase from some Desulfovibrio species (D. gigas, D. desulfuricans) contains nickel and iron–sulfur clusters (Teixeira et al., 1983; Krüger et al., 1982), and carbon monoxide dehydrogenase from Clostridium thermoaceticum and other Clostridium species has also been shown to contain nickel and iron–sulfur proteins (Ragsdale et al., 1983). Aerobic CO dehydrogenase from Pseudomonas carboxydehydrogena and P. carboxydovorans, however, contains molybdenum rather than nickel.

The nickel in most of the enzymes cited is found as Ni(III) in the aerobically isolated inactive states; that this is so can readily be demonstrated by EPR (Teixeira et al., 1985). The Ni(III) in hydrogenase obtained from D. gigas seems to be reduced to Ni(II) with a reduction potential of -0.22 V (Teixeira et al., 1983); however, a more recent study (Teixeira et al., 1985) attributes the loss of the Ni(III) EPR signal upon reduction to an antiferromagnetic coupling between the Ni(III) and a reduced paramagnetic Fe-S cluster. There is some indication (Lindahl et al., 1984) that certain of the Ni ligands in methanobacterial hydrogenase are sulfur atoms. These observations are consistent with the notion that Ni constitutes a part of an iron–sulfur protein, just as does Mo in the Mo-Fe protein of nitrogenase (Section 7.1). It has also been suggested that the Ni in these clusters acts as the binding site of H_2 in the case of hydrogenase and perhaps of CO in the case of CO dehydrogenase. A Fe-Ni-C(O) complex formation has recently been demonstrated for the latter enzyme from C. thermoaceticum (Ragsdale et al., 1985).

The last suggestion may deserve a little more elaboration. Ni, as well as its congeners Pd and Pt, is a well-known catalyst for hydrogenation reactions in which the metal binds and activates H_2. Ni(0) is also known to readily bind CO to form $Ni(CO)_4$. The suggestion is consistent with these facts, which means then that the Ni-Fe-S cluster is to keep the virtual oxidation state of the Ni low, perhaps 0 or I, in order to use the

Ni center as the H_2 or CO binding center. This idea is very much analogous to an idea about the Mo-Fe-S cluster found in nitrogenase (see Section 7.1).

3.3 Peculiarity of Molybdenum

There are more than two dozen essential elements (Fig. 1-1). Most of them are light elements, with atomic number less than 34 (selenium). Molybdenum and iodine are exceptions. This, then, is the first of the peculiarities of molybdenum. Enzymes containing molybdenum are involved in several oxidation–reduction reactions, including nitrogen fixation (a reduction reaction). Most oxidation–reduction reactions are carried out with the aid of other metals, such as iron and copper found in a variety of metalloenzymes and proteins, as discussed above. One likely reason is that iron and copper are more abundant; nevertheless, despite its rarity, molybdenum has been specifically selected for use in certain kinds of oxidation–reduction enzymes. This is the second peculiarity of molybdenum. Finally, molybdenum is unique in its ability to constitute the essential site of nitrogenase; no other element seems to be able to replace it (although vanadium or tungsten has been reported to show some activity in certain cases). This is the third peculiarity (uniqueness) of molybdenum. I will discuss the first and second of these characteristics in this section and defer the discussion of nitrogenase to Chapter 7.

3.3.1 Availability

Since it belongs to the second series of transition metals, molybdenum has a greater tendency to assume higher oxidation states than its counterpart of the first transition series, chromium. Thus, Mo(VI), as MoO_4^{2-} or MoO_6 (as in polyacids), is the predominant oxidation state, whereas Cr(VI) is highly oxidative and unstable (in a thermodynamic sense). This is equivalent to saying that Mo(VI) has a very low reduction potential, the very reason that MoO_4^{2-} could have been prevalent in the earlier stages of the evolution of the earth, as discussed in Chapter 1.

MoO_4^{2-} behaves more or less like nonmetallic oxyanions such as SO_4^{2-} or SeO_4^{2-}. Because of the stability and solubility of MoO_4^{2-}, the abundance of molybdenum in seawater has been rather high over most of the time life has existed on earth. Its concentration in seawater today is comparable to that of iron despite the fact that its abundance in the earth's crust is much lower than that of iron.

3.3.2 Oxidation–Reduction and Other Characteristics of Molybdenum in Aqueous Medium

The oxidation–reduction reactions of molybdenum, even those of simple compounds in aqueous media, are not very well understood. The most common oxidation states are VI, V, IV, and III. Mo(VI), which exists as $H_m MoO_4^{-2+m}$ or as polymerized forms in aqueous media, would be reduced first to Mo(V) (MoO_2^+ or $MoOCl_5^{2-}$) and then to Mo(III) in strongly acidic solutions, particularly in concentrated HCl. Mo(IV), too, seems to be unstable in aqueous medium; thus, even if it were produced in the course of reduction of Mo(VI) or Mo(V), it would disproportionate to Mo(V) and Mo(III). That is:

$$2Mo(IV) \rightarrow Mo(V) + Mo(III) \quad \text{in acidic aqueous medium} \qquad (3\text{-}3)$$

It is to be noted that this disproportionation reaction would be prevented under such conditions that two Mo(IV) entities are prevented from coming close together, as, for example, in an enzyme in which one Mo atom is enclosed in a chain of a large polypeptide and spatially isolated from other such species. A mononuclear MoO_x [$x = 1$ for Mo(IV) and $x = 2$ for Mo(VI)] model compound of this type has recently been devised and prepared (Berg and Holm, 1984).

At neutral and basic pH values, $Mo^{IV}O_2$ is the thermodynamically stable species. Thus, $Mo^{VI}O_4^{2-}$ would under these conditions be reduced to Mo(IV). MoO_2, once formed, is a very inert solid. It may be assumed that MoO_4^{2-} is first reduced to form a basic oxide of the type $Mo^{IV}O_m(OH)_n$, probably not of definite composition, and the latter eventually turns into MoO_2. The reduction potential line given in Fig. 3-3 is thought to be that for the couple $Mo^{VI}O_4^{2-}/MoO_2$. The reduction potential for $MoO_4^{2-}/MoO_m(OH)_n$, the latter being presumed to be the primary reduction product, may be slighty lower than the one indicated by the line in the diagram.

The chemistry of molybdenum, particularly in its higher oxidation states, is dominated by oxo derivatives. The simplest Mo(VI) compound in aqueous media, MoO_4^{2-}, is stable in neutral to basic solutions but tends to oligomerize at lower pH, resulting in various so-called isopolyacids whose fundamental units are the octahedral MoO_6. Examples include $Mo_7O_{24}^{6-}$ and $Mo_8O_{26}^{4-}$. A similar tendency is also observed with WO_4^{2-}. Halides such as $MoCl_6$ exist, but they can only be prepared from molybdenum metal and the molecular halogens. The reaction of MoO_3 with the strong halogenating agent SO_2Cl_2 leads to $MoCl_4$, whereas the dissolution of MoO_4^{2-} in 12 M HCl results in the formation of $[MoO_2Cl_4]^{2-}$. The reaction of MoO_4^{2-} with chelating agents such as $(CH_3)_2NCS_2^-$, or that of MoO_2Cl_2 with bidentate chelating agents such as acetylacetone or

8-hydroxyquinoline, leads to the formation of dioxomolybdenum(VI) complexes of the type $Mo^{VI}O_2L_2$. These complexes have an approximately octahedral structure:

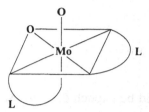

X-ray crystallographic study has shown that the bond length Mo—O is rather short, 163–171 pm, suggesting some double bond character (Spivack and Dori, 1973). As can be inferred from the description up to this point, the Mo—O bond persists under various reaction conditions. The only known simple monoxo entity is the halide $MoOCl_5^{2-}$; all other species contain either two or even three oxygens, i.e., MoO_2L_2 or MoO_3L.

It will be useful to discuss briefly the nature of the bond Mo—O, formally to be regarded as Mo^{VI}—O^{2-}. Because the d shell in Mo^{VI} is empty (d^0), the σ bond is supplied by a lone pair from the oxygen atom. As noted above, the bond distance is short, and hence there must be additional bonding as well, most likely a π-type interaction between a d_π orbital of the molybdenum and p orbitals of the oxygen atom (Fig. 3-6). There are two equivalent d_π-p_π bonds perpendicular to each other. Thus, the bond may be expressed by Mo≡O; but the extent of the observed π bonding is not so high that the overall character of the Mo—O bond seems to be about the same as that of a double bond. An important conclusion to be drawn from this description is that the electron density on oxygen is much less than the nominal value of $2-$, because of the donation of oxygen electrons to the vacant orbitals of the molybdenum atom. In fact, the description Mo=O^0 is probably fairly accurate as far as the effective charge on each atom is concerned. This in turn implies

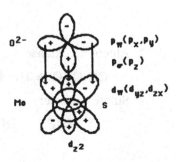

Figure 3-6. Bonding scheme in Mo–O moiety.

that the oxygen atom in the Mo—O bond may be fairly electrophilic, i.e., it would tend to accept a pair of electrons from a nucleophilic agent.

Mononuclear oxo complexes of Mo(V) are scarce; instead various compounds of the type

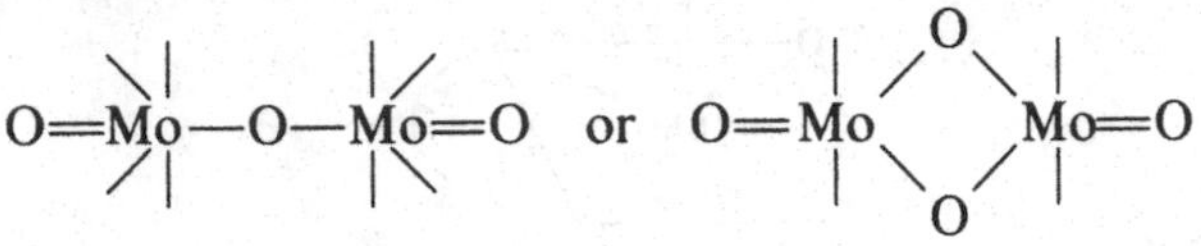

$$O{=}\overset{|}{\underset{/}{Mo}}{-}O{-}\overset{|}{\underset{|}{Mo}}{=}O \quad \text{or} \quad O{=}Mo\overset{O}{\underset{O}{\diagup\diagdown}}Mo{=}O$$

are well known. As would be expected, these Mo—O bonds, too, have double bond character.

Oxo Mo(IV) compounds are known but they are less numerous and less stable than those of Mo(VI) and Mo(V). The only dioxo complex known is $[Mo^{IV}O_2(CN)_4]^{4-}$ (Cotton, 1973); all the others are monooxo complexes of the type $MoOX_2$ (X = bidentate ligand). This is the major difference between Mo(IV) and Mo(VI), and it is due to the weakened d_π-p_π bond between Mo(IV) and O^{2-}. Since Mo(IV) is a d^2 species, the electron-withdrawing effect of Mo(IV), i.e., its ability to accept electron from O^{2-}, should be much weaker than that of Mo(VI).

3.3.3 The Types of Oxidation–Reduction Reactions Catalyzed by Molybdenum Enzymes

The known reactions catalyzed by enzymes containing molybdenum are as follows:
1. Xanthine oxidase:

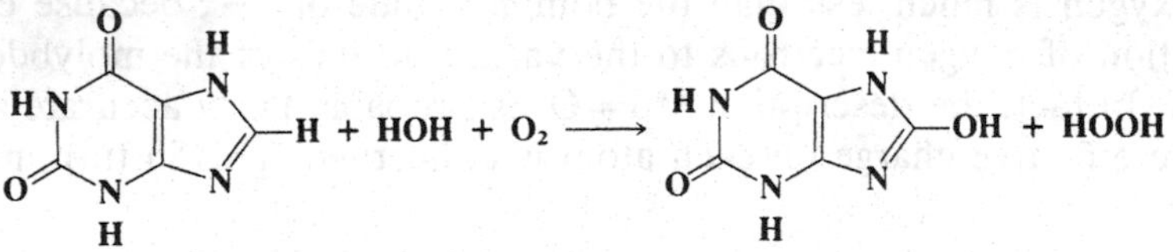

$$(3\text{-}4)$$

2. Aldehyde oxidase:

$$CH_3CHO + HOH + O_2 \rightarrow CH_3COOH + HOOH \qquad (3\text{-}5)$$

3. Sulfite oxidase:

$$SO_3^{2-} + HOH \rightarrow SO_4^{2-} + 2H^+ + 2e \qquad (3\text{-}6)$$

4. Nitrate reductase:

$$NO_3^- + 2H^+ + 2e \rightarrow NO_2^- + HOH \qquad (3\text{-}7)$$

It should also be noted that two electrons are involved in each of these reactions. In the first two cases, the reaction can be regarded as the removal of a hydride ion from the substrate; thus,

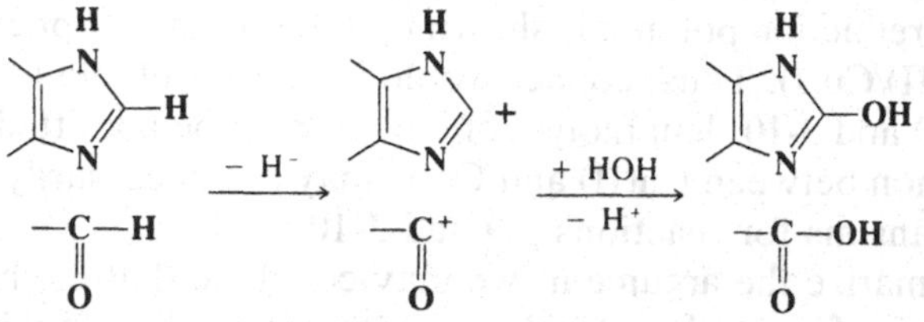

(3-8)

The reduction potentials of these processes are fairly low: E_0' (uric acid/ xanthine) $= -0.36$ V and E_0' (CH_3COOH/CH_3CHO) $= -0.59$ V. Therefore, an entity with low reduction potential, one in the range of, say, -0.3 to -0.5 V, would be suitable to carry out the oxidations. As an inspection of Fig. 3-3 will indicate, the lower portion of the Mo(VI)/Mo(IV) system, the Fe(III)/Fe(II) system, and the Sn(IV)/Sn(II) system are all apparently suitable for this job. Provided that the assumption of a two-electron process is correct, the system Fe(III)/Fe(II) can be rejected.

Likewise the reduction potential of SO_4^{2-}/SO_3^{2-} is rather low, -0.33 V, again falling in the range of the Mo(VI)/Mo(IV) system. However, it may be that the potential of Mo(VI)/Mo(IV) is much higher in sulfite oxidase than in the enzymes mentioned above. The reason is that the likely electron acceptor for Mo(IV) in sulfite oxidase is cytochrome b_5, which has a rather high potential ($E_0' = 0.0$ V), unlike the FAD found in the other two oxidases. In these three cases (reactions 3-4–3-6) molybdenum acts as an oxidizing center.

The reduction potential NO_3^-/NO_3^- is quite high, $+0.94$ V, and hence it is likely that the reduction potential of the molybdenum in nitrate reductase might lie in the highest portion of the Mo(VI)/Mo(IV) potential range.

The overall argument may become a little clearer if one compares these enzymatic reactions with some of those catalyzed by copper enzymes. The following reactions catalyzed by the so-called nonblue copper-enzymes are similar to reactions 3-4 and 3-5:

$$RCH_3NH_2 + HOH + O_2 \xrightarrow{\text{amine oxidase}} RCHO + NH_3 + HOOH \qquad (3\text{-}9)$$

$OHC(CH(OH))_4CH_2OH + HOH + O_2$

$$\xrightarrow{\text{galactose oxidase}} OHC(CH(OH))_4CHO + HOOH \quad (3\text{-}10)$$

The reduction potentials of reactions of this type would correspond to that of CH_3CHO/C_2H_5OH, which is -0.2 V: higher than those of uric acid/xanthine or CH_3CHHO/CH_3CHO. Correspondingly, the catalysts required should be ones with reduction potentials higher than -0.15 V, in the range of reduction potentials shown by the ordinary copper (complex) couples Cu(II)/Cu(I). Thus, copper would be more suitable for catalyzing reactions 3-9 and 3-10 than molybdenum. It must be admitted, however, that alternation between Cu(II) and Cu(I) may not necessarily play a part in the mechanisms for reactions 3-9 and 3-10.

To summarize the argument, we may conclude that the basic fitness of molybdenum for xanthine oxidase, aldehyde oxidase, sulfite oxidase, and nitrate reductase can be established by reference to the redox potentials of the Mo(VI)/Mo(IV) couple and also by matching the nature of the reactions with that of the molybdenum redox system, i.e., two-electron transfers are required. This implies that the prototype of contemporary molybdenum enzymes may have been MoO_4^{2-} or its simpler complex, such as $Mo^{VI}O_2(\text{amino acid})_2$ or $Mo^{VI}O(\text{proteinoid})$.

It has been well established that all of the molybdenum enzymes, with the exception of nitrogenase, contain a common factor, called "Mo cofactor" (Nason et al., 1971; Pateman et al., 1964). More recent studies (Johnson et al., 1984; Johnson and Rajagopalan, 1982) indicate that the cofactor contains a MoO_nS_2 ($n = 1$ or 2) unit, and the sulfur is a part of a pterin (urothione derivative).

3.3.4 Xanthine Oxidase and Aldehyde Oxidase

Reactions of the types 3-4, 3-5, 3-9 and 3-10 are accomplished by dehydrogenases dependent on NAD^+ or by flavoproteins containing either FAD or FMN. Thus,

$$\overset{|}{H}C{=}O + NAD^+ \rightarrow {}^+C{=}O + NADH \xrightarrow{\text{HOH}} \overset{OH}{\underset{|}{-}}\overset{|}{C}{=}O + NADH + H^+$$
$$(3\text{-}11)$$

$$-\overset{|}{C}H-OH + NAD \rightarrow -{}^{\pm}\overset{|}{C}-OH \xrightarrow{\text{NADH}} -\overset{|}{C}{=}O + NADH + H^+$$
$$(3\text{-}12)$$

$$\text{HOCH}_2\text{COO}^- + \text{O}_2 \xrightarrow{\text{glycolate oxidase}} \text{OHCCCOOO}^- + \text{HOOH} \quad (3\text{-}13)$$

$$\text{D-HOOCCH}_2\text{CH(NH}_2)\text{COOH} + \text{HOH} + \text{O}_2 \xrightarrow{\text{D-aspartate oxidase}}$$
$$\text{HOOCCH}_2\text{COCOOH} + \text{NH}_3 + \text{HOOH} \quad (3\text{-}14)$$

From this survey, it can be concluded that the most likely pathway for the reactions catalyzed by these molybdenum enzymes is the removal of a hydride from the substrates, as was assumed above (reaction 3-8). The other types of reactions represented by reactions 3-9, 3-10, 3-12, and 3-14 can occur either by hydride removal (as in rection 3-12) or by removal of two hydrogens of the type

$$-\overset{|}{\text{CH}}-\text{XH} \rightarrow -\overset{|}{\text{C}}{=}\text{X} + 2\text{H}$$

(as in all the other reactions). These two hydrogen atoms are accepted by FAD or FMN in flavoproteins.

A number of different mechanisms have been put forward for xanthine oxidase (and xanthine dehydrogenase) and aldehyde oxidase. Williams and Wentworth (1973) suggested that the hydrogen on the substrate is removed as an atom, i.e., $\text{Mo(VI)}{=}\text{O} + \text{H} \cdot \rightarrow \text{Mo(V)}-\text{OH}$, and they seem to have discarded the possibility of involvement of Mo(IV) as far as these two enzymes are concerned. Stiefel and Gardner (1973) emphasized the importance of hydrogen (proton) transfer in the course of reaction. They reasoned that molybdenum is important because of the deprotonation of a ligand bound to Mo(VI). Their proposed mechanism can be summarized as follows:

$$\text{E}-\overset{..}{\text{Mo(VI)}}-\text{NR} + \text{X}-\text{H(xanthine)} \longrightarrow$$

$$\left[\begin{array}{c} \text{X}{-\!\!-\!\!-\!\!-\!\!-}\text{H} \\ {}\vdots\uparrow \\ \text{E}-\text{Mo(VI)}-\overset{..}{\text{NR}} \end{array}\right] \xrightarrow{+ \text{OH}^-}$$

$$\left[\begin{array}{c} {}\overset{\displaystyle\curvearrowleft \text{OH}^-}{\text{X}{-\!-\!-\!-\!-}\text{H}} \\ {}\vdots\uparrow \\ \text{E}-\text{Mo(VI)}-\overset{..}{\text{NR}} \end{array}\right] \longrightarrow \text{E}-\text{Mo(IV)}-\text{NHR} + \text{X}-\text{OH} \quad (3\text{-}15)$$

Here it is assumed that the molybdenum possesses a deprotonated bound group of an amine or sulfide type, and that it is this neighboring group that accepts a proton from the substrate. Bray (1973) proposed a modification in which the important neighboring group is assumed to be a sulfide. His mechanism is as follows:

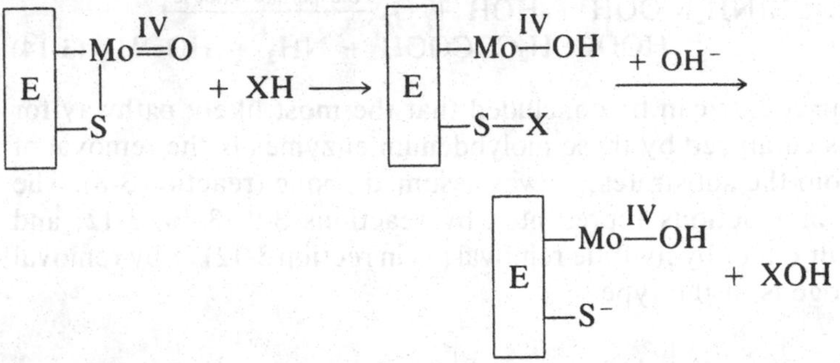

$$
\boxed{E \begin{array}{l} -\overset{IV}{Mo}{=}O \\ | \\ -S \end{array}} + XH \longrightarrow \boxed{E \begin{array}{l} -\overset{IV}{Mo}{-}OH \\ \\ -S{-}X \end{array}} \xrightarrow{+\ OH^-}
$$

$$
\boxed{E \begin{array}{l} -\overset{IV}{Mo}{-}OH \\ \\ -S^- \end{array}} + XOH \qquad (3\text{-}16)
$$

Olson *et al.* (1974) proposed the following mechanism, which appears to combine those developed by Stiefel and Gardner and by Bray.

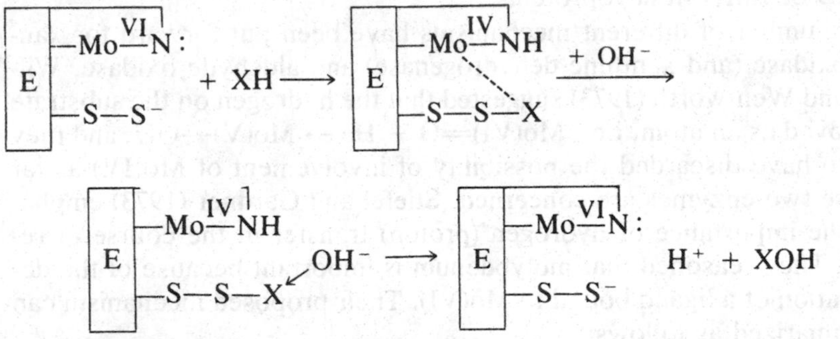

$$
\boxed{E \begin{array}{l} -\overset{VI}{Mo}{-}N: \\ \\ -S{-}S^- \end{array}} + XH \longrightarrow \boxed{E \begin{array}{l} -\overset{IV}{Mo}{-}NH \\ \\ -S{-}S{-}X \end{array}} \xrightarrow{+\ OH^-}
$$

$$
\boxed{E \begin{array}{l} -\overset{IV}{Mo}{-}NH \\ \\ -S{-}S{-}X \end{array}} \xrightarrow{\quad OH^- \quad} \boxed{E \begin{array}{l} -\overset{VI}{Mo}{-}N: \\ \\ -S{-}S^- \end{array}} H^+ + XOH \quad (3\text{-}17)
$$

Bray (1973) and Williams and Wentworth (1973) emphasized the importance of the Mo$=$O bond. This linkage has a double bond character, and the nature of the "O" in $Mo^{VI}{-}O^{2-}$ is far removed from that of O^{2-} and closer to O^0, which may act as a hydride acceptor. Thus, the possibility arises that the reaction proceeds in the following way (Ochiai, 1978a):

$$[Mo^{VI}{-}O^{2-} \Leftrightarrow Mo^{IV}{=}O] + XH \to Mo^{IV}{-}OH + X^+ \quad (3\text{-}18)$$

X^+ may have to be captured temporarily by a nucleophilic site (such as $-S-$ or $-S-S-$) before it reacts further with water. This set of events appears, on the surface, to be equivalent to the proposal of Bray (3-16). Mo(IV) then would be oxidized back to Mo(V) and then to Mo(VI) by a one-electron carrier, such as iron–sulfur unit, as presumed in the mechanisms advanced by Bray (1973) and Olson *et al.* (1974). The electrons (two per molecule of substrate) are eventually picked up by FAD; the

$FADH_2$ thus produced is oxidized by O_2 in a final step to regenerate FAD and form HOOH.

3.3.5 Sulfite Oxidase and Nitrate Reductase

Sulfite oxidase contains cytochrome b_5 as well as molybdenum. The electron transfer is considered to proceed as follows:

$$SO_3^{2-} \rightarrow Mo \rightarrow \text{cytochrome } b_5 \rightarrow \text{acceptor} \qquad (3\text{-}19)$$

The direction of electron transfer is the opposite of that in nitrate reductase:

$$NADH \rightarrow FAD \rightarrow \text{iron–sulfur protein} \rightarrow Mo \rightarrow NO_3^- \qquad (3\text{-}20)$$

In both cases, however, the molybdenum plays a role in substrate binding and perhaps in oxygen atom transfer as well. It is very likely that the Mo=O bond plays an important role in these enzymes. That is:

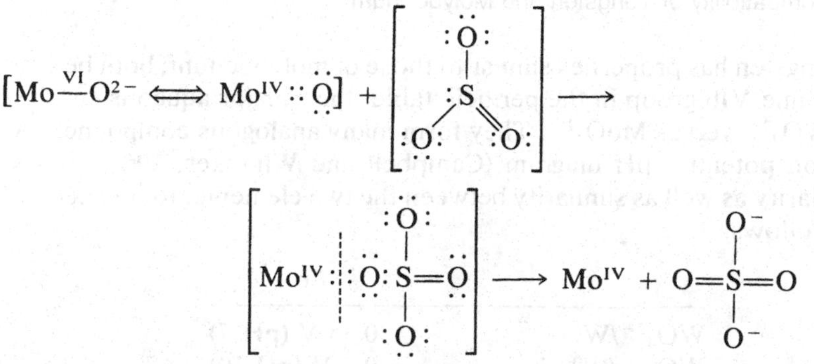

$$[Mo\overset{VI}{-}O^{2-} \Longleftrightarrow Mo^{IV} :: \ddot{O}] + \begin{bmatrix} :\ddot{O}: \\ | \\ :S \\ :\ddot{O}: \quad :\ddot{O}: \end{bmatrix} \longrightarrow$$

$$\begin{bmatrix} :\ddot{O}: \\ | \\ Mo^{IV} :: \ddot{O}: S = \ddot{O} \\ | \\ :\ddot{O}: \end{bmatrix} \longrightarrow Mo^{IV} + \begin{matrix} O^- \\ | \\ O = S = O \\ | \\ O^- \end{matrix} \qquad (3\text{-}21)$$

$$[Mo^{IV}:] + \begin{bmatrix} \ddot{O}: \\ :\ddot{O}-N \\ \ddot{O}: \end{bmatrix} \longrightarrow \begin{bmatrix} \ddot{O}: \\ Mo^{IV} :: \ddot{O}-N \\ \ddot{O}: \end{bmatrix} \longrightarrow$$

$$[Mo\overset{VI}{-}O^{2-} \Longleftrightarrow Mo^{IV} :: \ddot{O}] + :N\begin{matrix} O^- \\ \diagdown \\ O \end{matrix} \qquad (3\text{-}22)$$

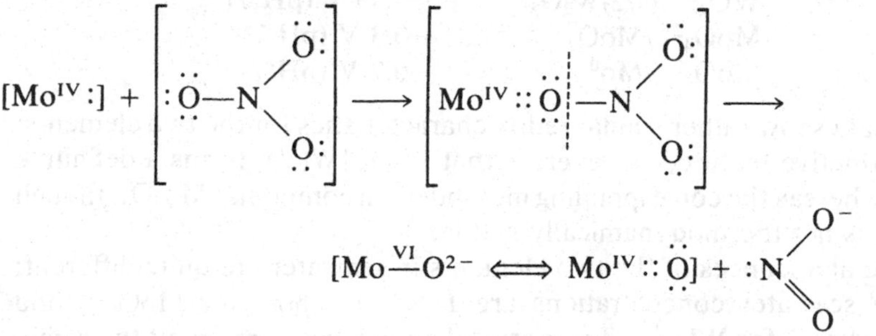

Whereas the oxygen atom in $Mo^{VI}O$ was postulated to be a hydride acceptor in aldehyde oxidase and xanthine oxidase, it is now assumed to be directly transferred between the molybdenum atom and the substrates in nitrate reductase and sulfite oxidase. An oxygen atom transfer mechanism has been suggested in view of the following known reactions (Schneider *et al.*, 1972; Chen *et al.*, 1976):

$$MoO_2(R_2dtc)_2 + P(C_6H_5)_3 \rightarrow MoO(R_2dtc)_2 + OP(C_6H_5)_3 \quad (3\text{-}23)$$

The basic concept that the "O" entity in $Mo^{VI}O$ is an electron-pair acceptor is common to both types of enzymes.

The reduction of NO_3^- to NO_2^- was assumed to be a two-electron reaction in the above mechanism; this does not necessarily have to be the case, however. NO_3^- may be reduced by two successive one-electron transfers in which case molybdenum as a two-electron mediator appears to lose its essentiality. In fact, a few other nitrate-reducing enzymes are known that contain FAD and cytochromes or iron–sulfur proteins but not molybdenum.

3.3.6 Compatibility of Tungsten and Molybdenum

Tungsten has properties similar to those of molybdenum; both belong to the same VIb group in the periodic table, their major aqueous species being WO_4^{2-} versus MoO_4^{2-}. They form many analogous compounds. A reduction potential–pH diagram (Campbell and Whiteker, 1969) shows dissimilarity as well as similarity between the two elements. Relevant data are as follows:

	Eo'
WO_4^{2-}/WO_2	-0.45 V (pH 7)
WO_4^{2-}/W^0	-0.7 V (pH 10)
$WO_4^{2-}/(1/2)W_2O_5$	-0.35 V (pH 7)
MoO_4^{2-}/MoO_2	-0.3 V (pH 7)
MoO_4^{2-}/Mo^0	-0.7 V (pH 7)

These data show rather similar redox characteristics for the two elements; one distinctive feature, however, is that W_2O_5 [W(V)] forms a definitive phase, whereas the corresponding molybdenum compound Mo_2O_5, though known, is not thermodynamically stable.

The abundances of the two elements in seawater are quite different; average seawater concentrations are 1×10^{-2} ppm for MoO_4^{2-} and 1×10^{-4} ppm for WO_4^{2-}. Their crustal abundances are about the same, however. The difference in seawater concentrations appears even more pronounced when expressed in terms of molar concentrations: 1×10^{-7} M (MoO_4^{2-}) and 5×10^{-10} M (WO_4^{2-}).

If our hypothesis regarding the use of the couple Mo(VI)/Mo(IV) is correct, and tungsten is assumed to work in a similar way, then the normal reduction potential of W(VI)/W(IV) must be raised slightly. This may or may not be possible; even if it is not, however, tungsten may find special use in some other enzymatic systems. Tungsten-substituted enzymes have been prepared as analogues for nitrate reductase (Guerrero and Vega, 1975; Lee *et al.*, 1974), sulfite oxidase (Johnson *et al.*, 1974b), and nitrogenase (Benemann *et al.*, 1973). These enzymes lacked activity, however, with the exception of weak activity shown by nitrogenase from *Rhodospirillum rubrum* (Paschinger, 1974) and hepatic sulfite oxidase (Johnson *et al.*, 1974b). Tungsten prevented the incorporation of molybdenum into xanthine oxidase without the tungsten itself being absorbed when rats were fed a diet containing both molybdenum and tungsten (Johnson *et al.*, 1974a). These observations indicate that tungsten is at best a poor substitute for molybdenum in these enzymes. This might be due to the slight difference in reduction potential between the two metals as outlined above, i.e., tungsten may be intrinsically less fit to do the required job compared with molybdenum. Another possible reason is that the enzymes (i.e., proteins) are so well adapted to molybdenum that they cannot accommodate tungsten in an optimum way, again because of subtle differences between the two elements despite their overall similarity (an application of the rule of evolutionary adaptation).

Formate dehydrogenase, which was not discussed in the preceding sections, has also been found to be a molybdenum- or, possibly, tungsten-dependent enzyme. Interestingly, this enzyme is also dependent on selenium (see Section 3.6). Formate dehydrogenase catalyzes the following reaction:

$$HCOOH - 2e \rightarrow CO_2 + 2H^+ \tag{3-24}$$

It seems to function in both the forward and the reverse sense, catalyzing either the dehydrogenation of HCOOH or the hydrogenation (reduction) of CO_2. An interesting observation has been made by Ljungdahl (1976): an interrelationship seems to exist between the preferred mode of physiological function of the enzyme (whether a dehydrogenation or reduction) and its metal requirement. The normal physiological role of formate dehydrogenase seems, in *Pseudomonas oxalaticus* and *E. coli*, to be the dehydrogenation of formate, whereas in *Clostridium* species (*C. pasteurianum*, *C. thermoaceticum*, and *C. formicoaceticum*) it is considered to be the reduction of CO_2. The enzyme in bacteria of the former type (which are, incidentally, facultative) contains 1 Mo, 1 Se, 1 heme, and 13–15 Fe and S per subunit of molecular weight 154,000 (for the *E. coli* enzyme). Tungsten acts as an antagonist of molybdenum in the *E. coli* enzyme. The enzymes in *C. thermoaceticum* and *C. formicoaceticum*, on the other hand, specifically incorporate tungsten from a medium that contains 100 times more molybdenum than tungsten. This tungsten-containing enzyme is quite active, its turnover number being 5 to 800 times that of most molybdenum enzymes. Thus, formate dehydrogenase from *C. thermoaceticum* and *C. formicoaceticum* is a true tungsten-containing enzyme, the first one to be found and so far the only one definitely known. It contains, in molar terms, 2 W, 2 Se, 36 Fe, and about 50 S in a species of molecular weight 340,000 (Yamamoto *et al.*, 1983).

Let us next turn to a possible explanation for this finding. The reduction potential of $CO_2/HCOOH$ is about -0.42 V. A molybdenum system based on the couple $Mo(VI)/Mo(IV)$ would not have a potential much lower than -0.35 V, as discussed above. Therefore, molybdenum may be suited to the oxidation of HCOOH. On the other hand, the reduction potential of the tungsten couple $W(VI)/W(IV)$ in a protein is expected to be a little lower than its normal value of WO_4^{2-}/WO_2 -0.45 V. Therefore, a tungsten enzyme may be more fit for the job of reducing CO_2, as opposed to the reverse reaction. The situation can be summarized as follows:

$$Mo(VI)/Mo(IV) \qquad CO_2/HCOOH \qquad W(VI)/W(IV) \qquad (3\text{-}25)$$
$$-0.35 \text{ V} \qquad\qquad -0.42 \text{ V} \qquad\qquad -0.45 \text{ V}$$

This argument, of course, presupposes that both tungsten and molybdenum are involved directly in the electron transfer to CO_2 or from HCOOH. The electron abstraction from HCOOH (probably in the form of hydride abstraction from the carbonyl carbon atom) is analogous to the reactions of aldehyde or xanthine oxidase discussed earlier. Thus, the assumption is probably not unreasonable. This appears also to be corroborated by a recent EXAFS data (Cramer *et al.*, 1985): the W enzyme lacks $W=O$ bonds, whereas the Mo enzyme contains $Mo=O$ bonds. As exemplified by equations 3-21 and 3-22, a reducing enzyme (the W enzyme in this case if the above argument is right) would not require $M=O$ bonds, whereas an oxidizing enzyme (the Mo enzyme in this case) would, in the present framework of mechanistic hypothesis. This is in agreement with the characteristics of the two enzymes as revealed by EXAFS.

3.4 Manifestation of Evolutionary Adaptation

Having adopted a specific inorganic element for a specific purpose, organisms (or protobionts) should have attempted to make increasingly better use of it. In other words, they would have tried to improve upon that specific function (any function for that matter), which included the element, particularly in regard to its specificity, selectivity, and efficiency. This process would have been genetic, involving mutations in genes, and must have taken place rather rapidly at the earlier steps of the formation of life as far as most of the basic enzymes and proteins are concerned. Little study has been made on this early evolutionary process, especially as it relates to metalloenzymes and metalloproteins.

What we do know are the results of such early evolutionary processes and the subsequent long history of evolution; that is, the evidence we have is the metalloproteins and metalloenzymes found in contemporary organisms. All we can do with these data is to attempt to rationalize why enzymes and proteins are in their present forms. I will discuss some of the manifestations of evolutionary pressure in the form of improved efficiency, selectivity, and specificity as they apply to the iron–sulfur proteins, cytochromes, and copper-containing enzymes and proteins dealt with in Section 3.2.

3.4.1 Iron–Sulfur Proteins

Most iron–sulfur proteins are involved in some type of electron transport system associated with FAD (FMN) and/or $NAD(P)^+$, and accordingly, they have fairly low reduction potentials, although their precise values can be subject to substantial variation over the range of known examples.

One type of iron–sulfur protein is involved in photosynthesis (see Fig. 3-7); in this case (ferredoxin) its role is to accept electrons from ionized chlorophyll and then transfer them to $NAD(P)^+$. The reduction potential of the ionized chlorophyll system is estimated to be about -0.6

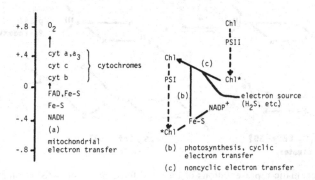

Figure 3-7. Electron transfer systems in (a) mitochondrion, and (b, c) in chloroplast.

V and that of NAD(P)$^+$ is about -0.3 V. Therefore, the ferredoxins must have reduction potentials in between -0.3 and -0.6 V. Ferredoxins from both plants and bacteria have indeed been found to show reduction potentials around -0.4 V, even though the potential of the simple couple Fe_2S_3/FeS is about -0.7 V. Binding an Fe_nS_m unit to a protein would certainly tend to raise its reduction potential, because the extremely low potential of Fe_nS_m is due to the very high affinity between S^{2-} and Fe(III), as well as to the lattice energy. Thus, a decrease in the number of S^{2-} ions bound to the Fe(III), coupled with disruption of the lattice energy of the solid, would reduce the stability of $Fe^{III}_nS_m$ relative to the corresponding Fe^{II} species, thereby raising the reduction potential of the Fe—S system. The probable way by which organisms have accomplished a change in the reduction potential is by isolating and incorporating into a protein two of the simplest subunits of the FeS solid, Fe_2S_2 and Fe_4S_4 unit. Ferredoxins containing Fe_2S_2 are found in cyanobacteria and plants, whereas Fe_4S_4 ferredoxins have been isolated from bacteria (Fig. 3-8). A few other special types of bacteria also seem to contain Fe_2S_2 proteins, including the recently characterized "red paramagnetic protein" from *Clostridium pasteurianum* (Meyer *et al.*, 1984).

The recent discovery of an Fe_3S_3-type cluster from *Azotobacter vinelandii, Desulfovibrio gigas,* and others (Emptage *et al.*, 1980; Stout *et*

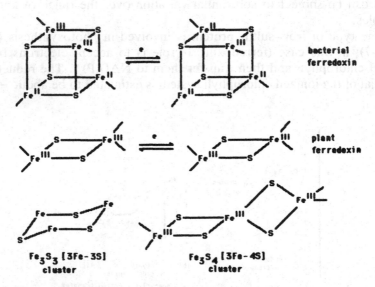

Figure 3-8. Schematic representations of bacterial ferredoxin ([4Fe–4S] cluster), plant ferredoxin ([2Fe–2S] cluster), and [3Fe–3S] and [3Fe–4S] clusters.

al., 1980; Stout, 1982; Munck, 1982) complicates the situation. The structure of the Fe_3S_3 cluster is essentially that of a twisted boat form of a six-membered ring, with Fe and S atoms at alternate positions. However, it is deemed highly likely (Johnson *et al.*, 1982) that in certain cases at least this Fe_3S_3 cluster is actually an artifact—an air-oxidation product of an Fe_4S_4 during an aerobic isolation process. A linear Fe_3S_4 cluster, three iron atoms arranged linearly and doubly bridged by sulfides, has also been identified as an inactive partially unfolded aconitase (Kennedy *et al.*, 1984; see Section 5.3.2 for aconitase). A recent development (e.g., see Morningstar *et al.*, 1985) seems to indicate that the Fe_3S_3 is a genuinely active unit in some other cases.

Another group of iron–sulfur proteins is that comprising the FAD (FMN)-dependent enzymes and proteins (involved in electron transfer; see Fig. 3-7). In these enzymes and proteins, which include adrenodoxin and putidaredoxin, as well as the iron–sulfur proteins of the mitochondrial respiratory chain, the job of the iron–sulfur unit is to accept electron(s) from or donate electrons to FAD(H) [or FMN(H)]. The reduction potential of FAD (FMN) is about -0.23 V; therefore, the reduction potentials of the Fe—S units in this class of enzymes should either be slightly lower or higher than -0.23 V. The measured reduction potentials of such Fe—S units from several organisms have been found to range from -0.3 to $+0.04$ V (Hall *et al.*, 1976). This rather high reduction potential may be induced by changing the conformation and the environment of the Fe—S unit, especially elongating the Fe—Fe distance and slightly bending the Fe—Fe bond at its center.

Next we consider a group of iron–sulfur units of the $Fe(cysteine)_4$ type; here the iron atom is surrounded by four cysteine sulfur atoms, each with one unit of negative charge. One such unit is found in rubredoxin. The Fe(III)-stabilizing effect of $S-\!-\!CH_2CH(NH_2)COOH$ is much less than that of the S^{2-} ion. Thus, rubredoxin's reduction potential ranges from -0.06 to 0 V. This group of iron–sulfur proteins is considered to be involved in rather special types of electron transfer, such as ω-hydroxylation of fatty acids (in *Clostridium, Pseudomonas, Micrococcus*, and others).

Those iron–sulfur proteins that show the highest reduction potentials, about $+0.35$ V, are called "high-potential iron–sulfur protein" (HiPIP); they are isolated from the anaerobic photoautotroph *Chromatium* and others. The structure of the Fe–S core of HiPIP turns out to be surprisingly similar to that of the Fe_4S_4 core in bacterial ferredoxin. The tricks that have been employed in forming such a high reduction potential out of the Fe_4S_4 core have been debated (see Sheridan *et al.*, 1981). Each of the iron atoms in the Fe_4S_4 core could be in either the Fe(II) or the Fe(III) state; or the electrons can be assumed to be delocalized among the four iron atoms and four sulfur atoms and hence each iron atom could carry a fractional electric charge. There are possibly five different overall oxidation states envisioned for Fe_4S_4 if we assign a discrete electric charge to each iron atom: $Fe^{III}_4S_4$, $Fe^{III}_3Fe^{II}S_4$, $Fe^{III}_2Fe^{II}_2S_4$, $Fe^{III}_3Fe^{II}S_2$, and $Fe^{II}_4S_4$. These ought to be described as $(Fe_4S_4)^{4+}$, $(Fe_4S_4)^{3+}$, $(Fe_4S_4)^{2+}$, and $(Fe_4S_4)^+$, if electron delocalization is taken into consideration. It is believed (Carter *et al.*, 1972) that bacterial ferredoxin operates with the couple $(Fe_4S_4)^{2+}/(Fe_4S_4)^+$, whereas HiPIP utilizes the couple $(Fe_4S_4)^{3+}/(Fe_4S_4)^{2+}$. That is:

$$Fe^{III}_3Fe^{II}S_4 \xrightleftharpoons[\text{HiPIP}]{e} Fe^{III}_2Fe^{II}_2S_4 \xrightleftharpoons[\text{ferredoxin}]{e} Fe^{III}Fe^{II}_3S_4 \qquad (3\text{-}26)$$

The function and the reason for the high reduction potential of HiPIP are not well understood. It might be noted, however, that an enzyme known as aconitase has been found to contain a rather unique Fe_4S_4 core, which appears to have some resemblance to that of HiPIP (see Section 5.3.2.2).

All of the iron–sulfur proteins discussed above have been subject to a long history of evolution in the course of which organisms have adopted them to meet specific needs within certain environmental constraints. This has led to modification of the Fe–S units and of the proteins themselves, leading to systems ideally suited for the required job. We are here presented with an excellent example of the interrelationship of thermodynamic effects and evolutionary pressures.

3.4.2 Heme Proteins

3.4.2.1 Porphyrins

There are a vast number of heme-containing enzymes and proteins. Oxygen carriers (hemoglobin, myoglobin, leghemoglobin) and enzymes involved in the reactions of oxygen and its derivatives will be more fully discussed in Chapter 4. Here we concentrate on cytochromes.

A special cytochrome, P-450, presents an interesting example of the effect of the evolutionary pressure. Resting P-450 is in a low-spin Fe(III) state, and has a reduction potential of -0.38 V (see Fig. 3-3). The Fe(III) in this state cannot be reduced by NAD(P)H. Only when P-450 binds a substrate does it become reducible by NAD(P)H, because its reduction potential is raised to -0.17 V. This potential is higher than that of putidaredoxin, one of the electron carriers in the P-450-dependent monooxygenases. If this were not the case, i.e., it had a high reduction potential,

P-450 would exist in its reduced state, i.e., Fe(II), while resting. This state [Fe(II)] has an affinity for O_2, and the product of the O_2 binding would lead to undesirable reactions such as the oxygenation of its own proximate amino acid residues. Thus, P-450 underwent modifications such that it becomes reactive only when the enzyme binds a substrate, which can block the attack of the protein's own amino acids by the active entity. This fine adjustment of the reduction potential of P-450 involves a conformational change of the protein and is, no doubt, a result of evolutionary adaptation.

Other cytochromes play important parts in the following three major electron transfer processes: mitochondrial respiration, the cyclic portion of photosystem I, and the electron transfer chain leading to photosystem I, starting either from an electron source compound (such as H_2S in the case of bacteria) or from photosystem II (in the case of cyanobacteria and eukaryotic plants). The thermodynamic factors (the reduction potentials) of these systems are illustrated in Fig. 3-7. The figure suggests the following:

1. Those cytochromes involved in the mitochondrial respiratory chain should have reduction potentials lying between about 0 and $+0.8$ V.
2. The reduction potentials of cytochromes involved in photosynthetic cyclic electron transfer (coupled with phosphorylation) should range from about -0.4 V to $+0.4$ V.
3. The electron transfer systems connecting PSI with PSII or some electron source (e.g., H_2S, CH_3COOH) would contain cytochromes whose reduction potentials range up to $+0.4$ V, starting at about $+0.2$ V in the case of PSII or much lower, about -0.2 V, in the case where H_2S is the electron source.

Other types of electron transfer systems occur more rarely in special organisms, including those that utilize SO_4^{2-}, CO_2, or NO_3^- as their ultimate electron acceptor. The reduction potential of NO_3^- to NO_2^- is $+0.94$ V, very close to that of oxygen ($+0.82$ V). Therefore, the reduction potentials of cytochromes involved in this electron transfer (NO_3^- reduction) should be similar to those in aerobic respiration. The other two electron acceptors, SO_4^{2-} and CO_2, have fairly low reduction potentials (-0.4 V). It is known that CO_2 reducers, such as methane-forming bacteria, lack cytochromes altogether; they instead utilize iron–sulfur proteins and cobalamin. SO_4^{2-} reducers, including *Desulfovibrio*, contain cytochromes with very low reduction potentials.

Not many cytochrome structures have been elucidated. Therefore, the conventional classification scheme for cytochromes relies not on struc-

Table 3-2. Reduction Potentials of Representative Cytochromes[a]

Type	Name	Source	E_0' (V)	Function (possible)[b]
A	a_3	Mitochondria	+0.35	AR (cytochrome oxidase)
	a	Mitochondria	+0.28	AR (cytochrome oxidase)
B	b	Mitochondria	+0.035	AR
	b_1	E. coli	−0.34	
	b_2	Yeast	+0.34	AR
	b_5	Microsomes	+0.02	P-450 system
	b_7	Aram (plant)	−0.03	AR
	b_6	Chloroplasts	−0.06	AR or PC
	b_{562}	Bacterium anitratum	+0.13	AR
	b_{556}	Saccharomyces cerevisiae	−0.02	AR
C	c	Mitochondria	+0.25	AR
	c_1	Mitochondria	+0.22	AR
	c_4	Azotobacter vinelandii	+0.30	AR
	c_{552}	Micrococcus denitrificans	+0.32	NR, AR
	c_{555}	Crithidia fasciculata	+0.28	AR
	c_{553}	Nitrosomonas europaea	+0.50	AR (NH_3 oxidation)
	cc′	Chromatium D	−0.01	PC–PNC (with H_2S, H_2)
	c_{553}	Chromatium D	+0.33	PC
	c_3	Desulfovibrio sp.	−0.2	SR
	$c_{555}(f)$	Chloroplasts	+0.3–+0.4	PSI–PSII
Cytochrome P-450			−0.38	
Hemoglobin			+0.17	
Horseradish peroxidase			−0.17	
Catalase			−0.42	

[a] Compiled from T. Yamanaka and K. Okunuki, in Microbial Iron Metabolism (J. B. Neilands, ed.), pp. 349–400, Academic Press (1974); R. E. Dickerson and R. Timkovich, in The Enzymes, 3rd ed. (P. D. Boyer, ed.), Vol. XI, Part A, pp. 397–547, Academic Press (1975); H. R. Mahler and E. H. Cordes, Biological Chemistry, Harper & Row (1971); G. R. Moore and R. J. P. Williams, Coord. Chem. Rev. 18:125–197 (1976).

[b] AR, aerobic respiratory electron transfer system; NR, nitrate-reducing electron transfer system; SR, sulfate-reducing electron transfer system; PC, photosynthetic, cyclic electron transfer system; PNC, photosynthetic, noncyclic electron transfer system; PSI–PSII, electron transfer system connecting PSI and PSII.

ture but on the characteristic absorption spectra; cytochromes are thus classified into types A, B, C, D, E, F, O, and so forth. Comprehensive lists of cytochromes are found in Yamanaka and Okunuki (1974) and Dickerson and Timkovich (1973). A few selected examples are given in Table 3-2.

As seen in Table 3-2, the reduction potentials of cytochromes (and of heme proteins in general) range from −0.2 to +0.5 V with the exceptions of the low potentials of cytochromes b_1, c_3, P-450 and catalase (as well as peroxidase). Each individual cytochrome was presumably created to meet a specific need (or needs) of a particular organism (evolutionary pressure). As exemplified by Athiorhodaceae (*Rhodospirillum* and *Rhodopseudomonas*), a single electron transfer system containing cytochromes may be utilized for both the cyclic phosphorylation system of photosynthesis and a portion of the same organism's respiratory electron transfer system.

How then can organisms modulate the reduction potential of cytochromes so as to meet specific needs? There are many factors that control the reduction potential of a cytochrome. Major factors include: (1) the nature of the peripheral substituents of the porphyrin ring, (2) the types of fifth and/or sixth ligands coordinated to the iron atom in the porphyrin ring, (3) the positions of the ligands (their distances and whether they are exactly axial or slightly displaced), and (4) the environment of the heme group (e.g., whether or not it is hydrophobic).

The structures of some representative heme groups are shown in Fig. 3-9. In general, electron-withdrawing groups or groups with conjugating double bonds (such as a vinyl group) on the peripheral positions of a porphyrin tend to stabilize lower oxidation states, thus raising the reduction potential. The rather high reduction potentials of various species of cytochrome a result partly from the fact that the porphyrin ring of this group bears an aldehyde carbonyl group, an electron-withdrawing group. However, as exemplified by the derivatives of porphyrin IX, including type b cytochromes, catalase, and hemoglobin, factors other than the

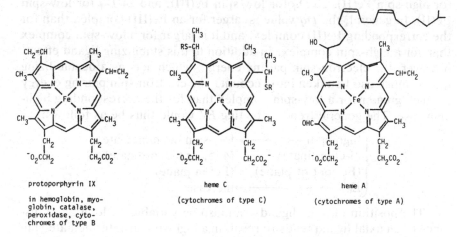

protoporphyrin IX

in hemoglobin, myo-
globin, catalase,
peroxidase, cyto-
chromes of type B

heme C

(cytochromes of type C)

heme A

(cytochromes of type A)

Figure 3-9. Structures of representative heme (porphyrin) groups.

peripheral substituents seem to exert greater effects on the reduction potential. These proteins, all of which contain the same porphyrin, show a wide range of reduction potentials: from +0.17 V (hemoglobin) to below −0.42 V (catalase).

Negatively charged ligands bind to a metal of higher oxidation states more strongly than to one in its lower oxidation states, thus tending to stabilize the higher oxidation states, which in turn lowers the reduction potential. A neutral ligand, on the other hand, does not affect the reduction potential significantly. The low reduction potential of P-450 is considered to be due to the fact that its fifth ligand is a S^- of cysteine (see p. 59). This can be compared with cytochrome c, which has a methionine (with a neutral S atom) as its fifth ligand and, as a result, a rather high potential. The extremely low potential of catalase is presumably due to its fifth ligand, which is believed to be a negatively charged carboxylate group.

One additional complicating factor in any consideration of the electronic structure (and, hence, of the reduction potential of a heme protein) is that its iron atom can be in various states: it may lie in the plane of the porphyrin ring or it may be out of the plane; furthermore, it may be in either a high- or a low-spin state. The Fe(II) in hemoglobin, for example, is located slightly above the center of the porphyrin ring with an imidazole group attached to it as a fifth ligand and it is in the high-spin state ($S = 2$). The iron atom of cytochrome c, on the other hand, is in the plane, is six-coordinated, and takes a low-spin state regardless of whether it is Fe(III) or Fe(II).

The ligand field stabilization energy (LFSE) for iron can readily be calculated for octahedral symmetry; it is 0 Dq for high-spin Fe(III), 4 Dq for high-spin Fe(II), 20 Dq for low-spin Fe(III), and 24 Dq for low-spin Fe(II). In general, the Dq value is larger for an Fe(III) complex than for the corresponding Fe(II) complex, and it is larger for a low-spin complex than for a high-spin complex. In addition to this stabilizing ligand effect, however, the electron-spin pairing energy, which acts as a destabilizing factor, must also be taken into account. The electron-spin pairing energy is much greater for a low-spin complex than for the corresponding high-spin case. The general trends affecting E_0' would thus be as follows:

$$\begin{cases} \text{high-spin} \longrightarrow \text{low-spin or immediate} \\ \text{(5-coordinate)} \quad \text{(4, 5, or 6-coordinate)} \\ \text{(Fe: out of plane)} \quad \text{(Fe: in plane)} \\ E_0' \longrightarrow \text{increase} \end{cases}$$

The position of axial ligands can also be variable. A long bond distance to an axial ligand tends to result in a high-spin or intermediate-spin state $[S = 3/2$ for Fe(III) and 1 for Fe(II)]. In addition, the ligand may be slightly offset from the axis; this, too, affects the reduction potential, as shown in Fig. 3-10. A slight displacement of the axial ligand from the strictly axial position reduces its destabilizing effect on the d_{z^2} orbital (the d_{z^2} level being lowered as a result) and also, though to a much lesser extent, destabilizes either d_{yz} or d_{zx} orbital. Figure 3-10 depicts the situation as it pertains to the Fe(III) oxidation state; in order for this to be reduced, an electron must be added to the lowest possible level. In the case of low-spin complexes, this level will have been destablilized by off-axis ligand displacement. Therefore, a displacement of this kind tends to decrease the reduction potential. With other spin states, however, the effect would not be as great, although the tendency would still be toward a slight decrease of reduction potential with off-axis displacement. The effects of structure on the reduction potentials of heme groups are summarized in Table 3-3.

A hydrophobic environment enclosing solvents of high dielectric constant, such as water, would have a stabilizing effect on more polarized entities. A strictly hydrophobic environment, in contrast, would favor less polarized species. A lower oxidation state would, therefore, be preferred in a hydrophobic environment, thus leading to an increase in reduction potential. This intuitive argument has been substantiated by a theoretical calculation showing a direct correlation between decreased polarity of the environment and the reduction potentials of type c cytochromes (Kassner, 1973). An inverse correlation was found between the fraction (%) of a heme group exposed to solvent and reduction potential in a series of heme proteins in a similar nonpolar environment (Stellwagen, 1978), which further corroborates the intuitive conclusion.

Organisms all of which utilize the same basic porphyrin–iron complex as a prosthetic group have nevertheless created a great number of diverse heme proteins and enzymes with a wide range of reduction potential, with

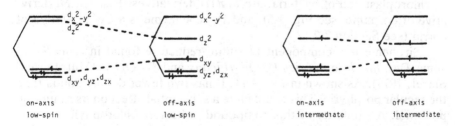

Figure 3-10. Effects of an off-axis ligand on the electronic structure of a heme group.

Table 3-3. The Effects of Structure on the Reduction Potential of Heme

Position of Fe	Coordination number	Spin state	E_0'	Position of 5th ligand	E_0'
Out-of-plane	5	High	Tend to decrease	Axial	Tend to increase
				Nonaxial	Tend to decrease
In-plane	4	Intermediate	Tend to increase		
	5	Low or intermediate	Tend to increase	Axial	Tend to increase
				Nonaxial	Tend to decrease
	6	Low	Tend to increase	Axial	Tend to increase
				Nonaxial	Tend to decrease

each one being particularly suited to a specific requirement. This must be seen as one more result of evolutionary pressure.

3.4.2.2 Porphyrin-like Compounds

Several porphyrin-like structures are known in nature. They include dihydro derivatives (chlorophyll a, b), tetrahydro derivatives (bacteriochlorophyll and sirohydrochlorin), corphinoid (as in F_{430}) and corrinoid (as in B_{12} coenzyme). The last two correspond to octahydro derivatives. In these compounds the porphyrin electronic structure is drastically changed due to the lessening of the degree of conjugation in the ring system as well as alterations in the peripheral substituents and the ring planarity. Except for siroheme (the iron derivative of sirohydrochlorin), all these porphyrin derivatives are used not as iron complexes. Chlorophylls, both of chloroplast and of bacteria, are Mg(II) derivatives, F_{430} is a Ni derivative of corphine (see Fig. 3-5), and B_{12} coenzyme is a Co derivative of corrin (see Section 7.2).

Siroheme is a component of sulfite reductase found in *E. coli* and some *Desulfovibrio* species (*D. nigrificans* and *D. gigas*) (Murphy and Siegel, 1973). As shown in Fig. 3-11, it has two fewer double bonds than the regular porphyrin. This could have a substantial effect on its reduction potential. A similarity of this compound to bacteriochlorophyll in terms of the degree of conjugation indicates some evolutionary relationship between these two types of compound.

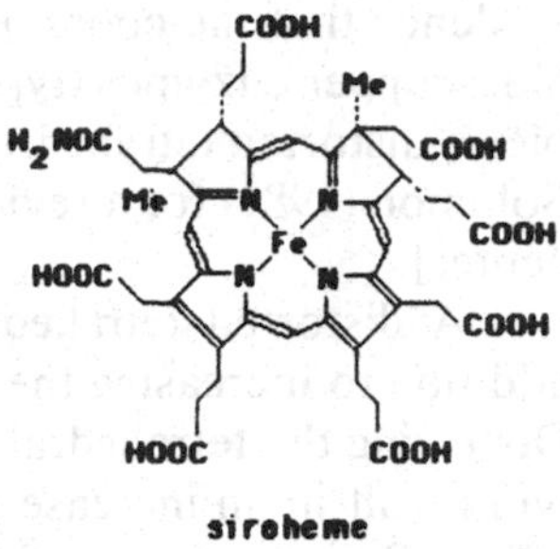

Figure 3-11. Siroheme (iron complex of sirohydrochlorin),
a unique factor found in sulfite reductase. (After Murphy
and Siegel, 1973.)

3.4.3 Blue-Copper Proteins and Enzymes

The structures of the blue-copper centers in the so-called blue-copper proteins (such as plastocyanin and azurin) and blue-copper enzymes (such as laccase and ceruloplasmin) seem to provide another example of the manifestation of the evolutionary adaptation rule. Here, both kinetic and thermodynamic consequences can be demonstrated. One characteristic feature of copper that must be borne in mind is that its stable coordination structures dramatically depend on whether the element is in the Cu(II) or Cu(I) state. Thus, the favorable coordination structure of Cu(II) is square-planar, whereas that of Cu(I) is tetrahedral.

In terms of absolute rate theory, the activation energy of a process can be regarded as the potential energy difference between an initial and a transition state. A significant portion of the activation energy (i.e., free energy of activation) comes from the energy required to reorganize the structure of the initial state so that it takes on the structure of the transition state. Therefore, the closer the structure of the initial state is to that of the transition state, the less the activation energy becomes, hence the higher the reaction rate (see also Section 5.1). The reduction of a square-planar Cu(II) species would require a significant rearrangement in the structure about the metal center, because the preferred structure of the reduced state, Cu(I), is tetrahedral. Hence, one would not expect the reduction rate of a square-planar Cu(II) complex to be high. One way to reduce the "cost" of this process (the large amount of rearrangement required in the coordination structure) would be to start with a compound whose structure was already intermediate between regular square-planar and regular tetrahedral (Ochiai, 1978b).

X-ray crystallographic analysis of a plastocyanin (Collman *et al.,* 1978) revealed that the coordination structure about its copper atom is that of a distorted tetrahedron with two imidazole (histidine) nitrogen atoms and two sulfur atoms (of cysteine and methionine) as ligands. There is now evidence that the coordination structure about the copper atoms in other blue-copper enzymes (type I center) and blue-copper proteins also resembles a distorted tetrahedron [Adman and Jensen (1981); see also Gray and Solomon (1981) for a review of the structure of the type I (type a) copper center].

A distorted tetrahedral structure serves other useful purposes, in addition to increasing the electron transfer rate, which is a kinetic effect. Distorting the tetrahedral structure will reduce the stability of Cu(II); this will result in an increase in reduction potential; this is a thermodynamic effect. Reduction potentials can also be manipulated by the proper choice of ligands. Sulfur atoms (such as two in plastocyanin, see above) stabilize the Cu(I) state, whereas nitrogen ligands (e.g., the two histidines in plastocyanin) tend to stabilize the Cu(II) state. Thus, the choice of ligands in plastocyanin seems to have been made in the interest of providing a subtle adjustment to the compound's reduction potential.

3.5 Oxidation–Reduction of Sulfur Compounds

3.5.1 Reduction of Sulfate

Nitrate (NO_3^-) is reduced directly to nitrite (NO_2^-) by nitrate reductase, whereas sulfate (SO_4^{2-}) cannot be directly reduced to sulfite (SO_3^{2-}) enzymatically. Instead, it must first be activated by combining with adenosine triphosphate (ATP) to form adenosine-5'-phosphosulfate (APS). APS is subsequently reduced by APS reductase to form AMP and SO_3^{2-}, i.e.,

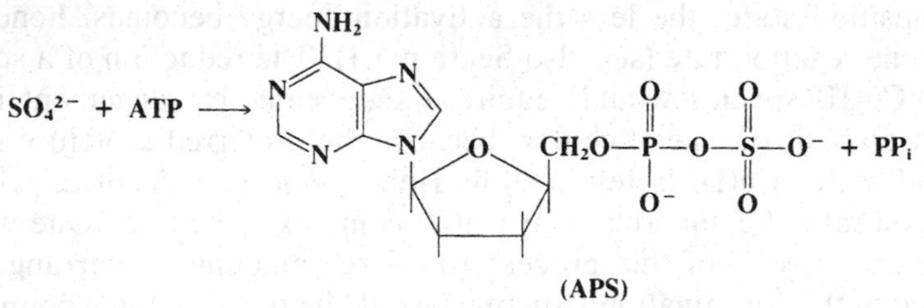

$$SO_4^{2-} + ATP \longrightarrow$$

(APS)

(3-27)

$$APS + 2e \xrightarrow{\text{APS reductase}} AMP + SO_3^{2-} \qquad (3\text{-}28)$$

In some systems, the ultimate electron source is H_2 or H_2S, whereas it is NADPH in others. In systems of the latter type, APS is further converted to 3'-phosphoadenosine-5'-phosphosulfate (PAPS) before it is reduced:

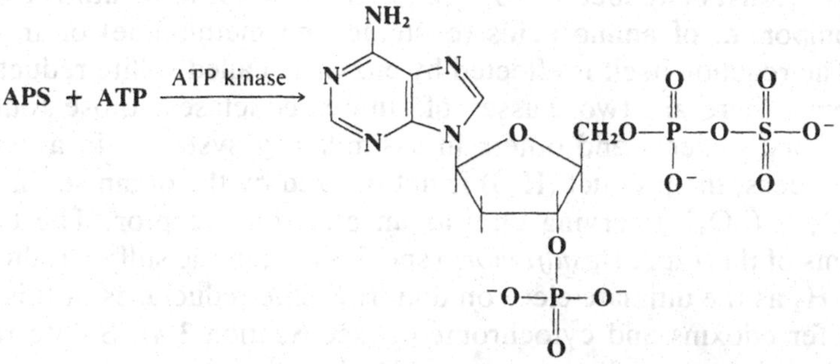

$$\text{APS} + \text{ATP} \xrightarrow{\text{ATP kinase}}$$

(PAPS)

(3-29)

The question to be examined here concerns why SO_4^{2-} has to be combined with AMP in order to be reduced. As shown in Fig. 3-3, the reduction potential of SO_4^{2-} to SO_3^{2-} is fairly low (-0.33 V), being almost equivalent to that of H^+. This means that SO_4^{2-} cannot possibly be reduced by H_2 or NAD(P)H. Would the binding of SO_4^{2-} to AMP significantly change its reduction potential? Let us estimate the reduction potential, which would be virtually the same as that of PAPS. We will consider the following process:

$$\text{APS} + \text{HOH} \rightarrow \text{AMP}(2H^+) + SO_4^{2-} \qquad (3\text{-}30)$$

It would be reasonable to assume that the free energy change of reaction 3-30 is similar to that of hydrolysis of ADP to AMP known to be -30.5 kJ/mole. Addition of reaction 3-31 to reaction 3-30 would give rise to reaction 3-28. Therefore, the free energy change ΔG of reaction 3-28

$$2H^+ + 2e + SO_4^{2-} \rightarrow SO_3^{2-} + \text{HOH} \qquad \Delta G = -2EF \quad (3\text{-}31)$$

is given by:

$$\Delta G(3\text{-}28) = -2E_0'(SO_4^{2-})F + (-30.5 \text{ kJ/mole}) \qquad (3\text{-}32)$$

In terms of reduction potential:

$$E_0'(\text{APS}) = E_0'(SO_4^{2-}) + 0.16 \text{ V} \qquad (3\text{-}33)$$

That is, the reduction potential of SO_4^{2-} can be raised by about $+0.16$ V upon binding with AMP to produce APS or PAPS. The reduction potential for APS can thus be estimated at about -0.22 V, making APS (PAPS) reducible by H_2 or NAD(P)H as seen in Fig. 3-3.

3.5.2 Oxidation–Reduction of Other Sulfur Compounds

SO_3^{2-} must be reduced to S^{2-} in order for sulfur to be utilized either as a component of amino acids (cysteine and methionine) or in other ways. The reaction itself is effected by enzymes called sulfite reductases. In general, there are two classes of sulfite reductases: those found in dissimilatory systems and others in assimilatory systems. In a dissimilatory process, the product (H_2S) is not utilized by the organism in question, SO_3^{2-} (SO_4^{2-}) serving only as an electron acceptor. The typical organisms of this type, *Desulfovibrio* species (anaerobic sulfate reducers), employ H_2 as the ultimate electron donor. Sulfite reductases of this class contain ferredoxins and cytochrome c_3 (see Section 3.4). Sulfite reductases from the other class (ones involved in the assimilatory process) contain FAD, FMN, and iron–sulfur proteins, and these utilize NAD(P)H as their ultimate electron source, with the product H_2S being incorporated into amino acids.

The oxidation of H_2S, S^0, or other low-oxidation-state sulfur compounds, such as $S_2O_3^{2-}$, is effected by two different classes of microorganisms. One consists of the so-called purple or green sulfur bacteria, which employ H_2S as a source of electrons in photosynthesis. Purple sulfur bacteria oxidize H_2S to S^0 and further to SO_4^{2-}, whereas green sulfur bacteria deposit the product sulfur (S^0) extracellularly. Both are anaerobes. By contrast, the second group of organisms are strictly aerobic. Species in this category oxidize H_2S, S^0, $S_2O_3^{2-}$, and others, using oxygen and then capture the energy released by the reaction. Examples of organisms of this type include *Thiobacillus thio-oxidans* and *Ferrobacillus ferro-oxidans*.

Thiols are the principal physiological compounds whose action is dependent on the oxidation–reduction chemistry of sulfur; examples include glutathione and lipoic acid:

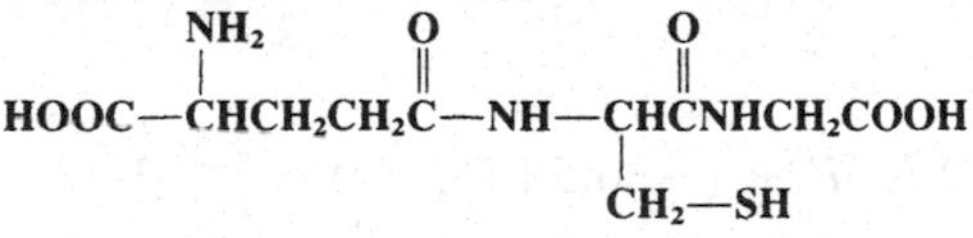

glutathione 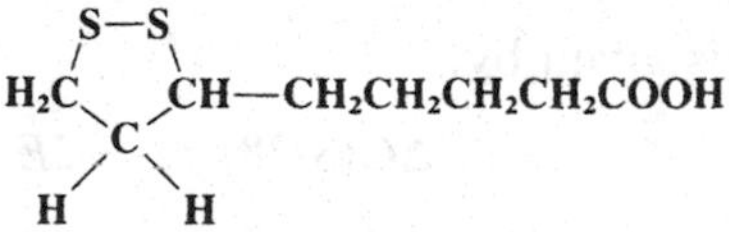   lipoic acid

The mode of their oxidation–reduction is represented by:

$$RSSR + 2H^+ + 2e \rightarrow 2RSH \qquad (3\text{-}34)$$

The persulfide bond is formed intramolecularly in the case of lipoic acid and intermolecularly in the case of glutathione. The reduction of the amino acid cystine to form 2 moles of cysteine falls in the same category. The reduction potentials of these reactions range from -0.29 (lipoic acid) to -0.23 V (glutathione). These values coincide with the range of reduction potentials of such important biomolecules as NAD(P)$^+$ and FAD (FMN); it is little wonder that thiols play important roles in biological oxidation–reduction processes. Glutathione peroxidase (see Section 3.6) and pyruvate dehydrogenase (which is dependent on lipoic acid) are typical examples of enzymes catalyzing persulfide–thiol intermediates.

3.6 Selenium, Its Oxidation–Reduction, and Enzymes Dependent on It

The oxidation–reduction of selenium compounds is quite different from that of sulfur compounds. Selenate, SeO_4^{2-}, is much more readily reducible than sulfate: the E_0' (pH 7) (SeO_4^{2-}/SeO_3^{2-}) is $+0.34$ V, which should be compared to an E_0' (SO_4^{2-}/SO_3^{2-}) of -0.33 V. The reduction of SeO_4^{2-} is considered to proceed in the same way as that of SO_4^{2-}; i.e., by way of the Se analogue of APS, although activation of selenate by ATP would hardly seem to be necessary based on the reduction potentials alone. The reduction of SeO_3^{2-} to Se^0 (E_0' $+0.2$ V) is also much easier than that of SO_3^{2-} (E_0' -0.11 V). On the other hand, HSe^- is a stronger reductant than HS^-: E_0' (Se^0/HSe^-) $= -0.7$ V versus E_0' (S^0/HS^-) $= -0.4$ V. Thus, the reduction potentials of selenium compounds cover a much wider range than do the corresponding sulfur compounds. The amount of selenium in the earth's crust is 2×10^{-4} times that of sulfur, thereby restricting its availability to organisms. As far as its potential usefulness in oxidation–reduction is concerned, selenium would seem to be much superior to sulfur, however. Therefore, it is not surprising that selenium has come to be recognized as an essential factor in a few selected oxidation–reduction (including dehydrogenation) enzymes. Examples are glutathione peroxidase (Diplock, 1976), formate dehydrogenase (Stadtman, 1974), and glycine reductase (Stadtman, 1974), which catalyze the following reactions:

$$2GSH(glutathione) + HOOH\ (ROOH) \xrightarrow{\text{glutathione peroxidase}}$$
$$GSSG + 2HOH\ (ROH + HOH) \quad (3\text{-}35)$$

$$NH_2CH_2COOH + R(SH)_2 + ADP + P_i \xrightarrow{\text{glycine reductase}}$$

$$CH_3COOH + NH_3 + ATP + R{\overset{\displaystyle S}{\underset{\displaystyle S}{\diagdown\ |\ \diagup}}} \quad (3\text{-}36)$$

$$\text{HCOOH} \xrightarrow{-2\text{H}^+, \text{ formate dehydrogenase}} \text{CO}_2 \qquad (3\text{-}37)$$

See Stadtman (1980) and Shamberger (1983) for recent reviews of selenium and selenium-dependent enzymes.

3.6.1 Glutathione Peroxidase

Glutathione peroxidase consists of four subunits, each of which has a molecular weight of 22,000 and contains 1 gram atom of selenium. The essentiality of selenium has been well established, but the mode of the participation of selenium in the catalysis process and even the state of selenium present in the enzyme are still obscure. The reduction potentials of substrates are $+1.32$ V for $\text{HOOH}/2\text{H}_2\text{O}$ and -0.23 V for $\text{GSSG}/2\text{GSH}$. Therefore, the catalytic entity's reduction potential would need to lie in between these two extremes. One proposed mechanism (Ganther *et al.*, 1974) is as follows:

$$\text{E—Se—H} + \text{ROOH} \rightarrow \text{E—Se—OH} + \text{ROH}$$

$$\text{E—Se—OH} + \text{GSH} \rightarrow \text{E—Se—SG} + \text{HOH}$$

$$\text{E—Se—SG} + \text{GSH} \rightarrow \text{E—Se—H} + \text{GSSG} \qquad (3\text{-}38)$$

This mechanism presumes that the selenium is formally in the oxidation state of Se($-$II). Evidence to the point exists in the sense that the only selenium found to be present in a reduced subunit is in the form of selenocysteine (Stadtman, 1980; Shamberger, 1983). This is oxidized in a two-electron step by ROOH (or HOOH), formally leading to the Se(0) oxidation state; that is:

$$\text{E}^+\text{—Se}^{-\text{II}}\text{—H}^+ + \text{ROOH} \rightarrow \text{E}^+\text{—Se}^0 + \text{ROH} + \text{OH}^-$$

$$\text{E}^+\text{—Se}^0 + \text{OH}^- \rightarrow \text{E}^+\text{—Se}^0\text{—OH}^- \qquad (3\text{-}39)$$

This Se^0 is assumed then to be reduced back to Se($-$II) by GSH; thus:

$$\text{E}^+\text{—Se}^0\text{—OH}^- + 2\text{GSH} \rightarrow \text{E}^+\text{—Se}^{-\text{II}}\text{—H}^+ + \text{GSSG} + \text{HOH} \qquad (3\text{-}40)$$

The latter reaction can be broken down into two component parts:

$$\text{E}^+\text{—Se}^0\text{—OH}^- + 2e + 2\text{H}^+ \rightarrow \text{E}^+\text{—Se}^{-\text{II}}\text{—H}^+ + \text{HOH} \qquad E_0' = x$$

$$\text{GSSG} + 2e + 2\text{H}^+ \rightarrow 2\text{GSH} \qquad E' = -0.23 \text{ V} \qquad (3\text{-}41)$$

from which it can be seen that, in order for reaction 3-40 to proceed spontaneously, the value x must be higher than -0.23 V. In terms of formal oxidation states, the first reaction of 3-41 corresponds to the couple of Se(0)/Se(−II), the reduction potential of which is −0.7 V. Bond formation between Se(−II) and E$^+$ (enzyme, probably resembling a carbonium) would stabilize the state E$^+$—Se^{-II}—H$^+$ relative to Se^{-II}—H$^+$. If the species E$^+$—Se^{-II}—H$^+$ can be reasonably approximated by H$^+$—Se^{-II}—H$^+$, then the value x should resemble the potential of Se(0)/HSeH, which is about −0.5 V at pH = 0. Thus, while the real value of x may be slightly higher than −0.5 V, it may not be as high as (or higher than) −0.23 V. It should be noted that this estimation process suggests that reaction 3-40 may not be spontaneous, even though it would be required in the mechanism proposed by the authors (Ganther *et al.*, 1974). A further question might be raised as well. It seems highly unlikely that the highest oxidation state of selenium in this enzyme would be a formal zero given the presence of such a strong oxidizing agent as ROOH.

An alternative mechanism is based on the premise that the selenium in the enzyme functions as the couple Se(IV)/Se(−II). That is, it is suggested that selenium oscillates between the selenite and selenide states:

$$E—O—Se^{IV}O^{2-} + 6GSH \rightarrow E—Se—H + 3GSSSG + 3HOH \quad (3\text{-}42)$$

The resulting selenide derivative is then reoxidized by ROOH:

$$E—Se—H + 3ROOH + 3HOH \rightarrow E—OSeO_2^- + 3ROH + 6H^+ \quad (3\text{-}43)$$

The reduction potential of the couple Se(IV)/Se(−II) is expected to be slightly higher than that of SeO$_3^{2-}$/SeH$^-$, estimated to be about −0.1 V. Therefore, the potential of the proposed redox couple comprising the catalytic center would appear to fall between the reduction potentials of the substrates, glutathione and ROOH. Such a system seems intuitively more plausible than one involving the formal couple Se(0)/Se(−II), and an analogous mechanism based on oxidation of R-SeH to RSeO$_3$H by 3ROOH has been suggested by Sies *et al.* (1982). An interesting sidelight to this whole issue is the fact that a non-Se-containing glutathione peroxidase appears to be present in mammalian liver (Sies *et al.*, 1982). The nature and mode of action of this enzyme are unclear.

3.6.2 Glycine Reductase

In order to understand this system, we require first an estimate of E_0' for the reaction:

$$NH_2CH_2COOH + 2H^+ + 2e \rightarrow CH_3COOH + NH_3 \quad (3\text{-}44)$$

This has been estimated to be about +0.52 V, based on the known reduction potential of CH$_3$OH/CH$_4$ and appropriate bond energy data. The reduction potential of persulfide (3-45) is known to be about -0.3 V. Therefore,

$$R\begin{matrix} S \\ | \\ S \end{matrix} + 2H^+ + 2e \longrightarrow R(SH)_2 \qquad (3\text{-}45)$$

the reduction of glycine by $R(SH)_2$ would be amply exothermic: $(2 \times 23.06) \times (+0.52 + 0.3) = -38\,\text{kcal} = -158\,\text{kJ}$. In the case of *Clostridium sticklandii*, the resulting energy is utilized to phosphorylate ADP to ATP. If the role of the enzyme system is to mediate the flow of electrons from $R(SH)_2$ to glycine, then the reduction potentials of the enzyme components must be between -0.3 and $+0.52$ V. The enzyme is known to contain several (three or four) components, one of which is a selenoprotein (Stadtman, 1974). In the reduced form of this selenoprotein, the selenium was found (Cone *et al.*, 1976, 1977) to be in the form of selenocysteine. The selenoprotein has a low molecular weight (12,000), as acidic and heat-stable, and contains 2 moles of cysteine residues in addition to 1 mole of the selenocysteine. Whether the selenium is involved directly in the catalytic process is not known, although it has been suggested (Stadtman, 1974; Cone *et al.*, 1977) that the selenoprotein does have electron carrier functions.

3.6.3 Formate Dehydrogenase

Formate dehydrogenases from a number of sources, including *E. coli*, *Clostridium thermoaceticium*, and *Methanococcus vannielii*, are known to be stimulated by molybdenum or tungsten as well as selenium. The formate dehydrogenase reaction is basically similar to that catalyzed by aldehyde oxidase in that both can be described as involving hydride abstraction (from carbonyl in aldehyde oxidase and from a carboxylate group in formate dehydrogenase). Therefore, it is very likely that in formate dehydrogenase the molybdenum atom [as Mo(VI) or an oxo entity bound to Mo], or its replacement, tungsten, acts as the requisite hydride abstractor, just as is the case in aldehyde oxidase or xanthine oxidase (see Section 3.3.4).

As for the role of selenium, nothing definite is known. One possibility is that selenium substitutes for the inorganic sulfur (i.e., selenide instead of sulfide) in the iron–sulfur unit of one of the electron carrier proteins. It has been demonstrated that sulfide and selenide are interchangeable chemically in adrenodoxin (Mukai *et al.*, 1973), ferredoxin (Fee and Palmer, 1971), and putidaredoxin (Tsibris *et al.*, 1968). However, little success has been made in substitution of selenium for sulfur in the case of glycine reductase (Stadtman, 1974).

3.6.4 Other Enzymes

A hydrogenase containing Se has been isolated from *Methanococcus vannielii* (Yamazaki, 1982). The chemical form of the selenium has been identified as selenocysteine.

Carbon monoxide oxidase from *Pseudomonas carboxydovorans* seems to be activated by selenium (Meyer and Rajagopalan, 1984).

Enzymes and Proteins in the Reactions of Oxygen and Oxygen Derivatives

We are here concerned with metalloproteins and metalloenzymes that participate in the reactions of oxygen and its derivatives. Oxygen (dioxygen), as O_2, is rather stable; that is, it is kinetically inert despite its biradical character. However, it is potentially a strong oxidant, having a reduction potential of $+0.8$ V at pH 7. The reduction potentials of oxygen and its derivatives are shown in Table 4-1 and Fig. 3-3. Hydrogen peroxide is also strongly oxidizing; so are the related hydroperoxides ROOH in which R is an organic residue.

The biological reactions of oxygen and its derivatives can be classified as follows:

1. Oxygen carrying or storing: O_2 binds reversibly to a metalloprotein.
2. Superoxide dismutase: $2O_2^- + 2H^+ \rightarrow O_2 + HOOH$.
3. Monooxygenase reactions (hydroxylation): $SH + O_2 + XH_2 \rightarrow SOH + HOH + X$.
4. Dioxygenase reactions: $SH + O_2 \rightarrow SHO_2$.
5. Dihydroxylation reactions: $S + O_2 + XH_2 \rightarrow S(OH)_2 + X$.
6. Reactions of hydrogen peroxide or hydroperoxides: $ROOH + XH_2 \rightarrow ROH + HOH + X$ (in catalase, $R = H$, $X = O_2$).

In this chapter we will be mostly concerned with the underlying problems associated with O_2 and its derivatives, rather than the minute details of the mechanisms debated in the contemporary literature.

Table 4-1. Reduction Potential (E_0' at pH 7)
of Oxygen and its Derivatives

	E_0'
$4H^+ + O_2 + 4e \rightarrow 2H_2O$	$+0.82$ V
$2H^+ + O_2 + 2e \rightarrow H_2O_2$	$+0.27$ V
$H^+ + O_2 + e \rightarrow HOO$	-0.45 V
$H^+ + HOO + e \rightarrow H_2O_2$	$+0.98$ V
$H^+ + H_2O_2 + e \rightarrow OH + H_2O$	$+0.38$ V
$2H^+ + H_2O_2 + 2e \rightarrow 2H_2O$	$+1.36$ V
$2H^+ + O_2^- + e \rightarrow H_2O + O$	-0.67 V
$2H^+ + O + 2e \rightarrow H_2O$	$+2.13$ V
$H_2O_2 \rightarrow H_2O + O$ at pH 7	$+35.7$ kcal
$H_2O_2 \rightarrow H_2O + O$ in gas phase	$+24.2$ kcal

4.1 Oxygen Carrying and Storing

4.1.1 Distribution

The majority of organisms on earth are aerobic and require oxygen as the ultimate electron acceptor. In small unicellular organisms, oxygen may simply diffuse through the cell membrane. Because of the organisms' small size, the rate of replenishment of oxygen by diffusion is adequate to obviate the need for any extra oxygen-carrying or -storing device. In large multicellular organisms, however, simple diffusion is not rapid enough to sustain metabolism; besides, the solubility of oxygen in cytoplasm is fairly low. These complex organisms, therefore, have to be equipped with two things: a supply of compounds with high affinity for oxygen, and an appropriate system to transport the oxygenated species throughout the body (e.g., circulating systems incorporating blood vessels). Our concern here are the compounds that have high affinities for oxygen and can bind it reversibly.

Hemoglobin is the most widespread example. It is even found in *Paramecium* (a unicellular protozoan), yeast (a unicellular fungus), and legume nodules. The hemoglobin in the legume nodule is called leghemoglobin and is slightly different from the other hemoglobins. Its function is considered to be oxygen-shielding for nitrogenase (see Chapter 13). Hemoglobin's oxygen-binding group is iron-protoporphyrin IX. Another protein containing an iron-porphyrin, chlorocruorin, is found in some families of polychaete (segmented worms). Other biological oxygen-carriers are hemerythrin and hemocyanin; despite their names they do not contain heme groups. Hemerythrin contains iron and is found in some

Table 4-2. Distribution of Naturally Occurring Oxygen-Carrying Proteins

Hemoglobin	All vertebrates: some very rare exceptions Invertebrates: most Annelida and endomostracan Crustacea; rare in arthropods, sporadic in molluscs; found in *Paramecium* Plants: found in yeast and in legume nodule (leghemoglobin)
Chlorocruorin	Found in four families of Polychaeta: Ampharetidae, Chlorhaemidae, Sabellidae, and Serpulidae (subunit M_r 16,000); (*Spirorbis borealis* possesses chlorocruorin, *S. corrugatus* has hemoglobin, and *S. millitaris* has no respiratory pigment)
Hemerythrin	Found in sipunculids (including *Sipunculus* and *Phascolosoma*), *Magelona* (a polychaete), the priapulids *Halicryptus* and *Priapulus*, and the brachiopod *Lingula*
Hemocyanin	Found in malacostracan Crustacea, decapods, and stomatopods; some chelicerates; *Limulus, Euscorpious,* and spiders; molluscs: chitons, cephalopods, and many gastropods including particularly the prosobranchs and the pulmonates

sea worms. Hemocyanin, a copper protein, is present in certain crustacea and molluscs, including cephalopods, gastropods, and others. As summarized in Table 4-2, the occurrence of these oxygen-carrying proteins seems to have no phylogenetic relationship. For example, within the genus *Spirorbis* (Polychaeta), *S. corrugatus* has hemoglobin, *S. borealis* uses chlorocurorin, and still another, *S. millitaris,* has no respiratory pigment at all. Thus, it seems that these respiratory pigments developed independently in different organisms depending on their situations, genetic apparatus, and niche. Hemoglobin and myoglobin are most widely distributed, undoubtedly because they arose from a mutational change of cytochromes, which are much more ancient and widespread than hemoglobin. Chlorocruorin is believed to have developed through a mutational modification of hemoglobin.

4.1.2 Application of Basic Fitness Rule

In hemoglobin, myoglobin, and chlorocruorin, 1 mole of oxygen binds with 1 mole of iron in the heme group:

$$Fe^{II}P + O_2 \rightleftharpoons O_2FeP \tag{4-1}$$

In the case of hemerythrin and hemocyanin, one molecule of oxygen binds with two atoms of the metal:

$$Fe^{II}_2 + O_2 \rightleftharpoons O_2Fe_2 \text{ (hemerythrin)} \tag{4-2}$$

$$Cu^I_2 + O_2 \rightleftharpoons O_2Cu_2 \text{ (hemocyanin)} \tag{4-3}$$

No other metallic species have been found to perform this function in nature. It is well known, however, that some laboratory-made Co(II) complexes and others can bind oxygen reversibly. The question then arises as to why only iron or copper is utilized for this purpose by organisms. As outlined in Chapter 2, a few basic rules can be invoked in order to provide an answer. First, the rule of abundance asserts that iron and copper are preferable to cobalt and others because these elements are more readily available, particularly in seawater (see Fig. 1-2).

Next let us apply the basic fitness rule to this problem. We assume the following processes:

$$[(HOH)M(+n)L] + O_2 \rightleftharpoons [(HOH)M(+n+1)L] + O_2^- \qquad \Delta G_1^a$$

$$[(HOH)M(+n+1)L] + O_2^- \rightleftharpoons [O_2^-M(+n+1)L] + HOH \qquad \Delta G_1^b$$
$$(4\text{-}4)$$

$$2[(HOH)M(+n)L] + O_2 \rightleftharpoons 2[(HOH)M(+n+1)L] + O_2^{2-} \qquad \Delta G_2^a$$

$$2[(HOH)M(+n+1)L] + O_2^{2-} \rightleftharpoons$$
$$[(LM(+n+1))_2 O_2^{2-}] + 2HOH \qquad \Delta G_2^b \quad (4\text{-}5)$$

The first step in these processes is a redox reaction and the second is the replacement of H_2O by O_2^- (or O_2^{2-}) on the $[(HOH)M(+n+1)L]$ entity. The overall free energy change ΔG_1 (ΔG_2) on oxygen binding is given by the sum of ΔG_1^a (ΔG_2^a) and ΔG_1^b (ΔG_2^b). An estimate for ΔG_1^b can be obtained from the equilibrium constant $K = 10^{-8.5}M$ for the following reaction:

$$(\text{metheme})Fe^{III}\text{-}(HOH) \rightleftharpoons (\text{metheme})Fe^{III}\text{-}OH^- + H^+ \qquad (4\text{-}6)$$

From the K value and K_W, we can estimate $(\Delta G_{OH} - \Delta G_{HOH}) = -42$ kJ where ΔG_{OH} and ΔG_{HOH} relate to the following equilibria:

$$(\text{metheme})Fe^{III} + HOH \rightleftharpoons Fe^{III}\text{-}HOH \qquad \Delta G_{HOH} \qquad (4\text{-}7)$$

$$(\text{metheme})Fe^{III} + OH^- \rightleftharpoons Fe^{III}\text{-}OH^- \qquad \Delta G_{OH} \qquad (4\text{-}8)$$

As both O_2^- and OH^- carry one unit of charge, ΔG_1^b would be similar to $(\Delta G_{OH} - \Delta G_{HOH})$. A similar value would be expected for the binding between O_2^- and M(III), if only the electrostatic interaction is taken into consideration. A double bond character that might exist between O_2 and Fe in hemoglobin is ignored in this discussion. We assume ΔG_2^b to be approximately -82 kJ. ΔG_1 and ΔG_2 values thus estimated in this way for various simple metal complexes and the naturally occurring oxygen carriers are given in Table 4-3.

For the processes 4-4 and 4-5 to be reversible, ΔG_1 or ΔG_2 should be in the range of roughly -8 to -50 kJ/mole (standard state $[O_2] = 1$

Table 4-3. Estimate of Free Energy Change in Oxygenation[a]

$M(+n)$	ΔG_1[a]	Approximate ΔG_1	ΔG_2[a]	Approximate ΔG_2
Ti(II)aq	+1.7	-8	-29.2	-50
Ti(III) (in 5F H_3PO_4)	+6.9	(-3)	-19.0	-40
V(II)aq	+4.4	-6	-24.0	-40
Cr(II)aq	+0.9	-9	-31.0	-50
Mn(II) (in 8F H_2SO_4)	+47.7	+38	+62.6	+40
Mn(II)/CN⁻	+5.2	-5	-22.4	-40
Fe(II) (pH 2)	+28.2	+18	+23.2	0
Fe(II) (pH 6)	+10.0	0	-11.8	-30
Fe(OH)$_2$	-2.6	-13	-38.0	-60
Fe(II)/CN⁻	+18.6	+9	+4.4	-15
Co(II)aq	+52.7	+43	+72.6	+50
Co(II)/NH$_2$	+12.6	+3	-7.6	-30
Co(II)/CN⁻	-9.0	-19	-50.8	-70
Cu(I)aq	+13.8	(+4)	-4.8	-25(-10[b])
Cu(I)/NH$_3$	+10.1	(0)	-12.6	-30(-15[b])
			experimental ΔG_1	experimental ΔG_2
Fe(II)-hemoglobin	+16.2	+6(-2[c])		-6[c]
Fe(II)-myoglobin	+14.3	+4(-4[c])	"	-8[c]
Cu(I)-hemocyanin	+22.9		+12.7	-7

[a] kcal/mol O_2 in medium unless otherwise stated. See text for the assumptions made for this estimate.

[b] ΔG_2[b] = -15 kcal is assumed, because of lower charges.

[c] Standard state = 1 torr P_{O_2}.

M; Ochiai, 1973a). Were ΔG_1 (ΔG_2) > -8 kJ/mole, oxygen binding would scarcely occur, whereas oxygen binding would be complete and hardly reversible (the reverse reaction goes only to a small extent) at ΔG_1 (ΔG_2) < -54 kJ/mole. With this criterion, we can say that Ti(III), V(II), Cr(II), some Mn(II) complexes, some Fe(II) complexes, and some Co(II) complexes could be suitable for reversible oxygen binding of the (1:1) type. Most of the metal ions listed in Table 4-3 have a ΔG_2 that is too large a negative value, implying that they are not suitable for reversible oxygen binding of the (2:1) type. Some possibility of reversible (2:1) oxygen adduct formation seems to exist with certain Mn(II), Fe(II), Co(II), and Cu(I) complexes. These conclusions are valid only in thermodynamic terms, however. In reality, there are other factors to be considered. One such factor is side reactions. Some relevant side reactions or further reactions of the oxygenated species are:

$$M(+n+1)O_2^- + M(+n) \rightleftharpoons$$
$$MO_2M \rightleftharpoons 2MO \xrightarrow{\ +M(+n)\ } 2\,(M(+n+1))_2O \quad (4\text{-}9)$$

$$M(+n+1)O_2^- + S(\text{solvent}) \rightleftharpoons M(+n+1)S + O_2^- \quad (4\text{-}10)$$

$$M(+n+1)O_2^- M(+n+1) + 2S \rightleftharpoons 2M(+n+1)S + O_2^- \quad (4\text{-}11)$$

If one of these reactions should occur before the desired reverse reaction of giving off O_2, the function of oxygen carrying (storing) would not be fulfilled. The formation of MOM (reaction 4-9) is strongly favored thermodynamically and takes place rapidly unless the bimolecular approach of $M(+n + 1)O_2^-$ and $M(+n)$ is blocked by some means. This type of reaction has been shown to be possible with $M(+n)$ having d^n ($n \leq 6$), thus including Fe(II), which is the active cation in hemoglobin and hemerythrin (Ochiai, 1974a). The rate of displacement of a ligand (O_2^- or O_2^{2-} in this case) by another ligand (a solvent molecule in this case) is known to be dependent on the central metal ion. It is particularly slow with Cr(III) and Co(III), but otherwise fairly fast. These considerations then indicate that all of the possible metal ions, Ti(II), V(II), Cr(II), Mn(II), and Fe(II), are disqualified for reversible oxygenation of the (1:1) type for the reason of the side reactions (4-9 and 4-10). As for (2:1)-type oxygenation, only Co(II) complexes seem to be suitable. Thus, we are led to the conclusion that no simpler transition metal complex except some Co(II) complexes are suitable for the job of reversible oxygen carrying and storing (of reaction type 4-5) in aqueous media. The process would thus have become feasible only after the metallic ion had been incorporated into a structure in which the interaction of MO_2 with M was blocked and water molecules were kept from freely coming to the active site. Needless to say, the presence of two proximate metallic sites would be essential for a smooth oxygen binding of the type found in hemerythrin and hemocyanin.

4.1.3 Genealogy of Proteins

The appearance of oxygen-carrying or -storing proteins in organisms is a fairly recent event in the long history of life. It occurred when multicellular organisms came into existence and the concentration of oxygen in the atmosphere became significantly high. Therefore, it is likely that by that time organisms had developed some proteins that could be adapted with a small genetic modification to acquire an oxygen-carrying (storing) function. Undoubtedly, some cytochromes, which are almost ubiquitous and much older than hemoglobin, myoglobin, and chlorocruorin, were the precursors for these oxygen carriers, as mentioned earlier.

It is difficult to identify the precursors for hemerythrin and hemocyanin. A possible precursor of hemerythrin is, as suggested by Neilands (1973), the so-called B2 protein of ribonucleotide reductase. There are two entirely different kinds of ribonucleotide reductase. One, found in *Lactobacillus leichmannii, Micrococcus denitrificans,* and some other prokaryotes, is dependent on vitamin B_{12} coenzyme. The other enzyme contains iron and is thought to be widely distributed. This enzyme, which converts ribonucleotide into the corresponding deoxyribonucleotide, should be essential to all organisms. Spectroscopic studies (Atkin *et al.,* 1973) have indicated a close similarity between the iron in B2 protein of ribonucleotide reductase and that in hemerythrin. More recent data on iron-dependent ribonucleotide reductase isolated from prokaryotes and eukaryotes are found in Thelander and Reichard (1979), Graslund *et al.* (1982), and Lammers and Follman (1983). The spectroscopic and magnetic data show that the iron atoms in B2 protein are antiferromagnetically coupled (Engström *et al.,* 1979). Since hemerythrin is found only sporadically in several genera, it is conceivable that mutation of the original protein B of ribonucleotide reductase in these organisms led to the creation of a new protein that happened to bind oxygen reversibly. This is a mere speculation; a more detailed comparative study of these two proteins is necessary to verify this suggestion.

It is well known that tyrosinase and hemocyanin have similar compositions and show similar spectral properties and both have the ability to bind oxygen. Tyrosinase is found in some fungi as well as animal tissues, and hence appears to be older than hemocyanin. Therefore, it is tempting to suggest that the precursor of hemocyanin could be tyrosinase. The recently determined primary structure of a tyrosinase (Lerch, 1982) from *Neurospora crassa* will give some information concerning this point, once the primary structure of a hemocyanin has been obtained.

4.1.4 The Active Site of Hemoglobin

We have one more basic question to deal with before leaving the subject of oxygen carriers: the chemical significance of the structures of the active sites in these metalloproteins. First, hemoglobin, myoglobin, and perhaps chlorocruorin have their Fe(II) atoms slightly (approximately 42 pm) out of the plane of the porphyrin ligand. This does not seem to be due to any special structural effect of the protein (globin), because five-coordinate porphyrin derivatives of Zn(II) as well as Fe(III) with an ordinary ligand at the apical fifth position also show such a structure. The fifth ligand is a histidine nitrogen at F8 in the primary structure of the protein. In six-coordinate porphyrin derivatives with a strong sixth ligand (including oxyhemoglobin and oxymyoglobin), the metal ion is known to lie in the porphyrin ring and at its center. Out of the plane, the metal ion is subjected to a less strong ligand field effect from the porphyrin ring. Thus, the Fe(II) or Fe(III) in five-coordinate derivatives takes a high-spin state. Does this have any significance in the function of hemoglobin?

First, let us consider some relevant facts. The metal–ligand interatomic distances usually decrease by about 12 pm when a complex changes from a high-spin state to a low-spin state (Stynes and Ibers, 1971). This applies to iron, cobalt, and nickel. The interatomic distance also decreases upon oxidation of the central atom, by approximately 40 pm, for example, in the conversion of Co(II) to Co(III) (Stynes and Ibers, 1971). In the case of hemoglobin, the Fe-histidine nitrogen distance is estimated to decrease by 15 pm upon oxygen binding (Gelin and Karplus, 1977); this in addition to the 42-pm decrease that is caused by the movement of the iron atom from the out-of-plane position to the in-plane position. Thus, there is a total displacement of the FB histidine residue of 57 pm upon heme iron's binding with oxygen.

There is as yet no definite proof that four-coordinate Fe(II)-porphyrin can bind oxygen reversibly. The four-coordinate porphyrin Fe(II) is in a spin state $S = 1$ (intermediate spin). There is also no proof for the existence of a low-spin five-coordinate Fe(II)-porphyrin complex. The en-

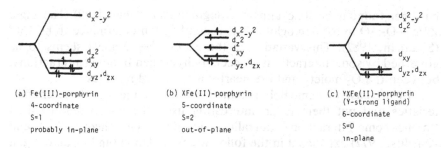

(a) Fe(III)-porphyrin
4-coordinate
S=1
probably in-plane

(b) XFe(II)-porphyrin
5-coordinate
S=2
out-of-plane

(c) YXFe(II)-porphyrin
(Y-strong ligand)
6-coordinate
S=0
in-plane

Figure 4-1. Spin states of Fe(II)/Fe(III)-porphyrin.

ergy level relationships of Fe(II)-porphyrin, therefore, seem to be as represented by Fig. 4-1. Because of a square-planar character, the d_{z^2} level lies low in the four-coordinate case and, within this limitation, the most probable state is $S = 1$, with the configuration $(d_{yz})^2(d_{zx})^2(d_{xy})^1(d_{z^2})^1$. The d_{z^2} level is destabilized upon binding a ligand at the apical, fifth position. This makes an $S = 1$ state such as (a) of Fig. 4-1 energetically unfavorable because of the unfavorable coulomb effect. This in turn forces the Fe(II) out of plane, resulting in a less strong field and $S = 2$, thereby relieving the system of the unfavorable coulomb effect. This still leaves an unpaired electron in the d_{z^2} orbital, however. The effect of the apical ligand is, thus, twofold: it destabilizes the d_{z^2} level and also leaves an unpaired electron. This is important in oxygen binding, because oxygen binding is regarded to be coupling of the unpaired electron in the d_{z^2} orbital of the metal and that in one of the π_g^* orbitals of the O_2 molecule (Ochiai, 1974b). This is true at least for the oxygenation of Co(II) complexes, and there is good correlation between the equilibrium constants for O_2-Co(II) systems and the corresponding binding strengths of the fifth ligands. These considerations would lead to the prediction that the rate of oxygen binding may be considerably lower with a low-spin five-coordinate complex than with a high-spin five-coordinate one in the case of Fe(II). An unpaired electron in d_{z^2} is, of course, available in the case of all Co(II) complexes regardless of whether they are in high- or low-spin states. This scheme (including the coupling of the unpaired d_{z^2} electron with an unpaired electron in π_g^* orbitals of the O_2), if applied to an Fe(II) system, would lead to an edge-on structure Fe—O—O for the Fe-O_2 binding with the angle $\angle$Fe—O—O being approximately 120°. Such an angle has been demonstrated in oxymyoglobin (Phillips, 1978) and an Fe-O_2 model compound (Collman et al., 1974). The Fe—O—O bond in oxyerythrocruorin has been found to be almost linear (Stiegman and Weber, 1979). The angle $\angle Fe—O—O$ in oxyhemoglobin has recently been determined to be 156° (Shaanan, 1982). This variation in the $\angle Fe—O—O$ angle seems to be governed by some interactions, especially hydrogen-bonding interactions, between the O_2 moiety and the nearby amino acid residues.

In the case of hemoglobin, nature took advantage of these characteristics and put them to an interesting use—the phenomenon called "heme–heme interaction" (an allosteric effect). One theory (Gelin and Karplus, 1977) explains it in the following way. Upon binding an oxygen molecule, the relevant histidine moiety moves closer to the porphyrin ring by as much as approximately 60 pm (see above). The histidine is then located asymmetrically with respect to the porphyrin ring, and the increased nonbonded van der Waals repulsion forces the porphyrin ring to swing away from the histidine group. This creates an unstable interaction between the porphyrin ring and the protein globin molecule. The protein molecule then relaxes this tension by changing the positions of various contacting residues slightly. This results in a change of location of crucial residues on the surface of one subunit where it interacts with an adjacent subunit. This change then would trigger a conformational change in this second subunit and, presumably, change slightly the position of the histidine residue on the second heme group, resulting in an enhancement of its oxygen affinity (see also Perutz, 1978, 1982).

4.1.5 The Active Sites of Hemerythrin and Hemocyanin

The mechanism of the formation of a $(2:1)$ binuclear μ-peroxo cobalt complex can be described as in Fig. 4-2 (Ochiai, 1973b). The formation of the binuclear complex, step (iii) in Fig. 4-2, would be facilitated if the

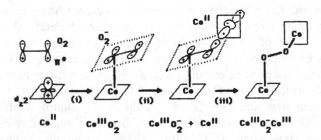

Figure 4-2. A model for the oxygenation of a Co(II) complex. (From Ochiai, 1973b.)

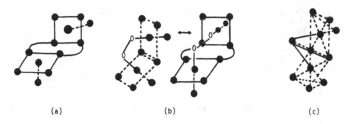

(a)   (b)   (c)

Figure 4-3. Models for a μ-peroxo (peroxo-bridged) complex.

second Co(II) entity were to be placed beside the first, with their planes of complex being more or less perpendicular to each other. Many devices can be put to use to juxtapose the two ligand planes. One possibility is to use two bidentate ligands to connect the two planes, as exemplified in Fig. 4-3a, bringing the two cobalt complexes into a suitable arrangement. This permits the oxygen molecule to fit properly in the bonding position, as shown in Fig. 4-3b. Some strain could result because of the exact nature of the bonding with O_2^{2-} molecule, particularly because of the existence of some π-bond character between the cobalt atom and the oxygen atom in certain cases. This strain can, however, be relaxed rather easily, for example, by choosing flexible bidentate ligands to bind the two complexes. Apparently this occurs with hemerythrin. An X-ray crystallographic study (Stenkampf *et al.*, 1976) has shown that the bridging groups in methemerythrin from *Thermiste dyscritum* are aspartic acid and glutamine and a μ-oxo group (Stenkampf *et al.*, 1985). If one regards the bidentate ligands as monodentate for the sake of simplicity, then the structure of methemerythrin appears as a composite of a trigonally distorted octahedron as shown in Fig. 4-3c, and this has, indeed, been verified by X-ray crystallography (Stenkampf *et al.*, 1976). It is likely that the coordination structure around the two iron atoms, and hence the conformation of its whole protein structure, would be altered upon binding oxygen. The necessary change involves a shortening of the coordination bonds between the iron atoms and their ligands, though the actual change turned out to be not too large (Stenkampf *et al.*, 1985). The protein structure as revealed by X-ray crystallography seems to be capable of accommodating a conformational change of this kind.

The latest X-ray crystallographic data (Stenkampf *et al.*, 1985) were interpreted to indicate that the dioxygen moiety bound to only one of the two iron atoms, the one with five ligands. It is described schematically as:

$$\underset{\text{glu}}{\overset{\displaystyle \underset{\displaystyle \text{his}}{\overset{\displaystyle \text{his}}{\underset{\displaystyle \text{Fe}}{|}}} \text{his}}{}} \quad \cdots \quad \text{O—H} + O_2 \longrightarrow \quad \cdots$$

(4-12)

This description implies that the coordinated O_2 moiety be a superoxide; this is in disagreement with the spectroscopic data. The resolution of the X-ray electron map was not too accurate so that further refinement may be needed to unequivocally determine whether the O_2 moiety is O_2^{2-} or O_2^{-} and whether both ends of the O_2 moiety bind to the iron atoms.

The copper atoms in hemocyanin are in a juxtaposition as demonstrated by a variety of methods. EXAFS and resonance Raman studies (Brown *et al.*, 1980; Larrabee and Spiro, 1980) indicate that three histidine residues bind to each of the two copper atoms in deoxyhemocyanin and that an O_2^{2-} moiety and probably a tyrosine residue bridge the two copper atoms in oxyhemocyanin. A more recent X-ray crystallographic determination of hemocyanin from spiny lobster (*Panulirus interruptus*) (Gaykema *et al.*, 1984) seems to confirm that each copper atom is coordinated

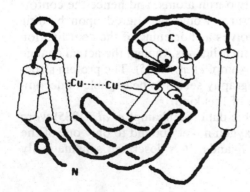

Figure 4-4. The structure of a hemocyanin active site (schematic). A cylinder represents an α-helical segment. (Adapted from Gaykema *et al.*, 1984.)

by three histidine residues and that two copper atoms are coupled at a distance of 380 pm, but could not confirm the presence of the bridging tyrosine residue. The distance 380 pm corresponds with 339 pm estimated for deoxyhemocyanin by EXAFS and 367 pm for oxyhemocyanin (Brown *et al.,* 1980). The structure of this hemocyanin, one of the smallest but nonetheless of substantial molecular weight (450,000), is shown in Fig. 4-4. Some recent progress in the chemistry of hemocyanin and other similar binuclear copper centers is reviewed by Solomon (1981), and hemocyanin, in particular, by Ellerton *et al.,* (1983).

4.2 Superoxide Dismutase

The reduction potentials of $O_2^-/HOOH$ and O_2/O_2^- are $+0.98$ and -0.45 V at pH 7, respectively. Suppose that the mechanism of the superoxide dismutase reaction is as follows:

$$M(+n) + O_2^- + 2H^+ \rightarrow HOOH + M(+n+1) \qquad k_a$$

$$M(+n+1) + O_2^- \rightarrow O_2 + M(+n) \qquad k_b \qquad (4\text{-}13)$$

Then it would be inferred that a redox system whose potential lies somewhere in the middle of the range -0.45 to $+0.98$ V would function as a catalyst for the superoxide dismutase reaction. Figure 3-3 indicates that appropriate systems are Fe(III)/Fe(II), Cu(II)/Cu(I), Mn(III)/Mn(II), and V(IV)/V(III). Indeed, superoxide dismutases from different sources contain copper, iron, or manganese (Fridovich, 1975, 1976). The enzyme dependent on copper (found in erythrocytes, mammaliin liver and brain) contains Zn(II) as well, which is considered to play a structural role. The reduction potential of the copper center of this enzyme has been reported to be $+0.42$ V (Rotilio *et al.,* 1973). Superoxide dismutase dependent on Mn(III) is found in mammalian mitochondria as well as bacteria. This fact is in accord with the thesis that the mitochondria of eukaryotic cells are derived from a bacterial symbiont (Roodyn and Wilkie, 1968; Margulis, 1981; Frederick, 1981). Iron-dependent superoxide dismutase is found in bacteria, but its location is different from that of the manganese enzyme. A number of reviews have recently been published on the subject; they include Fee (1980), Vallentine and Pantoliano (1981), and Rotilio *et al.* (1982).

Since the copper ion is assumed to oscillate between Cu(II) and Cu(I) in superoxide dismutase, the coordination structure about the copper atom would be expected to be a distorted tetrahedron, as discussed in Section 3.4.3 for the blue-copper centers. However, spectroscopic data on the copper in the superoxide dismutase are all compatible with a square-planar structure or one very close to it. Indeed, an X-ray crystallographic study (Richardson *et al.*, 1975) showed that the ligands about copper in a bovine enzyme are four histidine residues arranged in a slightly distorted square-planar manner. The same study also showed that the zinc atom is only 600 pm away from the catalytic copper atom, bridged by an imidazole ring. Rotilio *et al.* (1982) argue that the coordination structure about the copper atom becomes a distorted tetrahedron severing the bond to the bridging imidazole when it is in the Cu(I) state, and that thus the structure about the copper atom oscillates between a distorted square-plane (square-pyramid if a water molecule was counted) in the Cu(II) state and a distorted tetrahedron in the Cu(I) state.

The overall rate of the bovine enzyme-catalyzed reaction has been established to be expressed by $k[\text{enzyme}][O_2^-]$, where k is 2.3×10^9 M^{-1}/sec at pH 7 and 25°C. The rate constant of the spontaneous dismutation reaction (nonenzymatic) is reported to be $k_s = 8.5 \times 10^7 M^{-1}/\text{sec}$ with a rate dependence of $k_s[O_2^-]^2$ under similar conditions. This reaction is very much pH-dependent, however. The X-ray crystallographic study mentioned above showed that the fifth axial coordination site of the copper in the enzyme is especially open to solvent. This fifth coordination site might therefore accept an O_2^- entity. Several positively charged amino acid residues (two lysines and an arginine) are located near the active center, the copper atom (Cudd and Fridovich, 1982). It has been proposed that these groups aid in the introduction of O_2^- in the cavity where the catalytic atom is residing (Koppenol, 1981). A recent study of the electrostatic vector field surrounding the active site using computer graphics has shown that indeed the conduit (cavity) created in the protein seems to facilitate the movement of the negatively charged O_2^- toward the active site (Tainer *et al.*, 1983; Getzoff *et al.*, 1983).

A detailed kinetic and spectroscopic study on bovine superoxide dismutase (Fielden *et al.*, 1974) led the authors to conclude that only about 50% of the enzyme was involved in the reaction. If the mechanism can be expressed by equation 4-13, the steady-state concentration of Cu(II) during the enzymatic reaction should be the same irrespective of whether the reaction is initiated with the Cu(II) form of the enzyme or the Cu(I) form. However, the EPR and optical spectroscopic data obtained in this study showed that the steady-state concentration of Cu(II) was about 73% when the Cu(II) form was allowed to react with O_2^- whereas it was only about 25% when the Cu(I) form of the enzyme was employed. There was always a gap of about 50% between these two values (Fielden *et al.*, 1974). This led to the conclusion cited above. Nonetheless, no significant difference was observed spectroscopically between the supposedly active form of the enzyme and the supposedly inactive form (Fielden *et al.*, 1974). See Rotilio *et al.* (1982) for more details on this point.

If a mechanism such as 4-14 (see below) is assumed in addition to 4-13, it would be unnecessary to invoke such a hypothesis as the existence of active and inactive forms of the enzyme (present in roughly equal amount).

$$Cu^{II} + O_2^- \rightleftharpoons Cu^{II}O_2^-$$

$$Cu^{II}O_2^- + O_2^- \rightarrow Cu^{II} + O_2^- + O_2^- \qquad (4\text{-}14)$$

A similar reaction can be assumed to occur with the Cu(I) form. The addition of these reactions would not change the kinetics (i.e., the form of the rate equation) as long as the O_2^- concentration is fairly low (as in the pulse radiolysis experiments where a typical high value of $[O_2^-]$ is 2×10^{-4} M).

The results from a kinetic study on an Fe superoxide dismutase (Fee and McClune, 1978) are consistent with mechanism 4-13 with Fe in place of Cu. The structures of Fe superoxide dismutase from *Pseudomonas ovalis* (Ringe *et al.*, 1983) and from *E. coli* (Stallings *et al.*, 1983) have recently been determined (to a resolution of 290 pm and 310 pm) by X-ray crystallography. These two enzymes have an essentially identical structure. Fe atom is coordinated by four amino acid residues, one of which is His-26 in the case of the *P. ovalis* enzyme. There is, however, an indication that the fifth site is occupied by a water molecule. The coordination structure seems to be a distorted square pyramid. The global structure of a Mn superoxide dismutase from *Thermus thermophilus* HB8 has recently been shown to be very similar to that of Fe SOD (Stallings *et al.*, 1985). The ligands for Mn(III) include three histidine residues and one aspartate residue (Stallings *et al.*, 1985).

4.3 Dioxygenase Reactions

Dioxygenase incorporates both of the oxygen atoms of a dioxygen molecule into a substrate. There seem to be two types of dioxygenation reactions:

$$-\overset{|}{C}=\overset{|}{C}- + O_2 \rightarrow -C=O + O=C- \qquad (4\text{-}15)$$

$$-\overset{|}{C}=\overset{|}{C}- \ + \ O_2 + NADH + H^+ \rightarrow -\overset{\overset{\displaystyle HO}{|}}{C}-\overset{\overset{\displaystyle OH}{|}}{C}- \ + \ NAD^+ \ (4\text{-}16)$$

Examples of the first type of dioxygenase are tryptophan 2,3-dioxygenase, indoleamine 2,3-dioxygenase, 2,3-dihydroxybenzoate 2,3-dioxygenase, pyrocatechase, metapyrocatechase, protocatechuate 3,4-dioxygenase, 3,4-dihydroxyphenylacetate 2,3-dioxygenase, steroid dioxygenase, homogentisate dioxygenase, and cysteamine dioxygenase. The first two enzymes listed contain an iron-porphyrin as the catalytic site, and the third enzyme is believed to be a copper enzyme (Sharma and Vaidyanathan, 1975). All the other enzymes contain nonheme iron as the sole catalytic entity. An interesting feature of this group of dioxygenases is that the Fe(III) is coordinated by two tyrosine O atoms and one or two histidine N atoms (Que *et al.*, 1980; Que and Epstein, 1981; Roe *et al.*, 1984). 4-Hydroxyphenylpyruvate dioxygenase has recently been shown to be another example of ferric-tyrosinate enzyme (Bradley *et al.*, 1986).

Dioxygenation of the second type (4-16) is better described as dihydroxylation, and is exemplified by benzoate 1,2-dioxygenase, anthranilate hydroxylase, and benzene hydroxylase (Fujisawa *et al.*, 1976).

A variant of reaction type 4-15 is catalyzed by a copper-containing enzyme, quercetinase:

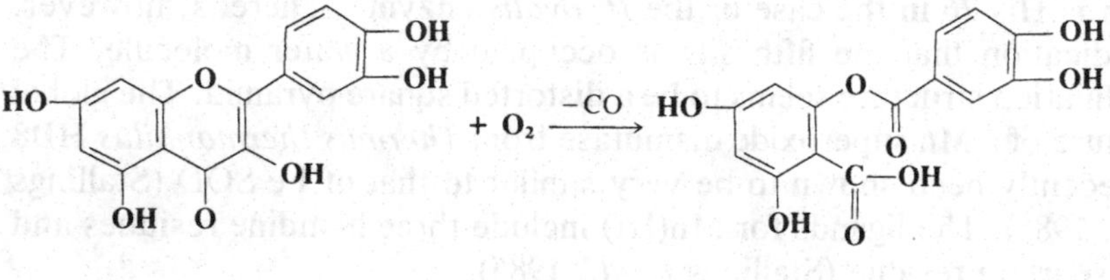

(4-17)

Thus, iron and copper again appear to be the essential elements in dioxygenases.

4.3.1 Naive Views—Dioxygenases

None of the mechanisms of the dioxygenase reactions has been elucidated unequivocally. The simplest conceivable mechanism for reaction 4-14 might be as follows:

$$M(+n) + O_2 \underset{(i)}{\rightleftharpoons} \underset{\text{Contract}}{\cdot O_2^- {-} M(+n+1)} \xrightarrow[\text{(ii)}]{\overset{\displaystyle -\overset{\displaystyle |}{C}=\overset{\displaystyle |}{C}-}{}}$$

(I)

$$\underset{\substack{|\nearrow\cdot\downarrow| \\ -\overset{|}{C}{-}\overset{|}{C}-}}{O{-}O{-}M(+n+1)} \xrightarrow{-M(+n)} 2 \overset{\displaystyle O}{\underset{\displaystyle \|}{-C-}} \qquad (4\text{-}18)$$

(II)

There are several pieces of experimental evidence for the involvement of the superoxide radical in the enzymatic reactions, at least for the heme-containing dioxygenases (Hirata and Hayaishi, 1975; Ohnishi et al., 1977; Hirata et al., 1977). Step (i) is reversible oxygen binding of the (1:1) type, and was discussed in Section 4.1. In the case of hemoglobin and myoglobin, there seems to be a considerable degree of double bond character between the metal and bonded oxygen; thus, the odd electron character is lessened. However, in the case of dioxygenases their odd electron seems to be required to attack the carbon–carbon double bond. It seems reasonable to conclude that the reduction potential of the metal in a dioxygenase of type 4-15 should be similar to those of hemoglobin and myoglobin, and that the electronic structure of the metal should be such that there will be no extensive coupling between the odd electron of O_2^- and one on the metal. This condition then dictates, as seen from Fig. 3-3, that iron and copper are basically suitable for these tasks. Step (ii) is the reverse of step (i), as far as the metal–oxygen bond is concerned, and the reversibility of step (i) would imply the facility of step (ii).

The O_2^- moiety on a cobalt complex of the (1:1) type, e.g., $[O_2^-Co^{III}(salen)]$, has a more pronounced odd electron character; its free radical character is manifested in its ability to initiate free radical polymerization. Such a cobalt complex should be able to mimic the dioxygenase reactions, and this is indeed the case. [Co(salen)] was found to catalyze the reactions normally effected by tryptophan 2,3-dioxygenase and quercetinase (Nishinaga, 1975; Nishinaga et al., 1974).

This mechanism involves the valence change of the catalytic element and a kind of reversible oxygenation; as such, it is consistent with the other kinds of oxygen-binding enzymes and proteins. The next section describes an entirely different mechanism in which the iron atom is supposed not to change its valence and not to be involved in the activation of oxygen.

4.3.2 An Alternative View—Dioxygenase

Spectroscopic (Que *et al.*, 1976) and kinetic (Que *et al.*, 1977) studies have shown that the iron, Fe(III) in protocatechuate 3,4-dioxygenase did not change its valence in the course of reaction. That it remained as Fe(III) upon forming the ES complex or the ternary complex ESO_2 was mainly ascertained from the EPR and Mössbauer spectral data. From kinetic studies it was concluded that only one of the hydroxyl groups in the substrate (protocatechuate) binds to the catalytic site, Fe(III). This coordination is then supposed to facilitate a keto–enol type of rearrangement in the substrate:

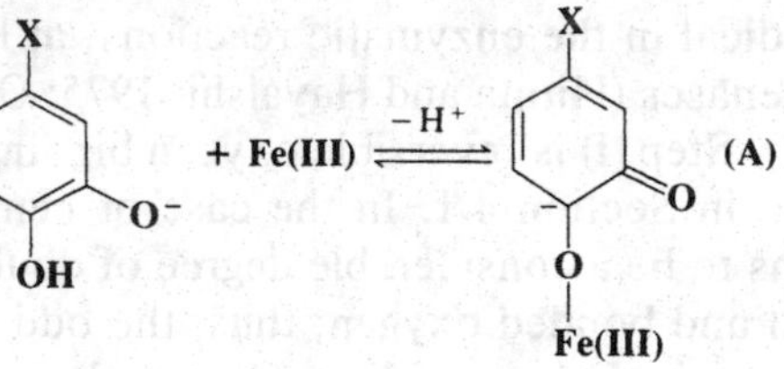

$$(4-19)$$

O_2 is then supposed to attack the carbanion:

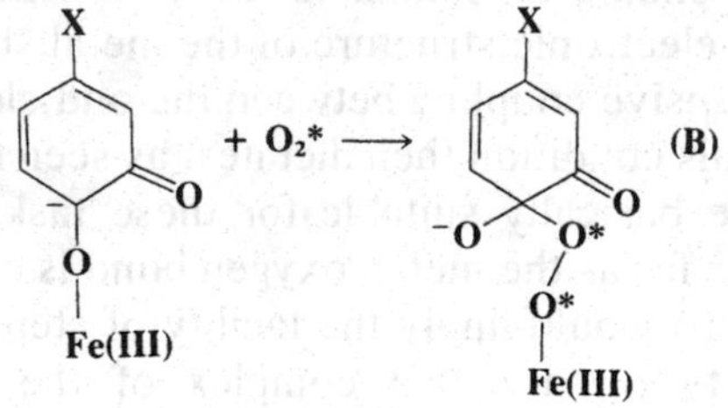

$$(4-20)$$

Subsequently a rearrangement occurs:

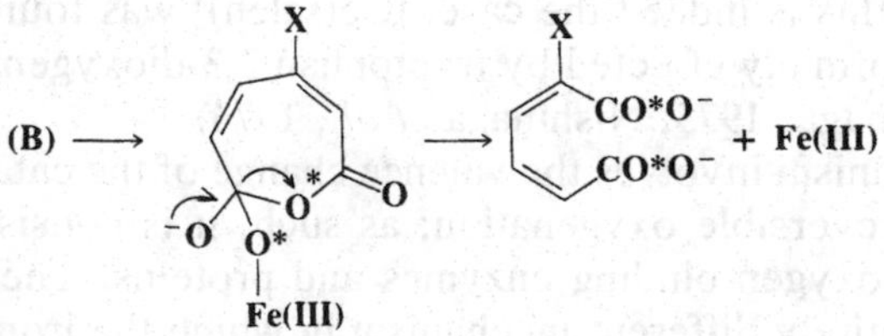

$$(4-21)$$

The spectral data revealed (1) that the Fe(III) in the native enzyme has a rhombic symmetry, (2) that the Fe(III) in the ES complex, which is supposed to be A in reaction 4-19, has an axial symmetry, and (3) that the ternary complex, which is supposed to be B, is in a field slightly off the axial symmetry.

These data certainly indicate different coordination structures around the Fe(III) in the native enzyme, the ES complex, and the complex in the presence of E, S, plus O_2, but would not necessarily prove that the structures proposed are the correct ones. One important point to be noted is that the spectral data were obtained at very low temperatures (4.2–50°K) and therefore may not accurately represent the situation at the usual active temperature (say, 298°K). Another point that has to be raised is that other dioxygenases, called "extradiol" oxygenases, which catalyze similar reactions and contain iron as the sole catalytic ion, have their iron in the +2 oxidation state (Arciero et al., 1983; for a review see Ochiai, 1977). An example is:

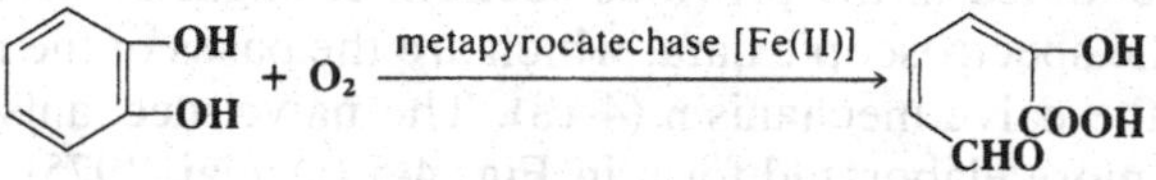

$$(4\text{-}22)$$

It is only natural to expect a similar mechanism here, too; that is:

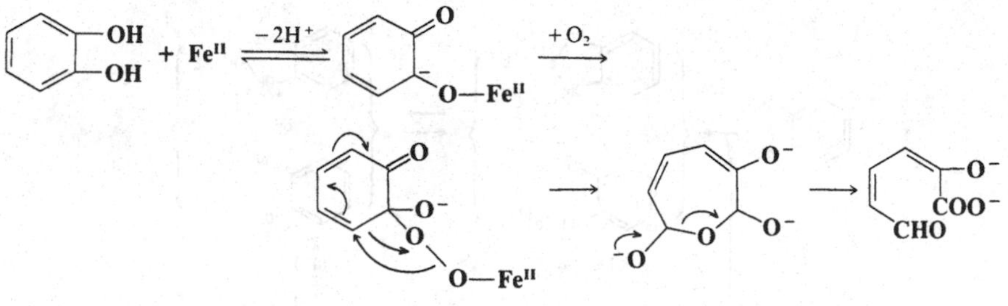

$$(4\text{-}23)$$

This is certainly possible on paper; but the question still has to be asked as to why nature here utilizes Fe(II) instead of Fe(III). Another basic question centers around whether keto–enol tautomerization like 4-19 can really be facilitated by Fe(III) [or Fe(II)] as in 4-23.

A simple ferric complex of nitrilotriacetate (NTA) has been shown to mimic a dioxygenase activity (Weller and Weser, 1982). The [Fe(NTA)] forms a relatively stable complex with 3,5-di-t-butyl catechol (dianion) as shown below. The complex slowly reacts with oxygen at an ambient temperature over days to form the expected products:

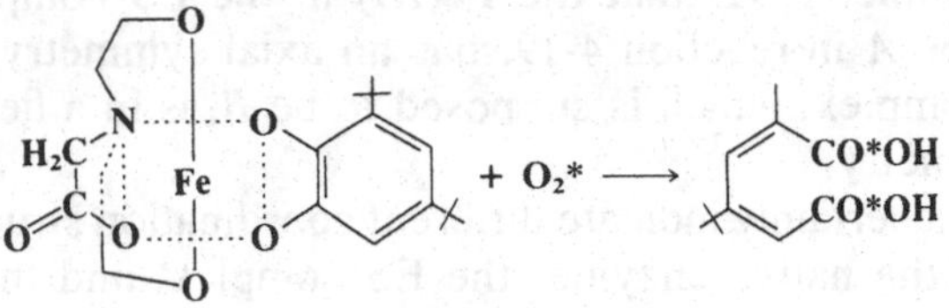

$$+ \; O_2^* \; \longrightarrow \qquad \qquad \text{(4-24)}$$

Studies with this or other model compounds could shed more light on the mechanism of the dioxygenase reactions.

4.3.3 Further Discussion on Dioxygenases

There are many questions that need to be answered regarding the mechanism outlined in the previous section. It might be useful to try to rationalize the spectroscopic data, which are the bases of the mechanism, in terms of the naive mechanism (4-18). The naive mechanism is shown in a slightly more elaborated form in Fig. 4-5 (Ochiai, 1975). It might be conjectured that intermediates **I** and **II** represent the ES complex, and intermediates **III** and **IV** represent the ES–O_2 complex detected in the

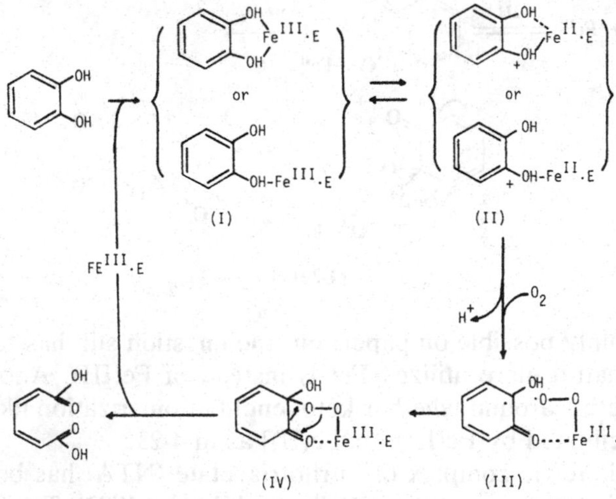

Figure 4-5. A mechanism of Fe(III)-dependent dioxygenase.

EPR and Mössbauer spectra referred to above. Intermediates **I** and **II** can be presumed here to be in equilibrium with each other:

$$\underset{(1-\alpha)}{(\mathbf{I})} \rightleftharpoons \underset{\alpha}{(\mathbf{II})} \qquad \Delta G, K; \; \alpha/(1-\alpha) = \exp(-\Delta G/RT) \qquad (4\text{-}25)$$

Therefore,

$$\alpha = \exp(-\Delta G/RT)/[1 + \exp(-\Delta G/RT)] \qquad (4\text{-}26)$$

If $\Delta G > 0$, as $T \to 0$, $\alpha \to 0$. Therefore, at the low temperatures at which the EPR and Mössbauer spectra were taken, only intermediate **I** will be observed. It is likely that intermediate **IV** is what was detected by EPR and Mössbauer at lower temperatures and that intermediate **III** is very short-lived. During the steady-state reaction condition at room temperature, the ES-O_2 complex was demonstrated to show a specific absorption band at 530 nm (Fujisawa *et al.*, 1972). This may be due to either **III** or **IV**. A Co(III) complex with an *ortho*-semiquinone ligand, which corresponds to **III**, has been synthesized and shown to have an absorption band at 512 nm with molar extinction coefficient = 1270 M^{-1}/cm (Wicklund and Brown, 1976; Wicklund *et al.*, 1976). Therefore, it is possible that the 530-nm band of the ES-O_2 complex is due to **III**, as suggested (Ochiai, 1975). A similar suggestion has more recently been made by Bull *et al.* (1981).

It is apparent that if ΔG in 4-25 has a very large positive value, the overall reaction would not proceed at all. Therefore, ΔG can be assumed to have a small positive value. When the reduction potential of E-Fe(III)/E-Fe(II) is very low, the value of ΔG, being related to the reduction potential, would be prohibitively high. The reduction potential of *o*-quinone/*o*-dihydroquinone is about $+0.35$ V; thus, the reduction potential of E-Fe(III)/E-Fe(II), very roughly speaking, must be about $+0.35$ V or perhaps slightly lower. If the potential E-Fe(III)/E-Fe(II) is much higher than $+0.35$ V, the substrate, catechol and similar compounds, would extensively reduce the Fe(III) in the enzyme, being oxidized itself to the semiquinone form even in the absence of O_2. This may lead to some other reaction paths. At an appropriate potential (perhaps slightly less than $+0.35$ V), the equilibrium **I** $\rightleftharpoons$ **II** would be pulled toward **II** only when **II** reacts with O_2, forming **III**. A similar argument can be made for copper-dependent 2,3-dihydroxybenzoate 2,3-dioxygenase. The effects of substituents on the catechol on the reaction rates have been studied (Walsh and Ballou, 1983; Walsh *et al.*, 1983). The electron-withdrawing substituents such as F, Cl, and Br reduced the oxygenation rate whereas the electron-donating substituents at C-4 position of the substrate accelerated the rate of formation of the intermediate ESO_2. These results are not inconsistent with this naive mechanism though interpreted in a different way by the authors.

A report by Brown *et al.* (1977) is quite interesting in this regard. A Cu(II) complex with a catechol as a ligand was shown to form a ring-cleaved product similar to the one found in the enzymatic reaction:

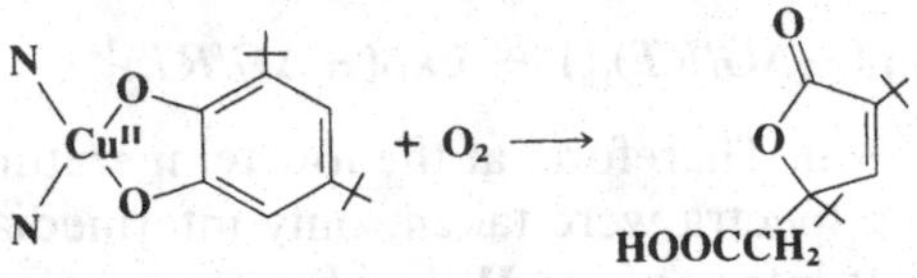

$$
\text{(V)} \quad + \; O_2 \longrightarrow
$$

(4-27)

This reaction cannot occur when the nitrogenous ligand is an aliphatic amine such as ethylenediamine; it occurs only when the ligand is either phenanthroline or bipyridyl. The latter are known usually to raise the reduction potential of transition metal cations. With ethylenediamine the reduction potential of complex V is too low, whereas with bipyridyl or phenanthroline it becomes appropriate for the reaction. Under oxygen-free conditions, complex V with either bipyridyl or phenanthroline remains stable indefinitely, apparently as a Cu(II) complex. These observations, as well as the spectroscopic data discussed above, are compatible with the mechanism of Fig. 4-5 and the interpretation above. In other words, the spectroscopic and kinetic data outlined in Section 4.3.2 would not necessarily require the mechanism described by 4-19–4-21.

The reaction mechanism proposed for the Fe(II)-containing enzymes (extradiol dioxygenase) is shown in Fig. 4-6. In this type of reaction, the most difficult step seems to be VII → VIII, which involves the reduction of Fe(III) to Fe(II). To facilitate this step, the reduction potential E-Fe(III)/E-Fe(II) must be raised. If this potential is too high, however, the step VI → VII would be hampered. Therefore, there must be an optimum range for the reduction potential, which appears to be slightly higher than the potential of

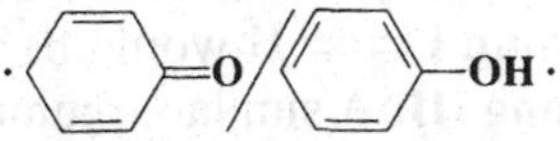

The potential of the latter process is likely to be higher than the potential of *o*-quinone/*o*-hydroquinone (catechol). Thus, it is very likely that the reduction potential Fe(III)/Fe(II) in the extradiol oxygenases is higher than that in the intradiol enzymes. This would explain the fact that the iron in the intradiol enzymes is Fe(III) in the resting (native) state whereas that in the extradiol enzymes is Fe(II). A recently discovered 4-hydrox-

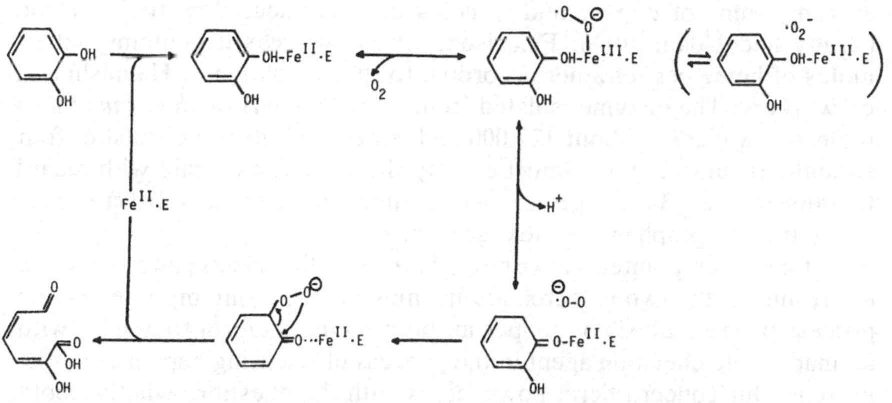

Figure 4-6. A mechanism of Fe(II)-dependent dioxygenase.

yphenyl pyruvate dioxygenase from *Pseudomonas* species contains Fe(III), which must be reduced to Fe(II) for activation (Lindstedt and Rundgren, 1982). All these data indicate that the essential part of dioxygenase activity is due to a superoxo–Fe(III) complex formation from Fe(II) and O_2.

These arguments would suggest that suitable elements for this kind of dioxygenase should have a redox couple of $M(+n+1)/M(+n)$ approximately in the potential range $+0.2$ to $+0.4$ V. Iron and copper, then, would be the most suitable. As in the case of superoxide dismutase, Mn species could function in a similar fashion, because of its similarity to iron species. Indeed, a Mn-containing dioxygenase has been isolated from *Bacillus brevis* (Que *et al.*, 1981).

4.3.4 Heme-Containing Dioxygenases

Known heme-containing dioxygenases are two very similar enzymes, tryptophan 2,3-dioxygenase and indoleamine 2,3-dioxygenase. There has been a serious dispute as to whether they contain only heme group(s) as the catalytic site or both heme and copper. According to Brady and Feigelson's group (Feigelson and Brady, 1974; Brady and Udam, 1976; Brady, 1976; Feigelson, 1976), the enzymes contain both heme and copper, whereas Hayaishi and his co-workers (Hirata *et al.*, 1976; Ishimura and Hayaishi, 1973, 1976) maintained that they contain only heme as the essential group.

Tryptophan 2,3-dioxygenase from rabbit liver has a molecular weight of about 167,000 and a subunit composition of $\alpha_2\beta_2$, i.e., it is a tetramer. One mole (of tetramer) of the enzyme has been estimated to contain

2 gram atoms of copper and 2 moles of heme according to Feigelson (Brady and Udam, 1976; Feigelson, 1976), whereas it contains only 2 moles of heme per tetramer according to the estimation of Hayaishi and co-workers. The enzyme isolated from *Pseudomonas acidovorans* has a molecular weight of about 122,000 and is believed also to consist of four subunits (Brady, 1976). Almost exactly the same can be said with regard to indoleamine 2,3-dioxygenase, which simply has a lower substrate specificity than tryptophan 2,3-dioxygenase.

It has been pointed out (Brady, 1976) that this discrepancy between the results of the two research groups may be due to an improper ashing process in the analysis of copper in the protein, and/or a treatment with an inadequate chelating agent in the process of releasing copper from the protein. Our concern here, however, is with the question: whether both copper and the heme are essential to the enzyme and, if essential, in what way they might work. This question has not been answered unequivocally.

4.3.5 Dihydroxylases

Reaction 4-16 represents an entirely different type of dioxygenation, distinct from reaction type 4-15; this may be better described as a "dihydroxylase" reaction. An example of such enzymes is benzoate 1,2-dioxygenase (Fujisawa *et al.*, 1976) which catalyzes the following reaction:

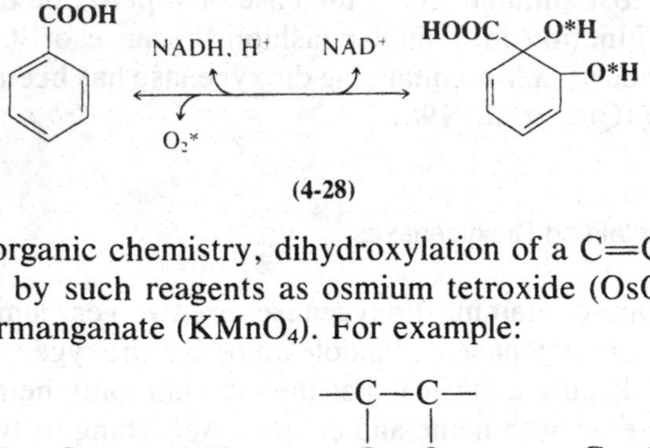

(4-28)

In synthetic organic chemistry, dihydroxylation of a $C{=}C$ bond can be accomplished by such reagents as osmium tetroxide (OsO_4) or alkaline potassium permanganate ($KMnO_4$). For example:

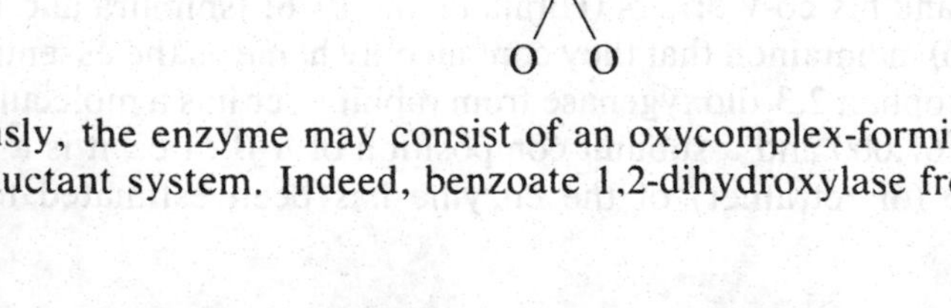

(4-29)

Analogously, the enzyme may consist of an oxycomplex-forming center and a reductant system. Indeed, benzoate 1,2-dihydroxylase from *Pseu-*

domonas arivilla consists of two proteins, a terminal oxygenase system containing nonheme iron and a reductase system containing FAD and nonheme iron (Fujisawa *et al.*, 1976). Therefore, a mechanism similar to reaction scheme 4-29 may be postulated for the enzymatic reaction:

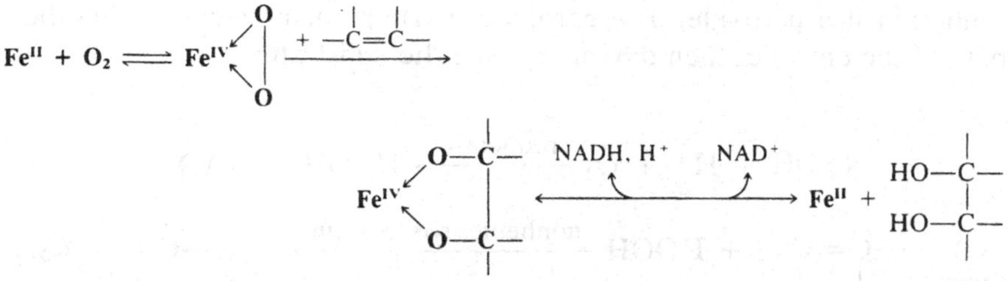

(4-30)

The second step of this mechanism may proceed as follows:

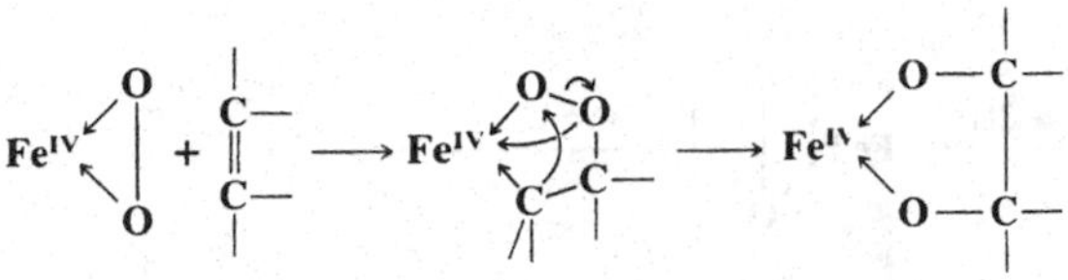

(4-31)

This is in accord with an established mechanism of the following type (Valentine, 1973):

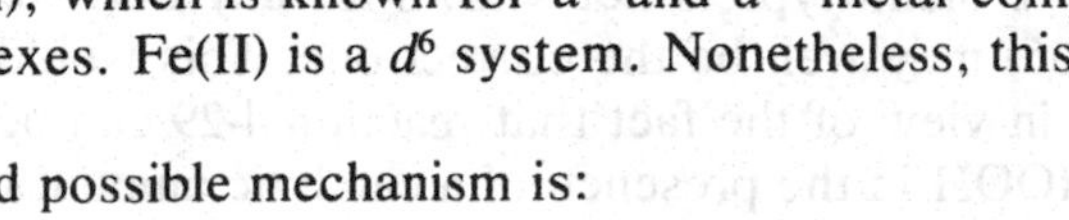

(4-32)

Such a mechanism presumes an oxygen complex of the so-called π-olefin type (side-on), which is known for d^8 and d^{10} metal complexes but not for d^6 complexes. Fe(II) is a d^6 system. Nonetheless, this mechanism is a possibility.

A second possible mechanism is:

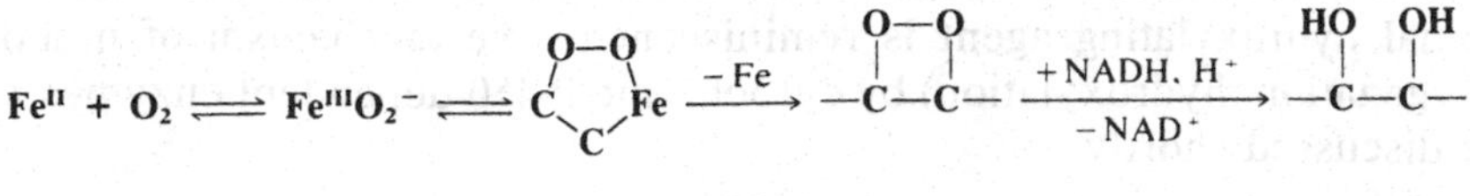

(4-33)

The problem here is that the dioxetane intermediate postulated, being so unstable, should decompose into two carbonyl compounds before it undergoes the hydrogenation. Therefore, this mechanism does not appear to be very feasible.

A third possibility is that the reductase system reduces O_2 first, thus rendering it a peroxide. The peroxide, perhaps being aided by another part of the enzyme, then dihydroxylates the substrate:

$$NADH + H^+ + O_2 \xrightarrow{\text{reductase}} HOOH + NAD^+$$

$$\begin{array}{c} -C=C- \\ | \quad | \end{array} + HOOH \xrightarrow{\text{nonheme iron protein}} \begin{array}{c} -C-C- \\ | \quad | \\ HO \quad OH \end{array} \quad (4\text{-}34)$$

The second step of 4-34 may proceed as follows:

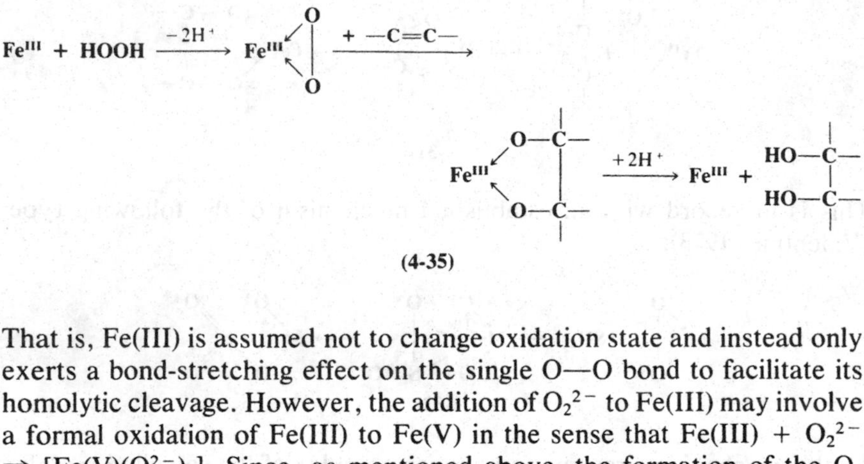

$$Fe^{III} + HOOH \xrightarrow{-2H^+} Fe^{III}\overset{O}{\underset{O}{\Big\langle}} \ + \ -C=C- \longrightarrow$$

$$Fe^{III}\overset{O-C-}{\underset{O-C-}{\Big\langle}} \xrightarrow{+2H^+} Fe^{III} + \begin{array}{c} HO-C- \\ | \\ HO-C- \end{array}$$

(4-35)

That is, Fe(III) is assumed not to change oxidation state and instead only exerts a bond-stretching effect on the single O—O bond to facilitate its homolytic cleavage. However, the addition of O_2^{2-} to Fe(III) may involve a formal oxidation of Fe(III) to Fe(V) in the sense that Fe(III) + O_2^{2-} ⇌ [Fe(V)(O^{2-})$_2$]. Since, as mentioned above, the formation of the O_2 complex of the π-olefin type is not known for any Fe(II) complex, this mechanism (4-35) may well be the correct one. This would also appear more plausible in view of the fact that reaction 4-29 can occur between an olefin and HOOH in the presence of a catalytic amount of OsO_4. The notion that HOOH is formed by the reductase system and becomes the actual hydroxylating agent is reminiscent of the mechanism of monooxygenation (hydroxylation) by cytochrome P-450-dependent enzymes to be discussed shortly.

4.4 Monooxygenase Reactions (Plus Peroxidases and Catalase)

The overall reaction of this type can be described by:

$$SH + O_2 + 2H^+ + 2e \rightarrow SOH + HOH \qquad (4\text{-}36)$$

This indicates that the reaction formally involves a two-electron reduction of O_2 and an insertion of an oxygen atom into a C—H bond of the substrate. There are apparently three different categories of enzymes that catalyze reactions of this type. One group of enzymes is dependent on P-450, which is believed to constitute the active site for the enzymatic reaction. Under physiological conditions the enzymes of this group utilize NADPH (or NADH) as the electron source. The second group comprises a variety of enzymes that contain nonheme iron and use α-ketoglutarate, pterin, or NADPH as the electron source (for review see Ochiai, 1977; Ullrich and Duppel, 1975; Udenfriend and Cardinale, 1982). The enzymes of the third category are dependent on copper at the catalytic site; known cases are tyrosinase and dopamine-β-hydroxylase (Ullrich *et al.*, 1977).

4.4.1 Cytochrome P-450-Dependent Enzymes

The enzymes of this class have recently attracted much attention for a variety of reasons: e.g., the versatility of the enzymes and their genetic relationship among themselves. Our main concern here is with their mechanism of oxygen activation. The following reaction scheme has been well established:

$$E\text{-}Fe^{III} + SH \rightleftharpoons E\text{-}Fe^{III}—SH \xrightarrow{e} E\text{-}Fe^{II}—SH \xrightarrow{O_2}$$
$$E\text{-}FeO_2—SH \xrightarrow{e} E\text{-}Fe^{III}(O_2{}^{2-})\text{-}SH \xrightarrow{2H^+} E\text{-}Fe^{III}$$
$$+ SOH + HOH \qquad (4\text{-}37)$$

There is a minor dispute as to whether the two electrons are added to the P-450 system simultaneously or, as shown here, separately. The microsome system of liver, which oxidizes NADPH and metabolizes exogenous drugs, is known to produce hydrogen peroxide. This fact is in accord with the reaction scheme, although there is some question as to the mode of hydrogen peroxide formation. The scheme above indicates that HOOH forms from $E\text{-}Fe^{III}(O_2{}^{2-})$—SH, whereas some experimental evidence (Estabrook and Werringloer, 1977) suggests that $O_2{}^-$ released from $E\text{-}Fe^{III}(O_2{}^-)$—SH disproportionates into O_2 and HOOH.

The most interesting and crucial point is the last step of reaction scheme 4-37. The peroxide entity on the Fe(III) of P-450 must be cleaved, one of the oxygen atoms being converted into HOH and the other being inserted into a C—H bond of the substrate. There are three possible modes of cleavage:

$$[P\text{-}Fe^{III} \leftarrow |\overline{O}\text{—}\overline{O}|] \rightarrow [P\text{-}Fe^{III} \leftarrow |\overline{O}] + |\overline{O}| \; (+2H^+ \rightarrow HOH) \quad (4\text{-}38)$$

$$[P\text{-}Fe^{III} \leftarrow |\overline{O}\text{—}\overline{O}|] \longrightarrow [P\text{-}Fe^{III} \leftarrow |\overline{O}\cdot]$$
$$+ \cdot\overline{O}| \; (+H^+ + H \rightarrow HOH) \quad (4\text{-}39)$$

$$[P\text{-}Fe^{III} \leftarrow |\overline{O}\text{—}\overline{O}|] \longrightarrow$$
$$\overline{\underset{\cdot}{O}}| + [P\text{-}Fe^{III} \leftarrow |\overline{O}|](+2H^+ + P\text{-}Fe^{III} \rightarrow HOH) \quad (4\text{-}40)$$

In the above equations, P is the porphyrin ligand of P-450. Accordingly, the hydroxylating agent is $P\text{-}Fe^{III} \leftarrow |\overline{O}$ in mode 4-38, $P\text{-}Fe^{III} \leftarrow |\overline{O}\cdot$ in 4-39 and $|\overline{O}$ in 4-40. In this regard it is useful to compare the above schemes with the reaction modes of peroxidases and catalase. It has been well established that in these enzymes a complex termed "complex I" forms, and that it has the formal chemical formula $[\cdot P^+\text{-}Fe^{IV}]$. This can arise in the following ways:

$$[P\text{-}Fe^{III} \leftarrow |\overline{O}\text{—}\overline{O}\text{—}R](R = H \text{ or alkyl}) \xrightarrow{+H^+} [P\text{-}Fe^{III} \leftarrow |\overline{O}] + ROH$$

$$[P\text{-}Fe^{III} \leftarrow |\overline{O} \Leftrightarrow \cdot P^+\text{-}Fe^{IV} \leftarrow |\overline{O}|] \xrightarrow{+2H^+} [\cdot P^+\text{-}Fe^{IV}] + HOH \quad (4\text{-}41)$$

or

$$[P\text{-}Fe^{III} \leftarrow |\overline{O}\text{—}\overline{O}\text{—}R] \longrightarrow [P\text{-}Fe^{III} \leftarrow |\overline{O}\cdot] + \cdot OR$$

$$[P\text{-}Fe^{III} \leftarrow |\overline{O}\cdot \Leftrightarrow P\text{-}Fe^{IV} \leftarrow |\overline{O}|] \xrightarrow{+ \cdot OR} [\cdot P^+\text{-}Fe^{IV} \leftarrow |\overline{O}|] + ROH$$

$$[\cdot P^+\text{-}Fe^{IV} \leftarrow |\overline{O}|] \xrightarrow{+2H^+} [\cdot P^+\text{-}Fe^{IV}]$$
$$+ HOH \quad (4\text{-}42)$$

Reaction scheme 4-41 corresponds to reaction mode 4-38 whereas 4-42 is analogous to 4-39. Cleavage mode 4-40 has no counterpart in the case of a hydroperoxide bound to Fe(III).

It has been shown (Nordblom et al., 1976; Coon et al., 1977) that a hydroperoxide such as cumene hydroperoxide can hydroxylate the usual substrate in the presence of Fe(III)-cytochrome P-450 and in the absence of NADPH and the reductase system. In view of this result, mode 4-40 has to be discarded, although the hydroxylation reaction by O_2 + the complete system may be different from that in the hydroperoxide + P-450 system and may thus proceed in manner 4-40. Quantum mechanical considerations argue, indeed, that O—O bond cleavage in Fe^{III}—O—O) most likely corresponds to mode 4-40, rather than 4-38 or 4-39 (Ochiai, 1975).

In order for either mode 4-38 or 4-39 to become feasible, it appears that the terminal oxygen would need to be bound to an electrophilic entity before the O–O bond cleavage. Thus, for reaction 4-39:

$$[\text{P-Fe}^{III} \leftarrow |\overline{O}—\overline{O}|] \xrightarrow{+X^+} [\text{P-Fe}^{III} \leftarrow |\overline{O}—\overline{O}—X]$$
$$\longrightarrow [\text{P-Fe}^{III} \leftarrow |\overline{O}\cdot] + \cdot OX \quad (4\text{-}43)$$

X^+ may usually be H^+, in which case the reaction is essentially the same as 4-42. XO or RO radical does not necessarily abstract a hydrogen atom from the porphyrin ring; it may abstract a hydrogen atom from a nearby protein residue. The latter appears to happen with a peroxidase, cytochrome c peroxidase; in this case $[\text{P-Fe}^{IV}]$ and an organic radical are formed. If it is to be assumed that all peroxidases should have a single, same mechanism, mode 4-42 may be preferred to mode 4-41 as the probable mechanism of peroxidases and catalase (Ochiai, 1977).

If there is available a device that effectively protonates the terminal oxygen further, O—O bond cleavage mode 4-38 may become feasible; that is:

$$[\text{P-Fe}^{III} \leftarrow |\overline{O}—\overline{O}|] \xrightarrow{+2H^+} [\text{P-Fe}^{III} \leftarrow |\overline{O}—\overline{O}^+—H] \rightarrow$$
$$\overset{\displaystyle |}{\underset{\displaystyle H}{}}$$
$$[\text{P-Fe}^{III} \leftarrow |\overline{O}] + HOH \quad (4\text{-}44)$$

In the case of a hydroperoxide as the hydroxylating agent,

$$[\text{P-Fe}^{III} \leftarrow |\overline{O}—\overline{O}—R] \xrightarrow{+H^+} [\text{P-Fe}^{III} \leftarrow |\overline{O}—\overline{O}—R] \rightarrow$$
$$\overset{\displaystyle |}{\underset{\displaystyle H}{}}$$
$$[\text{P-Fe}^{III} \leftarrow |\overline{O}] + ROH \quad (4\text{-}45)$$

This is very similar to scheme 4-41. P-Fe^{III}—O can be regarded as the "ferryl ion," a term coined by Mason (1957). That this oxygen may have a character of $O(^1D$, the first excited state of the O atom) is indicated here by the description of O as $|\overline{O}$ and is in accord with the fact that a catalytically active ferryl moiety (iron-oxo complex) can be obtained from a reaction of iodosylbenzene and Fe(III)-porphyrin (Lichtenberger et al.,

1976; Groves and Nemo, 1983), though some authors describe it as $Fe^V{=}O$. A tentative conclusion then would be that the mode of O—O bond cleavage in P-450-dependent enzyme systems can be regarded as scheme, 4-38, 4-44, 4-45, provided that some moiety is located near the heme center, which can strongly and rapidly protonate the terminal oxygen atom of the coordinated O_2; that is:

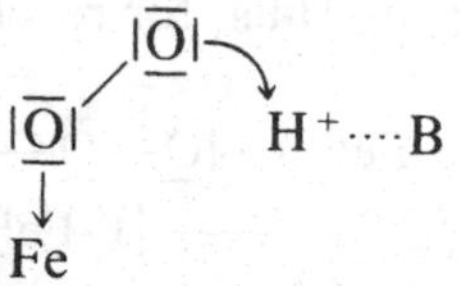

$$
\begin{array}{c}
\overset{\displaystyle |\overline{O}|\searrow}{} \\
|\overline{O}|^{\nearrow} \qquad H^+ \cdots B \\
\downarrow \\
Fe
\end{array}
\qquad\qquad (4\text{-}46)
$$

There are some indications that such readily exchangeable protons are present near the heme iron (White and Coon, 1982; Griffin and Peterson, 1975; Philson *et al.*, 1979; LoBrutto *et al.*, 1980).

The next question then is how P-Fe^{III}—O would hydroxylate a C—H bond. There are two possible modes. In mode A, the oxo ligand functions as an activated oxygen atom $O(^1D)$, the first excited state of atomic O. $O(^1D)$ is known to perform the following types of reaction, which are exactly the same as the types of reaction catalyzed by the enzymes dependent on P-450:

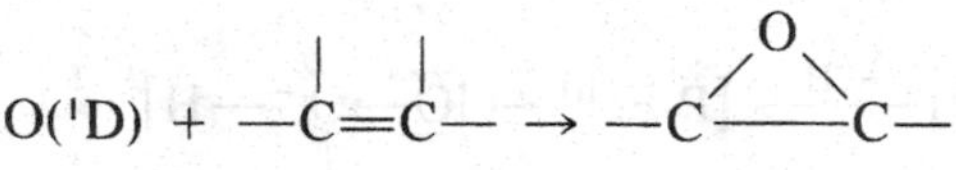

$$
O(^1D) + \underset{|}{\overset{|}{-}C}{=}\underset{|}{\overset{|}{C}}{-} \rightarrow -\overset{\displaystyle O}{\overset{\displaystyle \diagup \diagdown}{C\text{———}C}}- \qquad\qquad (4\text{-}47)
$$

$$
O(^1D) + \underset{|}{\overset{|}{-}C}{-}H \rightarrow \underset{|}{\overset{|}{-}C}{-}O{-}H \qquad\qquad (4\text{-}48)
$$

These reactions may be regarded as involving electrophilic attack of $O(^1D)$ on the C=C bond or the C—H bond. $O(^1D)$, when bound to Fe(III), would be activated in the sense that its electrophilicity should be increased by the electron-withdrawing effect of Fe(III). The results with a Mn(III)-containing model compound that olefins which are electron-rich and have a *cis* geometry interact more efficiently with the oxo complex (Collman *et al.*, 1984) are consistent with the idea that the O atom bound to the metal center acts as an electrophile.

Another possibility, mode B, assigns a free radical character to the oxo ligand (Groves *et al.*, 1978; Groves, 1980). In an experiment with norborneol and a microsomal enzyme system, an isotope effect k_H/k_D of as high as 11.5 ± 1 was obtained:

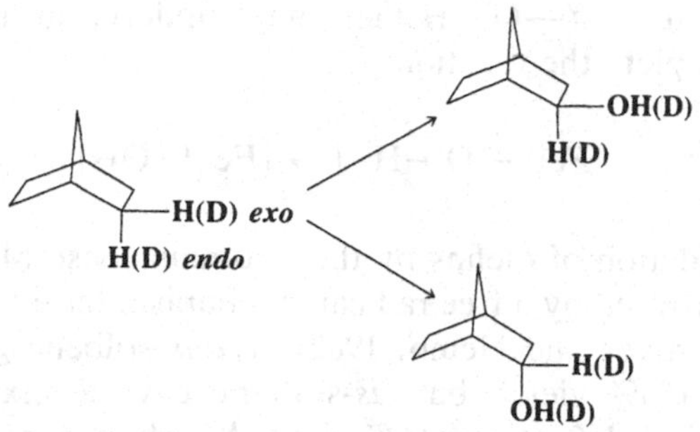

(4-49)

This large isotope effect value is similar to those observed for alkane oxidations by oxo complexes of manganese and chromium. In addition to the large isotope effect, a significant (14–18%) degree of epimerization (*endo* $\rightleftharpoons$ *exo*) was observed. The cumene hydroperoxide-mediated hydroxylation of 1,2-dideuterocyclohexane resulted in a significant allylic rearrangement (Groves and Nemo, 1982). The reaction of norcarane with $Mn^{III}(TPP)Cl$ and iodosylbenzene gave a mixture of 2-norcaranyl and 3-cyclohexenyl methyl derivative (Groves and Nemo, 1982). In this case, $Mn^V{=}O$ is assumed to abstract a hydrogen from the substrate to form a norcaranyl free radical that rearranges to a 3-cyclohexenyl methyl radical. From these observations the conclusion was drawn that the hydroxylation occurs by a homolytic hydrogen abstraction and readdition of the resulting OH onto the carbon radical. That is:

$$(Fe^{III}{-}O)^{3+} + H{-}\overset{|}{\underset{|}{C}}{-} \rightarrow \left[Fe^{IV}{-}OH^- \cdot \overset{|}{\underset{|}{C}}{-} \right] \rightarrow$$

$$Fe^{III} + HO{-}\overset{|}{\underset{|}{C}}{-} \quad (4\text{-}50)$$

$(Fe^{III}{-}O)^{3+}$ can be regarded as $(Fe^V{\leftarrow}|\overline{O}|^{2-})$; there could be some degree of double bond character in an $Fe^V{-}O^{2-}$ bond, as in, e.g., the vanadyl VO^{2+}. To explain the high degree of reactivity of the $(Fe{-}O)^{3+}$ shown in 4-50, one has to assume that the oxo ligand has some degree of free radical character; that is:

$$(Fe^{III}{-}O)^{3+}({\equiv}Fe^{III} \leftarrow |\underline{O}) \leftrightarrow [Fe^{IV} \leftarrow |\underline{O} \cdot] \quad (4\text{-}51)$$

Then

$$[Fe^{IV} \leftarrow |\underline{O} \cdot] + H{-}\overset{|}{\underset{|}{C}}{-} \rightarrow \left[Fe^{IV}{-}O{-}H^- \cdot \overset{|}{\underset{|}{C}}{-} \right] \quad (4\text{-}52)$$

The intermediate Fe^{IV}—O—H then must undergo an internal electron transfer to complete the reaction:

$$[Fe^{IV}—O—H^-] \rightarrow [Fe^{III} \cdot OH] \qquad (4\text{-}53)$$

The epoxidation of olefins by the oxomanganese Mn^v=O was also suggested to proceed by a free radical mechanism based on the following observation (Groves and Nemo, 1982). *Trans*-stilbene gave only *trans*-stilbene oxide (53% yield), but *cis*-stilbene gave a mixture of *cis*- and *trans*-stilbene (in 1:1.6 ratio) in 88% yield. A carbon–carbon bond rotation in an intermediate Mn(IV) carbon radical species could explain this result as shown below:

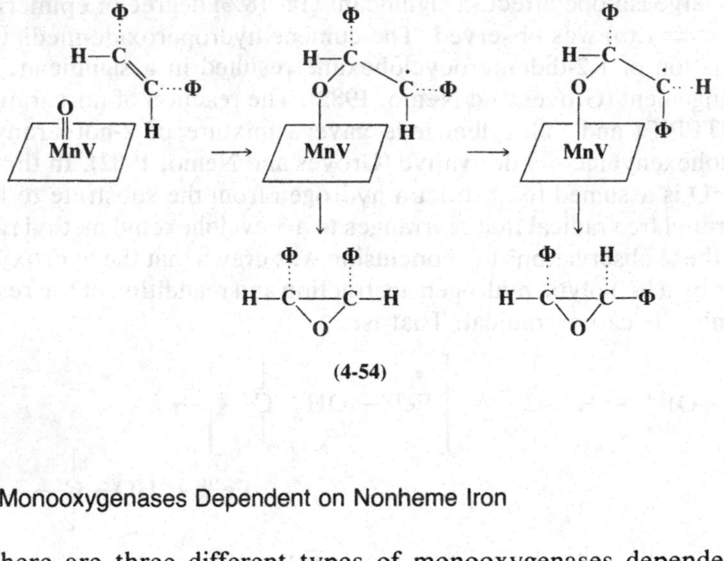

(4-54)

4.4.2 Monooxygenases Dependent on Nonheme Iron

There are three different types of monooxygenases dependent on nonheme iron (Ullrich and Duppel, 1975). The first type is the pterin-dependent monooxygenase and is exemplified by phenylalanine hydroxylase, tyrosine hydroxylase, and tryptophan hydroxylase. The second type requires α-glutarate and ascorbate as well as Fe(II), and includes prolyl hydroxylase, lysyl hydroxylase, and thymine 7-hydroxylase. The third type is exemplified by 4-methoxybenzoate O-demethyl monooxygenase and contains iron–sulfur cores. Squalene epoxidase, which catalyzes a reaction of type 4-47, seems to require NADPH-cytochrome P-450 reductase and FAD, but apparently does not contain a significant amount of P-450 (Schroepfer, 1982). This may well represent a fourth type of monooxygenase, but it is also possible that this is an FAD-dependent monooxygenase described in Section 4.7.

The reaction sequence of the pterin-dependent hydroxylases is:

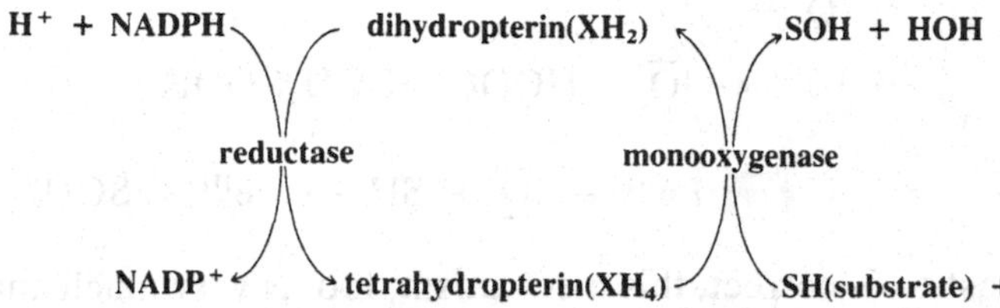

$$H^+ + NADPH \qquad dihydropterin(XH_2) \qquad SOH + HOH$$

$$reductase \qquad\qquad monooxygenase$$

$$NADP^+ \qquad tetrahydropterin(XH_4) \qquad SH(substrate) + O_2$$

$$(4\text{-}55)$$

As far as the overall process is concerned, this system is very similar to that of the P-450-dependent monooxygenase. The details of the mechanism have not been unveiled, but it is very tempting to suggest a mechanism similar to 4-38, 4-44. That is:

$$Fe^{III} + XH_4 \rightarrow Fe^{II} + XH_3 + H^+$$

$$Fe^{II} + XH_3 + O_2 \rightarrow Fe^{III}O_2^- + XH_2 + H^+$$

$$Fe^{III}O_2^- + 2H^+ \rightarrow Fe^{III}O + HOH$$

$$Fe^{III}O + SH \rightarrow Fe^{III} + SOH \qquad\qquad (4\text{-}56)$$

The α-ketoglutarate-dependent enzymatic reactions are represented by:

$$SH + HOOCCH_2CH_2CCOOH + O_2 \rightarrow$$
$$SOH + HOOCCH_2CH_2COOH + CO_2 \quad (4\text{-}57)$$

The reaction sequence and mechanism are not known. However, the following scheme is a possibility:

$$\overset{\displaystyle O}{\overset{\displaystyle \|}{HOOCCH_2CH_2CCOOH}} \xrightarrow{\ enzyme\ }$$

$$\overset{\displaystyle O}{\overset{\displaystyle \|}{HOOCCH_2CH_2C^+}} + CO_2 + H^+ + 2e$$

$$Fe^{III} + e \rightarrow Fe^{II}, \ Fe^{II} + O_2 + e \rightarrow Fe^{III} \leftarrow |\overline{O}\!-\!\overline{O}|^{2-}$$

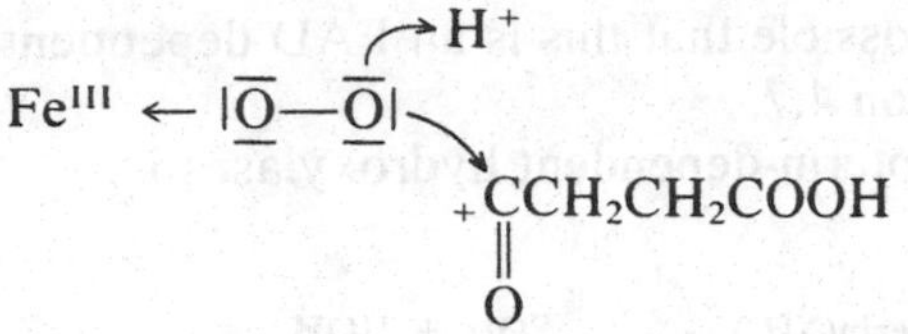

$$Fe^{III} \leftarrow |\overline{O}-\overline{O}| \underset{\underset{O}{\overset{\parallel}{+CCH_2CH_2COOH}}}{\overset{\curvearrowright H^+}{\diagdown}}$$

$$\rightarrow Fe^{III} \leftarrow |\overline{O} + HOOCCH_2CH_2COOH$$

$$Fe^{III} \leftarrow |\overline{O} + SH \rightarrow Fe^{III} + SOH \qquad (4\text{-}58)$$

If this suggestion is correct, this mechanism, too, is very much analogous to that of P-450-dependent monooxygenases. The initial reduction of Fe(III) to Fe(II) may be accomplished by the reducing agent (ascorbic acid) instead of a reductase system inherent in the enzyme.

The reaction of the iron–sulfur-containing enzyme, 4-methoxybenzoate O-demethyl monooxygenase, can be described as follows:

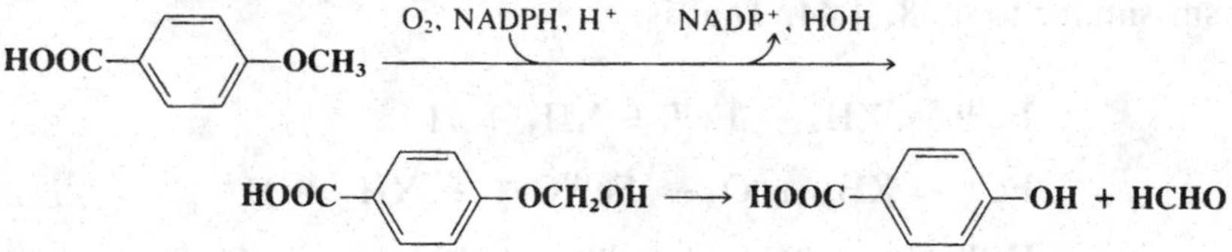

$$HOOC-\langle\text{ring}\rangle-OCH_3 \xrightarrow[]{O_2, NADPH, H^+ \qquad NADP^+, HOH}$$

$$HOOC-\langle\text{ring}\rangle-OCH_2OH \longrightarrow HOOC-\langle\text{ring}\rangle-OH + HCHO$$

$$(4\text{-}59)$$

This is again a typical monooxygenation; the mechanism then could be similar to that of monooxygenases outlined in the previous section. The catalytic center seems to contain an iron atom, but its nature is not known.

4.4.3 Monooxygenases Containing Copper

The similarity between tyrosinase and hemocyanin has been well established. Therefore, it is likely that tyrosinase forms $Cu^{II}-O_2^{2-}-Cu^{II}$ (see Section 4.1.5). As in the case of P-450, $Cu^{II}-O_2^{2-}-Cu^{II}$ may react in the following ways:

$$Cu^{II}-O_2^{2-}-Cu^{II} \rightarrow \overset{\displaystyle |\overline{O}|}{\underset{\downarrow}{Cu^{II}\cdots Cu^{II}}} \xrightarrow{+2H^+} \overset{\displaystyle |\overline{O}|}{\underset{\downarrow}{Cu^{II}\cdots Cu^{II}}} + HOH$$

$$\overset{\displaystyle |O|}{\underset{\downarrow}{Cu^{II}\cdots Cu^{II}}} + SH \rightarrow Cu^{II}\cdots Cu^{II} + SOH \qquad (4\text{-}60)$$

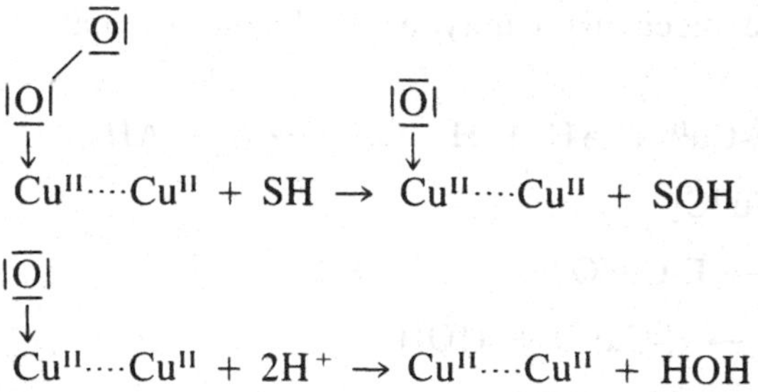

$$
\begin{array}{c}
\overset{\displaystyle |\overline{O}|}{\overset{\displaystyle /}{\underset{\displaystyle \downarrow}{|\overline{O}|}}} \\
Cu^{II}\cdots Cu^{II} + SH \rightarrow \overset{\displaystyle |\overline{O}|}{\underset{\displaystyle \downarrow}{Cu^{II}}}\cdots Cu^{II} + SOH
\end{array}
$$

$$
\overset{\displaystyle |\overline{O}|}{\underset{\displaystyle \downarrow}{Cu^{II}}}\cdots Cu^{II} + 2H^{+} \rightarrow Cu^{II}\cdots Cu^{II} + HOH \tag{4-61}
$$

Reaction scheme 4-60 corresponds to scheme 4-38 and 4-61 to 4-40. In this enzyme system, the substrate itself provides the reducing agent; that is:

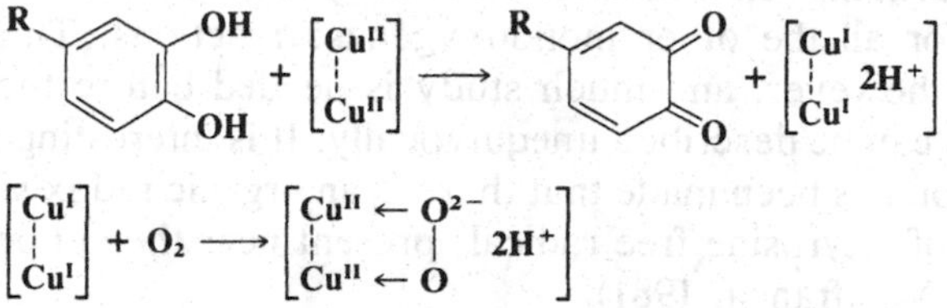

$$\tag{4-62}$$

Therefore, this enzyme acts as an oxidase (catechol activity) for the substrate and also as a monooxygenase (cresolase activity).

Dopamine-β-hydroxylase requires ascorbic acid as the electron source. The overall reaction is:

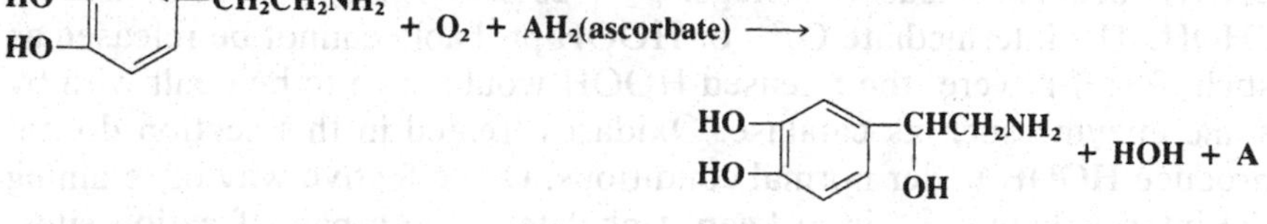

$$\tag{4-63}$$

The enzyme probably contains one copper atom per subunit of molecular weight of about 36,000. Whether multiple copper atoms are arranged in juxtaposition as in hemocyanin and tyrosinase or a mononuclear copper center constitutes the catalytic site is not known. If the catalytic site contains a dinuclear copper system, then the mechanism may be expected to be similar to that of tyrosinase. If the catalytic site is composed of a single copper atom, the mechanism may be analogous to that at the P-450 site. That is:

$$E\text{-}Cu^{II} + AH_2 \rightarrow E\text{-}Cu^{II} + AH + H^+ \quad (2AH \rightarrow A + AH_2)$$

$$E\text{-}Cu^{I} + O_2 \rightarrow E\text{-}Cu^{II}O_2^-$$

$$E\text{-}Cu^{II}O_2^- + AH_2 \rightarrow E\text{-}Cu^{II}O_2^{2-} + AH + H^+$$

$$E\text{-}Cu^{II}O_2^{2-} + 2H^+ \rightarrow E\text{-}Cu^{II}O + HOH$$

$$E\text{-}Cu^{II}O + SH \rightarrow E\text{-}Cu^{II} + SOH \tag{4-64}$$

The catalytic intermediate postulated here, $Cu^{II}O$, is the so-called "cupryl ion." Other mechanisms can be formulated, but if one accepts the fundamental notion that nature chooses simple, common ways of effecting apparently somewhat dissimilar reactions, then mechanism 4-64 may well be preferable, in the sense that it is essentially analogous to the mechanism postulated for all the other monooxygenase reactions. This is a mere speculation, however, and much study is needed before these reaction mechanisms can be described unequivocally. It is interesting to note that the suggestion has been made that there is an organic redox site, possibly in the form of a tyrosine free radical, present near the copper site (May et al., 1981; Villafranca, 1981).

4.5 Oxidases Effecting Reduction of Oxygen to Water

The overall reaction in this category may be written as:

$$\text{reduced substrate(S)} + O_2 \rightarrow \text{oxidized S} + 2HOH \tag{4-65}$$

This involves the addition of four electrons to O_2. The reduction of O_2 to 2HOH occurs in at least two steps: $O_2 + 2e \rightarrow O_2^{2-}$; $O_2^{2-} + 2e + 4H^+ \rightarrow 2HOH$. The intermediate O_2^{2-} or HOOH probably cannot be released as such, for, if it were, the released HOOH would have to be dealt with by some enzyme such as catalase. Oxidases treated in this section do not produce HOOH under normal conditions. One effective way of retaining the intermediate O_2^{2-} is to keep it chelated onto a pair of cation sites. This necessitates the presence of a dinuclear center in these enzymes. Indeed this seems to be a general feature common to all such oxidases. That is:

$$\begin{bmatrix} M_a(+n) \\ M_b(+m) \end{bmatrix} + O_2 \rightarrow \begin{bmatrix} M_a(+n+1) \leftarrow O \\ M_b(+m+1) \leftarrow O \end{bmatrix}^{2-} \tag{4-66}$$

The addition of two electrons to this site completes the reduction of O_2 to 2HOH; that is:

$$\begin{bmatrix} M_a(+n+1) \leftarrow O \\ \vdots \\ M_b(+m+1) \leftarrow O \end{bmatrix}^{2-} \xrightarrow{+4H^+,2e} \begin{bmatrix} M_a(+n+1) \\ \vdots \\ M_b(+m+1) \end{bmatrix} + 2HOH \quad (4\text{-}67)$$

$M_a(+n+1)$ and $M_b(+m+1)$ must then be reduced back to $M_a(+n)$ and $M_b(+m)$, respectively, in order to permit the next cycle of reaction. Therefore, there needs to be a source of four electrons per cycle. The four electrons come ultimately from the substrate(s) and it would be convenient if the enzyme proteins contained four electron-depositing centers. This is exactly what is found in cytochrome c oxidase and the so-called blue-copper oxidases.

4.5.1 Cytochrome c Oxidase

Cytochrome c oxidase contains 2 moles of copper and 2 moles of heme (cytochrome a and cytochrome a_3) per mole. The two copper atoms have different characters Cu(v) and Cu(s). Cu(v) in the oxidized state, $Cu^{II}(v)$, shows an absorption band at 830 nm, and $Cu^{II}(s)$ gives rise to an EPR spectrum that is very sensitive to saturation effects by power increase. It has been suggested from a variety of experimental data that there is a kind of magnetic interaction between Cu(s) and cytochrome a_3. Several lines of evidence, including kinetic results (Chance et al., 1975; Antonini et al., 1977), have indicated that O_2 binds to the couple Cu(s)–cyt a_3; that is:

$$O_2 + [Cu^I(s)\text{----}cyt\ a_3(Fe^{II})] \rightleftharpoons [Cu^{II}(s)\text{----}O_2^{2-}\text{----}cyt\ a_3(Fe^{III})] \quad (4\text{-}68)$$

If the other sites, cyt a and Cu(v), are in the reduced state, they can supply two more electrons to the Cu(s)–cyt a_3 site, thus reducing the O_2^{2-} further down to 2 molecules of water. An EXAFS study (Powers et al., 1981) indicates an RS^- bridge between Fe of cyt a_3 and Cu(s) in the oxidized state; the distance between the $Fe^{III}(a_3)$ and $Cu^{II}(s)$ was estimated to be 375 ± 5 pm.

The electrons from the reductant, cytochrome c under physiological conditions, enter cytochrome c oxidase through cyt a. They are rapidly transferred to the second electron reservoir, i.e., Cu(v). The reaction sequence then can be depicted as:

$$\begin{bmatrix} \text{cyt a(Fe}^{\text{III}}) & \text{cyt a}_3(\text{Fe}^{\text{III}}) \\ \text{Cu}^{\text{II}}(v) & \text{Cu}^{\text{II}}(s) \end{bmatrix} \xrightarrow{e} \begin{bmatrix} \text{cyt a(II)} & \text{cyt a}_3(\text{III}) \\ \text{Cu}^{\text{II}}(v) & \text{Cu}^{\text{II}}(s) \end{bmatrix} \longrightarrow$$

$$\begin{bmatrix} \text{cyt a(III)} & \text{cyt a}_3(\text{III}) \\ \text{Cu}^{\text{I}}(v) & \text{Cu}^{\text{II}}(s) \end{bmatrix} \xrightarrow{3e} \begin{bmatrix} \text{cyt a(II)} & \text{cyt a}_3(\text{II}) \\ \text{Cu}^{\text{I}}(v) & \text{Cu}^{\text{I}}(s) \end{bmatrix} \xrightarrow{O_2} \quad (4\text{-}69)$$

$$\begin{bmatrix} \text{cyt a(II)} & \text{cyt a}_3(\text{III}) \\ \text{Cu}^{\text{I}}(v) & \text{Cu}^{\text{II}}(s) \end{bmatrix} O_2{}^{2-} \xrightarrow{4H^+} \begin{bmatrix} \text{cyt a(III)} & \text{cyt a}_3(\text{III}) \\ \text{Cu}^{\text{II}}(v) & \text{Cu}^{\text{II}}(s) \end{bmatrix} + 2\text{HOH}$$

4.5.2 Blue-Copper Oxidases

Included in this group are laccase, ascorbate oxidase, and ceruloplasmin. These enzymes all have a very similar copper composition: they contain four copper atoms per functional unit. Ceruloplasmin, which is now called ferroxidase, contains six copper atoms, two of which appear to be catalytically inactive, though. One copper, Cu(a) (or type I), is responsible for the blue color of the enzymes, thus designated "blue copper." Another copper atom, Cu(b) (or type II), has no conspicuous character, but is essential to the enzymatic activity. The remaining two copper atoms are in the form of a dimeric structure, $Cu_2(c)$ (type III). It is now believed based on various experimental studies that Cu(a) is the electron acceptor site from the substrate and $Cu_2(c)$ is the oxygen-binding site. Several lines of evidence also suggest that there is a rapid electron transfer within the protein, perhaps by the route $Cu(a)–Cu(b)–Cu_2(c)$.

A likely reaction sequence, therefore, is very analogous to that in cytochrome c oxidase; that is:

$$[\text{Cu}^{\text{II}}(a)\text{-}\text{Cu}^{\text{II}}(b)\text{-}\text{Cu}^{\text{II}}{}_2(c)] \xrightarrow{e(\text{from substrate})}$$

$$[\text{Cu}^{\text{I}}(a)\text{-}\text{Cu}^{\text{II}}(b)\text{-}\text{Cu}^{\text{II}}{}_2(c)] \xrightarrow{\text{rapid}}$$

$$[\text{Cu}^{\text{II}}(a)\text{-}\text{Cu}^{\text{I}}(b)\text{-}\text{Cu}^{\text{II}}{}_2(c)] \xrightarrow{\text{rapid}}$$

$$[\text{Cu}^{\text{II}}(a)\text{-}\text{Cu}^{\text{II}}(b)\text{-}\text{Cu}^{\text{II}}(c)\text{---}\text{Cu}^{\text{I}}(c)] \xrightarrow{3e(\text{from substrate})}$$

$$[\text{Cu}^{\text{I}}(a)\text{-}\text{Cu}^{\text{I}}(b)\text{-}\text{Cu}^{\text{I}}(c)\text{---}\text{Cu}^{\text{I}}(c)] \xrightarrow{O_2}$$

$$[\text{Cu}^{\text{I}}(a)\text{-}\text{Cu}^{\text{I}}(b)\text{-}\text{Cu}^{\text{II}}(c)\text{---}\text{Cu}^{\text{II}}(c)] \xrightarrow{4H^+}$$
$$\overset{\diagdown\quad\diagup}{O_2{}^{2-}}$$

$$[\text{Cu}^{\text{II}}(a)\text{-}\text{Cu}^{\text{II}}(b)\text{-}\text{Cu}^{\text{II}}{}_2(c)] + 2\text{HOH} \quad\quad (4\text{-}70)$$

4.5.3 Tyrosinase–Catecholase Activity

The functional unit of this enzyme consists of two juxtaposed copper atoms (see Sections 4.1.5 and 4.4.3). However, it does not contain two additional electron pooling sites as found in the blue-copper oxidases. As suggested above, the substrate appears to give the two additional electrons directly to O_2^{2-} bound with the dinuclear copper site. That is:

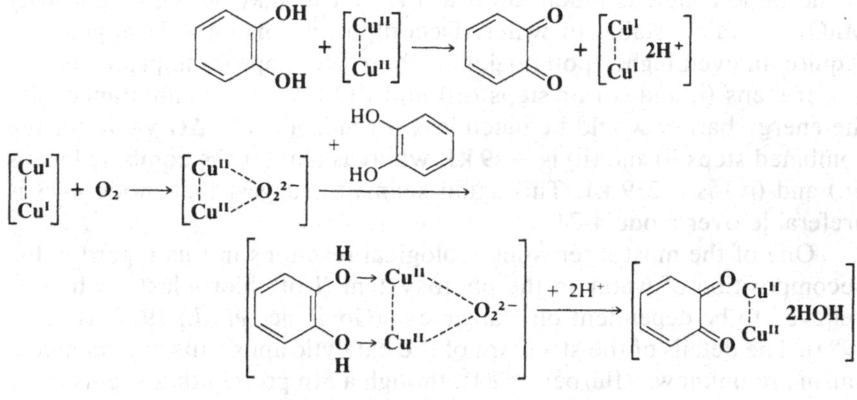

$$(4-71)$$

4.6 The Reverse of Oxidase Reactions—Decomposition of Water

The oxidative decomposition of water appears simple:

$$2HOH \rightarrow 4H^+ + 4e + O_2 \qquad (4-72)$$

This is exactly the reverse of the reactions catalyzed by oxidases as discussed in the previous section. Despite its apparent simplicity, however, its mechanism does not seem to have been elucidated fully in any system. Thermodynamically speaking, any oxidizing agent whose reduction potential at pH 7 is higher than $+0.82$ V should be able to decompose water; however, the reaction is usually very slow. For example, MnO_4^-, which is capable of oxidatively decomposing water, remains stable in pure water. CoF_3, on the other hand, rapidly decomposes water.

There are several possible pathways for the oxidative decomposition of water. Two of them are:

$$2HOH \xrightarrow{(i)} 4H^+ + 4e + 2[O], \quad 2[O] \xrightarrow{(ii)} O_2 \qquad (4-73)$$

$$2HOH \xrightarrow{(iii)} 2H^+ + 2e + 2[OH], \quad 2[OH] \xrightarrow{(iv)} HOOH,$$

$$HOOH \xrightarrow{(V)} 2H^+ + 2e + O_2 \qquad (4-74)$$

Estimated ΔG values (at 25°C, pH 7) are (i) $+577$ kJ, (ii) -497 kJ, (iii) $+426$ kJ, (iv) -166 kJ, and (v) $+52$ J; the corresponding reduction potentials for the reverse processes (i.e., reduction) are (i) $+1.49$ V, (iii) $+2.43$ V, and (v) $+0.27$ V. Therefore, if the reaction path is to be represented by 4-73, the minimum required reduction potential for the oxidant is not $+0.82$ V as cited above, but rather $+1.49$ V. The E_h^0 value of Co(III) is $+1.87$ V; thus, it may be able to decompose water according to mode 4-73. In an acidic medium the $E_h^0(MnO_4^-/Mn^{II})$ is $+1.51$ V, which is about the same as the minimum required potential. The potential of the same couple is much lower at pH 7. This may be the reason why MnO_4^- is fairly stable in water. Decomposition mode 4-74 appears to require an even higher potential and, therefore, appears improbable.

If steps (i) and (ii) or steps (iii) and (iv) take place simultaneously, the energy barrier would be much less formidable. The ΔG value for the combined steps (i) and (ii) is $+79$ kJ, whereas that for the combined steps (iii) and (iv) is $+259$ kJ. This again seems to suggest that mode 4-73 is preferable over mode 4-74.

One of the most interesting biological reactions in this regard is the decomposition of water in the photosystem II of chloroplasts, which is believed to be dependent on manganese (Govindjee et al., 1977; Barber, 1984). The details of the structure of the catalytic apparatus and its mechanism are unknown (Barber, 1984), though a Mn protein that seems to be essential for the water decomposition has recently been isolated and purified (Abramowicz et al., 1984). The protein contains two Mn atoms per 34,000 molecular weight, and exhibits an EPR spectrum that indicates a binuclear Mn center. There seems to be another pair of Mn atoms involved in the process, which has characteristics different from the first pair (see Kambara and Govindjee, 1985).

The discussion above may shed some light on the problem. As suggested above, a feasible mechanism would be one in which steps (i) and (ii) in mode 4-73 occur simultaneously. A likely mechanism may, thus, be as follows:

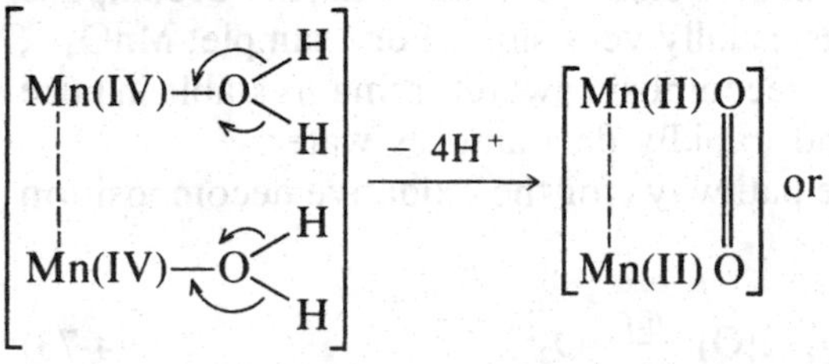

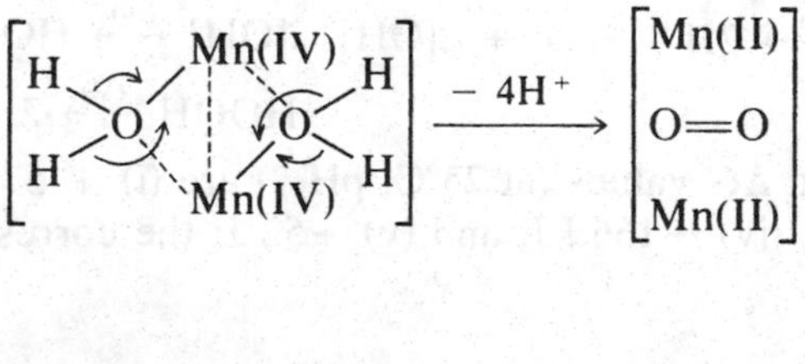

(4-75)

Two manganese atoms are assumed to be in juxtaposition, and they are supposed to oscillate between Mn(II) and Mn(IV). A similar reaction scheme involving Mn $\begin{array}{c} O \\ \diagdown \diagup \\ \diagup \diagdown \\ O \end{array}$ Mn has been proposed (Sawyer and Bodini, 1975; Lawrence and Sawyer, 1978). The four electrons stored as 2Mn(II) are then carried to the reaction center P_{680} of the photosystem II.

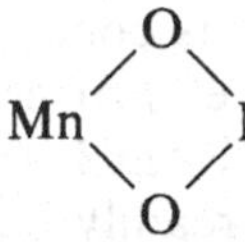

Kambara and Govindjee (1985) recently proposed an elaborate model for the photosynthetic water oxidation, in which the manganese pair in the protein of 34,000 molecular weight function as a couple of Mn(IV)/Mn(III) and the other Mn pair provides water-binding or -mediating sites.

4.7 Oxidases That Produce Hydrogen Peroxide

There are a number of oxidases that produce hydrogen peroxide as one of the end products instead of water. Some are metal-free, whereas others are metal-dependent. Not all metal-free enzymes involved in reactions with oxygen produce HOOH; an example is a group of FAD-dependent monooxygenases that catalyze reactions of type 4-36. Many FAD (or FMN)-dependent oxidases, however, do produce HOOH as the end product. This is because FAD (or FMN) functions in the following way:

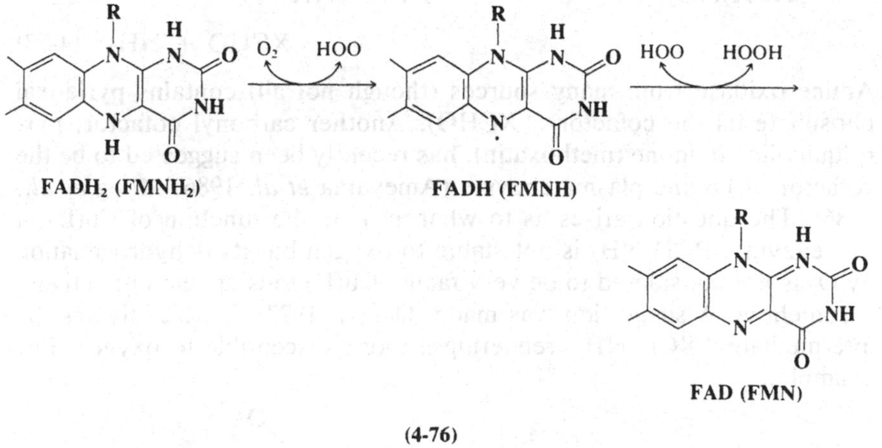

FADH₂ (FMNH₂)   FADH (FMNH)

FAD (FMN)

(4-76)

Many FAD (or FMN)-dependent oxidases utilizes FAD (FMN) itself as the dehydrogenating center, but some of them have other dehydrogenating or electron-accepting centers, such as molybdenum in xanthine oxidase. Electrons (of hydrogen atoms) thus captured are transferred to FAD (FMN), the latter being reduced to $FADH_2$ ($FMNH_2$). $FADH_2$ is then oxidized by oxygen as shown above. Metals are involved only indirectly in these enzymes, as far as the reaction with oxygen is concerned.

The reason that reactions like 4-76 ever proceed smoothly is the stability of the intermediate oxidation state, FADH. Usually the reaction $O_2 + H^+ + e \rightarrow HO_2$ would not proceed readily because of the unfavorable free energy change. This is the very reason that transition metals, especially Fe(II) and Cu(I), as well as $FADH_2$ are utilized by nature to activate O_2 or facilitate the initial step of O_2 reactions that usually involve superoxide formation (in a formal sense).

Several Cu-containing oxidases appear to differ from the above-mentioned ones in the sense that they do not contain FAD or FMN. Examples include amine oxidases, urate oxidase, and galactose oxidase. In these enzymes, copper appears to act as an activator for the substrate rather than for oxygen. The mechanism of amine oxidase reactions has been suggested (Ochiai, 1977) to be:

$$
\begin{array}{c}
\overset{\displaystyle H}{\underset{\displaystyle |}{^+XC}}{=}O + H_2NCH_2R \xrightarrow{\;-HOH\;} \overset{\displaystyle H}{\underset{\displaystyle |}{^+XC}}{=}NCH_2R \xrightarrow{\;-H^+\;}
\end{array}
$$

(cofactor)   (substrate)

$$
X{=}CH{-}N{=}CHR \xrightarrow{\;+H^+\;} {^+XCH_2N}{=}CHR \xrightarrow{\;+HOH,\,-RCHO\;}
$$

$$
^+XCH_2NH_2 \xrightarrow{\;+O_2,\,-HOOH\;} {^+XCH}{=}NH \xrightarrow{\;+HOH,\,-NH_3\;}
$$

$$
^+XCHO + NH_3 \quad (4\text{-}77)
$$

Amine oxidase from many sources (though not all) contains pyridoxal phosphate as the cofactor (^+XCHO). Another carbonyl cofactor, pyrrolquinoline quinone (methoxatin), has recently been suggested to be the cofactor of bovine plasma enzyme (Ameyama et al. 1986; Moog et al., 1986). The question arises as to what may be the function of Cu(II) in this enzyme. RCH_2NH_2 is not stable to oxygen but its dehydrogenation by O_2 is not considered to be very facile. Cu(II) thus appears to activate it somehow. A suggestion was made (Ochiai, 1977) that it activates the intermediate $^+XCH_2NH_2$, rendering it more susceptible to oxygen. For example:

$$
\underset{\displaystyle \lfloor\!-\!CH_2}{Cu^{II} \longleftarrow NH_2} \xrightarrow[\;]{\;-H^+\;} \underset{\displaystyle \lfloor\!-\!CH_2}{Cu^{I} \longleftarrow NH} \underset{fast}{\overset{\displaystyle O_2}{\rightleftharpoons}} \underset{\displaystyle \lfloor\!-\!CH_2}{\overset{\displaystyle \overset{O\,\cdot}{\diagup}}{\underset{\displaystyle \downarrow}{O}}Cu^{II} \longleftarrow NH} \xrightarrow[fast]{\;}
$$

$$
\begin{array}{c}
\overset{\displaystyle\text{OH}}{\overset{\diagup}{}}\\[-2pt]
^-\text{O}\\
\text{Cu}^{II} \longleftarrow \text{NH} \xrightarrow{\ +\text{H}^+\ } \text{Cu}^{II} \leftarrow \text{NH} + \text{HOOH} \quad (4\text{-}78)\\
\underset{\displaystyle\text{CH}}{\big\lfloor\rule{1.2cm}{0pt}} \qquad\qquad \underset{\displaystyle\text{CH}}{\big\lfloor\rule{0.8cm}{0pt}}
\end{array}
$$

This is analogous to the mechanism postulated for a nonheme iron-containing dioxygenase (Fig. 4-6) in the formal sense. In the absence of O_2, the first step, equilibrium between Cu^{II} and Cu^{I}, is established far toward the Cu^{II} state, and usual physicochemical techniques would not be able to detect the postulated Cu^{I} state.

A similar substrate activation mechanism could be suggested for urate oxidase (see Ochiai, 1977). An alternative mechanism suggested by Knowles *et al.* (1982) and Yadav and Knowles (1981) involves the attacking by OH^- (on Cu^{II}) of $^+XCH_2NH_2$, thus rendering the C—H bond susceptible to attack by O_2. O_2 abstracts the H atom in the form of hydride; thus:

$$
\begin{array}{c}
\diagdown\diagup O^-\ \diagdown\ \overset{\displaystyle H_2N}{}\ \overset{\displaystyle H}{|}\\
Cu^{II} + C + {}^-OO \longrightarrow\\
\diagup\diagdown\ \overset{\displaystyle |}{\underset{\displaystyle X^+}{}}\diagdown H
\end{array}
$$

$$
\begin{array}{c}
\diagdown\diagup\\
Cu^{II} + \text{HO}\!-\!\overset{\displaystyle NH_2}{\underset{\displaystyle X^+}{\overset{|}{\underset{|}{C}}}}\!-\!\text{H} + \text{HOO}^- \quad (4\text{-}79)\\
\diagup\diagdown
\end{array}
$$

The resulting ^+X—CH(OH) (NH$_2$) deaminates spontaneously to lead to the formation of $^+XCH{=}O$. This mechanism does not appear to be very feasible from the organic chemical point of view.

Galactose oxidase is quite unique; it catalyzes the reaction:

$$
RCH_2OH + O_2 \rightarrow RCH{=}O + HOOH \qquad (4\text{-}80)
$$

The substrate can be nearly any primary alcohol, including galactose but excluding methanol and ethanol. No unusual cofactor has been found to be involved except for a single Cu(II) atom per molecule of molecular weight of about 68,000 (Bereman *et al.*, 1977). A proposal for the mechanism assumes an oscillation between Cu(I) and Cu(III) (Dykacz *et al.*, 1976; Hamilton, 1980). The reaction involves a dehydrogenation of the group —CH$_2$—OH. Here, too, a mechanism similar to 4-78 may be feasible. It could, for example, explain the fact that the copper(II) in the enzyme does not seem to change appreciably its valence upon addition of the substrate. Yet another mechanism has been proposed, in which an

involvement of an organic free radical in the active center is postulated
(Ettinger and Kosman, 1980).

4.8 Summary I—Basic Fitness of Iron and Copper (and Manganese)

This survey clearly indicates that iron and copper are best suited to
the enzymatic reactions that involve oxygen (and its derivatives) as one
of the substrates. The known types of these enzymatic reactions are
summarized as follows:

1. Reversible O_2 addition (O_2 transport and storage): Fe-heme, Fe-
   nonheme, Cu
2. Superoxide dismutase: Fe, Cu, Mn
3. Dioxygenase: Fe-heme, Fe-nonheme, Cu, Mn
4. Dihydroxylase: Fe-nonheme
5. Monooxygenase (hydroxylase): Fe-heme, Cu
6. Catalase and peroxidase: Fe-heme, Mn, (Se)
7. Oxidase: Cu, Cu-Fe-heme

The suitability of iron, copper, and manganese for the first two classes
was discussed in the respective sections.

The essential requirement for a dioxygenase seems to be a reversible
oxygenation of the type: $O_2 + M(+n) \rightarrow O_2^- + M(+n+1)$, if the naive
mechanism is to be believed. Thus, iron, copper, and perhaps manganese
are appropriate for this class of reactions for the same reasons as those
cited for the reversible O_2 addition. This explains the fact that some
dioxygenases may use Mn(II) (instead of nonheme iron), as shown to be
the case in *Bacillus brevis* (Que *et al.*, 1981).

Dihydroxylases appear to involve a nonheme Fe(III). The mechanism
is still obscure. However, if mechanism 4-35 in Section 4.3.5 is to be
believed, the main function of the catalytic metal cation is to promote the
addition of HOOH across a $C{=}C$ bond, which involves a homolytic
cleavage. This mechanism may require a formal oxidation of Fe(III) to
Fe(V). Whether or not the redox potential of the iron in this class of
enzymes is appropriate for this kind of oxidation is not known. It might,
however, be possible that the enzymatic reaction does not require the
oxidation postulated in scheme 4-35. In that event, the criterion for the
basic fitness may be the ability of Fe(III) to cleave the HO—OH bond
homolytically. Fe(III) seems to be appropriate for the purpose, though
Fe(III) may not necessarily be uniquely fit for it. Perhaps Fe(III) bound
to porphyrin would not be suitable for this kind of reaction for steric
reasons.

The reactions of monooxygenases consist of four major steps: (a) O_2 addition to Fe(II) [or Cu(I)], (b) one-electron addition to form O_2^{2-}—Fe(III), (c) cleavage of the O—O bond in O_2^{2-}—Fe(III) to form O—Fe(III), and (d) transfer of the "O" on O—Fe(III) to the substrate. The basic fitness of iron and copper for step (a) has been established above. The second step (b) has little to do with the problem of basic fitness. The driving force of step (c) might be rather facile transfer of two electrons from Fe(III) onto O_2^{2-}. This process may be assisted by the addition of two protons to O_2^{2-}, as discussed in Section 4.4.1. The low reduction potentials of Fe(III) in P-450, catalase, and peroxidases are compatible with the hypothesis that the O—O bond cleavage may involve a formal (in the quantum mechanical sense) oxidation of Fe(III) to Fe(V). It is not certain whether this kind of formal oxidation is feasible in the case of copper, i.e., O_2^{2-}—Cu(II) being oxidized to form $O^{2-}(H_2O)$ and O^{2-}—Cu(IV).

Most of the oxidases that involve the reduction of O_2 to 2HOH rely on a binuclear moiety. Copper is appropriate for the formation of complexes of the type $O_2^{2-}M_2$. A similar reaction is, of course, possible with Cu(I)—Fe(II) as in cytochrome c oxidase. The enzymes of this class require, in addition, a source of two more electrons, which are needed to complete the formation of 2HOH from $O_2^{2-}M_2$.

Oxygen appeared in the atmosphere rather late in the overall history of the evolution of life (see Chapter 1). Therefore, the enzymes and the proteins involved with O_2 and its derivatives can be considered to be latecomers among all the enzymes and the proteins. By the time oxygen became amply available to organisms, copper had become rather abundant in the marine environment. Thus, copper could easily have been picked up by the organisms to deal with O_2, and happened to be basically fit for the jobs.

By this stage of the evolution of life, heme compounds had become widely available and were utilized by almost all organisms. Only minor genetic alterations may have been necessary to convert existing cytochromes to entities useful in dealing with oxygen. The evolution of nonheme iron enzymes and proteins is rather obscure.

4.9 Summary II—Differential Functions of Heme Proteins: Evolutionary Adaptation

It is an amazing fact that the functions of heme proteins are so widely varied, from simple electron carriers to oxygen carriers, monooxygenases, dioxygenases, and catalase and peroxidase. Each of these different heme-containing enzymes and proteins should have its own unique reduction potential appropriate for its function. The variation of reduction potential as a result of evolutionary pressure was discussed in Section 3.4 and will not be further discussed here, though it is a major contributor to the differential functions of heme proteins.

A second important factor is substrate specificity. A cytochrome involved in the electron transfer systems does not have to bind a chemical substrate. All it has to do is to accept and release an electron, or in other words, to change the oxidation state of its heme iron, $Fe(III) \rightarrow Fe(II)$. Therefore, it can function without a substrate-binding site, and hence most of the electron-transferring cytochromes are six-coordinate, without leaving open a site to bind a substrate directly to the heme group. All the other heme proteins and enzymes have to be able to bind substrates, and hence either the heme iron is five-coordinate with the sixth position vacant (to be occupied by the substrate) or, if six-coordinate, the sixth ligand should readily by displaced by the substrate. An example of five-coordinate heme is found in myoglobin and hemoglobin. The sixth ligand in catalase and peroxidase in the native state seems to be a water molecule, which appears to be readily replaced by their substrates. P-450 appears to be six-coordinate in the resting state but seems to change to five-coordinate upon binding a substrate. This would be brought about by a conformational change in the protein due to the substrate binding.

The substrate specificity and reaction specificity are determined mainly by the specific protein structure around the active site and perhaps a subtle difference in the electronic structure of the heme group, which in turn is controlled by the axial ligand, the environment (e.g., amino acid residues nearby), and other more subtle factors. An example of the latter factor (subtle difference) can be found in catalase, peroxidase (horseradish, HRP), and cytochrome c peroxidase (CCP). The relevant reactions are:

catalase: $HOOH + Fe\text{-heme} \rightarrow Cat\text{-I} \xrightarrow{+HOOH}$

$$Fe\text{-heme} + 2HOH + O_2 \quad (4\text{-}81)$$

peroxidase $HOOH(ROOH) + Fe\text{-heme} \rightarrow HRP\text{-I} \xrightarrow{+AH_2}$

$$HRP\text{-II} + AH \quad (4\text{-}82)$$

CCP: $HOOH + Fe\text{-heme} \rightarrow ES\ complex \xrightarrow{+2cyt\ c^{II}}$

$$Fe\text{-heme} + 2cyt\ c^{III} \quad (4\text{-}83)$$

HRP and CCP do not decompose HOOH as catalase does. The initial reaction, however, appears to be analogous in all three systems, i.e., a two-electron oxidation of the enzyme. It leads to Cat-I, HRP-I, and ES complex, whose characteristics are subtly different from one another. Cat-I and HRP-I both can be expressed as Fe^{IV}—P^+ (P = porphyrin), but their absorption spectra are not identical. The ES complex of CCP appears to be a composite of Fe^{IV}—P and a free radical (other than the porphyrin radical P^+).

The difference in the absorption spectra of Cat-I and HRP-I is attributed to the different ground state of P^+ in the two systems (Dolphin and Felton, 1974). It has been identified as $^2A_{1u}$ in Cat-I and $^2A_{2u}$ in HRP-I. Under appropriate reaction conditions, catalase may function as a peroxidase, but HRP cannot function as a catalase. If it is supposed that the second step of the catalase reaction (4-81) is a simultaneous removal of two electrons from HOOH by Cat-I, then the two oxidizing centers in Fe^{IV}—P^+ must be within a specified distance of each other, presumably within the H $\cdots$ H distance of HOOH. If the distance between Fe^{IV} and the positive charge center on the porphyrin ring is much larger than the H $\cdots$ H distance of HOOH, the reaction of a second HOOH with Fe^{IV}—P^+ might not materialize. Perhaps the distance between the two oxidizing centers matches that of the H $\cdots$ H of HOOH in the case of Cat-I, but it is much larger in the case of HRP-I. This may in turn be a result of the different ground level, $^2A_{1u}$ and $^2A_{2u}$ as suggested. The cause of this subtle difference between the two enzymes may be due to the difference in the fifth ligands and the different environments of the heme.

In the case of CCP, the reaction of the ES complex with a second HOOH must be prevented and the two electrons lost from the enzyme (to HOOH) must be replenished by two molecules of cytochrome c. Whether or not there is a direct contact between the two heme groups (of CCP and cytochrome c) is not known. However, nature designed CCP in such a way that the electrons may come through an amino acid residue in CCP, which is presumably closer to the heme center in cytochrome c when CCP (rather than ES complex) binds with the cytochrome c.

Another peroxidase, chloroperoxidase, catalyzes the oxygenation of Cl^- to ClO^-. It appears that the mechanism involves O—Fe^{III}, as in P-450. The reaction in this case could better be described as an electrophilic reaction of an O atom toward Cl^- rather than a free radical reaction, though.

The reason for the difference between O_2-carrying heme proteins and heme-dependent dioxygenases may be the extent of double bond character of the bond between Fe and O in Fe^{III}—O—O^-. When the extent of electron coupling between Fe^{III} and O_2^- (i.e., the double bond character) is large, as in hemoglobin and myoglobin, the reactivity of O_2^- on Fe(III) is severely reduced. A decreased coupling between Fe(III) and the un-paired electron on O_2^- would leave O_2^- highly reactive, permitting O_2^- to abstract H from a substrate. This seems to be the case with the heme-dependent dioxygenases. The difference between these two classes of proteins could be due to differences in the fifth ligand.

Enzymes in Acid–Base Reactions

5

The term "acid–base reaction" is taken here in the broadest sense; that is, it includes any reaction involving a heterolytic cleavage and/or formation of covalent bonds, without an apparent transfer of an odd number of electrons. This definition covers most of the enzymatic reactions other than oxidation–reduction reactions. Enzymes included in this category are the majority of transferases, hydrolases, lyases, isomerases, and ligases (synthetases).

The role of metal ions in these enzymatic reactions is rather poorly understood. Two major opposing views have been put forward in order to explain the enormous activities of the metalloenzymes or metal-activated enzymes of this category. They are the "entatic" theory (Vallee and Williams, 1968) and the "superacid" concept (Dixon *et al.*, 1976). The former theory maintains that the structure about the metal ion in a metalloenzyme (or metalloprotein) is poised and strained so that the structure of the enzyme–substrate complex approximates that of the transition state in the enzymatic reaction. In the "superacid" concept, such a poised structure is considered unnecessary and, instead, the ubiquitous hydrophobic nature of the active site structure in such an enzyme is invoked to explain the fact that the metal ionic center acts as a superacid.

The reality of enzymatic reactions, however, seems to be far more complicated than the pictures evoked by either of these theories. Jencks (1975) has given a lucid account of the rate-enhancing effects and the specificities of enzymes. According to his view, nature utilizes the substrate–enzyme binding energy, covalent or otherwise, to effect the enhancement of the enzymatic reaction rate and to produce the specificity. He coined the term "Circe effect" for this kind of effect, and meant by it a much wider concept than, for example, the "induced fit effect" of Koshland. In the same article he emphasized the importance of entropy effects. We will discuss the proposed "Circe effect" briefly, stressing those points relevant to our discussion of metalloenzymes or metal-activated enzymes.

5.1 General Mechanisms of Rate Enhancement by Enzymes

A very general mechanism for enzymatic reactions is given by:

$$E + S(\text{substrate}) \underset{k_{-1}}{\overset{k_1}{\rightleftharpoons}} ES \quad \Delta G_1, \quad ES \overset{k_2}{\rightarrow} E + P \quad (5\text{-}1)$$

The corresponding reaction rate r can be expressed by:

$$r = k_2 [E]_0 [S]/((k_{-1} + k_2)/k_1 + [S]) = V_{max} [S]/(K_m + [S]) \quad (5\text{-}2)$$

where

$$V_{max} = k_2 [E]_0 \qquad\qquad\qquad (5\text{-}3)$$

$$K_m \text{ (Michaelis–Menten constant)} = (k_{-1} + k_2)/k_1 \qquad (5\text{-}4)$$

If $k_2 \ll k_{-1}$, then K_m can be regarded as the dissociation constant K of the ES complex, or the inverse of the association constant $K_1 = k_1/k_{-1}$ of the enzyme and the substrate. The free energy of association of E and S, ΔG_1, is related to K_1 in the following way:

$$\Delta G_1 = -RT\ln K_1 = \Delta H_1 - T\Delta S_1 \qquad (5\text{-}5)$$

The rate constant k_2 can be written as:

$$k_2 = A\exp(-E^*/RT) = \kappa\exp(-\Delta G^*/RT)$$
$$= \kappa\exp[-(\Delta H^* - T\Delta \hat{S}^*)/R\bar{T}] \quad (5\text{-}6)$$

E^*, ΔG^*, ΔH^*, and ΔS^* are the activation energy, activation free energy, activation enthalpy, and activation entropy, respectively; A is called a preexponential factor and is related to the collision frequency in the "collision theory" of rate law, and κ is the transference constant. The "Circe effect" asserts that k_2 is affected by ΔG_1 or, in other words, that a portion of ΔG_1 is utilized to raise the value of k_2. There are a number of ways in which ΔG_1 can affect k_2.

According to the absolute rate theory, an activation parameter such as ΔG^*, ΔH^*, or ΔS^* is given by the difference in its value between the initial state and the transition state; that is:

$$\Delta G^* = G_{\text{transition state(t)}} - G_{\text{initial state(i)}} \qquad (5\text{-}7)$$

$$\Delta H^* = H_t - H_i \qquad\qquad\qquad (5\text{-}8)$$

$$\Delta S^* = S_t - S_i \qquad\qquad\qquad (5\text{-}9)$$

Table 5-1. Rate- and Specificity-Enhancing Effects on Enzymes or
Enzyme–Substrate System

Mutual effect; "Circe effect" in the presence of E and S	Other names	Nonmutual effect
Effect on enzyme		Effect on enzyme
(1) on initial state (H_1, G_1)	"Induced fit" (determines specificity)	(6) entatic effect: effect on the initial state of
(2) on transition state (H_t, G_t)	—	E (G_1)
Effect on substrate		(7) specific environment of the active site of
(3) on initial state (H_1, G_1)	Strain, destabilization (raises k_2)	enzyme: hydrophobicity, etc.;
(4) on transition state (H_t, G_t)	—	effect on the initial state and/or the
(5) on entropy of initial state (S_1)	(a) Bringing together two substrates in close proximity	transition state of substrates
	(b) Orbital steering	(8) specific factor: cofactor, metal ion, etc.

Therefore, the effect of a change in ΔG_1 or the binding energy finds its expression in the overall reaction through either the initial state or the transition state or both. Moreover, both the initial state and the transition state are composed of two components, the enzyme and the substrate. The effect of ΔG_1, thus, may involve either the enzyme or the substrate or both. Table 5-1 is a summary of effects on enzyme or enzyme–substrate system, that contribute to the enhancement of enzymatic reaction rates. There are two categories of such effects. One includes the "mutual effects" that become operative when both enzyme and substrate are present together, and an effect in this category is what Jencks (1975) means by "Circe effect." The other is "nonmutual effects" on enzyme alone. Such things as "entatic effect" and the hydrophobicity of the active site of enzymes belong to this category. These latter effects are supposed to be present irrespective of the presence or absence of substrate but should have been created by evolutionary pressure in anticipation of dealing with certain specific substrates effectively. Obviously, the effects in the first category, i.e., mutual effects, are also the products of evolutionary pressures.

Since $\Delta G = -RT\ln K$ and likewise $\Delta G^* = -RT\ln k_2$, a change in ΔG (ΔG^*) of 5.85, 11.7, 23, and 46 kJ/mole (1.4, 2.8, 5.5, and 11 kcal/mole) would at 300°K cause a change in K (k_2) of 10-, 100-, 10^4-, and 10^8-fold, respectively. Note that a relatively small change in ΔG (ΔG^*) will cause a large difference in K (k_2) values.

The binding forces between enzymes and substrates are of various natures, including covalent, electrostatic, and van der Waals (hydrophobic). The strengths of these interactions are heavily dependent on the distances involved. Thus, a specific interaction is required between an enzyme and a substrate in order to obtain a maximal binding energy; a slight mismatch may lead to a significantly lower binding energy and hence to a large decrease in the enzymatic reaction rate. On the other hand, a large binding energy may not necessarily induce an activating change in the conformation of the enzyme or the substrate in certain situations. This appears to happen in cases where inhibitors act as the "substrate."

A few examples of these effects will now be presented, mostly drawn from the article of Jencks (1975). Hexokinase catalyzes the following reaction:

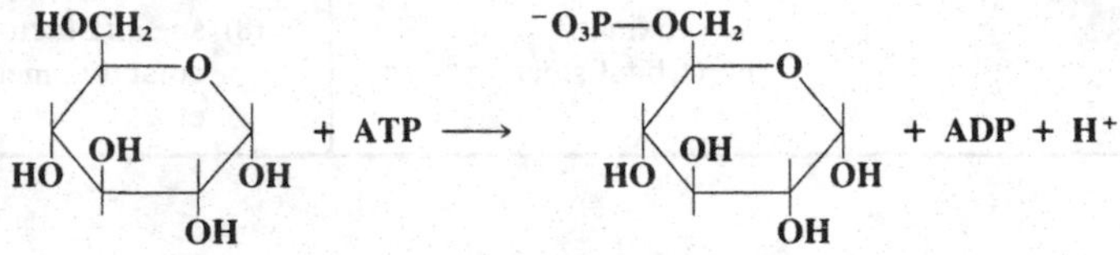

HOCH$_2$ ... $+ ATP \longrightarrow$ ... $^-O_3P{-}OCH_2$ ... $+ ADP + H^+$

$$(5\text{-}10)$$

That is, it is a reaction of the type:

$$ATP + R{-}OH \rightarrow ADP + R{-}O{-}PO_3^{2-} + H^+ \qquad (5\text{-}11)$$

The enzyme actually can hydrolyze ATP, in which case ROH = HOH. However, the rate constant of ATP hydrolysis is 4×10^4 times smaller than that of the phosphorylation of glucose. HOH is small enough to fit in the glucose-binding site of the enzyme but it does not contain the specific pyranose ring of glucose that is supposed to interact specifically with the enzyme. A portion of the specific binding energy associated with the pyranose ring is thought to be utilized to cause a conformational change of the enzyme, a change from the inactive form to the active form. That is, it is considered to cause an "induced fit" [effect (1) in Table 5-1]. The estimated binding energy ΔG of the active form of the enzyme to glucose is -46 kJ/mole (-11 kcal/mole), whereas the experimentally determined overall binding energy of glucose to the enzyme is -21 kJ/mole (-5.1 kcal/mole). Therefore, the difference of -25 kJ/mole (-5.9 kcal/mole) may have been used to cause the conformational change. These values were, by the way, obtained in the presence of the specific activator, Mg(II); Mg(II) may be substituted by Mn(II). The divalent cation is considered to be involved in binding ATP to the enzyme. This subject will be discussed later.

Elastase is a protein-hydrolyzing enzyme (proteinase) in which a serine residue plays a central role. Proteinases will be discussed in more detail later, but elastase provides a useful illustration here. Using a series of polypeptides as the substrate, the binding energies to elastase and their effects on the hydrolysis rate were estimated. For example, in the case of the hydrolysis of an oligopeptide analogue Ac—Pro—Ala—Pro—Ala—$CONH_2$ in which the bond CO—NH_2 is to be split, the interaction with the enzyme of even the peripheral fourth and fifth peptide units, Pro and Ac, seems to produce as much as 11.7 kJ/mole (2.8 kcal/mole) of binding energy. This energy is utilized to destabilize the bound substrate [effect (3) in Table 5-1] and thus to enhance the rate by a factor of 10^2.

The magnitude of the effect of bringing two reactants (substrates) together in proximity and in a correct orientation [effect (5) in Table 5-1] is remarkable in certain cases. The following three nonenzymatic reactions have, for example, relative rates of 1, 6×10^4, and 3×10^{15}, respectively:

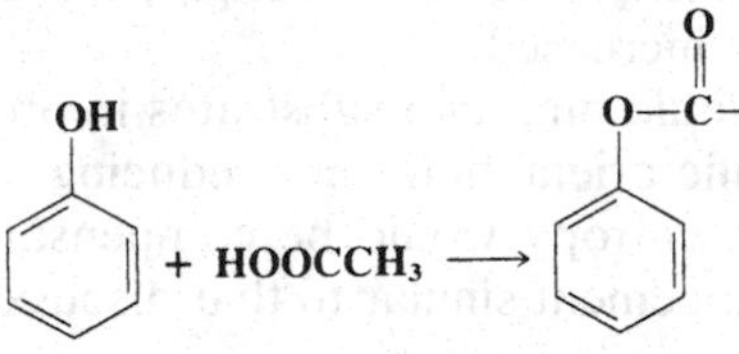

(5-12)

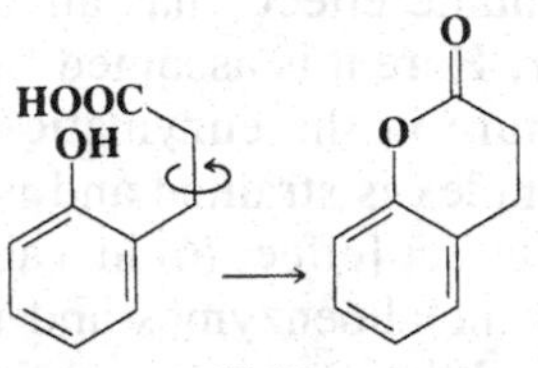

(5-13)

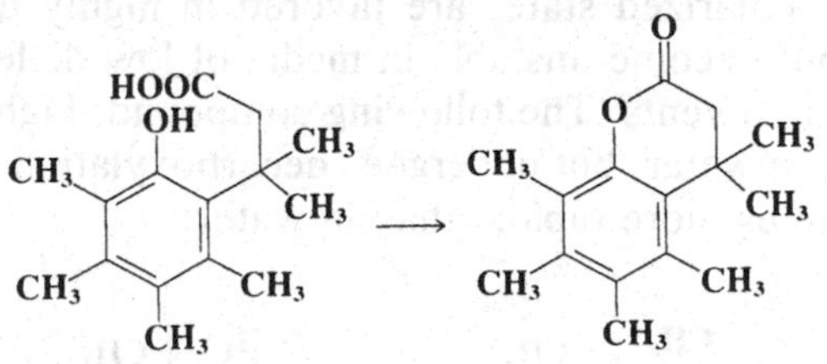

(5-14)

The first of these, a bimolecular reaction, occurs only when the two reactants happen to come in contact; only then do they form a transition state complex. In binding with each other, they lose their transitional entropies and some of the rotational entropies. In addition, they have to align themselves in the correct relative orientation and thus lose a portion of their vibrational entropies in forming the transition state. It has been estimated that ΔS^* is on the order of -35 to -45 e.u. in most bimolecular reactions. This amounts to an increase of 42–59 kJ/mole (10–14 kcal/mole) in ΔG^*, or a 10^8–10^{10}–fold decrease in the rate constant. In reaction 5-13, the two reactants have been incorporated into a single molecule; thus, the reaction becomes intramolecular. The reaction starts from a point at which the translational and part of the rotational freedoms of the reactants have already been lost. Therefore, the ΔS^* loss will be much less than that in reaction 5-12. This enhances the reaction rate. However, the bond C_α—C_β can still rotate as shown in reaction 5-13 and a rotation may take the reacting COOH group away from the OH group, which lessens the chance of a successful reaction. In reaction 5-14, this rotation, too, is frozen by attaching bulky substituents and, as expected, the rate is enormously enhanced. The discrepancy between the estimated factor 10^8–10^{10} and the observed 10^{15} may be attributed to various electronic factors; in the methyl-substituted compound, for example, the basicity of the OH group should be greatly increased.

An ideal enzyme would bind two substrates in proper juxtaposition, and in the correct specific orientation, thus reducing the entropies of the substrates. This loss in entropy would be compensated by the binding energy, and a rate enhancement similar to that discussed above could be expected.

The definition of "entatic effect" has already been given in the introduction to this chapter. Here it is assumed that the evolutionary pressure has created a structure in the enzymatic active site such that the initial state of the ES complex is strained and is close to the structure of the transition state. This effect [effect (6) in Table 5-1] has been invoked particularly in respect to metalloenzymes and metalloproteins, and will be discussed more fully a little later.

Electrically polarized states are favored in highly dielectric media such as water but become unstable in media of low dielectric constants such as nonpolar solvents. The following compound, highly polarized, is relatively stable in water, but undergoes decarboxylation in ethanol (less polar) 10^4–10^5 times more rapidly than in water:

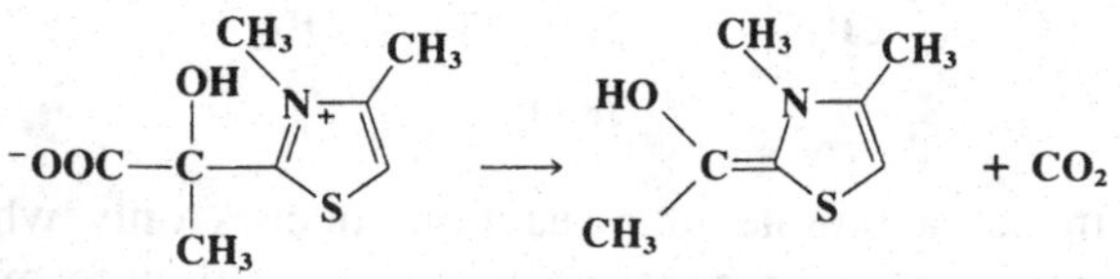

(5-15)

the decomposition is faster still in aprotic solvents. The space surrounding the active site of an enzyme is hydrophobic in many cases. This factor may help in producing nonionic products from a polarized substrate in analogy to the nonenzymatic example mentioned above. The hydrophobicity of the space near the active site gives rise to another effect. The potential energy created by a charged spherical entity with a charge z and a radius r is proportional to z/Dr, where D is the dielectric constant of the medium. The electric charge effect on the overall potential energy will be reduced when the substance is introduced into a medium of high dielectric constant such as water, but will be increased if the substrate is introduced into a region of low dielectric constant [e.g., D(benzene) = 4.6]. The hydrophobic space in an enzyme, having a lower dielectric constant (than the water medium), increases the effect of an electric charge such as a cation. This is one of the important factors that was cited in proposing the "superacid" concept (Dixon *et al.*, 1976). Namely, a metal ion in a metalloenzyme would show a Lewis acid character in the hydrophobic medium to a much greater extent than is true in water medium.

At this point it is useful to discuss the mechanisms of several more important enzymes and, in particular, the essentially or the modes of function of metal ions in those enzymes.

5.2 Proteinases and Peptidases: Metalloenzymes and Metal-Free Enzymes

There are a number of metal-dependent proteinases and peptidases; examples include carboxypeptidase A and B obtained from mammalian pancreas, and some proteinases of microbial origins. However, there are a great many proteinases and peptidases that do not require metal ions for activity. Then a legitimate question is: why are metal ions required for certain of these enzymes? Put another way, why could organisms not have done without the metal ions in all cases if they succeeded in inventing so many enzymes that do not require metal ions? In other words, what is really the specific function of a metal ion that makes it essential to certain organisms (or certain enzymes)? This is the sort of question posed in the discussion of enzymes and proteins involved in oxidation–reduction, electron transfer, and oxygen reactions (Chapters 3 and 4). The questions in these latter cases proved rather easily answerable. The same question in the present context, however, seems more difficult to answer, and the effects of metal ions or the reasons for their essentiality appear to be, at best, rather subtle.

First let us survey proteinases and peptidases and their characteristics. Table 5-2 lists some of the prominent proteinases and peptidases.
Proteins or polypeptides are described by:

$$
\begin{array}{cccc}
R_1 & R_2 & R_3 & R_n \\
| & | & | & | \\
\end{array}
$$

HOOCCH-NH-CO-CH-NH-CO-CH-NH---CO-CH-NH₂

(C-terminal)   (N-terminal)

The unit —CO—CHR—NH— represents one amino acid residue; the
end with a free carboxyl group is called the C-terminal and the other end,
with a free amino group, is the N-terminal. Heterotrophs all have to digest
available proteins and peptides into their component amino acids from
which they then construct their own essential proteins; therefore, it is not
surprising that they have a great number of proteinases and peptidases
capable of hydrolyzing a variety of peptide bonds.

5.2.1 Mechanisms of Metal-Free Proteinases

Now let us take a closer look at the mechanisms of typical proteinases. The mechanism of the serine proteinases has been fairly well elucidated (Stroud, 1974). These enzymes' catalytic site consists of serine,
histidine, and aspartic acid, with the serine playing a vital role; thus, they
are called "serine proteinase." Figure 5-1 shows the mechanism of a
serine proteinase as established so far. The nature of the pocket into
which the group R fits determines the specificity. Chymotrypsin, chymotrypsin C, and *Meribium* proteinase A accommodate rather bulky R
groups. Elastase's pocket seems to be rather small and only small aliphatic
nonionic R groups such as those found in glycine and alanine can fit into
it. The other enzymes in this class appear to have negative charges in the
pocket which interact with the positively charged protonated amino group
of lysine or the guanidinium group of arginine.

The acidic hydrogen from the OH group of the catalytic serine is first
associated with the histidine. The O^--serine thus produced attacks the
carbon atom of the carbonyl group of the peptide bond, forming a tetrahedral structure about this carbon atom. The hydrogen ion attached to
the histidine residue is then transferred to the suitably situated NH group

Table 5-2. Proteinases and Peptidases[a]

Enzyme (EN or EC number)	Substrate, preferential cleavage point[b]
Metal-free enzyme	
Serine proteinase	
Chymotrypsin (3.4.21.1)	Tyr-, Trp-, Phe-, Leu-
Chymotrypsin C (3.4.21.2)	Leu-, Tyr-, Phe-, Met-, Trp-, Gln-
Meribium proteinase A (3.4.21.3)	Tyr-, Phe-, Leu-
Trypsin (3.4.21.4)	Arg-, Lys-
Thrombin (3.4.21.5)	Arg-
Plasmin (3.4.21.7)	Arg-, Lys-
Kininogenin (3.4.21.8)	Arg-, Lys-
Elastase (3.4.21.11)	Uncharged small nonaromatic side chain
Subtilisin (3.4.21.14)	No clear specificity
Sulfhydryl proteinase	
Papain (3.4.22.2)	Arg-, Lys-, Phe-X-
Ficin (3.4.22.2)	Phe-, Tyr-
Chymopapain (3.4.22.6)	Glu-, Asp-
Acid proteinase	
Pepsin A (3.4.23.1)	Phe-, Leu-
Pepsin B (3.4.23.2)	Gelatin
Pepsin C (3.4.23.3)	Hemoglobin
Chymosin (3.4.23.4)	Phe-, Leu-
Metalloenzyme or metal-activated enzyme	
Crotalus atrox proteinase (3.4.24.1) (Zn enzyme)	Leu-, Phe-, Val-, Ile-
Collagenase (3.4.24.3) (Zn enzyme?)	Helical region of collagen
Bacillus subtilis proteinase (3.4.24.4) (Zn enzyme)	-Leu, -Phe[c]
B. thermolyticus thermolysin (3.4.24.4) (Zn enzyme)	-Leu, -Phe[c] ·
Aminopeptidase (3.4.11.1) (Zn enzyme)	N-terminal residues
Proline aminopeptidase (3.4.11.5) [requires Mn(II)]	N-terminal proline only
Carboxypeptidase A (3.4.12.2) (Zn enzyme)	C-terminal, except Arg-, Lys-, Pro-
Carboxypeptidase B (3.4.12.3) (Zn enzyme)	C-terminal, Lys-, Arg-
Aspartate carboxypeptidase (3.4.12.9) [requires either Zn(II) or Co(II)]	Asp-
Glutamate carboxypeptidase (3.4.12.10) [requires Zn(II)]	Glu-
Aminoacyl-histidine dipeptidase (3.4.13.3) [activated by Zn(II) or Mn(II)]	Aminoacyl-His

[a] Compiled from M. Florkin and E. H. Stotz (eds.), *Comprehensive Biochemistry*, Vol. 13, 3rd ed., Elsevier (1973).
[b] Cleavage at the carboxyl end of the amino acid whose symbol is given.
[c] Cleavage at the amino end of the amino acid whose symbol is given.

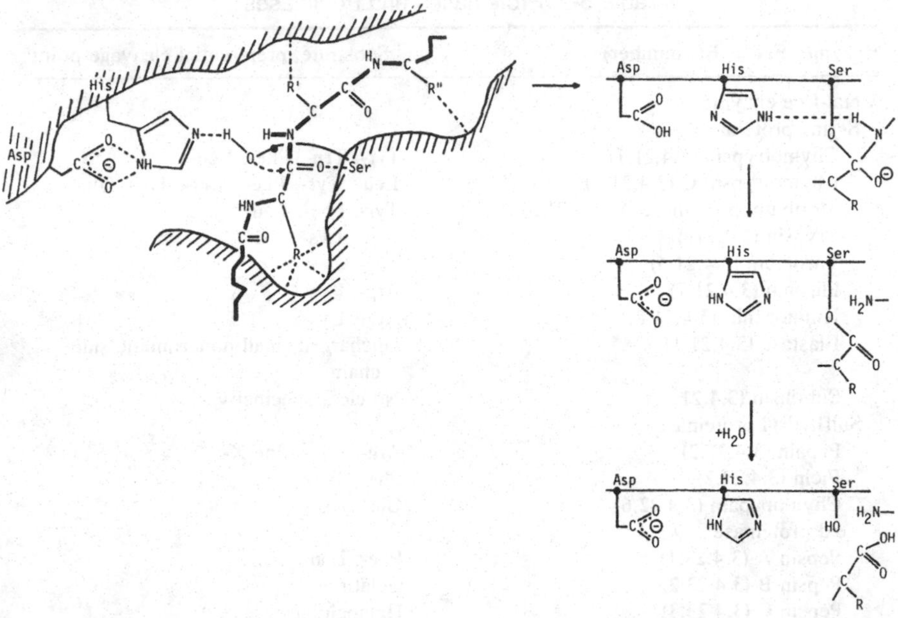

Figure 5-1. The reaction mechanism for a serine proteinase. (After Stroud, 1974.)

of the peptide, when the peptide bond splits, resulting in NH_2— and the acylated serine residue. The formation of such an acyl-serine unit has been established in a number of studies (Bell and Koshland, 1971). Finally, a water molecule attacks the acyl-serine bond, leading to the recovery of the serine—OH group and the formation of a carboxylic acid. Thus, the reaction has been completed. This mechanism may be considered as involving a general acid–base catalysis. In this mechanism the binding energy of the substrate to the enzyme is utilized to change the geometry around the peptide carbonyl carbon from sp^2 in its free state to something more like the sp^3 geometry that is supposed to be required in the transition state. This process presumably entails some loss of entropy and increase of enthalpy in the pretransition state, thus reducing ΔG^* as a result. The binding energy is, of course, also utilized to bring the reacting groups, the serine OH and the carbonyl group of the peptide bond, closer together.

Acyl-enzyme intermediates have also been postulated for the other classes of proteinases. The acyl group seems to bind to a sulfhydryl group of a cysteine residue in the sulfhydryl proteinases (Bell and Koshland, 1971). One or two aspartate carboxyl groups are believed to be involved in the formation of an acyl-enzyme intermediate in the case of the acid proteinases such as pepsin A (Fruton, 1976).

5.2.2 Mechanisms of Metalloenzymes

The mechanism of a Zn(II) enzyme, carboxypeptidase A, has been fairly well elucidated by use of X-ray crystallography (Quiocho and Lipscomb, 1971; Rees *et al.*, 1981). A diagrammatic sketch of the essential portion of an enzyme-substrate (analogue) complex is given in Fig. 5-2. The substrate specificity seems to be determined by the nature of the pocket into which the R group of the C-terminal fits. Carboxypeptidase A prefers a non-polar bulky group, particularly an aromatic group, such as a phenyl ring (phenylalanine, tyrosine) or an indole moiety (tryptophan). Carboxypeptidase B, whose structure has been less completely elucidated, prefers a positively charged R group such as those found in arginine and lysine. The essential Zn(II) ion is bound by two histidine residues (nitrogen), one glutamate residue (oxygen), and a water molecule in the substrate-free state. The carbonyl oxygen of the peptide bond to be split has been found to bind to the Zn(II) ion in the ES complex. The function of Zn(II) seems to be threefold. First, because of the binding energy, the Zn(II) ion can bring the substrate into a correct position with regard to the enzyme. Second, the Zn(II) ion polarizes the carbonyl group and renders its carbon atom more susceptible to a nucleophilic attack. A possible third function is as follows. The coordinating structure about the Zn(II) ion is very distorted from that of a regular tetrahedron, being more like a trigonal bipyramidal structure lacking an apical ligand, as illustrated in Fig. 5-3. This structure suggests that the Zn(II) ion may be able to bind yet another ligand at the apical position. The ligand would most likely be a water molecule, which could thus be brought into a position next to the peptide carbonyl group, thereby facilitating the addition of H_2O (or OH^-) to the carbonyl group. The reaction, if it in fact proceeds in this way,

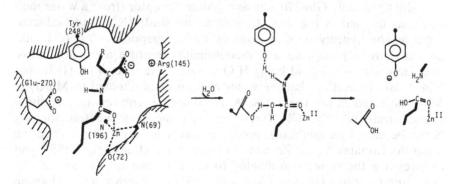

Figure 5-2. A proposed mechanism of carboxypeptidase A. (After Quiocho and Lipscomb, 1971.)

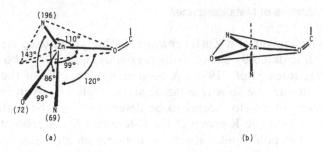

Figure 5-3. (a) The coordination structure of Zn(II) in carboxypeptidase A, and (b) a model for it.

approximates an intramolecular process. The advantages brought about by the intramolecularity of a reaction have already been emphasized in the previous section. The following data (Table 5-3) (Buckingham, 1977) on a model system illustrate the same point. The polarization of the carbonyl group by the coordination provides a rate increase of 10^4 over that found in the absence of a metal ion, whereas the intramolecular attack of OH^- or H_2O on the carbonyl group brings about a tremendous rate enhancement of 10^9–10^{11}. A study (Groves and Dias, 1979, 1983; Groves and Chambers, 1984) with Co(II)-, Cu(II)-, or Zn(II)-mediated amide hydrolysis indicates that the coordinated OH^- attacks the amide carbonyl group, which is a part of the ligand. This intramolecular attack of the OH^- on the peptide carbonyl is about 10^5 times faster than the corresponding bimolecular reaction. This oriented coordination of a water molecule is, then, assumed to be the third crucial function of the Zn(II) ion. The proposed mechanism (Quiocho and Lipscomb, 1971) (Fig. 5-2) does not take this effect of the Zn(II) ion into consideration.

Most probably Glu-270 acts as a proton acceptor (from a water molecule) as implied in Fig. 5-2. A more recent study (Kuo and Makinen, 1982) on the hydrolysis of an ester by carboxypeptidase A [Co(II)-substituted] strongly suggests a five-coordinate structure about the active metal ion with one ligand being H_2O or OH^-. Nakano et al. (1979) obtained data indicating that the water molecule coordinated to Mn(II) in Mn-substituted carboxypeptidase A would not be replaced upon addition of a substrate, Gly-Tyr. Gly-Tyr, which is a rather poor substrate (or may better be called an inhibitor), however, has been shown to replace the water coordinated to the Zn atom in the enzyme (Rees et al., 1981) and also replace the water coordinated to the Co atom in the case of Co-substituted enzyme (Kuo and Makinen, 1982). Apparently the NH group of Gly-Tyr displaces the water molecule (Rees et al., 1981). This seems to be the reason why Gly-Tyr is an inhibitor or a poor substrate. The

Table 5-3. Effect of the Coordination and the Intramolecularity on the Rate Constant of Hydrolysis of Glycine Amide[a]

Reaction	Rate constant (k)	Relative value
$NH_2CH_2CONH_2 + OH^- \rightarrow NH_2CH_2COO^- + NH_3$	$2.2 \times 10^{-3}\ M^{-1}\ sec^{-1}$	1
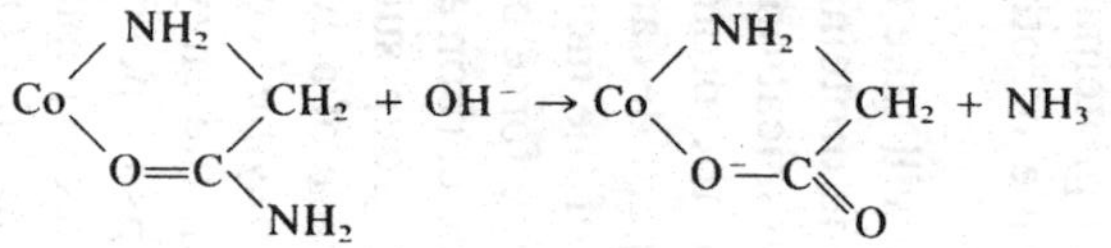 Co complex (Co—NH_2—CH_2, $O{=}C$—NH_2) $+ OH^- \rightarrow$ Co complex (CH_2, $O{=}C{=}O$) $+ NH_3$	$35\ M^{-1}\ sec^{-1}$	10^4
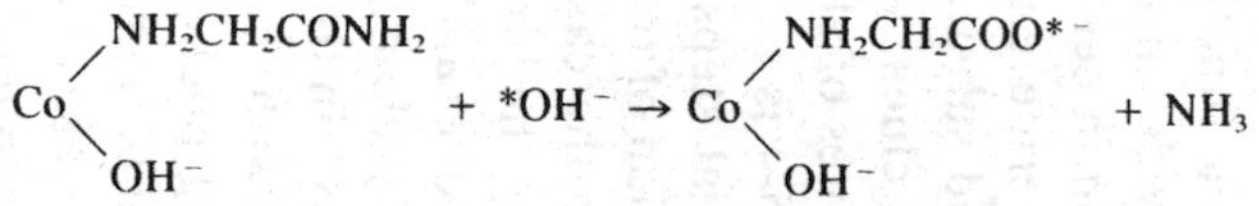 Co complex ($NH_2CH_2CONH_2$, Co—OH^-) $+ {}^*OH^- \rightarrow$ Co complex ($NH_2CH_2COO^*$, Co—OH^-) $+ NH_3$	$1.24\ M^{-1}\ sec^{-1}$	10^3
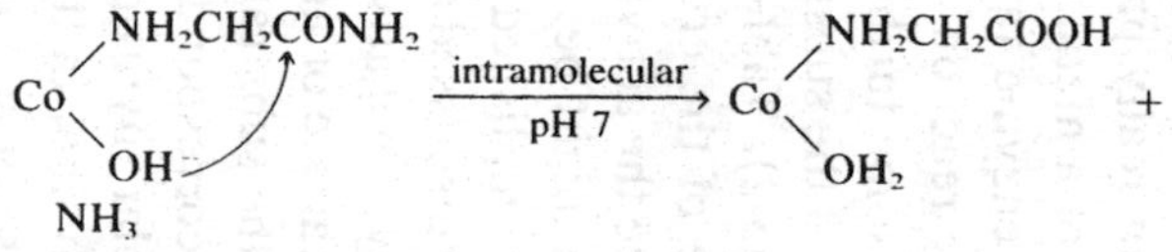 Co complex ($NH_2CH_2CONH_2$, Co—OH^-) $\xrightarrow[\text{pH 7}]{\text{intramolecular}}$ Co complex (NH_2CH_2COOH, Co—OH_2)	$1.8\ sec^{-1}$	10^{10}
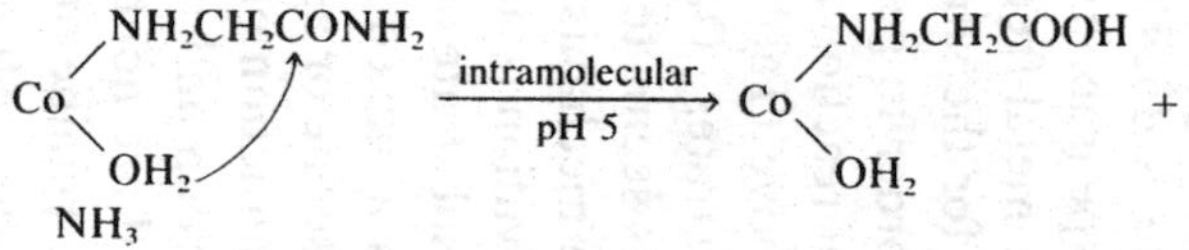 Co complex ($NH_2CH_2CONH_2$, Co—OH_2) $\xrightarrow[\text{pH 5}]{\text{intramolecular}}$ Co complex (NH_2CH_2COOH, Co—OH_2)	$3.5 \times 10^{-3}\ sec^{-1}$	10^9

[a] Source: D. A. Buckingham, in *Biological Aspects of Inorganic Chemistry* (A. W. Addison et al., eds.), pp. 141–196, Wiley–Interscience.

implication is that a true substrate may be one that would not displace the water (or OH$^-$) molecule coordinated to the catalytic cation.

Another similar enzyme, thermolysin, and its adduct with a dipeptide have been studied by X-ray crystallography (Kester and Mathews, 1977). The structure around the active site, again containing Zn(II), and the binding mode of the peptide turned out to be essentially the same as those in carboxypeptidase A. The authors (Kester and Mathews, 1977) proposed a mechanism analogous to that shown in Fig. 5-2; in thermolysin Glu-143 takes the place of Glu-270 of carboxypeptidase A. An argument for a mechanism of the intramolecular type (as made above for carboxypeptidase A) may be made here analogously. Other possible mechanisms will be discussed later (Section 5.2.5). Coleman (1984) gives a balanced review on the mechanisms of carboxypeptidase, thermolysin, and carbonic anhydrase.

5.2.3 Need for a Metal Cation

The foregoing discussion of the functions of the metal ion in carboxypeptidase A does not really provide an answer to the basic question of the indispensability of a metal ion in certain proteinases and peptidases. Could a metal-free enzyme (a serine enzyme or otherwise) have been devised for the same reaction and substrate?

In order to get some further clues, we will take a closer look at the modes of reaction and the substrates of all the proteinases and peptidases listed above (Table 5-2). Table 5-4 is a classification of reaction modes and substrates. One of the crucial steps in the proteinase or peptidase reactions seems to be the activation of the peptide carbonyl group, whatever the mechanism might be. In the case of the metal-free proteinases, this activation is accomplished by the binding force between the enzyme and the substrate (as discussed above) and the carbon atom of the carbonyl group is attacked by a negatively charged group such as an O$^-$ in the case of serine or aspartate or an S$^-$ in the case of a cysteine residue. The main interaction of the substrate with the enzyme is established between the R group and a corresponding pocket in the enzyme. The R groups preferred are nonpolar (hydrophobic) groups or positively charged ones such as arginine or lysine. In the case of carboxypeptidases, on the other hand, the substrates carry a negatively charged COO$^-$ group near the peptide bond of the reacting center. In carboxypeptidase A, this negatively charged COO$^-$ group binds electrostatically to a positively charged arginine residue (No. 245) in the enzyme. In the case of carboxypeptidase B, the preferred R group (see Table 5-4) is positively charged; this implies

Table 5-4. A Classification of Substrate Specificity for Proteinases and Peptidases

Enzyme	Substrate specificity and mode of cleavage	Reaction center
(1) Serine proteinases Sulfhydryl proteinases[a] Acid proteinases	$-\text{NH}-\text{CO}-\overset{\mid}{\underset{\text{R}}{\text{CH}}}-\text{NH}-$ R = nonpolar (aromatic or aliphatic) or positively charged	serine cysteine carboxyl
		$-\text{O}^{-},\ -\text{S}^{-},$ $-\text{COO}^{-}$
(2) *Crotalus atrox* proteinase	$-\text{NH}-\text{CO}-\overset{\mid}{\underset{\text{R}}{\text{CH}}}-\text{NH}-$ R = nonpolar bulky	Zn(II)
(3) *Bacillus* proteinase	$-\text{CO}-\overset{\mid}{\underset{\text{R}}{\text{CH}}}-\text{NH}-\text{CO}-$ R = nonpolar (aromatic or aliphatic)	Zn(II)
(4) Carboxypeptidases	$^{-}\text{OOC}-\overset{\mid}{\underset{\text{R}}{\text{C}}}-\text{NH}-\text{CO}-$ R = nonpolar bulky or positively charged	Zn(II)
(5) Aminopeptidases	$-\text{NH}-\text{CO}-\overset{\mid}{\underset{\text{R}}{\text{CH}}}-\text{NH}_2$ N-terminal	Zn(II)
(6) Zn(II)-, Co(II)-, or Mn(II)-activated enzymes 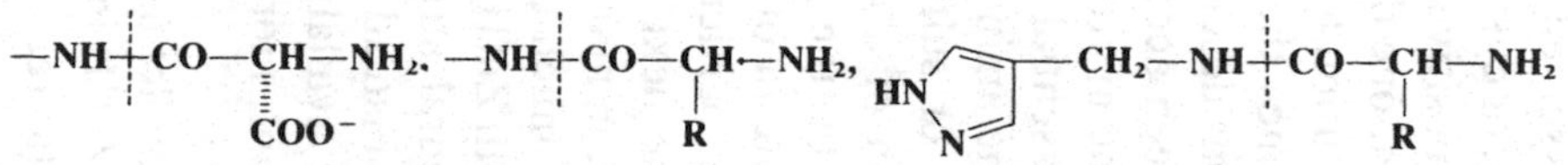	$-\text{NH}-\text{CO}-\overset{\mid}{\underset{\text{R}}{\text{CH}}}-\text{NH}_2,\ -\text{NH}-\text{CO}-\overset{\mid}{\underset{\text{COO}^{-}}{\text{CH}}}-\text{NH}-\text{CO}-\overset{\mid}{\underset{\text{R}}{\text{CH}}}-\text{NH}_2$	

[a] An exception: chymopapain, R = negatively charged.

that the pocket (in the enzyme), which must accommodate the positively charged R group, is likely to be negatively charged. For both carboxy-peptidases, the approach of the proper substrate, which contains a COO^- group, would be difficult because of an electrostatic repulsion if the active site possessed the usual negatively charged catalytic center (serine, aspartate, or cysteine). Instead, the substrate would prefer a catalytic center that is positively charged. Placing a metal ion such as Zn(II) in a proper location, one that would allow it to interact with carbonyl oxygen, would be a solution to this problem, and it is a good one. The metal ion not only overcomes the electrostatic barrier problem but also renders the three functions described. The reaction of proteinases 2 and 3 in Table 5-4 could have been accomplished with serine- or sulfhydryl enzymes, but apparently the microorganisms took advantage of the benefits of incorporating a nonessential metal ion. What about enzymes in groups 5 and 6 in Table 5-4? The amino peptidase reaction could presumably be carried out by a serine- or sulfhydryl-type peptidase. On the other hand, one possible function of a cation in the enzymes of these two groups of peptidase would be the metal's ability to bind the terminal amino group of the substrate in order to bring it to a proper position. This may turn out to be true, particularly in those cation-activated enzymes (group 6).

The next question is why the Zn(II) ion was selected. As is well known, in many Zn(II) enzymes Zn(II) can be replaced by Co(II) without loss of much enzyme activity. This can easily be explained by the fact that both Zn(II) and Co(II) are capable of adopting tetrahedral or distorted tetrahedral structures. Thus, organisms could, in principle, have used Co(II) instead of Zn(II). Zn is, however, much more abundant, more readily available in seawater; the organisms simply selected the more abundant, more readily available element (rule of abundance, see Chapter 2).

Now let us summarize the discussion given above for the essentiality of the Zn(II) ion in certain types of proteinases and peptidases. The fundamental reason for the essential role of a metal ion seems to be that since certain substrates carry a negatively charged group (such as COO^-) near the reaction center (CO—NH), it is necessary to have a positive charge at the catalytic site. The rather negatively charged enzymatic center (O^- of serine or S^- of cysteine) in metal-free proteinases would not effectively bind such substrates. The functions of a metal ion in metalloproteinases or metallopeptidases are (1) binding the substrates in a proper position relative to the enzyme, (2) polarizing the peptide-bond carbonyl group to make it more susceptible to a nucleophilic attack, and (3) bringing the other substrate, water, into position next to the peptide carbonyl group. This last function requires that the coordination structure about the metal be a fairly distorted tetrahedron, indeed closer to trigonal bipyramidal, and this structure can be regarded as a manifestation of the "entatic effect" caused by evolutionary pressure. Finally, the reasons that Zn(II) was chosen are (1) abundance and (2) the fact that Zn(II) can readily assume the required poised structure. Other potentially available elements such as Ni(II), Fe(II), and Cu(II) would not be very suitable for this kind of job because they are unable to satisfy the second criterion. Besides, Zn(II) is among the most strongly polarizing cations if the polarizing effect is measured by the z_{eff}/r value. z_{eff}/r values decrease in the order Ni(II) > Cu(II) > Zn(II) > Co(II) > Fe(II) > Mn(II) > Mg(II) > Ca(II) (see Section 2.2). It might also be pointed out here that, as discussed in Section 1.4, Cu(II) was not available to the primitive organisms when they needed a metal ion for those important functions (such as RNA polymerization). This leaves Zn(II) as the appropriate species for this kind of biological function.

5.2.4 Consideration of the "Superacid" Theory

In an interesting set of experiments, the Zn(II) ion was removed from carboxypeptidase A by careful dialysis techniques, after which a number of different metal ions were added to the apoenzyme, and the respective catalytic activities were measured (Coleman and Vallee, 1966). The results were: Co(II) > Zn(II) $\approx$ Ni(II) $\gg$ Mn(II) $\sim$ Cu(II) $\sim$ Cd(II) $\sim$ Hg(II) = 0 for peptidase activity; and Cd(II) > Hg(II) > Zn(II) $\approx$ Co(II) $\approx$ Ni(II) > Mn(II) > Cu(II) = 0 for esterase activity. These results apparently contradict the conclusions drawn above. In particular, the structures of Ni(II) complexes are usually octahedral or square-planar and less often tetrahedral. These observations led to the development of the concept of the "superacid" (Dixon et al., 1976). The argument goes like this. Since Ni(II) is as active as Zn(II) or Co(II) in the peptidase action of carboxypeptidase, and since Ni(II) is reluctant to assume a tetrahedral or similar structure, we may conclude that the specific tetrahedral structure is not important and that the important factor is instead the Lewis acidity of the metal ion. The metal ion in the active site might act as a super Lewis acid because the space around the active site of these enzymes is often hydrophobic, and hydrophobicity is assumed to increase the acidic effect of a metal ion:

$$\text{acidic effect} \propto z_{eff}/Dr$$
(D becomes smaller with increase in hydrophobicity)   (5-16)

The coordination structure about Ni(II) depends not only on the nature of the atomic orbitals of the metal ion but also on how flexible the polypeptide chain and appropriate branches of the enzyme are. If the polypeptide chain is rather inflexible and the directional properties of the metal ion orbitals are not strong enough, then the metal ion Ni(II) may coordinate the active site of the enzyme only rather reluctantly and is forced to assume the normally disfavored distorted tetrahedral structure or a structure similar to it. If this argument is valid, even Ni(II) could assume a distorted tetrahedral structure and hence the "entatic effect" concept would still be valid. The binding constants of metal ions to carboxypeptidase A have been measured (Piras and Vallee, 1967). Log K_{app} was found to be 10.5 for Zn(II), 7.0 for Co(II), 8.2 for Ni(II), 10.6 for Cu(II), 4.6 for Mn(II), 10.8 for Cd(II), and 21.0 for Hg(II). These values can be compared with, e.g., log K values for M(II)-ethylenediamine complexes; they are 4 for Zn(II), 6 for Co(II), 8 for Ni(II), 11 for Cu(II), 3 for Mn(II). The difference in log K between Ni(II) and Co(II) should be at least 2, if the enzyme complexes are free from unusual strain, as is the case with the ethylenediamine complexes, whereas the difference in log K_{app} between Ni(II) and Co(II) for metal-carboxypeptidase A is, in fact, only 1.2. This implies that the coordination structure around Ni(II) in carboxypeptidase A is rather strained. The reported log K_{app} values, by the way, also indicate that the metal-binding site of the enzyme is very much adapted to the Zn(II) ion. The structure of the active site might be extensively altered when Cu(II) or Mn(II) binds with it, thus explaining why, in these cases, so much of the catalytic activity is lost. If, on the other hand, the Lewis acidity were the sole parameter determining the enzymatic activity, Cu(II) should be the best catalytic metal.

In the case of esterase activity the situation may be entirely different. This reaction might not require as much specific structure at the catalytic site as the peptidase reaction. The problem of the esterase activity will be deferred to the next section, in order that further aspects of the mechanism of carboxypeptidase A activity can be discussed.

The proposed effect of the environmental hydrophobicity on the Lewis acidity of a metal ion may merit a brief discussion. The effect of a high-dielectric-constant environment (e.g., water) is of a long distance nature. When the potential energy of an ion with a charge z and a radius r is expressed by z/Dr, the effect of the dielectric constant on an object is meaningful only if that object is separated from the ionic center by at least one or more intermediary molecules having the dielectric constant (e.g., water in the present example). If, however, the object is bound directly to the ionic center, the effect of the ionic center on the object should, to a first approximation, be unaffected by the nature of the wider environment. This direct effect would be determined solely by the nature of the metal ion and its other coordinating ligands. Thus, it seems unreasonable to postulate a special effect of the hydrophobic environment that will make a metal ion a super Lewis acid, at least in the cases in which a substrate is directly bound to the metal.

The influence of the hydrophobicity on long-distance effects of an electric charge, however, should have some substance. This may indeed be one of the most important consequences of the hydrophobic space often found in the enzymes. This effect would, for example, help the ionic center attract and position the substrates. Furthermore, if the substrate-accommodating pocket is surrounded by ionic residues, then the electrostatic effect of a catalytic cation will be dampened.

A secondary effect of the environment is the solvation effect on a complex. Solvation by water tends to stabilize a somewhat more relaxed state of a metal–ligand system, i.e., a metal–ligand bond in an aqueous environment could be elongated slightly. This more relaxed state is stabilized by the water molecules surrounding it due to its being more polarized. Conversely, it is conceivable that a metal–ligand bond would be slightly shorter in a hydrophobic environment than in a medium of high dielectric constant and that, therefore, the substrate could be more highly activated in a nonpolar enzyme pocket than in a polar pocket that is filled with water.

5.2.5 Other Mechanisms

Before leaving the subject of carboxypeptidase A and thermolysin, we must review other important alternative mechanisms. The peptidase activity of these enzymes is quite different from their esterase activity in many respects, as mentioned earlier. This difference probably stems from the facts that a rotation about the C—O bond in esters is rather free, whereas the rotation about the amide C—N bond to be broken in peptidases is rather restricted because of a significant degree of double bond character of the C—N bond (Cleland, 1977). Therefore, it may be possible to interpret the difference between the esterase and peptidase activities in terms of similar mechanisms that have different rate-determining steps (Cleland, 1977). In the hydrolysis of an ester, Makinen et al. (1976) detected an acyl intermediate at low temperatures. They assigned the acyl accepting group to Glu-270 residue. Thus,

$$\text{Glu—}\overset{\text{O}}{\overset{\|}{\text{C}}}\text{—O}^- + \text{R}'\text{—O—}\overset{\text{O}}{\overset{\|}{\text{C}}}\text{—R} \xrightarrow{-\text{OR}'} \text{Glu—}\overset{\text{O}}{\overset{\|}{\text{C}}}\text{—O—}\overset{\text{O}}{\overset{\|}{\text{C}}}\text{—R} \quad (5\text{-}17)$$

This is the acylation process; the subsequent deacylation (5-18) completes the esterase reaction:

$$\underset{\displaystyle\text{Glu}-\overset{\displaystyle O}{\overset{\|}{C}}-O-\overset{\displaystyle O}{\overset{\|}{C}}-R}{} \xrightarrow{+OH^-} \text{Glu}-\overset{\displaystyle O}{\overset{\|}{C}}-O^- + HO-\overset{\displaystyle O}{\overset{\|}{C}}-R \qquad (5\text{-}18)$$

The acylation process should involve the creation of tetrahedral geometry about the carbonyl carbon undergoing hydrolysis, as is also true with the serine and other metal-free proteinases (see Fig. 5-1):

$$\text{Glu}-\overset{\displaystyle O}{\overset{\|}{C}}-O^- + R-\overset{\displaystyle O}{\overset{\|}{C}}-OR' \rightarrow \text{Glu}-\overset{\displaystyle O}{\overset{\|}{C}}-O-\overset{\displaystyle R}{\underset{\displaystyle O^-}{\overset{|}{\underset{\displaystyle (-NHR')}{C}}}}-OR' \qquad (5\text{-}19)$$

$$(-NHR')$$

The formation of a tetrahedral structure of this type will be easier with an ester than with an amide. The pH profile of the rate of peptidase activity indicates that a group with pK = 6 is involved in the catalytic activity, and Cleland (1977) assigns pK = 6 to the Glu-270 residue. This assignment is somewhat doubtful, however, because the pK_a of carboxylic acid residues in a wide variety of proteins is known to range from 1.8 to about 4.7. Cleland proposes the following mechanism, based on these findings:

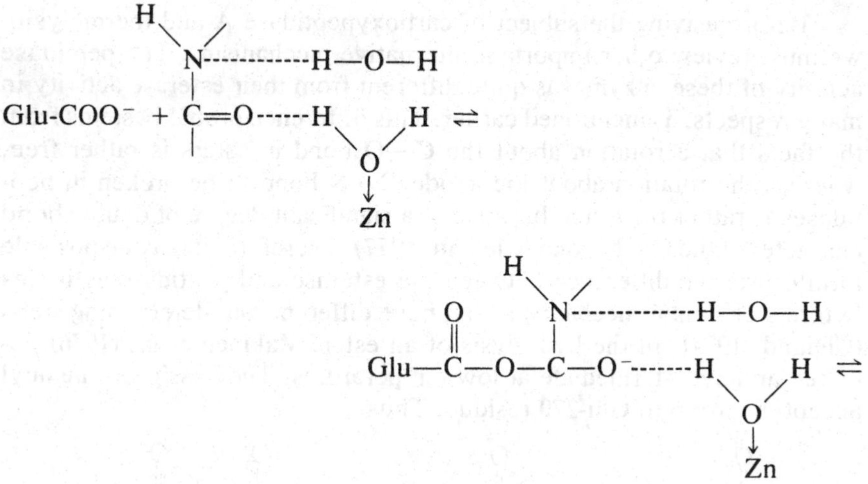

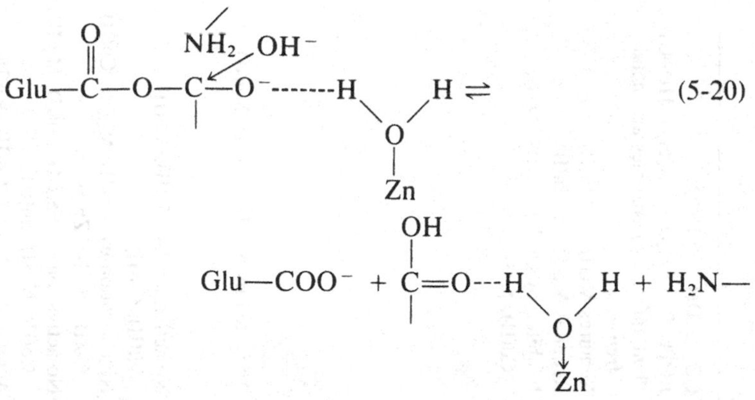

$$\underset{\text{Glu}-\overset{\overset{\displaystyle O}{\|}}{C}-O-\underset{\underset{\displaystyle |}{}}{\overset{\overset{\displaystyle N\overset{\nearrow}{H_2}}{}}{C}}\overset{\nwarrow}{\underset{}{O^-}}\text{------H}\underset{\underset{\displaystyle Zn}{|}}{\underset{\displaystyle O}{\diagup}}\text{H}}{} \rightleftharpoons \qquad (5\text{-}20)$$

$$\text{Glu}-COO^- + \underset{\underset{\displaystyle |}{}}{\overset{\overset{\displaystyle OH}{|}}{C}}=O\text{---H}\underset{\underset{\displaystyle Zn}{\downarrow}}{\underset{\displaystyle O}{\diagup}}\text{H} + H_2N-$$

The carbonyl group is assumed here to be polarized by a hydrogen bond to a water molecule coordinated to the Zn(II) ion. It should be noted, however, that the formation of an acyl intermediate with Glu-270 has yet to be detected and characterized in the case of an amide hydrolysis (peptidase activity). The recent result that $H^{16}OH$—$H^{18}OH$ kinetic isotope effect for the reaction of benzoylglycylphenylalanine with carboxypeptidase A is different from that of benzoylglycyl-β-L-phenylacetate has been interpreted to indicate that there is no common anhydride intermediate in these reactions (Breslow et al., 1983).

5.3 Metalloenzymes Other Than Proteinases and Peptidases

A few examples of other metalloenzymes catalyzing acid–base-type reactions are listed in Table 5-5. These are only the better characterized enzymes; there exist many other metalloenzymes in this category. Here we define a metalloenzyme as one that contains firmly bound metal atom(s) or has been shown to bind activating metal atoms to its own protein portion. It is often difficult, however, to make a clear distinction between a metalloenzyme so defined and a metal ion-activated enzyme.

5.3.1 Zn(II) [Co(II)] Enzymes

A detailed review on Zn enzymes has recently been published by Vallee and Galdes (1984). Readers are referred to that review for more up-to-date structural information on the individual enzymes.

Table 5-5. Examples of Metalloenzymes Other Than Proteinases and Peptidases

Enzyme	Metal	Note
Carbonic anhydrase	Zn(II) enzyme	$CO_2 + OH^- \rightleftharpoons HCO_3^-$
Alkaline phosphatase	Zn(II) enzyme	$ROPO_3^{2-} + H_2O \rightleftharpoons ROH + HOPO_3^{2-}$
Aldolase	Zn(II) enzyme (yeast and bacterial enzyme only)	Plant and animal enzymes are metal-free
DNA polymerase	Zn(II) enzyme	Requires Mg(II) or Mn(II)
RNA polymerase	Zn(II) enzyme	Requires Mg(II) or Mn(II)
AMP deaminase	Zn(II) enzyme	Inhibited by Cu(II), Fe(II), Ag(I), Cd(II), Ni(II)
3'-Nucleotidase	Zn(II) enzyme (from mung bean, wheat seedling only)	
Mannosephosphate isomerase	Zn(II) enzyme	
Transcarboxylase	Zn(II) (Co(II)) enzyme	
Fumarate hydratase	Zn(II) enzyme	
Pyruvate carboxylase	Mn(II) enzyme (Zn(II) in *Saccharomyces cerevisiae* enzyme)	
Oxaloacetate decarboxylase	Requires Mn(II) (not in all enzymes)	
Urease	Ni(II) enzyme (from jack beans)	Activated by Mn(II), Co(II), Mg(II)
Amidophosphoribosyl transferase	Fe enzyme (2Fe per subunit)	
Anthranilate synthetase	Fe (or Co) enzyme	
Lysin 2,3-aminomutase	Requires Fe(II), a pyridoxal enzyme	No activation by Mg(II), Mn(II), Zn(II), Ni(II), Cu(II)
L-Serine dehydratase	Requires Fe(II), a pyridoxal enzyme	No activation by Mn(II), Mg(II), Cu(II), Co(II), Ni(II), Zn(II)
Phosphotransacetylase	Requires Fe(II) (Mn(II), 50% active)	No activation by Fe(III), Zn(II), Mg(II), Co(II), Ni(II), Sn(II)
Histidine ammonia lyase	Fe(II) enzyme	Activated by Mn(II), Cd(II), Mg(II), Co(II), Ni(II), Cu(II), Zn(II)
Aconitase	Fe enzyme, requires Fe(II) [4Fe-4S] cluster	No activation by Ca(II), Mg(II), Mn(II), Fe(III), Co(II), Ni(II), Cu(II), Cd(II)
Citraconate hydratase	Requires Fe(II)	
Tartarate dehydratase	Requires Fe(II)	

5.3.1.1 Aldolase

This is the enzyme responsible for one of the pivotal steps in glycolysis. It catalyzes the following reaction:

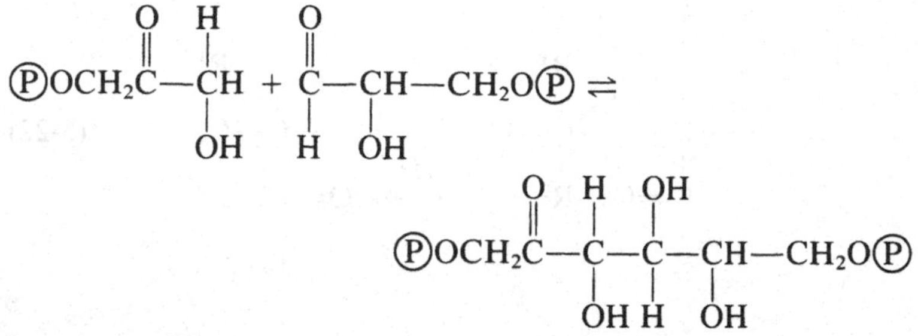

$$\text{(P)OCH}_2\overset{\overset{\displaystyle O}{\|}}{C}-\overset{\overset{\displaystyle H}{|}}{\underset{\underset{\displaystyle OH}{|}}{CH}} \;+\; \overset{\overset{\displaystyle O}{\|}}{C}-\overset{\underset{\underset{\displaystyle H}{|}}{}}{CH}-\overset{\underset{\underset{\displaystyle OH}{|}}{}}{CH}-CH_2O\text{(P)} \;\rightleftharpoons$$

$$\text{(P)OCH}_2\overset{\overset{\displaystyle O}{\|}}{C}-\overset{\overset{\displaystyle H}{|}}{\underset{\underset{\displaystyle OH}{|}}{C}}-\overset{\overset{\displaystyle OH}{|}}{\underset{\underset{\displaystyle H}{|}}{C}}-CH-CH_2O\text{(P)} \qquad (5\text{-}21)$$

There are two classes of aldolases, although enzymes of both classes catalyze exactly the same reaction. Enzymes from higher plants and animals (class I) are metal-free, whereas those from bacteria and yeast (class II) usually contain Zn(II) as an essential factor. There are, in addition, known aldolases that seem to contain Co(II) or Mn(II) instead of Zn(II). It is not known how closely the protein structures of enzymes of classes I and II resemble each other.

Aldolase is a cytoplasmic enzyme in a eukaryotic cell. The fact that the aldolase in the prokaryotes (bacteria) is quite different from that in the eukaryotes may suggest that aldolase I (of the eukaryotes) evolved independently of aldolase II. One evolutionary theory (see Chapter 1) maintains that mitochondria in the eukaryotic cells were derived from symbiotic bacteria. The gene for aldolase II in the symbiotic bacteria may have been lost in the process of the conversion of the bacteria to a cell organelle, the mitochondrion. If this theory is right, the host cells that were to become eukaryotic must have developed independently of the prokaryotes and must have been already equipped with some cytoplasmic enzymes including aldolase I as well as with a nucleus.

Reaction 5-21 is a kind of aldol condensation. In organic chemistry, the base-catalyzed aldol condensation is known to occur by the following mechanism:

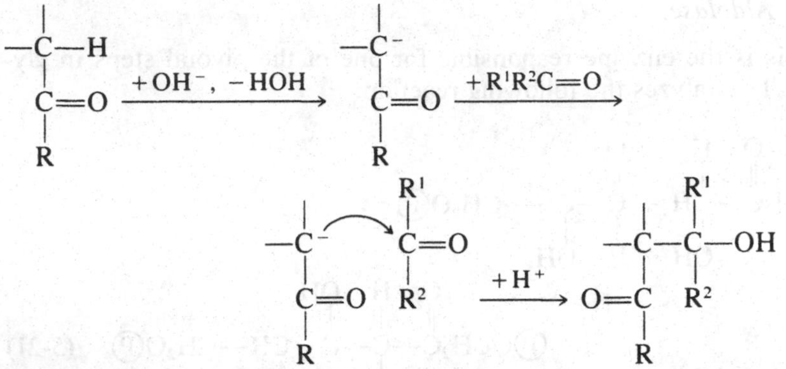

$$
\begin{array}{c}
-\overset{|}{\underset{\underset{R}{\overset{|}{C}}=O}{C}-H} \xrightarrow{+OH^-, \ -HOH} \quad -\overset{|}{\underset{\underset{R}{\overset{|}{C}}=O}{C^-}} \xrightarrow{+R^1R^2C=O}
\end{array}
$$

$$
-\overset{|}{\underset{\underset{\underset{R}{|}}{\overset{|}{C}=O}}{C^-}} \quad \overset{R^1}{\underset{R^2}{\overset{|}{C}=O}} \xrightarrow{+H^+} \quad -\overset{|}{\underset{O=C}{\overset{|}{C}}}-\overset{R^1}{\underset{R^2}{\overset{|}{C}}}-OH \quad (5\text{-}22)
$$

In this mechanism, the crucial steps are the deprotonation of the α carbon of the first reactant and the subsequent attack of the resulting anion onto the carbonyl group of the second reactant.

A proposed mechanism (Lai and Horecker, 1972) for class I aldolase involves a Schiff base formation and a rearrangement of the keto–enol type. That is:

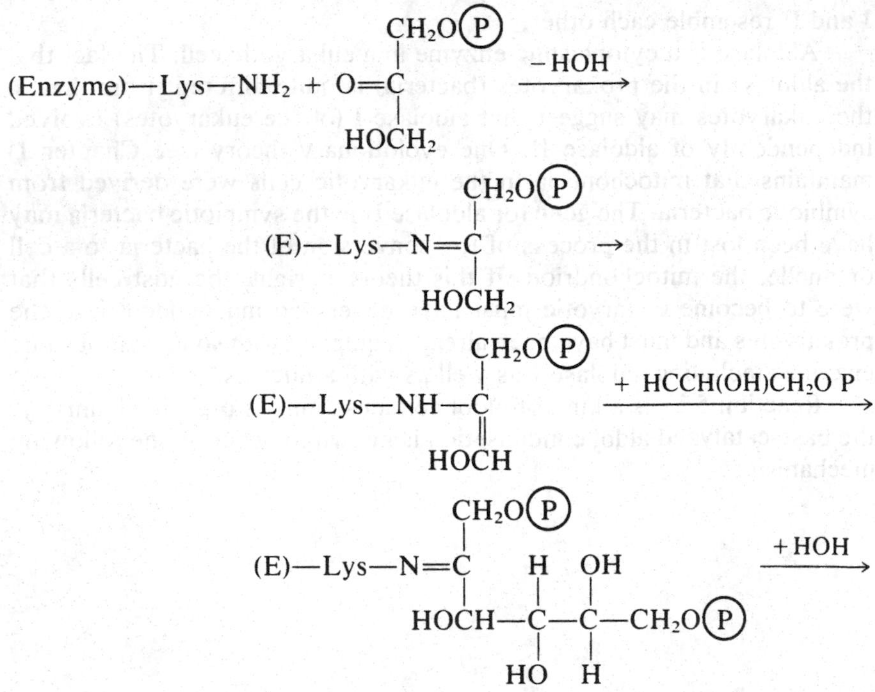

$$
\text{(Enzyme)}-\text{Lys}-\text{NH}_2 + O=\overset{\overset{\displaystyle CH_2O\,(P)}{|}}{\underset{\underset{\displaystyle HOCH_2}{|}}{C}} \xrightarrow{-HOH}
$$

$$
\text{(E)}-\text{Lys}-\text{N}=\overset{\overset{\displaystyle CH_2O\,(P)}{|}}{\underset{\underset{\displaystyle HOCH_2}{|}}{C}} \rightarrow
$$

$$
\text{(E)}-\text{Lys}-\text{NH}-\overset{\overset{\displaystyle CH_2O\,(P)}{|}}{\underset{\underset{\displaystyle HOCH}{\|}}{C}} \xrightarrow{+\ HCCH(OH)CH_2O\ P}
$$

$$
\text{(E)}-\text{Lys}-\text{N}=\overset{\overset{\displaystyle CH_2O\,(P)}{|}}{C} \quad \underset{\underset{\displaystyle HO \quad H}{\overset{\displaystyle H \quad OH}{|} \quad \overset{\displaystyle |}{}}}{\overset{}{HOCH}-\overset{|}{C}-\overset{|}{C}-CH_2O\,(P)} \xrightarrow{+HOH}
$$

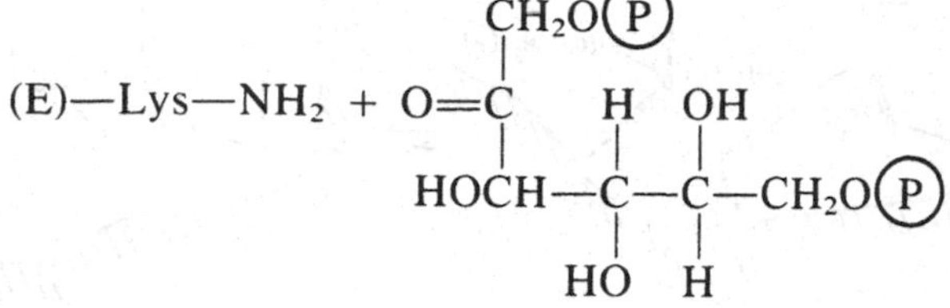

$$(E)-Lys-NH_2 \ + \ O=\overset{\overset{\displaystyle CH_2O\,\textcircled{P}}{|}}{C}$$

$$HO\overset{|}{C}H-\overset{\overset{\displaystyle H}{|}}{\underset{\underset{\displaystyle HO}{|}}{C}}-\overset{\overset{\displaystyle OH}{|}}{\underset{\underset{\displaystyle H}{|}}{C}}-CH_2O\,\textcircled{P} \quad (5\text{-}23)$$

Mn(II)-substituted aldolase can be produced by growing yeast in a medium containing Mn(II). An EPR study on this Mn(II) aldolase yielded data consistent with a Mn(II)-dihydroxyacetone phosphate structure as shown in Fig. 5-4 (Mildvan *et al.*, 1971). The coordination of dihydroxyacetone to the metal cation, Zn(II) in the native enzyme, facilitates the heterolytic cleavage of the C—H bond on the terminal carbon, and this process is assumed to be further facilitated by a proton-accepting group on the protein. This picture corresponds to the simple organic reaction mechanism (5-22). It is not known, however, whether the carbonyl group of the second substrate, glyceraldehyde phosphate, is coordinated to the same cation site on the enzyme. This would appear to have the advantage that the second substrate would be brought into close proximity and its carbonyl group would be activated. In view of the facts that the cation is fairly tightly bound in the enzyme and the ionic radius of Zn(II) is rather small, it does not seem very likely, however, that the second substrate directly binds to the cation center. Therefore, the activation of the second substrate may be accomplished by other active residues in the protein. However, a mechanism involving the simultaneous coordination of both substrates to the catalytic cation should not entirely be discarded at this stage. A possible mechanism is shown in Fig. 5-5. In the reverse mode of the reaction, the cation would help in bringing the substrate into the correct position relative to other activating sites such as A, B_1, and B_2.

5.3.1.2 Carbonic Anhydrase

The structure and mechanism of this enzyme are fairly well documented (Lindskog *et al.*, 1982; Bertini *et al.*, 1982; Tashian and Hewett-

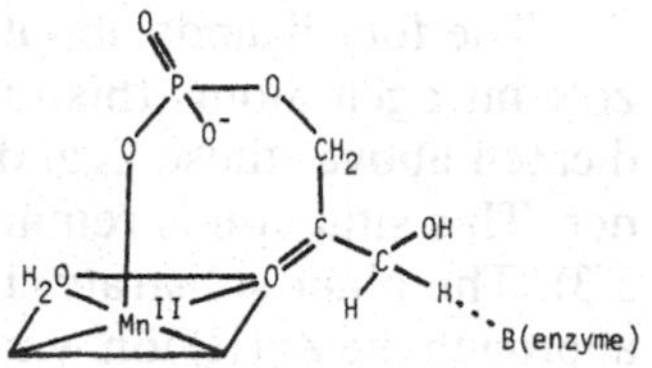

Figure 5-4. A proposed mechanism for the Mn(II)-aldolase. (After Mildvan *et al.*, 1971. *Biochemistry* 10:1191.)

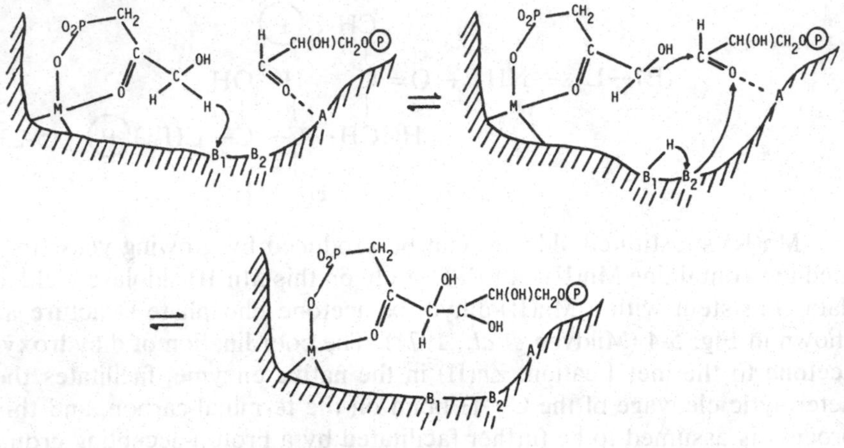

Figure 5-5. A possible model for the aldolase reaction mechanism.

Emmett, 1984). The coordination structure about the catalytic element Zn(II) has been established to be a distorted tetrahedron. A characteristic feature is the presence of water molecules in an icelike structure trapped in a hydrophobic pocket of the protein. The reaction catalyzed by this enzyme is the hydration of CO_2 and the dehydration of H_2CO_3; that is:

$$CO_2 + OH^- \rightarrow HCO_3^- \tag{5-24}$$

Thus, the substrates are CO_2, HCO_3^-, and OH^- (or HOH). CO_2 and HCO_3^- are rather hydrophobic; the nature of the pocket into which the substrates are supposed to enter and where the catalytic cation is located seems thus to be appropriate. What is the role of the water? Water molecules trapped in the pocket seem to be utilized. It may be that the icelike structure extends all the way to the surface of the protein where the water molecules come in contact with solvent water. If so, then the consumed water may not need to be replaced directly from the solvent but could arise instead through a series of rearrangements of the icelike structure whereby the entering water molecule from the solvent actually is incorporated at the extremity of the icelike structure.

The four ligands about Zn(II) in the resting enzyme are three imidazole nitrogen atoms (histidine residues) and one water molecule. As indicated above, these ligands are arranged in a distorted tetrahedral manner. This situation is reminiscent of that in carboxypeptidase A (see Fig. 5-3). The other substrate, being small and polarized, would then need to approach the Zn(II) ion whose coordination structure may include a some-

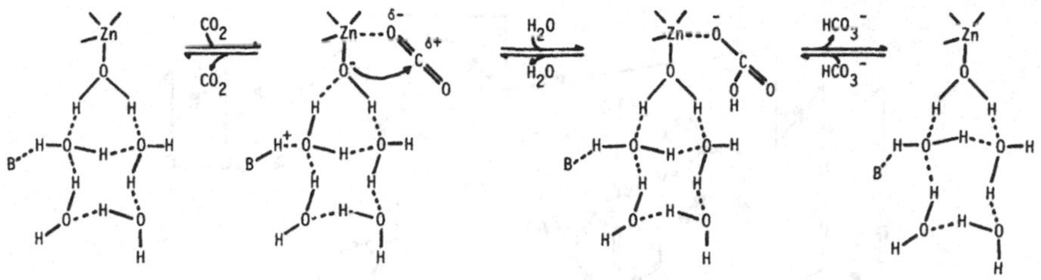

Figure 5-6. A proposed mechanism for carbonic anhydrase.

what open space as in the case of carboxypeptidase A. The approach of CO_2 to Zn(II) does not have to be of a coordinate bonding type, although the approach itself would tend to further polarize the $O{=}C{=}O$ bond. The mechanism suggested may be summarized as shown in Fig. 5-6 (Ochiai, 1977). A similar mechanism has been proposed by Lindskog et al. 1982), in which the presence of a proton acceptor (from the coordinated water molecule) is postulated near the zinc site.

5.3.1.3 DNA Polymerases

DNA polymerases as well as RNA polymerases from many sources have now been recognized to be Zn(II) enzymes (Mildvan and Loeb, 1981; Wu and Wu, 1981); it is quite possible that indeed all DNA polymerases in all kinds of cells are Zn(II) enzymes. The enzymes require, in addition to the firmly bound Zn(II), a divalent cation, usually Mg(II). Mg(II) can, however, be substituted by Mn(II) among other ions. This latter fact is common to many enzymes in which Mg(II) is the activator and will be discussed in Section 5.4.

DNA polymerase catalyzes the addition of a mononucleotide to a free end of DNA with the aid of a DNA template (Fig. 5-7). DNA polymerase I from E. coli is a single polypeptide chain of molecular weight 109,000 (Mildvan and Loeb, 1981; Loeb, 1974). It is more or less spherical with a diameter of about 6500 pm and contains a single atom of zinc per molecule. This 6500-pm diameter should be compared to the 2000-pm diameter of the double helix of a DNA molecule. A modification of a proposed mechanism is shown in Fig. 5-8 (Slater et al., 1972). Zn(II) binds the $3'{-}OH$ group while a proton acceptor on the protein picks up the OH group's proton, increasing the nucleophilicity of the oxygen atom, which in turn attacks the phosphorus atom of the incoming nucleotide. The divalent cation is supposed to be located nearby and bring the mononucleotide, especially its reacting phosphate group, into the correct

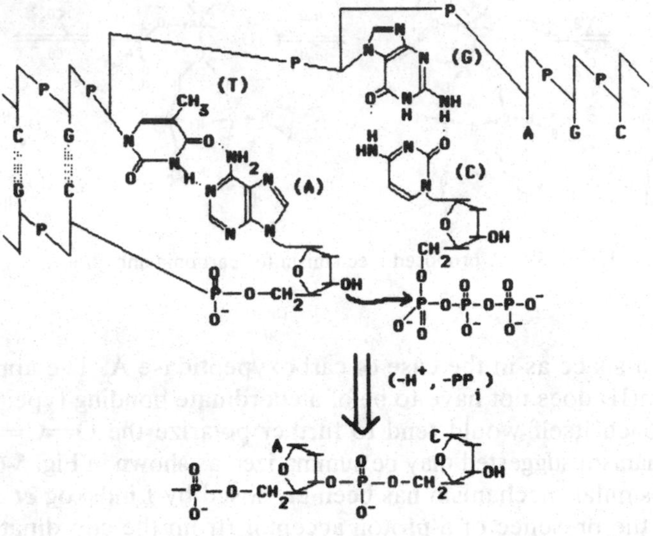

Figure 5-7. The reaction catalyzed by DNA polymerase.

position next to the free end of the polynucleotide. A more recent study (Mildvan and Loeb, 1981; Abbound *et al.*, 1978) with Mn(II) as the divalent cation indicated that Mn(II) binds the γ- and more loosely the α-phosphate oxygens, assisting the leaving of the pyrophosphate (Fig. 5-9).

Other divalent cations such as Be(II), Co(II), Cd(II), Cu(II), Ni(II), and Pb(II) as well as Mn(II) can replace the physiological cation Mg(II). They activate the polymerization function of the enzyme, but reduce the fidelity of the enzyme (Loeb and Mildvan, 1981; Loeb and Zakour, 1980; Sirover and Loeb, 1976). This is crucial, since the enzyme's task is to produce a polynucleotide chain that is exactly complementary to the template DNA. Even the native enzyme with the proper cation [Mg(II)] makes a mistake occasionally, but use of other cations in place of Mg(II) increases the chance of the enzyme's making an error. The correct substrate is usually selected through a matching of hydrogen bondings between the base of the template and the base of the incoming mononucleotide. The free energy difference between a correct and a wrong base pair is, however, not very large, being estimated to be about 8.2 kJ/mole (2 kcal/mole). This is the reason for the occasional mismatching. Different divalent cations bind with mononucleotides in different ways; Mg(II) binds with the pyrophosphate portion only, but most transition metal cations can bind with the base portion as well as the phosphate portion. Thus, incorporation

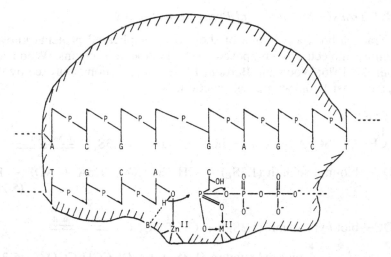

Figure 5-8. A proposed mechanism for DNA polymerase. (After Slater *et al.*, 1972. *J. Biol. Chem.* 247:6784.)

of different cations into the reaction system could cause a slight change in the conformation of the incoming mononucleotide as well as in the hydrogen bonding capacity of its base portion and hence could result in utilization of a mismatched base. This can be considered to be a major cause of the observed increase in infidelity in such cases. If this explanation is correct, then the divalent cation is not necessarily located near the Zn(II) center as implied in Fig. 5-8. This consideration would in turn lead to the hypothesis that the oxygen atom of the phosphate group that is to become the bridging phosphate would likely be coordinated (or at least be attracted) to the cation center Zn(II) so that the two reacting centers, i.e., OH and the phosphate, may be brought into close contact. This suggestion can be summarized as shown in Fig. 5-9.

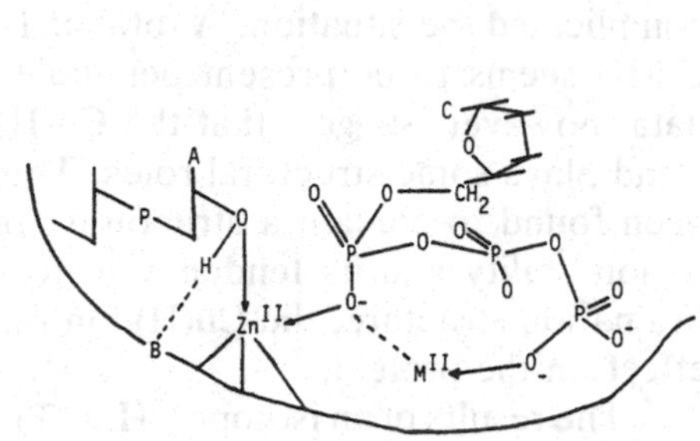

Figure 5-9. A proposed mechanism for the effect of a divalent metal ion on DNA polymerase. (After Sloan *et al.*, 1975. *J. Biol. Chem.* 250:8913.)

5.3.1.4 Transcarboxylase and Pyruvate Carboxylase

Transcarboxylase is one of the most complicated proteins known; the entire molecule is composed of 30 polypeptide chains (Wood and Zwolinski, 1976; Wood and Barden, 1977). The reaction catalyzed by this enzyme consists of two partial reactions:

$$\underset{\displaystyle CH_3CHCO-SCoA}{\overset{\displaystyle \overset{\textstyle COO^-}{|}}{}} + ATP + \text{biotinyl subunit } (1.3S_E) \xrightleftharpoons[]{12S_H \text{ subunit}}$$

$$^-OOC-\text{biotinyl subunit } (1.3S_E) + CH_3CH_2CO-SCoA + ADP + P_i$$
$$(5\text{-}25a)$$

$$^-OOC-\text{biotinyl subunit } (1.3S_E) + CH_3COCOO^- \xrightleftharpoons[]{5S_E \text{ subunit}}$$

$$\text{biotinyl subunit } (1.3S_E) + {}^-OOCCH_2COO^- \quad (5\text{-}25b)$$

As indicated in these equations, there are at least three different kinds of reactive subunits. The so-called $12S_H$ subunit (molecular weight 360,000) catalyzes the transfer of a CO_2 group from one of the substrates to the biotin residue on another subunit ($1.3S_E$). The latter serves as the CO_2 carrier. The second step, the transfer of CO_2 from the carrier to another substrate, is catalyzed by yet another subunit, $5S_E$ (molecular weight 120,000). The enzyme from propionic acid bacteria (*Propionibacterium shermanii*) contains varying amounts of Zn(II) and Co(II), but the total amount of Zn(II) and Co(II) is constant. The ratio of Zn(II) to Co(II) varies according to the content of these two metal ions in the organism's growth medium. It seems, therefore, that Zn(II) and Co(II) are completely interchangeable as is true in many other Zn(II) enzymes. These metal ions have been shown to be firmly bound to the $5S_E$ subunit. There are several lines of evidence indicating that the Zn(II) and/or Co(II) atoms in the $5S_E$ subunits are located near the pyruvate-binding sites.

A report by Fung *et al.* (1974) that $5S_E$ also contains Cu(II) has complicated the situation. A total of 12 gram atoms of Zn(II), Co(II), and Cu(II) seems to be present per mole of the enzyme. Several pieces of data, however, suggest that the Cu(II) is not catalytically active and instead plays some structural roles. This is the only protein that has so far been found to contain a structural copper. Because of the high coordination ability and its tendency to form a tetragonal (square-planar) coordination structure, the Cu(II) ion must exert a fairly specific structural effect on the protein.

The results of an isotope ($^3H = T$) distribution experiment were found to be consistent with the following mechanism for the carboxyl transfer step (5-25b) (Wood and Barden, 1977):

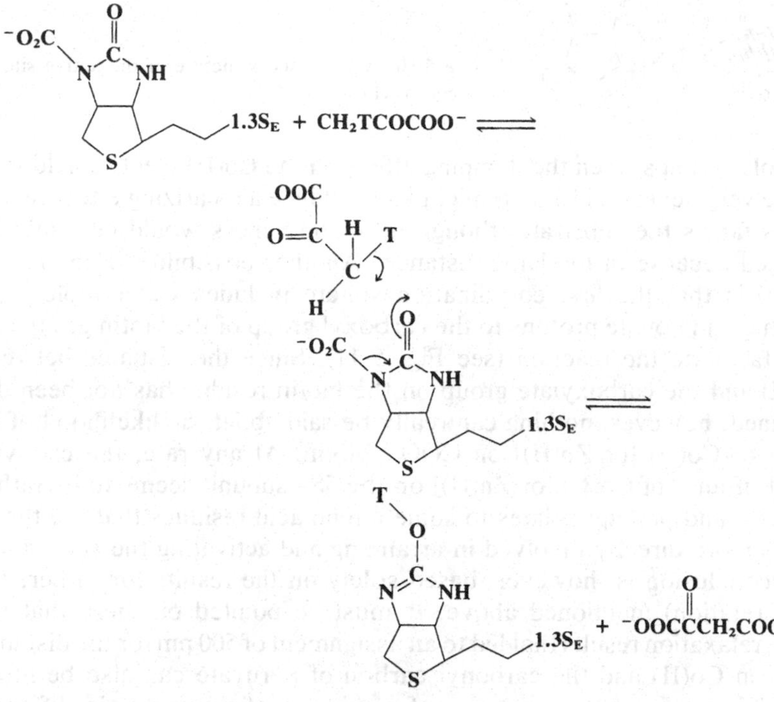

^-O_2C
O
C
N
NH
S
$1.3S_E$ + $CH_2TCOCOO^-$
OOC
O=C
H
T
C
H
O
^-O_2C
C
N
NH
S
$1.3S_E$
T
O
C
N
NH
S
$1.3S_E$
O
+ $^-OOCCCH_2COO^-$

(5-26)

The catalytic action of subunit $5S_E$ should comprise at least two functions: one is to bring pyruvate into the correct position relative to the biotin residue and another is to activate the C—H bond in pyruvate. Our concern here is then to consider what role Zn(II) and/or Co(II) might play in regard to these functions of the $5S_E$ subunit.

The NMR relaxation technique has been utilized to estimate the distances between the paramagnetic ion Co(II) and the magnetic nuclei on the substrates (Fung *et al.*, 1974, 1976a,b). A conclusion from these studies is that the distances from Co(II) to C-1 (i.e., COO^-), C-2 (i.e., CO), and C-3 (CH_3) of pyruvate are 630, 500, and 630 pm, respectively. These distances are too large to permit a direct (in the first coordination sphere) binding of pyruvate to Co(II). The situation is sketched in Fig. 5-10. The first coordination sphere of the Co(II) ion must thus be occupied by amino acid residues, if the interpretation of the NMR relaxation data is correct. It is impossible to determine whether or not Co(II) exerts some sort of polarizing effect on the substrate. If the intervening layer is composed of

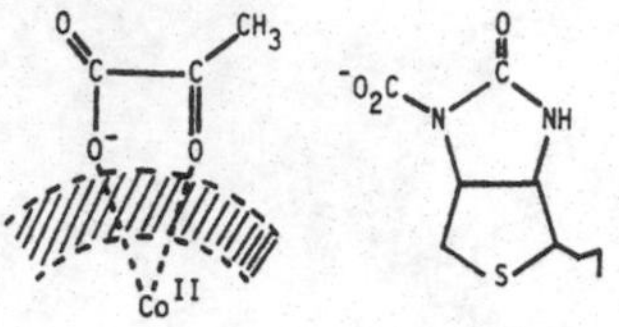

Figure 5-10. A proposed structure of the active site of transcarboxylase.

nonpolar groups, then the damping effect on the Co(II)-electric field may not be very severe and Co(II) might still produce a polarizing effect reaching as far as the substrate, though its effectiveness would certainly be reduced because of the large distance. Another possibility (Fung *et al.*, 1976b) is that the first coordination sphere includes water molecules, which could donate protons to the carboxyl group of the biotin group and thus facilitate the reaction (see Fig. 5-11). Since the distance between Co(II) and the carboxylate group on the biotin residue has not been determined, however, nothing can really be said about the likelihood of an effect of Co(II) [or Zn(II)] on COO$^-$-biotin. At any rate, the catalytic effect, if any, of Co(II) [or Zn(II)] on the $5S_E$ subunit seems to be rather indirect, and perhaps relates to some amino acid residues that are themselves more directly involved in arranging and activating the substrates. This conclusion is, however, based solely on the results (or rather, the interpretation) mentioned above. It must be pointed out here that the NMR relaxation results that led to an assignment of 500 pm for the distance between Co(II) and the carbonyl carbon of pyruvate can also be interpreted in terms of the existence of a mixture of pyruvates in different coordination states. For instance, as much as 2% of pyruvate could indeed be in the first coordination sphere while 98% is in the second coordination sphere, assuming that pyruvate molecules in the inner coordination sphere are rapidly exchanging with those in the outer sphere. Under these conditions, then, it might be that pyruvate in the inner sphere actually could be under the catalytic influence of the metal cation.

Pyruvate carboxylase is an enzyme very similar to transcarboxylase. It catalyzes the following reaction:

$$\text{E-biotin} + \text{ATP} + \text{HCO}_3^- \rightarrow \text{E-biotin}-\text{COO}^- + \text{ADP} + \text{P}_i \quad (5\text{-}27a)$$

$$\text{E-biotin}-\text{COO}^- + \text{CH}_2\text{COCOOH} \rightarrow \text{E-biotin} + \text{OOCCH}_2\text{COCOO}^- \quad (5\text{-}27b)$$

The second step is exactly the same as the corresponding step in the transcarboxylase reaction. There are three known types of pyruvate carboxylase (Wood and Barden, 1977). The first type (I) includes enzymes from various animal sources and *Arthrobacter globiformis* and *Bacillus stearothermophilus*. These enzymes are tetrameric, with subunits of molecular weight 110,000 to 130,000, and they contain Mn(II) (one atom per

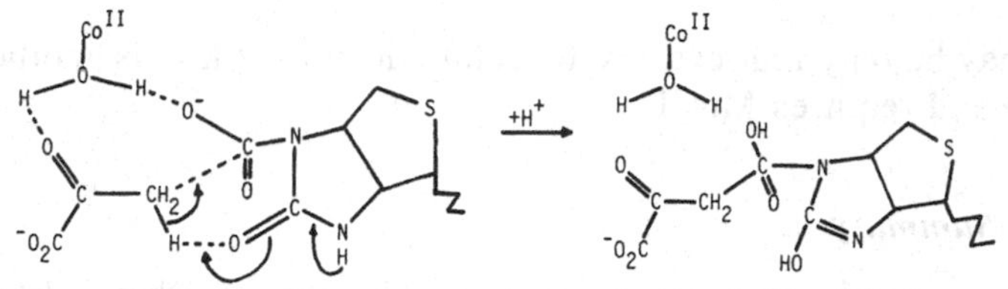

Figure 5-11. A proposed mechanism for transcarboxylase. (After Wood and Zwolinski, 1976. *Crit. Rev. Biochem.* 4:47.)

subunit); in some examples, Mn(II) can be replaced by Mg(II). For example, the enzyme from chicken liver contains only Mn(II), but that from calf liver contains both Mg(II) and Mn(II) in varying ratio. The second type (II) is obtained from yeast. It has a molecular weight of 446,000, a value very similar to that of the enzymes of type I, and it is also a tetramer. Each subunit contains one biotin unit and one Zn(II) ion. There is a third type (III) of the enzyme, obtained from *Pseudomonas citronellolis* and *Azotobacter vinelandii*. The enzymes of type III appear to be different from those of the other two types, but no structural details have yet been revealed.

It seems that a single polypeptide (subunit) of pyruvate carboxylase catalyzes both steps 5-27a and 5-27b unlike the case with transcarboxylase. It is not known which of the steps is catalyzed by the metal cation. If the mechanism of pyruvate carboxylase is similar at all to that of transcarboxylase, the second step (5-27b) may be the one to be catalyzed by the metal ion. NMR relaxation studies similar to those on transcarboxylase have been reported (Fung *et al.*, 1973; Reed and Scrutton, 1974). The distances between Mn(II) and the methyl, carbonyl, and carboxyl carbons of the bound pyruvate were estimated to be 620, 710, and 850 pm, respectively. These values are again too large for the direct coordination of pyruvate to Mn(II). As in the case of transcarboxylase, the metal cation does not seem to be involved in the direct binding of one of the substrates.

The similarity of type II enzymes to those of type I is very striking; this suggests a very similar protein structure. If so, then the function of Zn(II) in type II enzymes may be similar to that of Mn(II) [or Mg(II)] in type I enzymes. This would suggest that the requirement for a specific divalent cation in pyruvate carboxylase, as well as in transcarboxylase, is not very strict. Various divalent cations of similar ionic radii may function equally efficiently in the catalytic role. Since the nature of the catalytic function of the divalent cation has not been fully elucidated, as discussed above, this must, however, be taken as a tentative conclusion. The major function of the cation may well be structural, and the catalytic effect may be only indirect. Oxaloacetate decarboxylase is another biotin enzyme and requires Mn(II).

5.3.1.5 Summary

Two general functions of the catalytic site, whether it be a metal cation or simply comprised of amino acid residues, are (1) binding at least one of the substrates and (2) activation (polarization, deformation) of a specific bond where the reaction is to take place. The above survey of four or five enzymes seems to support these functional assignments to the divalent cation, especially Zn(II). However, the degree of the involvement of the catalytic cation in these functions varies from one enzyme to another. At one extreme, both of the reacting substrates may directly bind to the same single cation, thus being brought into close contact and simultaneously being activated. At the other extreme, the effect of the metal ion appears to be rather indirect, and in this case the divalent cation seems to be highly interchangeable, i.e., Mn(II), Zn(II), Mg(II), and Co(II) seem to be almost equally effective. Note that the electronic configuration of these cations are spherical (being d^0, d^5, or d^{10}; d^7 would be spherical in a tetrahedral field) and their z_{eff}/r values are similar (see Section 2.2).

5.3.2 Iron-Dependent Enzymes

5.3.2.1 Serine Dehydratase and Lysine 2,3-Mutase

As shown in Table 5-5, a number of enzymes require specifically Fe(II). Of the listed enzymes, lysine 2,3-mutase and L-serine dehydratase depend on pyridoxal as a coenzyme (Sagers, 1974). The reaction of serine dehydratase can be mimicked by a pyridoxal–metal ion system. The mechanism of this kind of reaction seems to be well established (see Ochiai, 1977):

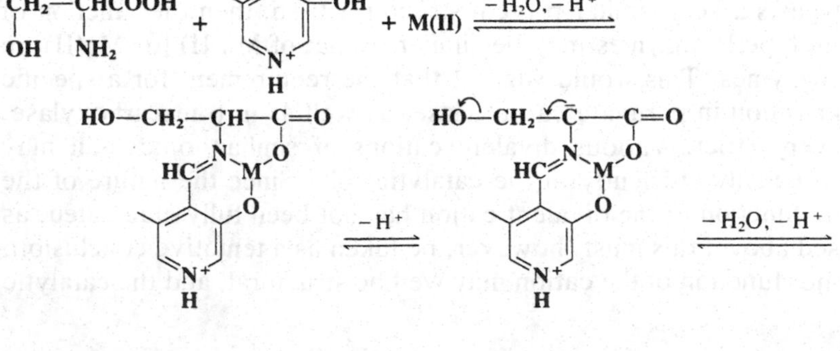

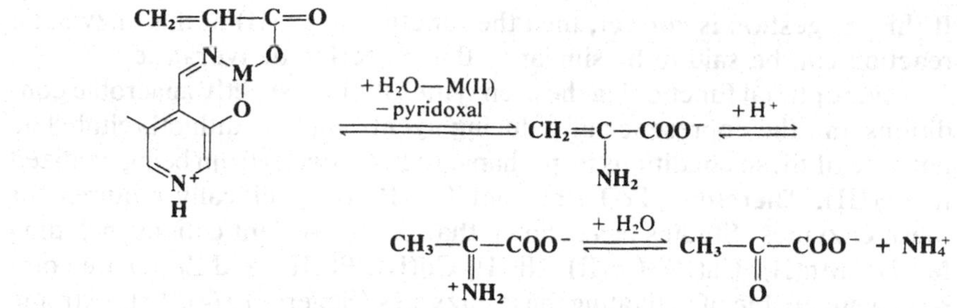

$$CH_2{=}CH{-}\underset{\overset{|}{\underset{O}{M}}}{\overset{N}{C}}{=}O$$

(structure with pyridoxal ring, N^+–H)

$$\xrightarrow[-\,\text{pyridoxal}]{+\,H_2O{-}M(II)} \quad CH_2{=}\underset{NH_2}{\overset{|}{C}}{-}COO^- \xrightleftharpoons{\;+H^+\;}$$

$$CH_3{-}\underset{\overset{+}{NH_2}}{\overset{|}{C}}{-}COO^- \xrightleftharpoons{\;+\,H_2O\;} CH_3{-}\underset{\overset{\|}{O}}{C}{-}COO^- + NH_4^+$$

$$(5\text{-}28)$$

A similar mechanism may be assumed also for the enzymatic reaction with Fe(II) as the M(II) in the scheme above. The function of Fe(II), then, is to bring two substrates, pyridoxal and a serine molecule, together in a specific arrangement. It may also facilitate some of the reaction steps, especially the second step in mechanism 5-28.

Lysine 2,3-mutase may possibly function by a similar mechanism:

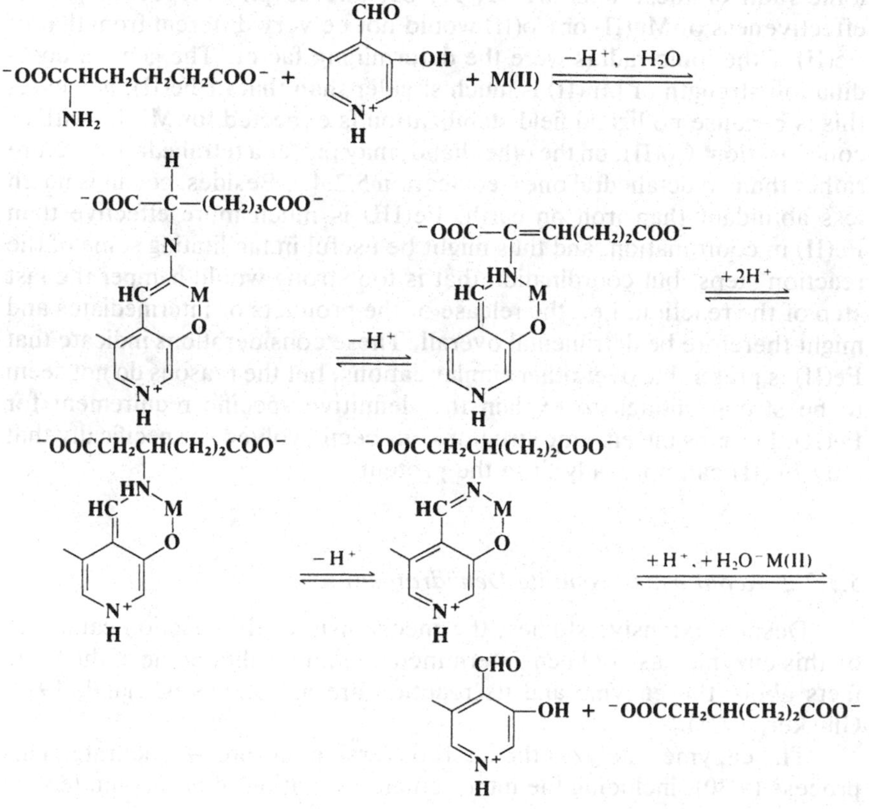

$$^-OOCCH\underset{NH_2}{\overset{|}{C}}H_2CH_2CH_2COO^- \;+\; (\text{pyridoxal ring, CHO, OH, } N^+{-}H) \;+\; M(II) \xrightleftharpoons{\;-H^+,\,-H_2O\;}$$

$$^-OOC{-}\underset{\overset{|}{N}}{\overset{\overset{H}{|}}{C}}{-}(CH_2)_3COO^- \qquad ^-OOC{-}C{=}CH(CH_2)_2COO^-$$

(pyridoxal-metal chelate structures, $HC{=}...M...O$, $N^+{-}H$) $\xrightleftharpoons{\;+2H^+\;}$

$$^-OOCCH_2CH(CH_2)_2COO^- \qquad ^-OOCCH_2CH(CH_2)_2COO^-$$

(pyridoxal-metal chelate structures, $HC{-}HN...M...O$, $N^+{-}H$) $\xrightleftharpoons{\;-H^+\;}$ (structure) $\xrightleftharpoons{\;+H^+,\,+H_2O{-}M(II)\;}$

$$\text{(pyridoxal ring, CHO, OH, } N^+{-}H) \;+\; ^-OOCCH_2CH(CH_2)_2COO^-$$

$$(5\text{-}29)$$

If this suggestion is correct, then the function of Fe(II) in this enzymatic reaction can be said to be similar to that in serine dehydratase.

For optimal functioning these enzymes require strictly anaerobic conditions and the copresence of reducing agents such as dithiothreitol. The purpose of these conditions is, perhaps, to keep Fe(II) from being oxidized to Fe(III). Therefore, Fe(II) but not Fe(III) is specifically required for these enzymes. Studies have shown that other divalent cations including Mg(II), Mn(II), Ca(II), Co(II), Ni(II), Cu(II), Pb(II), and Zn(II) are completely incapable of activating these enzymes (Sagers, 1974). This extreme specificity warrants further consideration. The mechanisms shown imply that at least two coordination sites on the divalent cation should be available for the substrates. The cation may also need to coordinate at least two or three amino acid residues in the protein. Therefore, a suitable divalent cation should have available five to six coordination sites, perhaps, in octahedral, trigonal-bipyramidal, or similar arrangement. This condition could be satisfied by Fe(II), Mn(II), Co(II), and Ni(II). The ionic radii of these ions are 92, 97, 89, and 83 pm, respectively. The effectiveness of Mn(II) or Co(II) would not be very different from that of Fe(II) if the ionic radius were the determining factor. The general coordination strength of Mn(II) is much smaller than that of Fe(II), however; this is because no ligand field stabilization is expected for Mn(II) with d^5 configuration. Co(II), on the other hand, may prefer a tetrahedral structure rather than an octahedral one (see Section 5.2.4.). Besides, cobalt is much less abundant than iron on earth. Fe(III) is much more effective than Fe(II) in coordination, and thus might be useful in facilitating some of the reaction steps, but coordination that is too strong would hamper the last step of the reaction, i.e., the release of the products or intermediates and might therefore be detrimental overall. These considerations indicate that Fe(II) is preferable over other similar cations, but the reasons do not seem to be strong enough to explain the definitive specific requirement for Fe(II). Perhaps the enzyme structure has been evolved so specifically that only Fe(II) can optimally fit in the protein.

5.3.2.2 Aconitase (Aconitate Dehydratase)

Despite extensive studies, the mechanism of the reaction catalyzed by this enzyme has not been determined unequivocally. Some of the basic facts about the enzyme and its reaction are as follows (Cleland, 1977; Glusker, 1971).

The enzyme catalyzes the interconversion: citrate → isocitrate. This process (5-30), including the intermediate formation of *cis*-aconitate, has been well established:

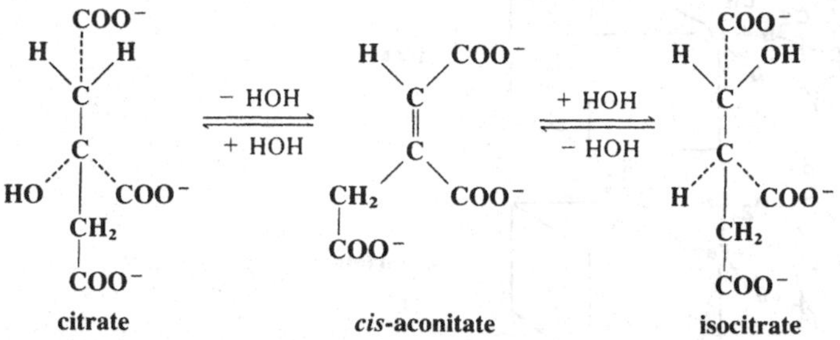

$$\text{(5-30)}$$

Mn(II), though not capable of activation, competes with Fe(II) for the binding sites on the apoenzyme. For optimal activation, a reducing agent must be added along with Fe(II). Other similar cations studied caused no activation, including Mg(II), Ca(II), Ba(II), Mn(II), Fe(III), Co(II), Ni(II), Cu(I, II), Zn(II), and Cd(II). Aconitase apparently exists in two conformations, one of which binds citrate, isocitrate, and their analogues, and the other binds cis-aconitase and its analogues. There is NMR evidence that carboxyl groups (at C-2 and C-3) of citrate as well as the 3-hydroxyl group are coordinated to an Fe(II), though a recent ESR study (Emptage et al., 1983) does not seem to confirm a strong binding of the carboxyl groups to Fe(II). However, the latter result may be questioned because of the O^{16}–O^{17} exchange between the solvent and the carboxyl oxygen atoms.

An ingenious mechanism, called the "ferrous wheel" model, was proposed, mostly based on X-ray crystallographic studies of citrate-Fe(II) and isocitrate-Fe(II) complexes (Glusker, 1971). However, the recent discovery that aconitase in its native inactive form contains "Fe_3S_4" cluster (Kent et al., 1982; Beinert et al., 1983) and that the activation by Fe(II) is in fact the formation of an active Fe_4S_4 cluster (Kent et al., 1982) forces us to reevaluate the whole problem of aconitase. The OH group of citrate has been demonstrated to bind to the Fe (designated as Fe_a), which is the labile one that is added to reconstitute the active cubane form (Emptage et al., 1983; Kent et al., 1985). Fe_a increases its ferrous character and its coordination sphere expands to five- or six-coordinate (from four-coordinate) upon binding the substrate. Obviously, Fe_a, very likely of Fe(II) character, is functioning as a Lewis acid here, unlike all the other Fe_nS_n core. A speculative mechanism is given in Fig. 5-12, in order merely to summarize what has been learned about aconitase.

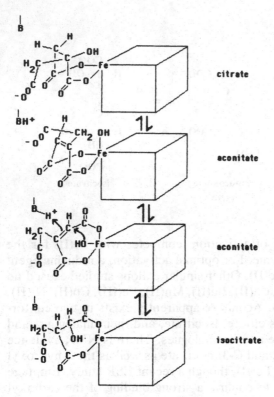

citrate

aconitate

aconitate

isocitrate

Figure 5-12. A proposed model for aconitase action.

5.3.2.3 Other Iron-Dependent Enzymes

Citraconate dehydratase and tartarate dehydratase appear to resemble aconitase in their reaction mechanisms, but no details are known. The mechanisms of other enzymatic reactions by iron-dependent enzymes are also obscure. One enzyme, phosphotransacetylase, may, however, deserve mention. It is activated by Fe(II) but also by Mn(II), albeit less effectively [about 50% as compared to Fe(II)]. It cannot be activated by any other metal ions studied, including Fe(III), Zn(II), Mg(II), Co(II), Ni(II), and Cu(II). As discussed earlier, Mn(II) is chemically very similar to Fe(II) and thus should be able to replace Fe(II) in a variety of enzymes (of acid–base type), were it not for evolutionary adaptation (see Section 3.3.2).

The so-called purple acid-phosphatases isolated from some mammalian, plant, and microbial sources are now believed to contain antiferromagnetically coupled Fe₂ (Davis and Averill, 1982). The oxidized form, which is inactive, can readily be converted to an active form by a mild reducing agent; the active form exhibits an unusual EPR spectrum centered at $g = 1.77$. Replacement of one of the Fe atoms produced an active FeZn enzyme that showed an EPR spectrum with $g = 4.3$, characteristic of Fe(III) (in high spin). These results suggest that an Fe(II)Fe(III) unit is active and that the Fe(II) can be replaced by Zn(II). Analogous enzymes isolated from sweet potatoes, soybeans, and other plant sources, however, have been reported to contain manganese as the sole catalytic element (Sugiura et al., 1981) (cf. Section 3.2.3).

5.3.3 Enzymes Dependent on Other Metals

Urease isolated from jack beans is the first reported Ni(II) enzyme (Dixon et al., 1976). Ni(II) is bound so firmly that attempts to remove it reversibly from the protein have been unsuccessful. This fact indicates that the Ni(II) ion is located in a position that is inaccessible to chelating agents. Pyruvate carboxylase, mentioned earlier, contains Mn(II) that also resists removal by chelating agents. Mn(II) in this enzyme appears to be shielded by at least one layer of ligands so that neither substrates nor solvent molecules can bind to it directly. Nothing concrete has yet been learned about the urease catalytic site containing Ni(II), but the difficulty of removing the Ni(II) suggests that it, too, is shielded by protein ligands, and hence that the substrates, urea and water, may not be able to bind directly with it. Another possibility is that the pocket that the substrates enter may be too small to permit the approach of most chelating agents. Because of this difficulty, the enzymatic activities of other metal ions have not been studied. A proposed mechanism (Dixon et al., 1976) involves the coordination of the carbonyl oxygen of urea to Ni(II). In this scheme, Ni(II) polarizes the C=O bond and renders it more susceptible to a nucleophilic attack.

Nickel is now known to occur in a number of proteins (Thauer et al., 1980). Nickel in redox enzymes and proteins has been discussed in Section 3.2.5.

Some Cu(II) enzymes in which Cu(II) exerts not an oxidizing effect but an electron-withdrawing effect are discussed in Chapter 4.

5.3.4 Concluding Remarks

It is rather difficult to make general statements with regard to the enzymatic activities of different metallic elements. The following list is an attempt to summarize a few of the more important features in this regard, however.

1. Catalytically active cations utilized in acid–base enzymatic systems are divalent. Prominent among them are Zn(II), Co(II), Mn(II), and Fe(II). Ni(II) and Cu(II) are utilized only in very special enzymes. Fe(III) seems to be utilized in several enzymes for structural purposes as well as for catalytic purpose in an enzyme—acid phosphatase. Cu(II) seems also to have structural functions, as for example in transcarboxylase.

2. Selection among Zn(II), Co(II), Mn(II), and Fe(II) seems to be made by the nature of the enzymatic active center; important factors include coordination number and structural requirements as well as characteristics of the reaction, such as the number of substrates that need to be brought onto the catalytic element.

3. Zn(II) and Co(II) prefer a tetrahedral or distorted tetrahedral structure and are thus utilized in those enzymes where a tetrahedral structure about the cation is required. Zn(II) and Co(II) are virtually interchangeable, but Zn(II) is preferred because it is more abundant than Co(II).

4. Fe(II) seems to require an octahedral structure and thus be employed by enzymes that require this geometry.

5. When the interaction between the metal ion and substrates is not very rigid (i.e., not direct), one metal ion can be substituted by another that has similar properties, including ionic radius and preferred geometry.

5.4 Metal Activation

A great number of enzymes are activated by the addition of metal ion(s); under physiological conditions the metal ion activation can indeed be regarded to be indispensable to these enzymes. A few examples of such metal ion-activated enzymes are given in Table 5-6.

The enzymes of the first category, kinases, catalyze the transfer of a phosphate group from a nucleotide, especially ATP to a substrate. For example, hexokinase catalyzes the reaction:

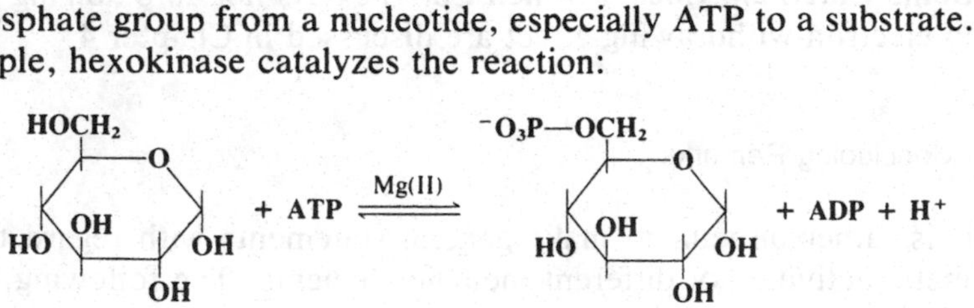

$$(5\text{-}31)$$

Mg(II) is the metal ion utilized by virtually every enzyme of this category. *In vitro*, Mg(II) can be replaced by Mn(II) in most enzymes, though Mn(II) is often less effective than Mg(II).

The enzymes of the second category in Table 5-6 are those that hydrolyze phosphate ester bonds:

$$R-O-\overset{\overset{\displaystyle O}{\|}}{\underset{\underset{\displaystyle O^-}{|}}{P}}-O^- + HOH \rightarrow R-OH + H-O-\overset{\overset{\displaystyle O}{\|}}{\underset{\underset{\displaystyle O^-}{|}}{P}}-O^- \quad (5\text{-}32)$$

The active metal ion is usually Mg(II) or Mn(II) in these enzymes, too.

The enzymes collected in the third group of Table 5-6 vary widely, but all involve ATP or other nucleotides in their reactions.

I shall discuss the effects of metal cations on these enzymes in order to shed some light on the questions of why the metal ions are required and why there seems to be a strong element specificity. It is useful first to consider the *in vitro* hydrolysis of ATP catalyzed by metal ions and complexes; this reaction provides some clues as to how metal ions bind with ATP.

5.4.1 Hydrolysis of ATP by Metal Ions

ATP has the structure shown below. Possible coordination sites on the molecule are N-7 of the adenine nucleus and the α, β, and γ phosphate oxygens.

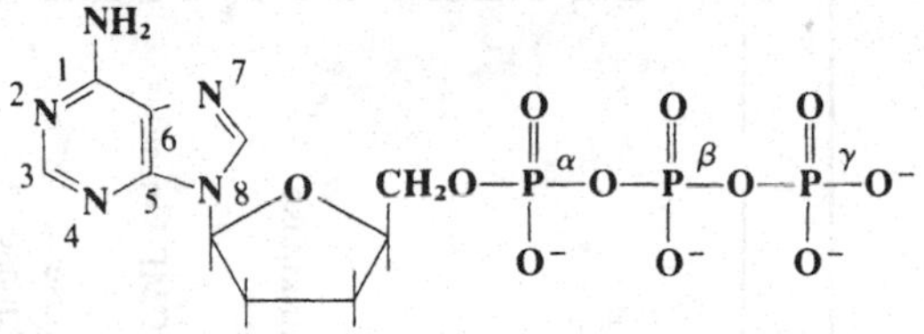

The binding mode depends on the metal ion. Mg(II) and Ca(II) bind only to pairs of phosphate oxygens: α and β or β and γ. For example,

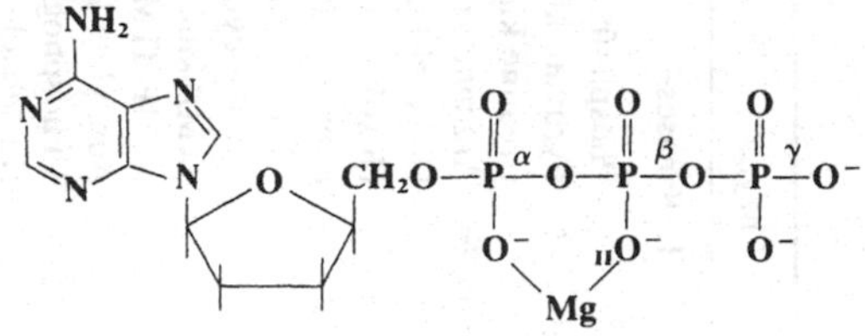

Table 5-6. Metal-Activated Enzymes[a]

Enzyme	Activated by	No effect (or inhibited by)[b]
I. Kinases		
Phosphofructokinase	Mg(II) > Mn(II),Co(II)	Zn(II)
Pyruvate kinase	Mg(II),Mn(II),Co(II) ≫ Ni(II),Cd(II)	Ba, Sr, Be, Ni, Cd, Zn
Creatine kinase	Mg(II),Mn(II),Ca(II),Co(II)	
Arginine kinase	Mg(II),Mn(II),Co(II),Ca(II),Sn(II)	
Glycerol kinase	Mg(II) ~ Mn(II) (Co(II)), Ni(II) in some cases)	
Hexokinase	Mg(II) ~ Mn(II) ≫ Ca(II)	
Phosphorylase kinase	Mg(II) (stimulated by Ca(II))	Ca(II)
AMP-dependent protein kinase	Mn(II) > Mg(II), Co(II) (in most cases)	
Uridine-cytidine kinase	Mg(II) > Mn(II),Fe(II),Co(II) ≫ Ca(II),Ni(II)	Ni, Zn, Cd, Ca
dTMP kinase	Mg(II) > Mn(II),Co(II)	
dAMP-dTMP-dGMP-dCMP kinase	Mg(II) ~ Mn(II) > Ca(II),Fe(II)	
Adenylate kinase	Mg(II) ≧ Ca(II) > Mn(II) > Ba(II)	
3-Phosphoglycerate kinase	Mg(II) ~ Mn(II)	Be, Fe(II), Zn
Nucleoside diphosphokinase	Mg(II) ~ Mn(II)	
II. Hydrolysis of phosphate ester or pyrophosphate		
ATPase		
Mitochondrial	Mg(II), H$^+$	
Chloroplast	Mg(II), H$^+$	Ni, Ca, Mn, Co, Sr
Sarcoplasmic reticulum	Ca(II) (or Sr(II)) and Mg(II) (or Mn(II))	

Enzyme	Activating cation[b]	Inhibitor
Nerve cell	Mg(II), K(I), and Na(I)	
5′-Nucleotidase		
Liver	Mg(II),Co(II) > Mn(II)	EDTA
Intestine	Mn(II),Mg(II),Co(II)	EDTA
Fish muscle	Mn(II) ~ Mg(II)	EDTA, Zn(II)
Ribonucleoside 2′,3′-cyclic phosphate diesterase	Co(II) or Mn(II) in some enzymes	Zn, Cu, Hg, EDTA
Nucleoside 3′,5′-cyclic phosphate diesterase	Mg(II) ~ Mn(II), Ca(II)	Zn(II)
Inorganic pyrophosphatase	Mg(II) > Mn(II) > Zn(II) > Co(II)	Ba, Ca, Cu, Fe, Ni, Fe(III)
Fructose-1,6-diphosphatase	Mn(II) > Mg(II)	
III. Other enzymes		
Asparagine synthetase	Mg(II) ~ Mn(II)	
ATP sulfurylase		
Bacterial	Mg(II)	
Mammalian	Mn(II)	
Penicillium chrysogenum	Zn(II)	
Methionine adenosyl transferase	Mn(II) and Mg(II), K(I), NH$_4^+$, Rb(I)	
B$_{12}$ adenosyl transferase	Mn(II) > Co(II),Mg(II),Zn(II); K(I) > NH$_4^+$	Ca, Cd
Glutamine synthetase	Mg(II)	
Staphylococcal nuclease	Ca(II) enzyme (can be replaced by Sr(II))	Zn, Hg, Cd
Deoxyribonuclease	Mg(II) and Ca(II) or Mn(II) alone	

[a] Compiled from P. D. Boyer (ed.), *The Enzymes*, 3rd ed., Vols. IV and VIII, Academic Press (1971, 1973).
[b] Unless otherwise indicated, divalent cation.

Mg(II) and Ca(II) also form [M₂ATP] complexes and Ca(II) seems to form a complex [M(ATP)₂] (Ochiai and Morand, 1985). Cu(II) binds strongly with N-7 of the adenine and at the same time with α and β phosphate oxygens, forming a tetragonal complex. Ni(II) seems to bind with N-7 as well as the α, β and γ phosphates; the overall structure of [Ni(ATP)(HOH)₂] would thus be close to an octahedron. The reason for the differences between the structures of Cu(II)-ATP and Ni(II)-ATP is the preferred coordination tendencies of the two metals. It was long thought that Mn(II) was also able to coordinate N-7 simultaneously with the α, β, and γ phosphates. A more recent study, however, has shown that one HOH molecule coordinated to Mn(II) is hydrogen-bonded to N-7 of adenine. It was further discovered that in a ternary complex, 8-hydroxyquinoline-Mn(II)-ATP, Mn(II) is not bound in any way to N-7, not even through a hydrogen bond (Hague *et al.*, 1972). That is, in a ternary complex, Mn(II) binds to only the phosphate groups.

    In detailed kinetic studies (Sigel and Amsler, 1976; Amsler and Sigel, 1976; Sigel, 1982), the metal ion activity for hydrolysis of ATP was found to decrease in the order Cu(II) > Zn(II) > Ni(II) > Mn(II). Judging from the postulated structure of the Cu(II)-ATP complex, it was suggested that coordination to N-7 of the base portion and to the α and β phosphates facilitates the hydrolysis of ATP. If the α and β phosphate portion chelates to a metal ion, that portion is stabilized by the chelate effect and the γ phosphate leaves more readily; that is:

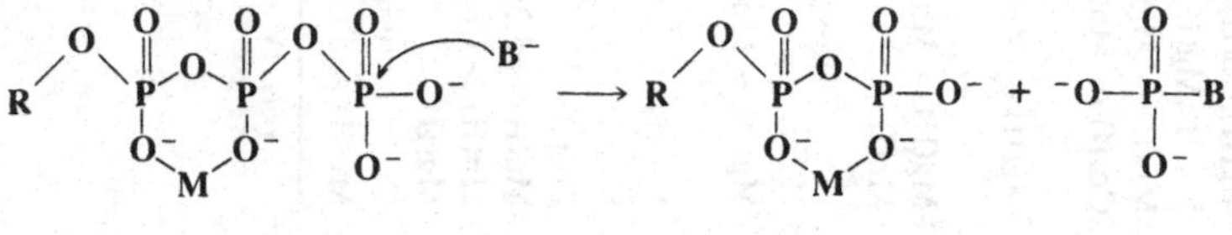

(5-33)

If, on the other hand, M(II) binds to the β and γ phosphate portion, the reaction would be:

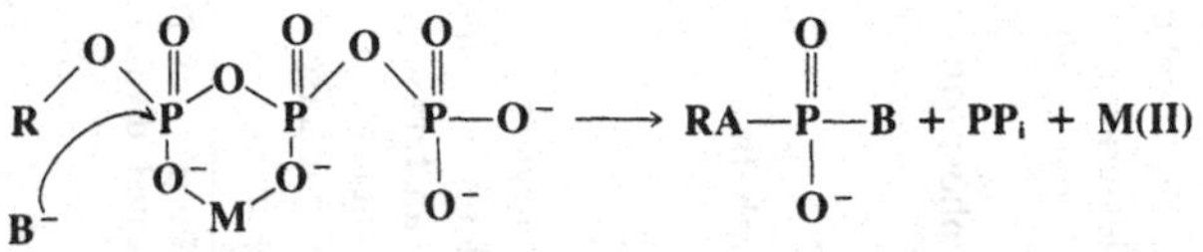

(5-34)

This type of reaction is observed in the enzymatic reaction of DNA polymerase (see Section 5.3.1.3). When all three phosphates are coordinated, the hydrolysis is not expected to be facilitated.

5.4.2 Some Pertinent Characteristics of Metal Cations

In the previous section, some differences among the coordination tendencies of various cations with ATP were mentioned. These appear to be important in understanding the differential activities of these ions in activating various enzymes. For this reason, some important basic characteristics of these elements are summarized below (cf. Section 2.2).

1. Ionic radius: Mg(II) (71 pm, coordination number = 4, tetrahedral) Cu(II) (71, 4, square-planar) < Zn(II) (74, 4, tetrahedral) < Co(II) (79, 4, tetrahedral) < Ni(II) (83, 6, octahedral) < Fe(II) (92, 6, octahedral) < Mn(II) (97, 6, octahedral) < Cd(II) (106, 6, octahedral) < Hg(II) (110, 4, tetrahedral; 116, 6, octahedral) < Ca(II) (114, 6, octahedral).
2. z_{eff}/r value (r value in units of nm; for the structure specified above): Co(II) (112) $\geqslant$ Cu(II) (111) $\geqslant$ Zn(II) (109) > Ni(II) (93) > Fe(II) (81) $\geqslant$ Mg(II) (78) $\geqslant$ Mn(II) (75) > Ca(II) (57)
3. Electronic configuration:
   a. d^0, d^5, d^{10}: Mg(II), Ca(II), Mn(II), Zn(II), Cd(II), Hg(II)
   b. d^n (n = 1–4, 6–9): Fe(II), Co(II), Ni(II), Cu(II)
4. Preferred coordination structures:
   a. Tetrahedral: Mg(II), Zn(II), Co(II), Hg(II)
   b. Tetrahedral or octahedral: Ca(II), Cd(II), Hg(II), Mn(II), Co(II)
   c. Octahedral: Fe(II), Ni(II)
   d. Tetragonal: Cu(II)
5. Irving–Williams series (this represents the general trend in the complex formation constants): Mn(II) < Fe(II) < Co(II) < Ni(II) < Cu(II) > Zn(II).
6. O-preferring: Mg(II), Mn(II), Ca(II), Fe(III). N-preferring: Fe(II), Co(II), Ni(II), Cu(II), Zn(II), Cd(II), Hg(II).

If the chelation to the phosphate portion of ATP is one of the important factors in ATP hydrolysis and in kinase reactions, then metal ions that prefer oxygen atoms would be most effective in the enzyme activation. Such metal ions would be Mg(II), Ca(II), and Mn(II). The polarizing effect (electron-withdrawing effect) is another relevant factor. If the value z_{eff}/r is a good measure of the polarizing effect, Co(II), Cu(II), and Zn(II) would be preferred over Fe(II), Mn(II), Cd(II), and Ca(II). If it is assumed that a coordination to the base portion is not desirable in these reactions, Ni(II), Cu(II), and Fe(II) may be inappropriate.

It should be noted that these conclusions would only apply rigorously to a situation where the metal cation exists in a more or less constraint-free state, whereas this is surely not an accurate description of a cation in an enzyme. Nevertheless, these considerations should be helpful in identifying the preferred cations for reactions involving ATP or other nucleotides as being Mg(II), Mn(II), and Co(II), and the others [Fe(II), Zn(II), Ni(II), and Cu(II)] would be poor choices.

5.4.3 Kinases

Having introduced the subject of ATP hydrolysis, I now return to a discussion of kinases.

Hexokinase consists of two subunits, which may be arranged in different ways in crystalline state. The BI form does not bind, whereas the BII form binds both substrates, sugar and mononucleotide (Blake, 1975). BII is a dimer that has the structure shown in Fig. 5-13 (Blake, 1975). Glucose and xylose bind in the deep cleft in the subunit. Glucose binds predominantly to one subunit (say, subunit A) and xylose to the other (subunit B). The presence of sugars causes extensive conformational changes (see Section 5.1) that seem to create the ATP-binding site between the subunits: in the absence of a sugar, ATP (or ADP) cannot bind to the enzyme. The detailed structure of the subunit was determined by X-ray crystallography at 300-pm resolution (Fletterick et al., 1973). However, the coordination mode of Mg(II) [or Mn(II)] has not been determined.

The major structural features shown in Fig. 5-13, i.e., the presence of a deep cleft and two globular domains in the subunit, seem to be common to phosphoglycerate kinase and pyruvate kinase. The NMR relaxation method has been employed to reveal the mode of binding of Mn(II) to pyruvate kinase and its substrates, ATP (ADP) and pyruvate (pyruvate phosphate) (Miller et al., 1968; Fung et al., 1973). The results were interpreted in terms of the mechanism shown in Fig. 5-14. This

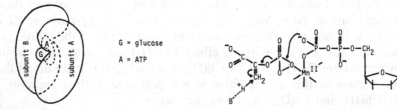

Figure 5-13. A structural model for hexokinase. (After Blake, 1975. *Essays Biochem.* 11:37.)

Figure 5-14. A proposed mechanism of action of pyruvate kinase. (After Reed and Scrutton, 1974. *J. Biol. Chem.* 249:6156.)

represents the reverse reaction. Mn(II) is here considered to assume an octahedral structure; two coordination sites are occupied by the substrates, two others by protein residues, and two water molecules complete the octahedron. Whether or not Mn(II) binds coordinatively to the protein residues cannot be determined directly by these methods, however. The protein structure itself, as shown in Fig. 5-14, seems to be capable of binding both substrates in a specific arrangement. In the forward reaction, a possible mechanism may be:

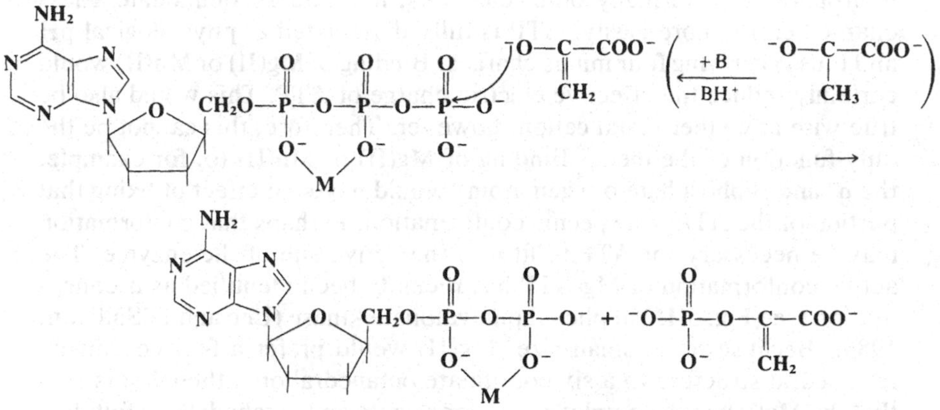

(5-35)

This is deduced from reaction 5-33. There is, however, a slight difference between the depicted forward and reverse mechanisms (i.e., that in Fig. 5-14 and reaction 5-35). In the reverse reaction (Fig. 5-14) both substrates are supposed to bind to Mn(II), whereas in the forward reaction (5-35) the cation binds only one of the substrates, ATP. The question arises as to whether or not the Mn(II)-binding mode could or should be different in the forward and the reverse reactions. It may be possible that the Mn(II)-binding mode is different in the two cases, but it seems more likely that the Mn(II)-binding mode is indeed the same for both the forward and reverse reactions. If so, mode 5-35 would appear to be the more likely one. In that event, M(II) cannot be considered to play a role in bringing the two substrates together, unlike the situation in some metalloenzyme systems discussed in the previous sections. It is, however, undeniable that M(II) could bring one of the substrates to the specific active site of the enzyme.

What, then, is the basis for the observed metal activation, other than the electron-withdrawing effect? Another conceivable role of metal ions in this type of enzymatic reaction is the neutralization of the negative

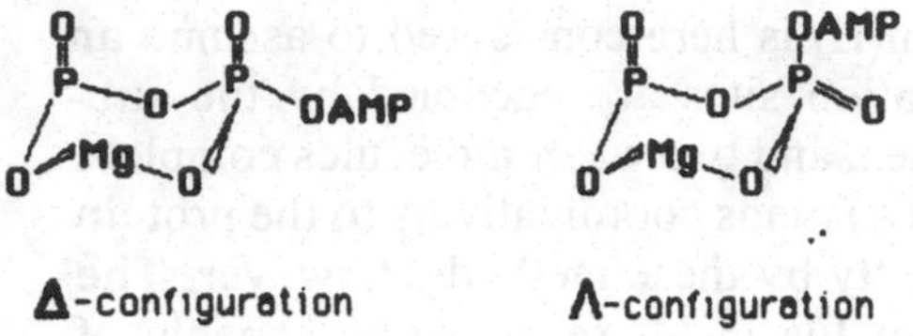

Δ-configuration   Λ-configuration

Figure 5-15. Configurations of Mg(II)-ATP complex.

charge on ATP. If the interior of the enzyme into which ATP enters is hydrophobic, as in many other enzymes, it would accommodate a less charged entity more easily. ATP is fully dissociated at physiological pH and thus is carrying four minus charges. Binding of Mg(II) or Mn(II) would certainly reduce the effective electric charge of ATP. This would also be true with any other metal cation, however. Therefore, this cannot be the only function of the metal. Binding of Mg(II) or Mn(II) to, for example, the α- and β-phosphate oxygen atoms would have the effect of fixing that portion of the ATP in a specific conformation. Perhaps that conformation may be necessary for ATP to fit into the active site of the enzyme. The active conformation of Mg-ATP has recently been identified as Δ configuration (see Fig. 5-15) in phosphomevalonate kinase (Lee and O'Sullivan, 1985). Because of its small size, Mg(II) would prefer a four-coordinate tetrahedral structure to a six-coordinate octahedral one, though it is said that the Mg(II)-aquo complex is six-coordinate and octahedral. In fulfilling this function, the size of the ion may not be particularly critical, if it is within a certain range and the ion can take a tetrahedral structure. Mn(II), being d^5, can take either an octahedral or a tetrahedral structure. As discussed in the previous sections, Co(II) is probably appropriate, because it prefers a tetrahedral structure. Indeed, in most kinase systems (*in vitro*) Mg(II) can be replaced by Mn(II) and in some cases, by Co(II) as well. Ni(II) and Fe(II) prefer an octahedral structure, and Cu(II) a tetragonal one. This may be one of the reasons why these cations are not usually effective in activating kinases. However, in certain enzyme systems this kind of restriction on the coordination structure of the divalent cation-ATP (or other mononucleotide) complex could be less severe. In such a case, other cations such as Fe(II) and Ni(II) could be utilized, even if much less effectively. Usually Ca(II) and Ba(II) are not effective as kinase activators. This is very likely to be due to their larger size.

Zn(II) is another cation that appears to have an appropriate size and a tendency to take on a tetrahedral structure. Zn(II), however, is often an inhibitor to kinases rather than an activator. This could be due to the fact that many, if not all, kinases contain a sulfhydryl group(s) in their active site and the fact that Zn(II) preferentially binds to SH groups.

5.4.4 Hydrolysis of Phosphate Ester Bonds

ATPase is a typical example of this class, but is one of the most complicated enzymes (see Chapter 10). The enzymes of this class are similar to kinases in that they involve ATP, mononucleotide, or other phosphate as one of the substrates. However, the second substrate is usually water, whereas it is a molecule much larger than water in the case of a kinase. This difference would be reflected in a difference in the active site structures of the enzymes of these two categories.

In the forward reaction of hydrolysis of a phosphate bond, the reaction scheme may be similar to that of the kinase reactions; that is, nucleophilic attack on the phosphorus atom of the phosphate group that is to leave:

$$X-O-\overset{\overset{O}{\|}}{\underset{\underset{O^-}{|}}{P}}-O^- \xrightarrow{\ OH^-\ } X-O^- + {}^-O-\overset{\overset{O}{\|}}{\underset{\underset{O^-}{|}}{P}}-OH \qquad (5\text{-}36)$$

Enzymes in this category may be subdivided into two groups: one group is involved in the hydrolysis of a pyrophosphate bond and the other in the hydrolysis of a phosphate ester bond. That is to say, the enzymes of the former group split a P—O—P bond, whereas those of the latter group cleave a C—O—P bond.

A divalent cation, Mg(II) or otherwise, would chelate preferentially with a pyrophosphate portion if one were available. Such chelation is, of course, possible in the case of ATPase (and in the hydrolysis of other triphosphate or diphosphate nucleotide) and also with inorganic pyrophosphatase. The function of the activating divalent cation in these enzymes may be regarded to be entirely analogous to that in the kinases discussed above. Such chelation is obviously not possible in the hydrolysis of a monophosphate ester or a cyclic ester. A divalent cation may chelate to these substrates in the following manner:

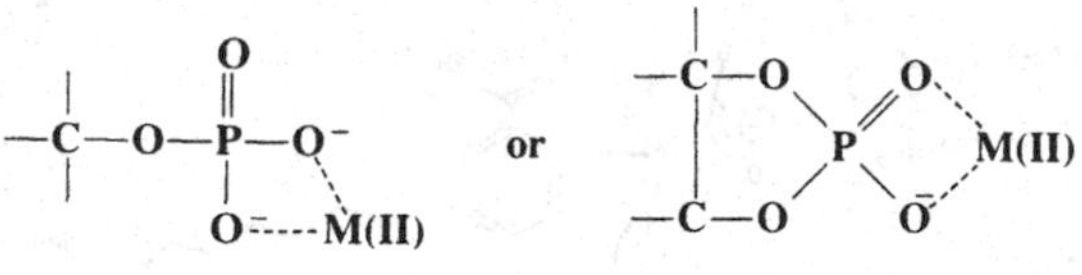

$$-\overset{|}{\underset{|}{C}}-O-\overset{\overset{O}{\|}}{\underset{\underset{O \cdots M(II)}{|}}{P}}-O^- \qquad \text{or} \qquad \begin{array}{c} -\overset{|}{C}-O \\ \overset{|}{C} \\ -\overset{|}{\underset{|}{C}}-O \end{array} \diagdown P \diagup\diagdown\begin{array}{c} O \cdots \\ \diagdown M(II) \\ O \end{array}$$

$$(5\text{-}37)$$

The binding of M(II) in this mode would appear to be rather weak and to be strained. Such a divalent cation would activate the substrate, however, in the sense that it withdraws electrons from the phosphate group, thereby rendering the phosphorus more susceptible to a nucleophilic attack. It will, of course, also function in neutralizing the electric charge of the substrate.

5.5 Other Metalloenzymes

Some oxidation–reduction enzymes contain metal atoms that function not as electron-transferring centers, but rather as Lewis acids. This is especially common with dehydrogenases dependent on NAD(P)$^+$; they usually contain either Zn(II) or Mn(II). Some other dehydrogenases are dependent on flavin and contain iron and/or molybdenum; the metals in these enzymes are functioning as electron carriers, however.

Of the very similar NAD(P)$^+$-dependent dehydrogenases, alcohol dehydrogenase is a Zn(II) enzyme, while lactate dehydrogenase, malate dehydrogenase, and D-glyceraldehyde-3-phosphate dehydrogenase are metal-free.

The mechanism of action of lactate dehydrogenase from dogfish has been established by use of X-ray crystallography (Adams *et al.*, 1973). The essential portion of the structure of the lactate–NAD$^+$–enzyme complex is schematically depicted in Fig. 5-16. Certain amino acid residues in the protein are arranged in such a way that the substrate (lactate) and the coenzyme's nicotine portion are brought into close proximity and in a specific arrangement. Arg-171, a positively charged residue, fixes the negatively charged substrate. The nicotine residue abstracts a hydride from the substrate, and a residue on the protein, Arg-109 in this case, would accept a proton from the substrate.

The structure of horse liver alcohol dehydrogenase was established by X-ray crystallography (Eklund *et al.*, 1974; Branden *et al.*, 1975; see

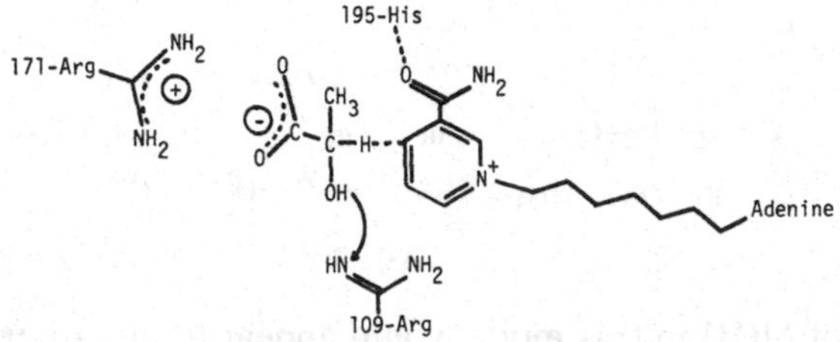

Figure 5-16. A schematic representation of the active site structure of the substrate–lactate dehydrogenase–NAD$^+$. (After Adams *et al.*, 1973. *Proc. Natl. Acad. Sci. USA* 70:1968.)

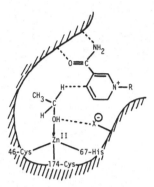

Figure 5-17. A schematic representation of the active site of the substrate–alcohol dehydrogenase–NAD$^+$.

also Zeppezauer, 1982). In the absence of the substrate (alcohol), the four approximately tetrahedral coordination sites of the Zn(II) ion are occupied by two cysteine sulfur atoms, one histidine nitrogen atom, and one water molecule. By the way, there is another Zn(II) ion in the protein, which appears to serve a structural role. It is very likely that the substrate displaces the water molecule coordinated to the catalytic Zn(II) ion (Branden *et al.,* 1975). Thus, in this case Zn(II) brings the substrate into the correct position relative to NAD$^+$ (see Fig. 5-17). That is, the Zn(II) ion is playing the role carried out by an arginine residue (Arg-171) in lactate dehydrogenase. Zn(II), by exerting its electron-withdrawing effect on the substrate, may also be facilitating the hydride and proton abstraction. A more recent study with Co(II)-containing liver alcohol dehydrogenase indicated a pentacoordinate structure around Co(II) in an NAD$^+$—enzyme—inhibitor complex (Makinen and Yim, 1981).

A legitimate question can then be asked: why does alcohol dehydrogenase require Zn(II) while other similar enzymes can do without a metal cation? The answer seems to be found in the difference in the natures of the substrates. Since the substrates for lactate dehydrogenase, malate dehydrogenase, and D-glyceraldehyde-3-phosphate dehydrogenase are negatively charged, a positively charged amino acid residue such as Arg or Lys may be sufficient to attract these substrates. Thus, these enzymes would be able to function without a metal cation. On the other hand, a neutral substrate such as alcohol would be difficult to attract to the active site. Therefore, it would be more convenient for an enzyme to have a coordinating metal cation to attract such a substrate.

If a cation acts as a coordination site for a substrate in dehydrogenases, it would not be surprising to find an Fe(II)-dependent dehydrogenase; such an enzyme has recently been isolated from *Zymmonas mobilis* (Scopes, 1983).

Structural Functions

Metal ions and compounds serve biological structural functions at both microscopic and macroscopic levels. The microscopic structural function is served when a metal ion plays a role in maintaining the tertiary or quaternary structure of a protein, polynucleotide, or other biological molecule. Metal compounds also constitute the important portions of skeletal or shell structures of organisms; this is their macroscopic structural function. Calcium and silicon play roles in maintaining the integrity of tissues; this task can be regarded to be intermediate between the macroscopic and the microscopic structural functions.

6.1 Effects on Molecular Structures

6.1.1 Effects on Protein Structures

Protein structure is conveniently characterized at four different levels. The primary structure is the amino acid sequence. The secondary structure is the conformation of the polypeptide chain, whether α-helix or β-sheet conformation. It is more or less dictated by the primary structure (Ramachandran rule; Ramachandran and Sasisekharan, 1968). The tertiary structure is the spatial configuration of a whole polypeptide chain. The quaternary structure is the way in which different polypeptide chains of a protein interact.

The tertiary structure is the one on which metal ions exert strong influence in certain proteins, although effects on the quaternary structures are not nonexistent. The spatial structure of a polypeptide chain is maintained by interactions of many different kinds among the chain's amino acid residues. These include electrostatic interactions between electrically charged residues (repulsive or attractive), hydrophobic interactions (at-

197

tractive), steric interactions (repulsive), and covalent bonds such as the disulfide bridge between two cysteine residues. A protein is considered to assume a minimum-energy conformation as a net result of these various interactions. Such an energy minimum is, however, considered to be relatively shallow in general, because of the great number of possible interactions. In other words, another energy minimum of a little higher energy may often be accessible through a slight conformational change. In general, there is only one optimal conformation for a protein. Any change in the tertiary structure may lead to a modification (usually deleterious) of the activity of the protein.

A metal ion could contribute to the maintenance of the tertiary structure through its coordination to specific amino acid residues. The available choice of metal ions for this kind of purpose would seem to be wide open, but it is restricted somewhat by the fact that nature would have shown a preference for more abundant metal ions (rule of abundance). In fact, Ca(II), Zn(II), Fe(III), and Mn(II) most often act as structural cations in various proteins and enzymes.

Ca(II) ion binds preferentially with the carboxyl group of aspartate and glutamate residues. This is due to the affinity of Ca(II) for carboxylate oxygen (see Chapters 8 and 9). The coordination structure of Ca(II) is usually octahedral in proteins. α-Amylase is an example of enzymes containing tightly bound Ca(II). Amylase from human saliva or hog pancreas contains 1 or 2 moles of Ca(II) per mole of the enzyme and has, in addition, disulfide bridges; amylase from *Bacillus subtilis*, however, has 3 or 4 moles of Ca(II) per mole and no disulfide bridge. Thus, in the latter enzyme, the tertiary structure is maintained partially by the bridging effect of the Ca(II) ions instead of by disulfide bridges (Sanders and Rutter, 1972). A thermally stable proteinase isolated from *B. thermolyticus* contains 1 mole of Zn(II) as the catalytic ion and 3 moles of Ca(II) per mole. Ca(II) ions were found to be bound to aspartate and glutamate residues (Mathews *et al.*, 1972). The specific structure of concanavalin A is maintained by Mn(II) and Ca(II). The ligands of the Mn(II) ion are a histidine nitrogen, three carboxylate oxygens, and two water molecules; they are arranged approximately in an octahedron. The ligands of the Ca(II) ion are four oxygen atoms from carboxylate and carbonyl groups and one water molecule (Hardman and Ainsworth, 1972).

Zn(II) ion is the most widely utilized metal ion in enzymes and proteins. It functions as a catalytic cation in enzymes of various types (see Chapter 5). It is likely that Zn(II) in these enzymes is also contributing to the maintenance of the entire structure of the proteins as well as to the catalytic activity. The preferred coordination structure about Zn(II) is tetrahedral or distorted tetrahedral, and Zn(II) binds preferentially with nitrogen, especially that of histidine imidazole, and with the sulfide of cysteine residue. There are other proteins and enzymes in which Zn(II) seems to play solely a structural role without being involved in the catalytic active site. Two of the four Zn(II) ions in alkaline phosphatase are believed to have only a structural effect. Insulin and the C-peptide fragment from proinsulin are packaged in Golgi vesicles, and Zn(II) ion is considered to maintain the orderly structure of the packaged insulin and C-peptide (Steiner *et al.*, 1974). Zn(II) in Cu(II)-dependent superoxide dismutase plays a structural role, although this Zn(II) may also play a catalytic role to a certain degree, as discussed in Section 4.2.

The effect of metal ions on the quaternary structure of a protein is exemplified by glutamate synthetase. The enzyme from *E. coli* (molecular weight 592,000) is composed of 12 identical subunits (molecular weight 49,000). The activity of this enzyme, which is allosteric, is inhibited by a number of substances, particularly the end products of glutamine metabolism; it is also dependent on the presence of cations (Stadtman, 1971). In the absence of any cation the enzyme takes the "relaxed" form, which is inactive. The addition of divalent cations such as Mg(II) and/or Mn(II) converts the "relaxed" form to the catalytically active "tightened" form. In this conversion process, the exposed SH groups become buried in the interior of the enzyme. The tryptophan and tyrosine residues that are exposed in the relaxed form become buried as well, the hydrophobic groups become exposed. This whole effect can be considered to be a combined result of the conformational change of each subunit and of changes in the quaternary structure of the protein. The more detailed structure of glutamate synthetase has recently been published (Almassy *et al.*, 1986). Two Mn(II) ions are located close by (580 pm from each other) and near the interface of subunits, which seems to be the catalytically active site. Other examples of allosteric effects of metal ions include Zn(II) effects on ornithine transcarbamoylase (Kuo *et al.*, 1982) and those on aspartate carbamoyltransferase (Hozantko *et al.*, 1981).

Another interesting example is the effect of Ca(II) on the quaternary structure of troponin, a component of muscle thin filament (Cohen, 1975); see Section 7.6 for more detail.

6.1.2 Effects on Polynucleotides

The phosphate backbone of polynucleotides is negatively charged at neutral pH. These negative charges tend to destabilize the double helix (or any other similar structure) of polynucleotides.

It has long been known that nucleic acid preparations from various sources usually contain metal ions of several kinds, including magnesium, calcium, barium, aluminum, chromium, manganese, iron, nickel, and zinc (Wacker and Vallee, 1959; Bryan, 1981). Metal ions in RNA are difficult to remove with chelating agents or by dialysis; they are considered to be playing a major role in stabilizing the conformation of RNA and maintaining the integrity of ribosomal RNA.

The effects of various metal ions on the double-helix-to-random-coil transition temperature of a pure DNA preparation were studied by Eichhorn and his co-workers (Eichhorn *et al.*, 1971; Eichhorn, 1981). Mg(II) raised the transition temperature, i.e., Mg(II) stabilizes the double-helix structure. This is attributed to the neutralization of the negative charges of the phosphate groups by the positive Mg(II) ion. Zn(II) also stabilizes the double-helix structure; it does so not only by neutralizing the negative charges but also by bridging the base pairs on the strands, thus increasing the interstrand cohesive force. Interstrand cohesive forces are mostly due to hydrogen bonds between the corresponding base pairs on the polynucleotides (adenine–thymine and guanine–cytosine in the case of DNA). If these hydrogen bonds were somehow disrupted, the double-helix structure would be destabilized. Cu(II) lowers the double-helix-to-random-coil transition temperature of DNA. This is interpreted as indicating that Cu(II) coordinates to the bases and thus disrupts the interbase hydrogen bonds.

In chromosomes, DNA molecules are present in supercoiled fashion. Negative charges on the backbone would be expected to destabilize such a structure. The apparent negative charges seem to be neutralized in part by metal ions, as described above, but also by basic proteins called histones. At neutral pH, histones are positively charged and intimately associated with the DNA in the chromosome.

RNAs in the group called "transfer" RNA are rather small; these assume the so-called "cloverleaf" structure. X-ray studies (Kim *et al.*, 1973; Rich and Raj-Bhandary, 1976; Teeter *et al.*, 1981) showed that a tRNA (for phenylalanine from yeast) is fairly compact, the cloverleaves being folded in a complicated manner as shown in Fig. 6-1 (Quigley *et al.*, 1978). The main binding interactions holding this structure in place are again the hydrogen bonds between the bases. Here hydrogen bonds not only form the basis for the bridged helical structure but also help to bring together some of the branches in the cloverleaf array, forming the appropriate folds. A more detailed X-ray crystallographic study (Quigley *et al.*, 1978; Teeter *et al.*, 1981) revealed that there were four Mg(II) atoms bound to the single tRNA molecule. The magnesium atoms are shown by black disks in Fig. 6-1. In the anticodon loop, in which three of the bases specify an amino acid, one Mg(II) seems to be contributing to maintaining this particular conformation of the loop, for there is no base–base pairing

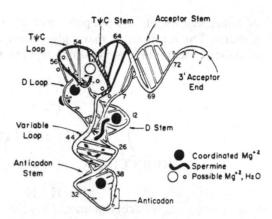

Figure 6-1. Structure of phenyl-
alanine RNA. (Reproduced from
Quigley *et al.*, 1978. *Proc. Natl.
Acad. Sci. USA* 75:64.)

in between nucleotides 32 and 38. The other magnesium atoms are also
located at points where extra help is required to maintain a specific, rather
strained conformation. The Mg(II) in the anticodon loop directly binds to
the phosphate oxygen of nucleotide 37. In addition, it is indirectly linked
to the other nucleotides by hydrogen bonds through coordinated water
molecules. One Mg(II) binds to the phosphate oxygens of nucleotides 20
and 21, and the other to the phosphate oxygen of nucleotide 19. The
fourth Mg(II) holds the rather sharp bend in the so-called D stem by
binding to the phosphate oxygens of nucleotides 8, 9, 11, and 12. In an
analogous manner to the histones in chromosomal DNA, a linear polyam-
ine molecule, spermine, was found to be bound to the groove just above
the anticodon loop (Quigley *et al.*, 1978; Teeter *et al.*, 1981). This is also
contributing to the stabilization of this particular conformation.

6.2 Effects on Intermolecular Interactions

6.2.1 Interactions with Polysaccharides and Their Derivatives

Polysaccharides do not seem to have the specific conformation of
structures (in the sense of the secondary and tertiary structures of protein).
The interactions of Ca(II) ions with polysaccharides or their derivatives,
however, play very important roles in maintaining intercellular bindings.
Bacterial cells have a network of polysaccharide fibers called the glycocalyx (Costerton *et al.*, 1978) outside the cell wall. The cells use the
glycocalyx to bind to other objects such as animal cells. Animal cells are
also known to have a glycocalyx coat. The glycocalices are extensions

of either cell-membrane lipopolysaccharides, glycoproteins, or peptido-glycans. The terminus of many glycoproteins of this type is a molecule of N-acetylneuraminic acid, which has the following formula:

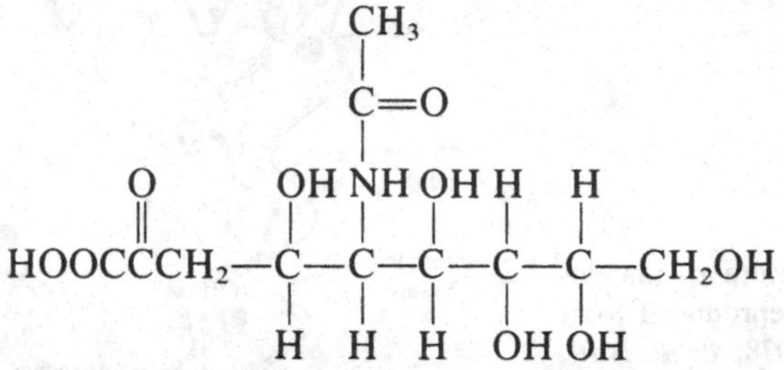

N-acetylneuraminic acid (sialic acid)

At neutral pH the carboxyl group is ionized; it is at this carboxylate site that Ca(II) binds. Ca(II) binds two terminal sialic acids of oligosaccharides from different cells and thus contributes to the intercellular binding. The fact that Ca(II) binds preferentially with carboxylate oxygen was previously encountered in the discussion of proteins (Section 6.1.1).

A component of the cell walls in plants is hemicellulose. Pectic substances, the major components of hemicellulose, contain polymers of galacturonic acid. The carboxylate groups of galacturonic acid units also bind with Ca(II); in this case Ca(II) bridges two molecules of pectic substances and contributes to the adhesion of adjacent plant cells.

A number of acidic mucopolysaccharides are found in connective tissues of animals and in the vitreous fluid of the eye. One example is hyaluronic acid whose structure is shown below. Hyaluronic acid forms a very viscous solution and acts as a lubricant in joints.

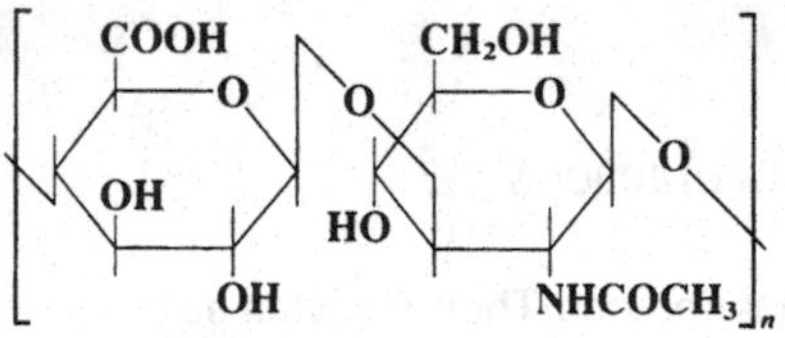

hyaluronic acid

It is also thought to hold cells together in its jellylike matrix. It is likely that here, too, Ca(II) plays a role in linking otherwise independent molecular strands. The interaction of Ca(II) ion with these acidic polysaccharides is electrostatic.

It has been found that many acidic mucopolysaccharides contain high levels of tightly bound silicon (Schwartz, 1973). Examples are 330–360 ppm Si reported in hyaluronic acid of human umbilical cord, 190 ppm in chondroitin-6-sulfate of human cartilage, and 430 ppm in heparan sulfate of cow lung. Silicon was also found in polyuronides of plants; for example, as much as 2580 ppm Si was reported to be present in the pectin of citrus fruit and 450 ppm in alginic acid from horsetail kelp. The tightly bound silicon in these materials was found to be covalently bonded to carbon through an oxygen atom. Thus, the complex system may be comprised of carbohydrate (alcohol) esters of silicic acid:

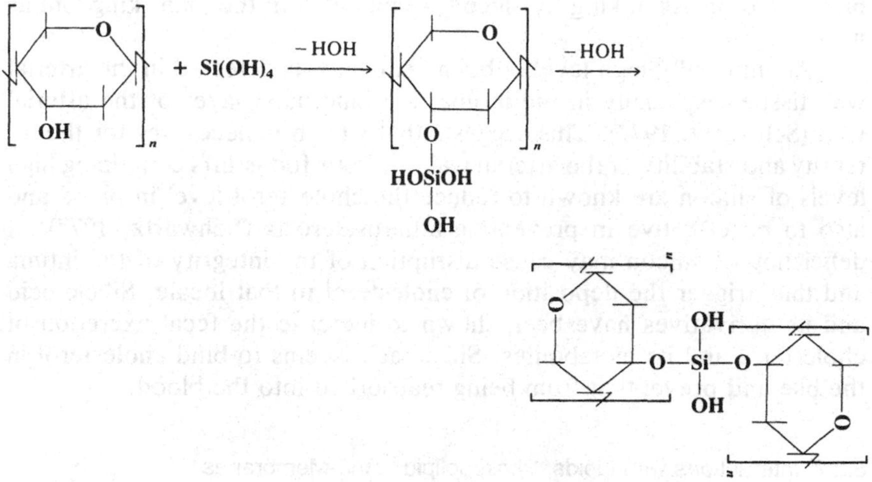
O
OH
n
+ Si(OH)4
−HOH
O
O
HOSiOH
OH
n
−HOH
n
OH
O—Si—O
OH
O
O
n

(6-1)

Alternatively, the linkages may be formed through carboxyl groups:

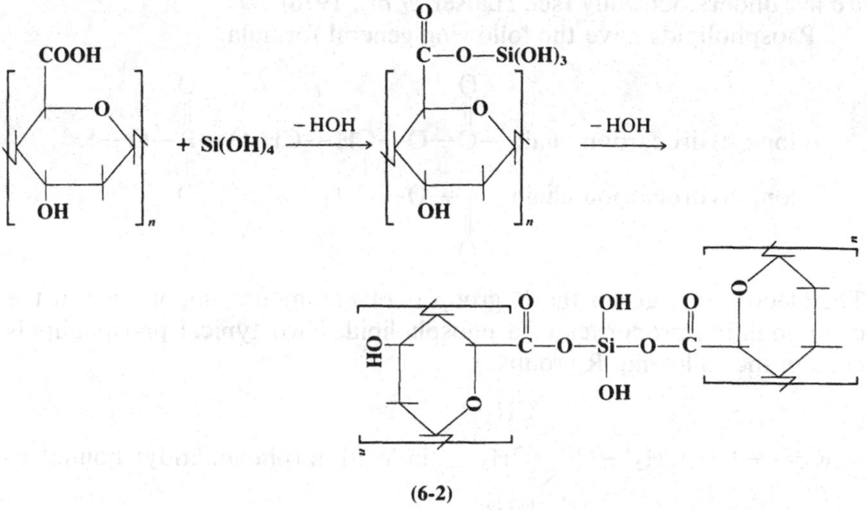
COOH
O
OH
n
+ Si(OH)4
−HOH
O
C—O—Si(OH)3
O
OH
n
−HOH
n
O
OH
O
HO
C—O—Si—O—C
OH
O
n

(6-2)

Whichever structure is correct, these equations illustrate the possibility that silicon can act as a bridge between two or more polysaccharide chains and thus contribute to the maintenance of the structure and integrity of connective tissues. The proteins in connective tissues, especially collagen, have also been found to contain bound silicon. Thus, the use of silicic acid as a cross-linking agent seems widespread. The fact that pectin and alginate contain high levels of tightly bound silicon suggests that the same principle of cross-linking by silicon is employed in the plant kingdom as well.

An unusually high level of bound silicon was detected in the arterial wall tissue, especially in the intima, the innermost layer of the arterial wall (Schwartz, 1977). This suggests that silicon is necessary for the integrity and stability of the arterial wall. Fibrous foodstuffs containing high levels of silicon are known to reduce the cholesterol level in blood and also to be effective in preventing atherosclerosis (Schwartz, 1977). A deficiency of silicon may cause disruption of the integrity of the intima and thus trigger the deposition of cholesterol in that locale. Silicic acid and its derivatives have been shown to increase the fecal excretion of cholesterol and its metabolites. Silicic acid seems to bind cholesterol in the bile and prevents it from being reabsorbed into the blood.

6.2.2 Interactions with Lipids, Phospholipids, and Membranes

The interactions of cations (and/or anions) with phospholipids and membranes are very complicated but are of great importance. Yet they are not understood fully (see Hauser $et\ al.$, 1976).

Phospholipids have the following general formula:

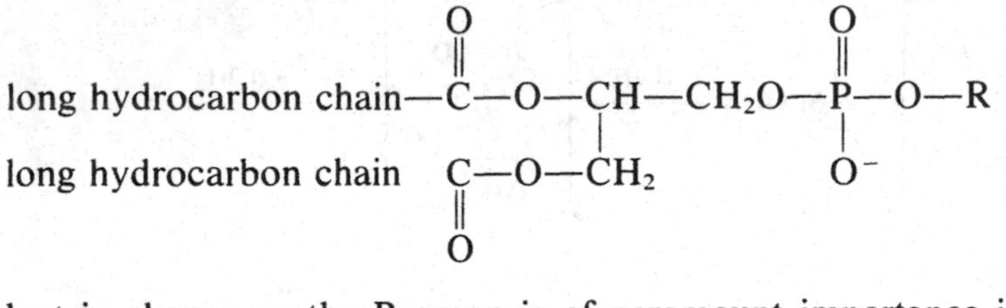

$$\text{long hydrocarbon chain}-\overset{\overset{\text{O}}{\|}}{\text{C}}-\text{O}-\underset{|}{\text{CH}}-\text{CH}_2\text{O}-\underset{|}{\overset{\overset{\text{O}}{\|}}{\text{P}}}-\text{O}-\text{R}$$

$$\text{long hydrocarbon chain}\quad\underset{\overset{\|}{\text{O}}}{\text{C}}-\text{O}-\text{CH}_2\qquad\qquad\text{O}^-$$

The electric charge on the R group is of paramount importance in the cation-binding properties of a phospholipid. Two typical phospholipids contain the following R groups:

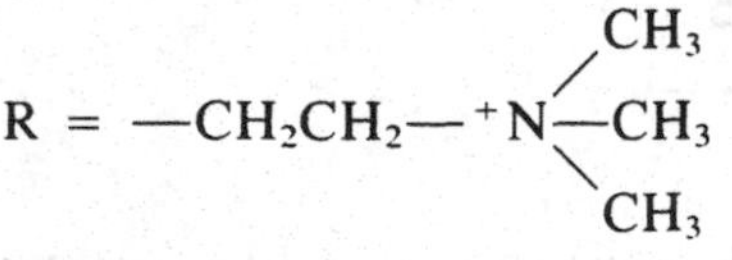

$$R = -CH_2CH_2-\overset{+}{N}\!\!\left\langle\begin{array}{l}CH_3\\CH_3\\CH_3\end{array}\right. \qquad \text{in lecithin (phosphatidylcholine)}$$

$$R = -CH_2CH(NH_2)COO^- \qquad \text{in phosphatidylserine}$$

The positive charge on the choline group in lecithin effectively counterbalances the negative charge of the phosphate group; thus, the hydrophilic head of lecithin is electrically neutral. Phosphatidylserine has two net units of negative charges and strongly binds with cations, particularly with Ca(II) and Ln(III) (Ln = lanthanide element). For example, Ca(II) ion added on both sides of the bilayer of phosphatidylserine stabilizes this artificial membrane, as manifested by an increased electrical resistance and a decreased water permeability (Triggle, 1972). Asymmetric addition of Ca(II) to the same bilayer, however, causes a reduction of the resistance, and the membrane becomes so unstable as to break when the difference in Ca(II) concentration between the two sides of the membrane is as little as 1 mM (Hauser and Phillips, 1973).

Another indication of the importance of cations on the membrane structure is the effect they have on the phase transition temperature of the lipid bilayers. Cations raise the transition temperature, their effectiveness decreasing in the order Ln(III) > Ca(II) > Na(I). The effect is quite significant in the case of a phosphatidylserine bilayer but only slight with a neutral lecithin bilayer (Hauser *et al.*, 1976).

The overall role of ions in membranes is more complex, however. For example, the positively charged choline group on the head of lecithin has been shown to be capable of binding anions. Such anion bindings are considered to affect in turn the cation-binding ability of the negative portion of the lecithin head. Moreover, various kinds of phospholipids are involved in a biological membrane. In addition, lipoproteins, glycoproteins, and other substances (including cholesterol) are present in the biological membranes. It is thus expected that the interactions of biomembranes with cations, especially the physiologically important Ca(II) ion, are very complicated. They are, however, of utmost importance in a great number of physiological phenomena (see Sections 7.6 and 9.4), including electrical conductance in nerve and muscle cell membranes, exocytosis and endocytosis.

6.3 Inorganic Substances in Macrostructures (Biomineralization)

Protective devices are advantageous to organisms, and mechanical rigidity is desirable for large organisms. Most organisms, therefore, have developed a great number of protective devices: chemical, physiological, and mechanical. The chemical and physiological protective devices will be discussed in Part V. Mechanical protection takes varied forms and includes such structures as cell wall, cuticle, shell, test, and frustule.

These structures are designed to cover the bodies of organisms and must be mechanically strong. Organic as well as inorganic substances are utilized for this purpose. Examples of the organic compounds include cellulose as the cell wall in algae and plants, and chitin [poly(N-acetyl)-D-glucosamine) as the shells or exoskeletons of insects and crustaceans, and as the cell wall of various fungi. Bacterial cell walls are mainly composed of peptidoglycans.

Inorganic compounds are often utilized in coverings and skeletons because of their inherent mechanical strength. The compounds to be selected must be relatively insoluble in water and must also be readily available. As a result of these requirements, certain calcium compounds and silica are the materials of choice for these structural purposes. In addition, a strontium compound is utilized in a special group of protozoa, the radiolarians. Examples are summarized in Table 6-1.

6.3.1 Silica as the Frustule of Diatoms

The cell walls of diatoms, a group of algae (mostly marine), consist of siliceous frustules encased in an organic coating. A major portion of the organic coatings is made up of proteins. The silica of diatoms is similar in chemical composition to a silica gel, $SiO_2 \cdot nH_2O$. The amino acid composition of the organic coating has been determined for several species (Hecky et al., 1973). The protein in the cell wall contains more serine,

Table 6-1. Shells and Skeletons Made of Inorganic Compounds

Compound	Plants	Invertebrates	Vertebrates
Silica	Diatoms (frustule), horsetail, grasses	Sponges, radiolarian	Connective tissue
Calcium carbonate	Cell inclusion, coccolithophores	Molluscs (shell), foraminifera	Egg shell (birds and reptiles)
Calcium oxalate	Some algae, mosses, and ferns	Occasionally found in insect eggs, larval cuticle	Soft tissue in association with bones
Calcium phosphate	Some bacteria	Some protozoa, coelenterate, arthropod, and brachiopod	Bone, dentin, and enamel
Strontium sulfate		Radiolarian	
Barium sulfate		Some coelentrate	
Magnesium carbonate		Coral reef	

threonine, and glycine, and less glutamic acid, aspartic acid, aromatic amino acids, and sulfur-containing amino acids than is found in the protein within the cell. Some selective data are given for *Cyclotella stelligera* in Table 6-2. Serine and threonine both have an aliphatic hydroxyl group. This finding led the authors to a hypothesis of a mechanism for silica incorporation, which includes a step where a silicic acid molecule undergoes a condensation reaction with two hydroxyl groups on the adjacent serine residues:

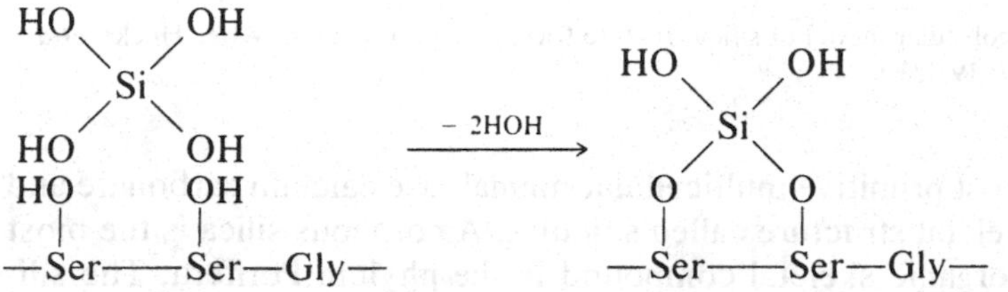

$$
\begin{array}{c}
\text{HO} \diagdown \quad \diagup \text{OH} \\
\text{Si} \\
\text{HO} \diagup \quad \diagdown \text{OH} \\
\text{HO} \quad\quad \text{OH} \\
| \quad\quad\quad | \\
\text{—Ser——Ser—Gly—}
\end{array}
\xrightarrow{\;-\,2\text{HOH}\;}
\begin{array}{c}
\text{HO} \diagdown \quad \diagup \text{OH} \\
\text{Si} \\
\text{O} \diagup \quad \diagdown \text{O} \\
| \quad\quad\quad | \\
\text{—Ser——Ser—Gly—}
\end{array}
\qquad (6\text{-}3)
$$

The initial layer of the bonded silicic acid may facilitate the further condensation or polymerization of silicic acid (Fig. 6-2). Perhaps this mechanism is an oversimplification, but it may nonetheless represent the fundamental process of the siliceous layer formation in diatoms (see Volcani, 1981, for a more recent review).

6.3.2 Silica in Other Organisms

Silica or silica gel is used as a skeletal material also in a large number of organisms (for a review see Simpson and Volcani, 1981). For example, an enormous number of a group of planktonic protozoa called radiolarians use opaline silica for their skeletons; the variation of their shapes is extraordinary (Riedel and Sanfilippo, 1981). A minor number of radiolarians use strontium sulfate instead of silica at the isospore stage; an example is *Sphaerozoum neapolitanum* whose isospores contain strontium sulfate, but its mature form bears a silica skeleton (see Anderson, 1981). Sponges,

Table 6-2. The Amino Acid Composition[a] of the Proteins in the Frustule of *Cyclotella stelligera*

	Asp	Glu	Cys	Met	Tyr	Phe	Gly	Thr	Ser
Cell wall	103	59	1	1	6	12	195	70	252
Cell content	119	109	2	20	29	46	110	54	98

[a] The number of amino acid residues per 1000.

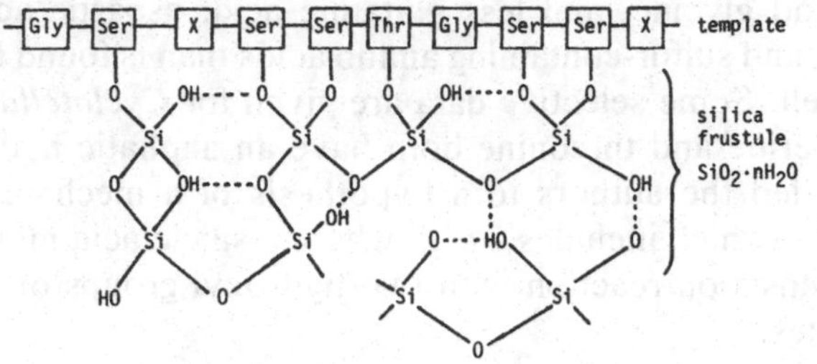

Figure 6-2. A molecular model of silica frustule formation of a diatom. (After Hecky *et al.*, 1973. *Mar. Biol.* 19:323.)

one of the most primitive multicellular animal, use calcium carbonate and silica in a skeletal structure called spicules. Amorphous silica is the most abundant inorganic skeletal compound in the phylum Porifera. The siliceous spicules are often bound together by a secondary deposit of silica or by a collagenous cement (Hartman, 1981; Garrone *et al.*, 1981).

Silica is often deposited as amorphous silica gel in the shoots of vascular plants, especially horsetails and grasses. Silicon is considered to be essential for the normal growth of these plants. For example, *Equisetum arvense* (a horsetail) has been shown to collapse when grown on silicon-free nutrient solutions but to be normally erect when sodium metasilicate is included in the growth medium (Kaufman *et al.*, 1981).

6.3.3 Formation of Calcium Carbonate Shell in Molluscs

Molluscan shells are generally composed of calcium carbonate (in either the calcite or the aragonite crystalline form) enclosed in an organic matrix. In order to gain clues as to the mechanism of the crystal formation, the composition of the organic matrix was studied. Since the organic matrix was observed to form prior to the mineralization, it was thought to provide a crystal nucleation site or matrix. The major component of the organic matrix is a glycoprotein containing high levels of acidic amino acids and acidic mucopolysaccharides. The amino acid composition of the glycoprotein was also determined for several species (Weiner and Hood, 1975). Some relevant data are shown in Table 6-3, which clearly shows the predominance of aspartate and glycine. Further study showed the predominance of the sequence Asp—X—Asp—X (X = mostly glycine or serine) in the protein matrix associated with calcite, with this sequence constituting between 18 and 38% of the whole protein in the samples studied. The corresponding number is lower, 8–14%, in protein

Table 6-3. The Amino Acid Composition[a] of the Organic Matrix of Molluscan Shells

	Crassostrea virginica	Crassostrea irredescens		Nortilus pompilius
Crystal form of $CaCO_3$:	Calcite	Calcite	Aragonite	Aragonite
Aspartate + asparagine	325	430	151	261
Glycine	247	260	159	236
Serine	215	119	84	79
Glutamate + glutamine	53	66	137	66
Proline	28	7	64	46
Subtotal	868	882	595	688
Other amino acids	132	118	405	312

[a] The number of amino acid residues per 1000.

associated with aragonite. If the Asp—X—Asp—X portion of the polypeptide is assumed to take a β-sheet form (this is quite likely), then the distance between two adjacent Asp residues would be about 690 pm (6.9 Å). This is the maximum Asp–Asp distance in a polypeptide. The Ca—Ca distance in calcite and aragonite is in the range of 300–650 pm (3–6.5 Å). It was proposed (Weiner and Hood, 1975; Weiner, 1979) that Asp—X—Asp—X acts as a template for the mineralization (Fig. 6-3). It should be noted that there is a definitive contrast between this Ca system and the SiO_2 formation system in diatoms, namely an Asp—X—Asp—X sequence for binding Ca(II), as opposed to a Ser–Ser sequence for binding SiO_4 units. A similar contrast is also found in the effects of calcium and silicon on the intermolecular interactions between polymeric substances as discussed in Section 6.2.2. A more recent review on the topic is found in Krampitz (1982).

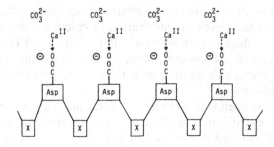

Figure 6-3. A molecular model of $CaCO_3$ shell formation.

6.3.4 Formation of Bone/Cartilage/Dentin

Cells that are responsible for the formation of these structures are called osteoblast (for bone), chondroblast (cartilage), and odontoblast (dentin). First the cells produce an organic matrix material and secrete it into their surroundings. The organic compounds used for the matrix are mainly proteins, such as collagen, elastin, a newly found "osteocalcin" (Martin, 1984) and "osteonectin" (see below), and mucopolysaccharides, such as chondroitin sulfate (see Section 11.5). Compounds of the latter type function as cementing agents, whereas collagen has been believed to provide the matrix onto which the calcium compound was deposited.

Calcification starts soon after the organic matrix material is laid down. The mechanism of calcification is now under intensive study. The major component of bone and dentin minerals is considered to be hydroxyapatite, which has the approximate formula $Ca(OH)_2 3Ca_3(PO_4)_2$. The required materials are $Ca(II)$ and PO_4^{3-}; *in vitro* calcium phosphate $Ca_3(PO_4)_2$ precipitates when the product $[Ca(II)] \times [PO_4^{3-}]$ is higher than $3.5 \times 10^{-6} M^2$.

The notion was once widely held that collagen might provide the crystallization site for the mineral. In fact, there exists a parallelism between the orientation of the hydroxyapatite crystallites and the organic matrix, collagen fiber. Collagen is a very special protein in that it contains about 35% glycine, 11% alanine, 12% proline, and 9% hydroxyproline. Because of the high content of proline and hydroxyproline, it takes on a special helical structure, which is believed to be triply stranded. Of the major components, only hydroxyproline might provide some sort of binding site for $Ca(II)$. In view of the nature of silicon or calcium binding to the matrices discussed in the previous sections, it seems highly unlikely, however, that the hydroxy group of hydroxyproline in fact provides a suitable binding site. Therefore, collagen itself probably is not the crystallization site of calcium phosphate.

A recently identified protein in lamellar bone has been demonstrated to bind both free calcium ions and synthetic apatite crystallites to the collagen and also facilitates the nucleation of calcium phosphate mineral on insolubilized collagen (Termine *et al.*, 1981; Eanes and Termine, 1983). This protein, coined osteonectin, a phosphate-containing glycoprotein, thus, seems to play a role as the nucleation site for the mineralization of lamellar bone tissues.

Emphasis has been shifted to a microstructure called variously "matrix vesicle," "extrusion," or "buds," as the probable site of actual mineralization (Hsu and Anderson, 1978). The matrix vesicles are submicroscopic vesicles (diameter 30–100 nm) found in the vicinity of cal-

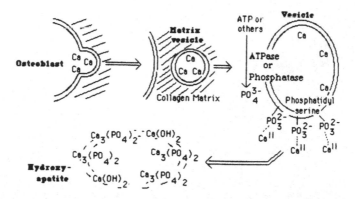

Figure 6-4. A model of bone (dentin) formation mechanism.

cifying cells at the beginning of calcification. They are membrane-bound structures in the early stages of mineralization, but the membranes disappear when calcification becomes more extensive. The membranes of the matrix vesicles seem to have derived from the cells (osteoblasts) but contain more sphingomyelin and phosphatidylserine than cellular fractions do. The vesicles have also been shown to contain ATPase, pyrophosphatase, and other phosphatases. It has been observed by electron microscopy that the first recognizable deposits of crystalline hydroxyapatite indeed occur in and on the matrix vesicles. These findings strongly suggest that the matrix vesicles are the site of mineralization. The chain of events in the mineralization may be as follows. Ca(II) ions are somehow accumulated in the calcifying cells. Ca(II) is then packed in vesicles, which may be derived from special parent cells. Perhaps this change in the condition of vesicles (from bound to off-state) may somehow activate phosphate-splitting enzymes such as ATPase and pyrophosphatase. These enzymes then increase the phosphate concentration in and around the vesicles until the value $[Ca(II)] \times [PO_4^{3-}]$ exceeds the critical value. This critical value was found to be smaller than the *in vitro* value of 3.5×10^{-6} M^2. ATP or GTP (or other nucleoside triphosphates) seems to have a stimulatory effect on the mineralization. It has been suggested that the phosphate group of phosphatidylserine in the vesicle membranes provides an adsorption site for Ca(II). This mechanism may be visualized as shown in Fig. 6-4. A recent review (Eanes and Termine, 1983) may be consulted for further details.

Miscellaneous Topics

7

7.1 Nitrogen Fixation

On the present earth, nitrogen gas (i.e., N_2) is the predominant form of the element. Most organisms cannot utilize this form of nitrogen. Fortunately, some organisms, including anaerobic heterotrophic, anaerobic autotrophic, aerobic bacteria, and cyanobacteria, can "fix" nitrogen (Postgate, 1974). They can convert nitrogen gas to ammonia. Ammonia is then utilized directly by these organisms. Certain plants, the legumes, which are themselves incapable of fixing nitrogen, have a kind of bacterium, rhizobium, attached to their roots. This symbiotic bacterium produces ammonia from nitrogen, and the host plant utilizes it. Ammonia is oxidized by various bacteria including *Nitrobacter,* to nitrate, which is utilized by other organisms; these latter organisms have mechanisms to reduce nitrate to nitrite and then to ammonia.

The enzyme that fixes nitrogen is called nitrogenase and consists of two proteins, "Mo-Fe" protein and "Fe" protein (Mortenson and Thorneley, 1979). Most of the iron in the enzyme exists in the form of iron–sulfur clusters of Fe_4S_4, and is involved in the transfer of electrons. The exact state of the molybdenum in the Mo-Fe protein is not fully understood. One model compound having a $MoFe_3S_4$ unit (Wolff *et al.*, 1978) mimics the spectroscopic behavior of the Mo-Fe protein (Cramer *et al.*, 1978a,b). In this model compound molybdenum atom substitutes for one of the iron atoms in the usual Fe_4S_4 structure. More recent information is found in Mazur and Chui (1982), Hausinger and Howard (1982), Shah and Brill (1981), Kurtz *et al.* (1979), Antonio *et al.* (1982), Averill (1983), and Burgmayer and Stiefel, (1985).

The detailed structure and mechanism aside, we are here interested in the question of why molybdenum is specifically suited to nitrogen fixation (Ochiai, 1978a). That is, we want to know the reasons for the basic fitness of molybdenum for the job.

7.1.1 The Basic Fitness of Molybdenum for Nitrogen Fixation

We will first consider the artificial synthetic process of ammonia for comparison. The catalyst in the well-known Haber–Bosch process for the industrial synthesis of ammonia from nitrogen and hydrogen is based on metallic iron. Let us assume the following mechanism for this synthesis (Makishima, 1960):

$$2K \; (K = catalyst) + N_2 \rightarrow 2(KN) \qquad \Delta H_1$$

$$2K + H_2 \rightarrow 2(KH) \qquad\qquad\qquad \Delta H_2$$

$$(KN) + 3(KH) \rightarrow 4K + NH_3 \qquad \Delta H_3 \qquad\qquad (7\text{-}1)$$

(KN) and (KH) can be regarded as nitride and hydride derivatives of the metallic catalyst K, respectively. The formation of ammonia would proceed smoothly and to an appreciable extent if none of the three enthalpy values (ΔH_1–ΔH_3) were exceedingly endothermic or exothermic. Appropriate ΔH_n values for common metallic elements are shown in Fig. 7-1. It is seen that the best catalyst is indeed iron; far behind iron are three elements, Ni, Mn, and Cr. These conclusions are in agreement with the experimental data. However, even with iron the reaction must be conducted at high temperature and pressure. This means that the activation energy of nitride and/or hydride formation with even the best catalyst,

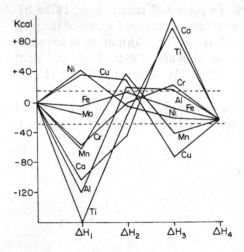

Figure 7-1. Enthalpy values for the following reactions: ΔH_1: $2K + N_2 \rightarrow 2(KN)$; ΔH_2: $2K + H_2 \rightarrow 2(KH)$; ΔH_3: $(KN) + 3(KH) \rightarrow 4K + NH_3$; ΔH_4: $N_2 + 3H_2 \rightarrow 2NH_3$. (From Ochiai, 1978a. *BioSystems* 10:329.)

iron, is quite high. It seems that especially the hydride formation step is subject to a high activation energy barrier in the case of iron. One of the most important differences between the biological nitrogen fixation and the industrial reaction is that molecular hydrogen gas is not directly utilized in nitrogen fixation by microorganisms. It is believed that nitrogen is given electrons by the enzymes and the necessary hydrogen is somehow supplied endogenously in the form of protons.

ΔH_1 represents the ability of a metal to bind N_2, cleave the N—N bond, and form a nitride. The ΔH_1 value for molybdenum is about -16 kcal/mole (-77 kJ/mole) of N_2, which is quite favorable for the fixation step. This suggests the possibility that, in early nitrogen-fixing organisms, the abundantly available (see Section 3.3.1) MoO_4^{2-} was reduced to the Mo^0 state and that the latter acted as the nitrogenase in these primitive organisms (or protobionts). The reduction potential of MoO_4^{2-}/Mo^0 at the pH of seawater (pH $= 8.1$) is estimated to be about -0.6 V. This value is based on crystalline Mo^0. In practice, the entity to be utilized might have been $Mo^{VI}O_mL_n$ (L = primitive protein or proteinoid); then the reduction potential would be slightly higher than -0.6 V. If this is the case, it would be possible to reduce the Mo(VI) to the requisite Mo(0) by the $Fe_2S_3/2FeS$ system ($E_0' = -0.7$ V) or primitive ferredoxin (see Section 3.2.1 and Fig. 3-3).

Thus, a possibility exists for the following process:

$$Mo^{VI}O_mL_n \xrightarrow[\text{ferredoxin(oxidized)}]{\text{ferredoxin(reduced)}} Mo^0L_{n'} \xrightarrow{+N_2} N_2Mo^0L_{n'} \qquad (7\text{-}2)$$

It is likely that the FeS entity remains associated with the product $Mo^0L_{n'}$; that is, $Mo^0L_{n'}$ may actually be an $MoFe_tS_mL_{n'}$. This may represent the precursor of the Mo-Fe protein (or the so-called Mo-Fe cofactor) of nitrogenase. The $N_2Mo^0L_n$ thus formed can react further in three different ways:

$$N_2Mo^{VI}L_n \longrightarrow \begin{array}{l} Mo^{VI}L_{n'} + 2NH_3 \qquad\qquad (7\text{-}3) \\[4pt] Mo^0L_{n'} + 2NH_3 \qquad\qquad (7\text{-}4) \\[4pt] Mo^{VI\text{-}n}L_{n'} + 2NH_3 \qquad\quad (7\text{-}5) \end{array}$$

In scheme 7-3 molybdenum oscillates between the oxidation states 0 and VI; this process corresponds to the nitride formation mechanism described earlier. In scheme 7-4 the electrons to reduce N_2 are supplied by another electron source, which may be another ferredoxin system. Scheme 7-5 is a combination of 7-3 and 7-4; portion of the electrons to be utilized to reduce N_2 come from both ferredoxin and molybdenum. If scheme 7-3 is correct, then the metal to be selected must readily interconvert between the oxidation states 0 and VI. Molybdenum and tungsten are two possible candidates. The tungsten redox system is similar to that of molybdenum. Chromium would not be suitable because it has the wrong redox potential, although it may function as a nitrate reductase (see Section 3.3). Tungsten (as WO_4^{2-}), however, is much less abundant than molybdenum in seawater, by a factor of about 10^3; thus, organisms would have preferred to incorporate molybdenum. It is noteworthy that *Nostoc* (a cyanobacterium) mutants, which require tungsten instead of molybdenum for growth on N_2 or NO_3^-, have been isolated (Singh *et al.*, 1978). Tungsten-substituted molybdenum enzymes, however, have generally been found to be inactive (Johnson *et al.*, 1974a,b; Nagatani and Brill, 1974). This suggests that organisms have fairly specifically adapted to molybdenum, which was their first choice. There may, however, be some exceptions where molybdenum can be replaced by tungsten, or even where tungsten is preferred to molybdenum (see Section 3.3.6). An active V-containing nitrogenase has very recently been reported (Robson *et al.*, 1986).

Iron cannot oscillate between the oxidation states 0 and VI readily; therefore, it cannot function in this way despite the favorable ΔH_1 value, and thus organisms did not find iron to be useful as a nitrogen-fixing agent. Iron in the form of iron–sulfur protein is involved in the related enzyme hydrogenase, however, as well as being present in the nitrogenase system.

That a Mo^0 complex binds N_2 and can yield NH_3 in the presence of a protic source (H_2SO_4 or CH_3OH) has been demonstrated (Chatt *et al.*, 1975). That is, Mo^0 can be oxidized by N_2 and H^+ to Mo^{VI}. This, however, cannot necessarily be taken to lend support to mechanism 7-3 in the actual nitrogenase system.

7.1.2 Alternative Views

The two other possibilities cited, 7-4 and 7-5, also deserve attention. Suppose that N_2 is reduced stepwise by two-electron processes as follows:

$$N_2 \xrightarrow{2e,\ 2H^+} N_2H_2 \xrightarrow{2e,\ 2H^+} N_2H_4 \xrightarrow{2e,\ 2H^+} 2NH_3 \qquad (7\text{-}6)$$

The reduction potential of the first step is not known. The reduction potential for an equivalent process, i.e., $N_2/2NH_2OH$, is, however, available: -2.4 V (at pH 7, 25°C). One reason for this large negative value is the inclusion of energy needed for the cleavage of the N≡N bond in the process (formation of NH_2OH). This bond energy is about 100 kcal (418 kJ). Even if this factor is taken into account, however, it seems likely that the potential of N_2/N_2H_2 would still be substantially negative. The reduction potentials for the subsequent processes are largely positive: about $+1.1$ V for $2NH_2OH/N_2H_4$ and $+0.7$ V for $N_2H_4/2NH_3$ at pH 7. Therefore, it is likely that the first step in this mode of reduction of N_2 presents a formidable barrier and may require an especially strong reducing mechanism. This might be related to the observation that nitrogenase requires a large amount of ATP for its functioning. The net reaction, $N_2/2NH_3$, is moderately exothermic at 25°C and should not require much energy. ATP may, therefore, be somehow involved in potentiating the reducing apparatus. This in turn presupposes a reducing or electron-donating center besides the nitrogen-accepting molybdenum center; that is, the picture so developed corresponds to either scheme 7-4 or 7-5.

In scheme 7-4 the molybdenum in the Mo-Fe protein would remain in the zero oxidation state and all the electrons are supplied by an electron donor, perhaps the Fe protein of the enzyme.

In scheme 7-5 the crucial first two electrons are probably supplied by the energized (by ATP hydrolysis) Fe protein, and the rest of the electrons needed, four, would then be provided by the molybdenum center of the Mo-Fe protein. The mid potential of the Fe protein of *Clostridium pasteurianum* (at pH 7.5) is -0.294 V and changes to -0.40 V when MgATP is bound to the enzyme (Zumft *et al.*, 1974). This implies that the binding of MgATP potentiates the Fe protein as hypothesized here. As a result, the molybdenum would be oxidized to the IV oxidation state; it would then need to be reduced back to the zero oxidation state by the iron–sulfur machinery of the Mo-Fe protein.

7.2 Vitamin B_{12} and B_{12} Coenzymes

The chemistry of B_{12} and B_{12} coenzymes, as well as the enzymatic reactions involving B_{12} coenzymes, are well documented (Pratt, 1972; Ochiai, 1977; Dolphin, 1982). It is not a topic to be covered fully by a few pages. However, certain essential facts must be reviewed in order to deal with our principal concern here, i.e., why is cobalt, not copper or iron or anything else, used specifically in these reactions?

I will first give a very brief account of the essential features of B_{12}, its coenzymes and the enzymatic reactions dependent on them.

7.2.1 B_{12}, B_{12} Coenzymes—Their Chemistry and Reactions

Vitamin B_{12} is a cobalt derivative of a porphyrin-like ring called "corrin." The structure of B_{12} is shown in Fig. 7-2a. The corrin ring

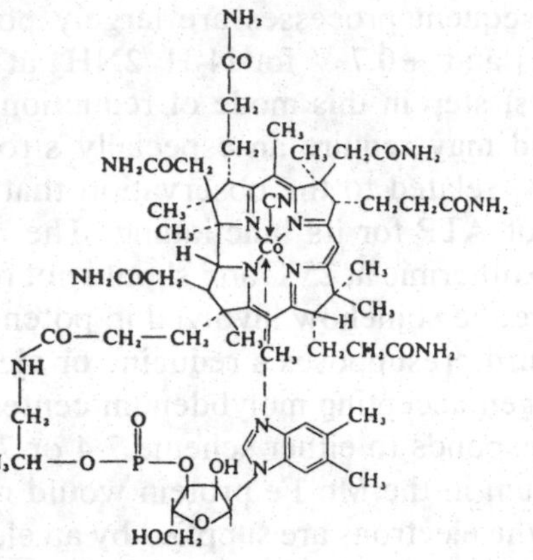

Figure 7-2. Structures of (a) vitamin B_{12} (cyanocobalamin), (b) its methyl derivative, and (c) B_{12} coenzyme (adenosylcobalamin).

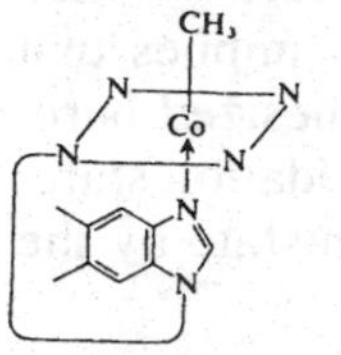

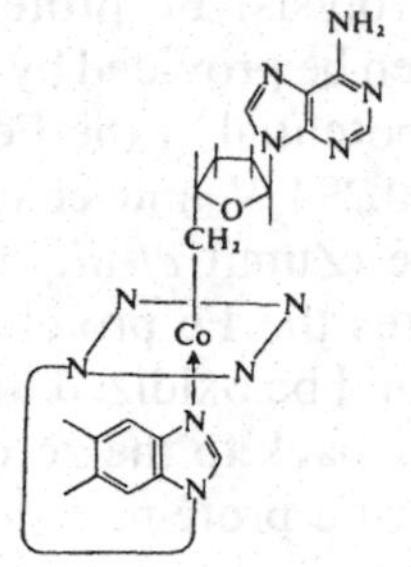

coordinates to the cobalt atom, with the cobalt atom and the four nitrogen atoms of the ring lying almost in the same plane. The fifth ligand (below the corrin ring) is dimethyl benzimidazole, and the sixth ligand is a cyanide anion in B_{12} itself. For this reason B_{12} is called cyanocobalamin. The sixth ligand can be varied. When it is water, the compound is called aquocobalamin.

The most interesting derivative is an alkyl derivative. An alkyl group can form a direct bond with the cobalt atom. Of all the alkyl derivatives, apparently only the methyl derivative (methylcobalamin or CH_3-B_{12}) and the adenosyl derivative (B_{12} coenzyme, adenosyl B_{12}) are physiologically active (see Fig. 7-2b,c).

The formation of an alkyl derivative is accomplished by a process called "oxidative addition" of an alkyl cation onto a Co(I) state. That is:

$$Co^I\text{-}B_{12} + R^+ \rightarrow R^-\text{—}Co^{III}\text{-}B_{12} \tag{7-7}$$

The formal oxidation state of cobalt is $+3$ in the alkyl derivatives, including B_{12} coenzyme and methylcobalamin, although the carbon–cobalt bond has a substantial degree of covalency. The cobalt in cyanocobalamin and aquocobalamin is also in the $+3$ state. In order for the alkylation reaction (7-7) to be effected, Co(III) must be reduced to Co(I). This can be done chemically with reducing agents such as BH_4^-, or it can be performed enzymatically. The formation of adenosyl B_{12} by an enzyme system in *Clostridium tetanomorphum* proceeds as follows:

$$Co^{III}(B_{12}) \xrightarrow{+2e(\text{from dithio protein})} Co^I(B_{12})$$

$$Co^I(B_{12}) + ATP \rightarrow \text{adenosyl } B_{12} + \text{triphosphate} \qquad (7\text{-}8)$$

Likewise, the formation of methylcobalamin is an oxidative addition:

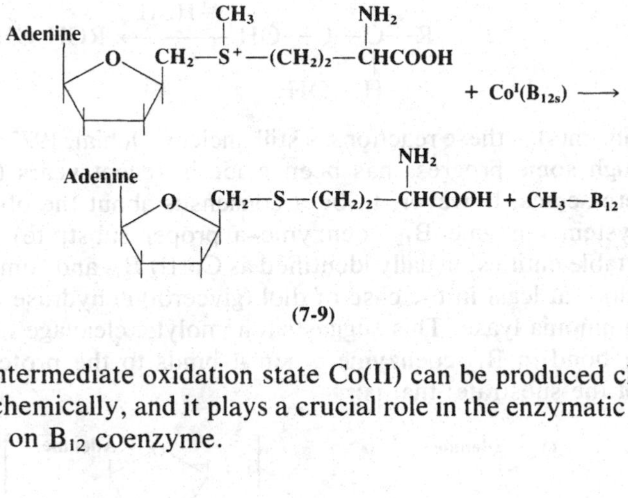

$$(7\text{-}9)$$

The intermediate oxidation state Co(II) can be produced chemically or electrochemically, and it plays a crucial role in the enzymatic reactions dependent on B_{12} coenzyme.

7.2.2 Enzymes Dependent on B_{12} Coenzyme

Some of the enzymes dependent on B_{12} coenzyme are glutamate mutase, L-β-lysine mutase, methylmalonyl CoA mutase, glycerol (diol) dehydrase, and ethanolamine ammonia lyase. The reactions catalyzed by these enzymes can be represented by:

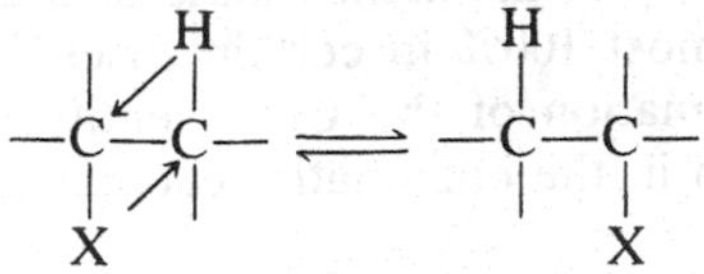

$$(7\text{-}10)$$

That is, they can be classified as 1,2-shift reactions. Two examples are:

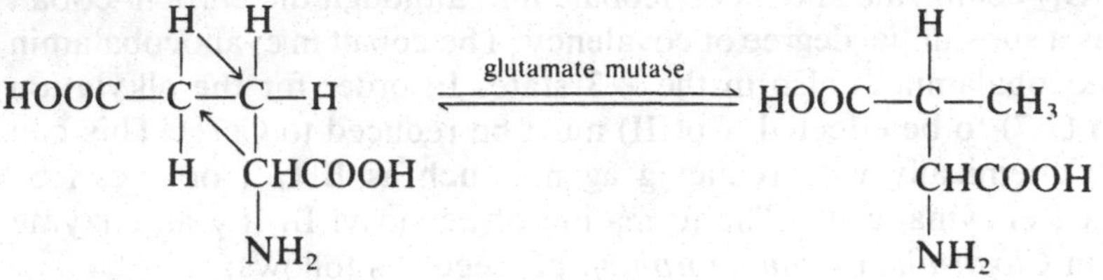

$$\underset{\substack{|\\ \text{NH}_2}}{\underset{\substack{|\\ \text{CHCOOH}}}{\text{HOOC}-\overset{\overset{\displaystyle H}{|}}{\underset{\underset{\displaystyle H}{|}}{C}}-\overset{\overset{\displaystyle H}{|}}{C}-H}} \quad \xrightleftharpoons{\text{glutamate mutase}} \quad \underset{\substack{|\\ \text{NH}_2}}{\underset{\substack{|\\ \text{CHCOOH}}}{\text{HOOC}-\overset{\overset{\displaystyle H}{|}}{C}-CH_3}} \qquad (7\text{-}11)$$

$$\underset{\substack{|\quad|\\ \text{HO}\;\;\text{OH}}}{R-\overset{\overset{\displaystyle H}{|}}{C}-\overset{\overset{\displaystyle H}{|}}{C}-H} \quad \xrightarrow{\text{diol dehydrase}}$$

$$\underset{\substack{|\quad|\\ \text{H}\;\;\text{OH}}}{R-\overset{\overset{\displaystyle H}{|}}{C}-\overset{\overset{\displaystyle H}{|}}{C}-OH} \quad \xrightarrow{-\text{HOH}} \quad RCH_2CHO \qquad (7\text{-}12)$$

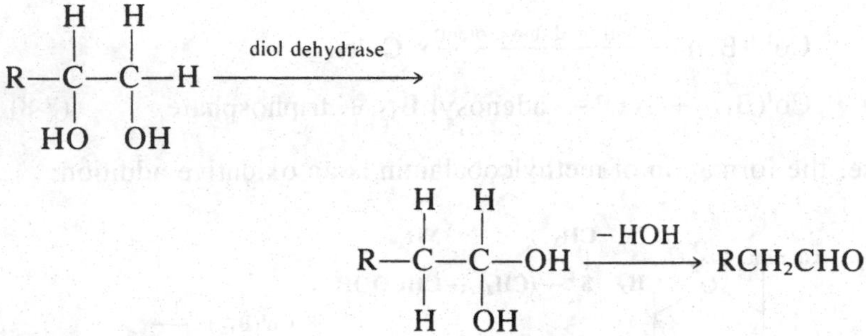

The mechanism(s) of these reactions is still unclear (Ochiai, 1977; Golding, 1982), though some progress has been made in recent years (Halpern, 1985). There seems, however, to be a consensus about the observation that the system (enzyme–B_{12} coenzyme–a proper substrate) produces EPR-detectable entities, usually identified as Co(II)-B_{12} and some organic free radical(s), at least in the case of diol (glycerol) dehydrase and ethanolamine ammonia lyase. This suggests a homolytic cleavage of the carbon–cobalt bond in B_{12} coenzyme when it binds to the protein in the presence of the substrate; that is:

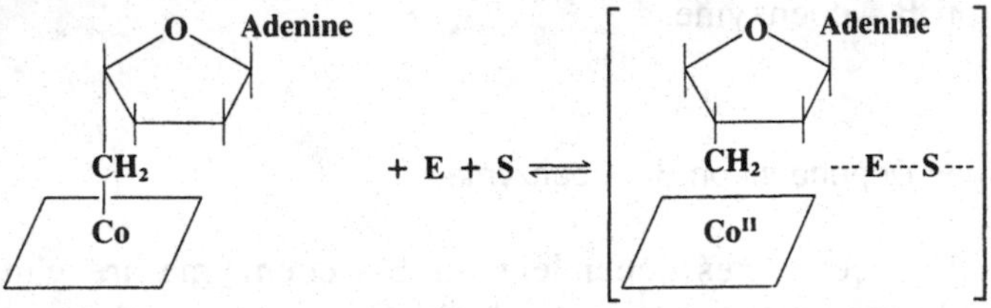

$$(7\text{-}13)$$

This homolytic splitting has also been detected by optical spectroscopy. The extent of the splitting in the steady state of the enzymatic reaction is quite substantial, almost 100% in certain cases. Therefore, it can be concluded that the formation of the Co(II) entity and an organic free radical is a crucial step in the enzymatic reactions, no matter what the actual mechanism of the 1,2-shift reaction might be (Halpern, 1985; Finke and Schiraldi, 1983; Finke *et al.*, 1983).

Other important enzymes dependent on B_{12} coenzyme include ribonucleotide reductase. Methane formation by methane bacteria is dependent on methylcobalamin. The mechanism of the methane formation seems to involve a methyl radical; this again suggests a homolytic cleavage of the methyl carbon–cobalt bond of the alkylcobalamin, forming a Co(II) intermediate.

7.2.3 The Basic Fitness of Cobalt

Cobalt is much rarer than other common transition metals such as iron, manganese, copper, and nickel, both in the upper crust of the earth and in seawater. Yet cobalt was selected for this particular job. Why?

From the discussion above, the important requirements that must be met by candidates for the job are clear, despite the fact that the mechanisms of the enzymatic reactions are far from well established. These requirements (Ochiai, 1978b) are: (1) the candidate should be readily able to take three consecutive oxidation states (e.g., I, II, III) in aqueous media; (2) the lowest oxidation state should be highly nucleophilic; and (3) the middle oxidation state should perhaps have an unpaired electron. Requirement 2 implies that the lowest oxidation state of the catalytic metal ion should have the d^8 or d^{10} configuration. This condition may be satisfied by Fe(0), Co(I), Ni(0), and Cu(I). Fe(0) and Ni(0) are not readily attainable, particularly in aqueous media. This leaves us Co(I) and Cu(I). Both Co(II) and Cu(II) can have one unpaired electron. However, Cu(I) is not very nucleophilic and Cu(III) is not readily attainable. Only the Co(III)/Co(II)/Co(I) system satisfies all of the requirements. This could explain why cobalt was selected (basic fitness). It must be pointed out, though inconsequential for this discussion, that the validity of the participation of Co(II) in the 1,2-shift reaction as postulated in requirement 3 (Ochiai, 1977; Golding, 1982) has been disputed by Finke and Schiraldi (1983), Finke *et al.* (1983, 1984), and Halpern (1985).

7.3 Chromium

The nutritional value of chromium has been recognized for higher animals (Baetjer, 1974). The only well-studied function of chromium is that of the so-called "glucose tolerance factor" (GTF); it potentiates insulin, which regulates the blood glucose level (Anderson and Mertz, 1977). Chromium has also been observed to have a tendency to accumulate in the nuclear fraction of tissues (Baetjer, 1974); RNA from beef liver has been reported to contain a fairly high level of chromium, 50–140 ppm (Wacker and Vallee, 1959; Bryan, 1981). I will review briefly the chemistry of chromium and the possible chemical basis for the physiological functions of chromium.

7.3.1 GTF

GTF can most conveniently be isolated from brewer's yeast (Anderson and Mertz, 1977). It is a rather small molecule with a molecular weight of about 500. It is water-soluble and relatively stable. Cr(III) was identified as a component of GTF, but otherwise GTF has defied a detailed elucidation of its composition and structure (Anderson and Mertz, 1977). It has been shown, however, that reaction of glutathione with chromium and nicotinic acid yields a stable, biologically active GTF-like complex (Anderson and Mertz, 1977). A synthetic complex containing only chromium and nicotinic acid was also said to be biologically active (Anderson and Mertz, 1977). These synthetic compounds bind tightly to insulin but can be dissociated from it at low pH. Therefore, it was believed (Anderson and Mertz, 1977) that GTF contained a dinicotinic acid–chromium–glutathione complex.

However, a recent development (Haylock et al., 1983) has cast a doubt on the validity of these earlier studies. Gonzalez-Veraga et al. (1982) synthesized a number of Cr–nicotinic acid complexes, none of which showed GTF activity. They indicated that Cr was associated with tryptophan moiety. Haylock et al. (1983) studied carefully Cr-containing fractions isolated from brewer's yeast grown in Cr-containing media. It was found that most of the fractions were actually artifacts formed as a result of direct reaction between chromium and components of the medium. It turned out that the GTF activity of the two biologically active Cr-containing fractions was cleanly separated from the Cr-containing material upon further purification. This led to a suggestion that chromium has nothing to do with GTF activity in brewer's yeast (Haylock et al., 1983).

It is still noteworthy that at lower concentrations, chromium has been found to be beneficial to many plant species, including algae, giant kelp, grapes, carrots, barley, cucumber, potatoes, peas, and beets (Baetjer, 1974). Chromium is toxic to plants at higher concentrations.

7.3.2 Other Possible Functions of Chromium

Almost half of the total cellular chromium in human liver is found to be concentrated in the nuclei (Baetjer, 1974). Human testis is known to accumulate chromium rapidly after injection of a trace dose. Along with these facts, the classical finding of a high level of chromium in RNA from beef liver suggests that chromium may play a role in the maintenance of the structure of some nuclear substances, particularly RNA. Chromate reacts with nucleic acid, forming a green complex (Baetjer, 1974). This is likely to be a Cr(III) complex. Cr(III) prefers an octahedral coordination structure. Mention was made in Section 6.1.2 of the structural role of M(II) in polynucleotides. M(II) binds the phosphate groups of polynucleotides via ionic bonding. Cr(III) then might well be involved in binding the base portions of polynucleotides, thus perhaps helping to maintain rather specific structures within the polynucleotides (e.g., the kinks in the polynucleotide chain). The suggestion has also been made that chromium may activate some enzymes (Anderson and Mertz, 1977).

7.4 Vanadium

Vanadium is said to be essential for a certain species of green alga and it is known to stimulate the growth of crop plants. However, definite evidence for the presence of a significantly high level of vanadium is found only in the toxic mushroom, *Amanita muscaria* (Biggs and Swinehart, 1976), among plant species. Its vanadium content ranges from 61 to 180 ppm (dry weight basis) and is about 100 times the mean content of any other plant.

An interesting group of animals, the ascidians (sea squirt), also contain high levels of vanadium, which is localized in a special set of cells called vanadocytes. In higher animals, vanadium has been reported to have some role in cholesterol metabolism and mineralization of tooth dentin and bone (Byerrum, 1974; Hopkins, 1974). Readers are referred to Chasteen (1983) and Kustin *et al.* (1983) for recent reviews. I will examine possible chemical reasons for these functions of vanadium.

7.4.1 Possible Oxidation–Reduction Functions

7.4.1.1 Vanadium in Ascidians

A major research interest has focused on the function of vanadium in ascidians. It was thought (Biggs and Swinehart, 1976) that vanadocytes also contained a high concentration of sulfuric acid, but it appears that this was an artifact and that the actual pH of the medium is 7.2 (Dingley *et al.*, 1982). Magnetic measurements and EPR studies (Biggs and Swinehart, 1976) and EXAFS studies (Tullius *et al.*, 1980) indicate that the vanadium in vanadocytes exists as V(III) under physiological conditions.

Reduction potentials were measured to be $E_0' = +0.33$ V at pH 0.5 (*Phallusia mammillata*) and $+0.37$ V at pH 0.42 (*Ascidia obliqua*) (Biggs and Swinehart, 1976). These values are close to the standard potential of $E_0' = +0.36$ V for VO^{2+}/V^{III} (pH 0). The hemolysate of the blood cells from *A. obliqua* showed an absorption spectrum having a band at 430 nm (molar extinction coefficient = 420) (Biggs and Swinehart, 1976). This spectrum is similar to that of $(VOV)^{4+}$, which has a peak at 425 nm.

These data suggest that the vanadium is in the $+3$ oxidation state under physiological conditions and that it may function as the couple V(IV)/V(III). What then is the function of the vanadium? It is still obscure. The vanadium in the vanadocyte is contained in small globules called vanadophores, and appears to be associated with small molecules, "tunichromes" (Macara *et al.*, 1979); the structure of a tunichrome, B-1, has recently been elucidated. It seems to function as a reducing agent for vanadium and perhaps as a complexing agent for the V(III) species (Bruening *et al.*, 1985).

Ascidians are the most primitive vertebrates, Urochordata, and thus intermediate between invertebrates and vertebrates. Their blood contains as many as ten morphologically distinct cells. Some of them are colored: green (vanadocyte), blue, and orange. Others are colorless and amoeba-like. Some contain iron as well as vanadium; iron either has not been detected or has not been sought in others. These facts suggest that the circulatory system in these primitive vertebrates contains all sorts of cells different from those found in vertebrates' respiratory (oxygen-carrying) cells as well as lymphocytic cells. The vanadocytes are among the most common cells in ascidian blood. This implies a respiratory role for the vanadocytes, i.e., $V(III) + O_2 \rightleftharpoons (VO_2)^{3+}$. As discussed in Chapter 4, this equilibrium in itself is favorable, but $(VO_2)^{3+}$ would easily be decomposed unless the complex was well protected from H^+ or water, perhaps by some specific feature of the protein present. Otherwise, likely decomposition pathways would include:

$$(VO_2)^{3+} + H^+ + HOH \rightarrow V^{IV}aq + HOO\cdot$$
or
$$(VO_2)^{3+} + HOH \rightarrow V^{IV}aq + {}^-OO\cdot \qquad (7\text{-}14)$$

It does not seem likely that the vanadium entity is well shielded in a protein. It has been shown (Macara *et al.*, 1979) that indeed the function of vanadium in ascidians is not as an "O_2 carrier."

Vanadium has been shown to have catalytic activity *in vitro* for the oxidation of amines (Byerrum, 1974; Hopkins, 1974). Can this be the function of vanadium in vanadocytes? If so, why is it localized in this specialized cell? Amines and other metabolites are present not in the vanadocytes but in other tissues. Therefore, it is not likely that vanadium is involved in the oxidation–reduction of organic metabolites.

Have we exhausted the possible functions? A few other general functions might be envisaged for vanadium. For example, it might participate in the decomposition of hydrogen peroxide or superoxide; that is, it might show catalase or superoxide dismutase activity. In the case of catalase action, vanadium would work in the following ways:

$$V^{III} + HOOH \xrightarrow{H^+} V^V + 2HOH$$

$$V^V + HOOH \longrightarrow V^{III} + 2H^+ + O_2 \qquad (7\text{-}15)$$

$$V^{III} + HOOH \rightarrow \left[V \underset{OH}{\overset{OH}{\diagdown}} \right] \xrightarrow{+HOOH} V^{III} + 2HOH + O_2 \qquad (7\text{-}16)$$

These processes are feasible in terms of reduction potentials. One possibility for superoxide dismutase activity might be as follows:

$$V^{III} + {}^-OO \cdot \xrightarrow{H^+} V^{IV} + HOOH$$

$$V^{IV} + {}^-OO \cdot \longrightarrow V^{III} + O_2 \qquad (7\text{-}17)$$

As mentioned earlier (Section 4.2), this reaction, too, is quite feasible (Ochiai, 1978b). The $E_0'(V^{IV}/V^{III})$ found for the vanadium in vanadocytes, $+0.33$–0.37 V, is close to the E_0' ($+0.42$ V) found for the copper in Cu/Zn superoxide dismutase. Perhaps oxygen is carried as dissolved in the blood plasma of those ascidians in which no blood iron has been detected or is bound to an iron complex in the blood of other ascidians in which iron has been detected. If so, vanadium could be functioning in a defense mechanism to protect against oxygen toxicity (see Chapter 13). Other possible effects (Kustin *et al.*, 1983) are (1) that vanadium may function as an antifeedant for tunicates, and (2) that vanadium may be involved in the tunic formation. The first suggestion is based on an observation that species whose outer coverings are rich in vanadocytes are shunned by predatory fish.

7.4.1.2 Other Oxidation–Reduction Effects

There is considerable evidence that vanadium is involved in lipid metabolism (Byerrum, 1974; Hopkins, 1974), particularly the metabolism of cholesterol. Vanadium has been reported to lower tissue cholesterol levels. It was found that vanadium inhibits one of the enzymes, squalene synthetase, in the cholesterol biosynthesis pathway. Squalene synthetase catalyzes the following reaction:

$$\text{Precursor of squalene} \xrightarrow[\text{pyrophosphate}]{\text{NADPH} \quad \text{NADP}^+} \text{squalene} \qquad (7\text{-}18)$$

It involves a hydride transfer (i.e., reduction) and, hence, the enzyme might contain electron transfer components. Since vanadium itself has electron-transferring capacity, it might interfere with the electron transfer components of the enzyme. Vanadium has also been reported to uncouple oxidative phosphorylation (Byerrum, 1974); the chemical basis of this effect should be similar to that of the previous one. The true physiological significance of these effects, however, is still unclear.

7.4.2 Other Effects

Vanadium is said to reduce the synthesis of cysteine and coenzyme A. CoA is the pivotal compound in lipid metabolism, hence this effect on CoA synthesis may be the basis of all of the known influences of vanadium on lipid metabolism. These effects, however, cannot be taken to represent the normal physiological functions of vanadium.

Another interesting effect is that vanadium mimics the action of insulin (see Chasteen, 1983). The incubation of rat adipocytes with 10–100 μM vanadyl(IV) or vanadate(V) stimulates the transport of glucose and 2-deoxyglucose into the cell. The maximal stimulation of hexose transport due to vanadium alone is virtually equivalent to that produced by insulin in the absence of vanadium.

One positive effect of vanadium is that on the mineralization of teeth and bones. Radioactive vanadium injected intravenously in mice was taken up by the teeth, bones, and fetus. Teeth and bones are mainly made of calcium phosphate. Vanadate, being analogous to phosphate, may be incorporated into the mineral part of the hard tissues. Looked at another way, calcium vanadate may promote the formation of the calcium phosphate crystals. Again, however, it is difficult to be sure that this finding represents a physiologically significant process. Vanadium may only be inadvertently taken up because of its similarity to phosphate. Nevertheless, vanadium appears to be essential to the proper growth of rats and chickens (Byerrum, 1974).

7.5 Tin

Tin was shown to enhance the growth of rats (Schwartz, 1974); the observed growth responses exceeded 50%. Nothing is known in detail about the physiological mechanism of the effect of tin. One or two interesting observations may, however, be mentioned. For example, tin levels in embryos and newborn animals are so low as to be hardly measurable, whereas the levels after birth seem to rise abruptly and then remain more or less constant (Schwartz, 1974). This suggests that tin may be deliberately excluded from the fetus, or that it cannot penetrate into the fetus. Another intriguing observation is the fact that organotin compounds such as tributyl tin chloride are as effective as such inorganic tin compounds as $Sn(SO_4)_2 2H_2O$ for growth enhancement. It was pointed out earlier (Section 3.2) that the reduction potential of Sn^{IV}/Sn^{II} ($E_0' = -0.13$ V) is such that Sn^{II} may function as a reducing agent in special cases. The effectiveness of the organotin compounds suggests that native tin may occur in fat tissues. The most provocative suggestion yet made is that the function of tin is to signal the birth to the animal (Schwartz, 1974). Animals lack many enzymes at the time of birth, e.g., the enzymes involved in glucose metabolism. This is because most nutrients have been supplied by the mother up until the time of birth. Therefore, a great burst in protein synthesis must take place immediately after birth. Perhaps tin, which was prevented from entering the fetus, now gets into the newborn animal and stimulates the burst of DNA/RNA transcription, translation, and protein synthesis (Schwartz, 1974).

7.6 Intracellular Ca(II)

Calcium is the fifth most abundant element (after C, H, O, N) in the human body. Ninety-five percent of the body calcium is in bones and tooth enamel, but is in dynamic equilibrium with the remaining pools of calcium. The remainder of calcium is widely distributed throughout the body, body fluid, and soft tissues.

The functions of Ca(II) in bones, shells, and other hard tissues, and in maintaining the integrity of cell membranes and the cell–cell adhesion were discussed in Chapter 6. The most important and intriguing functions of Ca(II), however, are carried out by small quantities of intracellular

Ca(II). This Ca(II) could come from the outside of a cell, through Ca(II) channels or active mechanisms, as well as from intracellular sources.

The functions of intracellular Ca(II) are diverse and numerous, and are being investigated intensively (Cheung, 1982; Campbell, 1983; Sigel, 1984; Spiro, 1983; Rubin et al., 1985). Intracellular Ca(II) regulates such cell activities as secretion of some hormones (e.g., insulin), release of neurotransmitter, chemotaxis, muscle contraction, exocytosis/ endocytosis, enzymatic activity, chemiluminescence, fertilization, and others. Some of these activities are elicited by some extracellular agents such as hormones, metabolites (e.g., glucose), and electrical signals; these are the so-called "primary messengers." Ca(II), along with cAMP (cyclic adenosine monophosphate) and other compounds as described below, are called "second messengers."

The mechanism of action of the second messengers are varied and complicated. They are discussed here only briefly, with emphasis being focused on the question: why calcium?

7.6.1 General Discussion of the Signal Transduction and Its Manifestation

In order for a signal, whether chemical or electrical, to have an effect, it must be transmitted to the effector (cellular). It has not been well delineated what types of signal transduction are operating in organisms, but the response time seems to be one of the important factors that determine the type of signal transduction. The response time and the duration of action range from minutes to hours and days in the regular hormonal actions, whereas those that involves cAMP and other second chemical messengers last from seconds to minutes. The fastest ones occurring in the millisecond range involve Ca(II) directly. In all three types, Ca(II) is involved but in different ways. In the first type, Ca(II) is involved only in indirect ways, but the processes of the second category involve Ca(II) in more direct and intimate ways. I exclude processes of the first type in this discussion.

The transduction of a signal (from a primary messenger), in order for its effect to be proper, should involve (1) transmittance of the information to the effector at the desired rate, (2) amplification of the signal, and (3) termination of the effects at the desired rate. That is:

$$\underset{\substack{\text{(primary} \\ \text{messenger)}}}{\textbf{signal}} \xrightarrow[\substack{\text{may be} \\ \text{amplified}}]{\substack{\text{across} \\ \text{membrane}}} \underset{\substack{\downarrow \\ \text{destroyed}}}{\textbf{second messenger}} \xrightarrow[]{\substack{\text{may be} \\ \text{amplified}}} \underset{\substack{\downarrow \\ \text{destroyed}}}{\textbf{effector}} \qquad (7\text{-}19)$$

There seem to be at least two different kinds of signal transduction in the second type as defined above. Primary messengers of kind A such as epinephrine, norepinephrine, and histamine exert their effects through the activation of adenyl cyclase, which catalyzes the formation of cAMP, which in turn activates protein kinase A. Protein kinase A then phosphorylates an enzyme, which will then, often, activate or deactivate specific enzymes that manifest the ultimate effect. The synthesis and decomposition of cAMP is often regulated by Ca(II). Ca(II) also activates the enzymes involved in some cases.

Primary messengers of kind B include adrenaline, noradrenaline, vasopressin, thrombin, glucose, angiotensin II, ACTH, plant lectin, and parathyroid hormone. The immediate effect of these messengers seems to be the activation of phospholipase C, which hydrolyzes phosphatidylinositides, especially triphosphoinositide (TPI), to diacylglycerol (DG, particularly stearyl-arachidonyl glyceride) and inositol-1,4,5-triphosphate (ITP) (Berridge, 1982, 1985; Takai et al., 1982). The function of DG is to activate protein kinase C, which then phosphorylates specific proteins to induce certain cellular responses. This latter process seems to be independent of the change in the cytosol [Ca(II)]. However, the activation of protein kinase C by DG seems to involve Ca(II) as well as some phospholipids. One of the other effects of a primary messenger of kind B is the increase of the cytosol [Ca(II)]; this is brought about by the plasma membrane inward transport and/or the release of Ca(II) from intracellular vesicles, especially the endoplasmic reticulum. The mechanisms of this influx of Ca(II) from the extracellular medium and the release of Ca(II) from intracellular vesicles are not well understood. A suggestion was made (Berridge, 1985) that DG can be converted to PA (phosphatidic acid) by DG kinase, and PA thus formed functions as a Ca(II) ionophore in a Ca(II) channel, thus facilitating the influx of Ca(II). ITP has been shown to promote the release of Ca(II) from the endoplasmic reticulum (Volpe et al., 1985). The mechanism of this process is unknown but would involve the opening of a gated Ca(II) channel located in the vesicle membrane. Ca(II) then regulates, through calmodulin [a Ca(II)-binding protein, see Section 9.4], cellular processes. One such process is considered to be the activation of phospholipase A_2, which hydrolyzes phospholipids such as phosphatidylinositides to form arachidonic acid. The latter is then converted to eicosanoids including prostaglandins, which are excreted to exert their effects. There is some indication that eicosanoids may activate cGMP (cyclic guanosine monophosphate) formation from GTP by guanosyl cyclase. These processes are summarized in Fig. 7-3.

Ca(II) influx caused by an electrical signal as at the presynaptic neural end or at muscle cells, triggers the fast reactions such as release of neurotransmitter or muscle contraction.

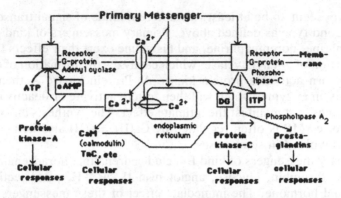

Figure 7-3. Functions of intracellular Ca(II) (see text for details).

7.6.2 Examples of the Action of Intracellular Ca(II)

A few examples will illustrate some of the principles operating in the action of Ca(II).

7.6.2.1 Epinephrine's Action on Glycogenolysis in Muscle Cells

This is one of the best understood systems (Alberts *et al.*, 1983). The process can be summarized as follows.

Epinephrine binds to a receptor at the surface of the muscle cell. The receptor protein, then, together with GTP-bound G protein, activates adenyl cyclase. This is catalytic in the sense that one molecule of receptor could activate more than two molecules of adenyl cyclase. This is the first of several amplifications involved. Adenyl cyclase catalyzes the formation of cAMP; this is the second amplification. One molecule of cAMP binds to one (designated as R) of the subunits of protein kinase A, releasing its active catalytic subunit. Protein kinase A catalyzes the phosphorylation of two subunits, α and β, of a phosphorylase kinase. Phosphorylation of the β subunit, a fast process, activates the catalytic subunit γ. In order for the γ subunit to become fully active, it must be bound with the δ subunit, which is identical to Ca(II)-loaded calmodulin (CaM). The activated protein kinase A then slowly phosphorylates the α subunit. This somehow makes subunit β susceptible to dephosphorylation by phosphoprotein phosphatase; this leads to the inactivation of the phosphorylase kinase. Further dephosphorylation of subunit α returns the entire protein to the original state.

Meanwhile, Ca(II), which has somehow increased in concentration, binds CaM and activates a phosphodiesterase that hydrolyzes and deactivates cAMP. In certain tissues including brain, Ca(II)-CaM seems to activate adenyl cyclase.

Because of the nature of the need, the intracellular concentration of the effecting agents such as cAMP and phosphorylase kinase must increase rapidly and then decrease quickly, once the need disappears. Cells attain this by removing or destroying the effecting agents at a rapid rate, as outlined above. This is one of the important characters of this kind of system.

7.6.2.2 Muscle (Skeletal or Smooth) Contraction

Skeletal muscle contraction involves Ca(II) binding to a subunit of troponin, Tn-C. A signal induces a sudden release of Ca(II) from the endoplasmic (called sarcoplasmic in this case) reticulum into the cytoplasm. Ca(II) immediately binds to Tn-C; this changes the relative positions of Tn-I and Tn-T (I and T subunit of troponin) and removes Tm (tropomyosin) from the blocking position to the nonblocking position (Potter and Johnson, 1982). Now actin is exposed and comes in contact with ATP-loaded myosin head, ATP being hydrolyzed as a result. This changes the conformation of the myosin, which is manifested as muscle contraction.

Ca(II) release would likely be carried out by opening of the Ca(II) channels. Ca(II)-ATPase, which pumps Ca(II) back into the sarcoplasmic reticulum or pumps Ca(II) out of the cellular membrane, is activated by Ca(II)-CaM. Therefore, as soon as Ca(II) bursts into the cytoplasm, it starts being pumped back into the endoplasmic reticulum or pumped out of the cell. Thus, the increase of $[Ca(II)]_c$ ($[Ca(II)]$ in the cytosol) would be only transient.

The contraction of smooth muscle involves a different set of chemicals and reactions. Excitation results in an increase of $[Ca(II)]$ in the cytoplasm. Ca(II)-CaM binds to and activates MLCK (myosin light chain kinase), which phosphorylates myosin using ATP. Phosphorylated myosin binds with actin; this composite (called actomyosin-P) hydrolyzes ATP and in so doing changes its conformation (contraction). A phosphatase is required to dephosphorylate and end contraction (Walsh and Hartshore, 1982). In this sense, it is similar to the mechanism of action of epinephrine on glycogenolysis as discussed above.

7.6.2.3 Release of a Neurotransmitter

An electrical signal causes opening of voltage-gated Ca(II) channels at the presynaptic terminal; it also causes release of Ca(II) from intracellular vesicles. This sudden rise and the subsequent rapid decrease of $[Ca(II)]_c$ occur in a manner similar to the triggering of muscle. Ca(II) binds to CaM; the Ca(II)-CaM enhances the fusion of the neurotransmitter-containing vesicle membrane and the membrane of the neuronal terminal, leading to release of the content into the junction space (DeLorenzo, 1982). This last step, the fusion of membranes and the subsequent release of the content of a vesicle, is quite common and is involved in the release or secretion of a number of chemicals including peptide hormones. This process always requires Ca(II). This function can be understood in terms of the structural effects of Ca(II) on phospholipid membranes as discussed in Chapter 6.

7.6.3 Why Ca(II)?

The free $[Ca(II)]_c$ in the resting cell is quite low, being usually about 10^{-7} M ($0.1\ \mu M$). This low concentration must be maintained despite the fact that $[Ca(II)]$ is quite high in the extracellular medium, typically at 10^{-3} M, because a persistent high concentration of free Ca(II) in the cytoplasm is quite deleterious, as might be inferred from the avidity of Ca(II) to bind a great variety of proteins (see Section 9.4). Free $[Ca(II)]_c$ is kept low usually by Ca(II) pumps; plasma membrane Ca(II)-ATPase extrudes Ca(II), and Ca(II) pumps and other mechanisms sequester Ca(II) into the endoplasmic reticulum at low $[Ca(II)]$ and then into mitochondria when $[Ca(II)]$ becomes relatively high ($> 10^{-6}$ M). This implies that there is a continuous influx of Ca(II) into the cytoplasm; this is partially due to imperfect shielding of the plasma membrane against Ca(II) invasion along the high concentration gradient. That is:

$$d[Ca(II)]_c/dt = \xi d[Ca(II)]/dx - k[Ca(II)]_c \qquad (7\text{-}20)$$

In the resting state, $[Ca(II)]_c$ is kept constant and low despite the relatively high rates of ξ (leak) and k (pump) terms, because they are more or less balanced. The chemical or electrical signal typically opens the gated Ca(II) channels, making ξ very high momentarily. This will lead to a very high $[Ca(II)]_c$ unless the k term is also activated. It usually increases because the Ca(II)-ATPase is activated by Ca(II), but not as much as to balance out the ξ effect completely. This k process, because of its dependence on $[Ca(II)]$, would lag slightly behind the ξ process. Therefore, $[Ca(II)]_c$ can increase quite a bit transiently, typically up to 10^{-5} M (i.e., up to a 100-fold increase), but then the k process would catch up and hence reduce $[Ca(II)]_c$ back to the resting value or near it. Meanwhile, the increase of

$[Ca(II)]_c$ could have exerted its effect as outlined above. The whole process can occur in milliseconds or lower ranges, because of high rates of ξ and k processes.

The situation is almost the same with respect to cAMP. cAMP in the resting cell is very low, indicating little adenyl cyclase activity. A primary messenger activates adenyl cyclase, which produces cAMP. The same messenger very likely induces an increase of $[Ca(II)]_c$, which combines with CaM and then activates, with a slight lag in time, phosphodiesterase; the latter enzyme deactivates cAMP. Therefore, this second messenger (cAMP)'s concentration, too, increases but transiently because of the high destruction rate. However, the whole process involves several steps of catalytic chemical reactions, and hence may not be as fast as that process above involving only Ca(II); these processes last typically from seconds to minutes.

The description above indicates that several specific characters are required for a second messenger. Some of those are (1) its intracellular concentration can be altered rapidly, and (2) it should be able to exert its effects rapidly by binding to its effector. As far as binding strength is concerned, Ca(II) is not particularly favorable; it has a rather special affinity for carboxylate oxygen, but its affinity is not spectacularly greater than those of similar cations such as Mg(II) or Zn(II). These latter cations, however, have more affinity toward nitrogen ligands. If one of these other cations had been chosen for a second messenger, organisms would have developed effectors that contain nitrogen ligand atoms. The only advantage Ca(II) may have in this regard seems to be that acidic amino acids, aspartic and glutamic acid, occur more frequently in nature than basic nitrogen-containing ones. If that is the case, Ca(II) would be more widely useful than the other cations.

The other relevant feature is the size of Ca(II), which is relatively large, 100–120 pm. Because of this, Ca(II)'s coordination is rather flexible, allowing Ca(II) to fit into different structures. Yet Ca(II) is more effective in causing conformational change in proteins, as compared to common cations of similar size such as K(I) or Na(I). In this regard, however, Ca(II) is less effective than Zn(II), Mg(II), and others. More important is the ease of ligand substitution about Ca(II); this is due to the relatively small value of the ionic potential (or charge density). The rates (in units of sec^{-1} at 25°C) of water exchange of metal aquo complexes have been determined to be about 10^8 for Ca(II), 2×10^7 for Zn(II) and Mn(II), 7×10^5 for Mg(II). That is, Ca(II) would be able to bind to and dissociate from a binding site (of a protein or otherwise) more rapidly, perhaps 100-fold or so, than Mg(II). Ca(II) among the common divalent cations, thus, best meets requirements (1) and (2) delineated above.

Because of widespread needs, cytosolic concentrations of Mg(II), K(I), and Cl⁻ are relatively high, being of the order of 10^{-3} M. Therefore, it would be difficult to rapidly change their concentrations by as much as 100-fold. These considerations, based on the ideas expressed by Blaustein (1985) and Levine and Williams (1982), seem to lead to a conclusion that Ca(II) is best fit for the task of a second messenger.

Chemical Principles of
Transport of the Elements

Chemistry of Uptake—
Thermodynamic and Kinetic
Factors in Passive Transport

8.1 Introduction

"Uptake" is used in a more general sense than simply absorption. It is taken to include such phenomena as the incorporation of a specific metal ion into a specific biological functional unit (such as chlorophyll or an enzyme), in addition to transport across a membrane (absorption). The problem of uptake of elements has many facets. Absorption entails the movement of molecules from one location to another across a boundary. The boundary can be a cell membrane; a membrane of a cytoplasmic organelle such as the mitochondrion, chloroplast, or nucleus; or a combination of a cell wall and a cell membrane. Some of the major aspects of absorption are illustrated in Fig. 8-1. There are two classes of transmembrane movement of a substance, whether it be a proton, a metal cation, a metal complex, an anion, or a neutral molecule. One class is known as "active" transport and the other as "passive." In passive transport, a substance moves along an electrochemical potential gradient; the motion occurs spontaneously without any added expenditure of energy. In active transport, a substance is forced to move against an electrochemical potential gradient; thus, the process requires an input of some sort of energy. In most cases this energy comes from the hydrolysis of ATP. A membrane-bound protein seems to be involved in this kind of transport. In other instances of active transport, the transmembranous movement of a substance is coupled with a second process of material transport that itself may or may not be active.

In the passive form of transport, the crucial factors are the nature of the boundary (membrane), the nature of the substance moved (hydrophobic, hydrophilic, or lipophilic) and the nature of the interaction between boundary and substance. Some cells may produce a special material to bind the substance to be transported, thus facilitating the uptake. The

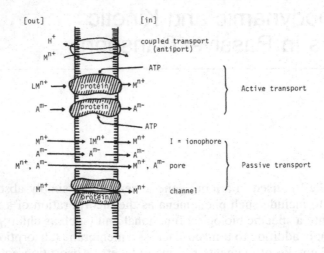

Figure 8-1. Typical transport phenomena across membrane system.

cells may even excrete a scavenger compound (ionophore) to capture the needed substance, particularly if it is not present in abundance. Pore opening of an appropriate size may be present or be created in the membrane to permit the passage of specific substances; these pores may in some cases even be lined with a specially created protein. Pores of the latter type are often referred to as "channels." The passive transport problems that are interesting from a bioinorganic standpoint are: (1) the state of the desired substance in the immediate surrounding of the cell; the surrounding may be seawater for many marine microbes and algae, soil for plant roots, plasma in the blood vessel, the gut lumen for the mucosal cells of the intestine, the tubule lumen for the tubular cells of the kidney, and so forth; (2) the selectivity: how a particular cell specifically absorbs the required substances or how a specific molecule incorporates a specific element; and (3) the structure and mechanisms of action of carriers and scavengers (ionophores). I will discuss some important general factors pertaining to problems 1 and 2 in this chapter; problem 3 will be dealt with in the next chapter.

8.2 Thermodynamic Considerations of Uptake of Metal Ions

Here I shall consider in general terms the thermodynamic factors involved in the uptake of metal ions (anions will be treated in a later section), focusing our attention on those factors that may determine the selectivity of uptake.

Let us consider a simple general case in which a metal (M) complex with a ligand (L) is picked by a receptor ligand L_1; that is:

$$ML + L_1 \rightleftharpoons ML_1 + L \qquad K_{L_1}(M) \qquad (8\text{-}1)$$

The amount, expressed as the concentration, of M taken up by the receptor is given by:

$$[ML_1] = K_{L_1}(M)\,[ML][L_1]/[L] \qquad (8\text{-}2)$$

8.2.1 From Seawater

When $L = (H_2O)_n$ and the medium is water, $K_{L_1}(M)$ is the ordinary metal complex formation constant (stability constant) and $[L]$ can be assumed to be equal to a constant C. Therefore,

$$[ML_1] = (K_{L_1}(M)/C)[ML][L_1] \qquad (8\text{-}3)$$

Next, we suppose that ML is present in a vastly large volume, corresponding to that of the sea, and that L_1 is localized at an organism or a single cell surface. In this case the uptake of the metal ion by the cell would not change the concentration of ML remaining in the medium. Therefore, the amount taken up is simply proportional to $K_{L_1}(M)[ML]$.

$K_{L_1}(M)$ values for typical ligands are illustrated in Fig. 8-2. The order of magnitude of $K_{L_1}(M)[ML]$ values of these and other ligands are summarized in Table 8-1, which indicates: (1) phosphate and polyphosphates tend to bind relatively strongly and seem to be suitable to take up Mg(II) and Ca(II). (2) Carboxylates, especially dicarboxylates, are suited to complex Ca(II). (3) The amine (and carboxylate) groups of amino acids and other nitrogenous ligands are suitable for the uptake of Cu(II). (4) Sulfur

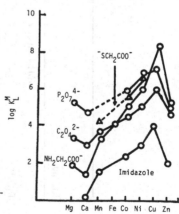

Figure 8-2. Formation constants of complexes of divalent metals with various ligands.

Table 8-1. $K_{L_1}^M$ and $K_{L_1}^M$ [ML] Values

Ligand		Cation						
		Na(I), K(I)	Mg(II)	Ca(II)	Zn(II)	Fe(II,III)	Cu(II)	Mn(II)
	[ML] (in sea)	10^{-1} M, 10^{-2} M	10^{-2} M	10^{-2} M	10^{-7} M	10^{-7} M	10^{-8} M	10^{-8} M
$P_2O_7^{4-}$	$K_{L_1}^M$	10^2	10^5	10^5	—	10^6	10^7	10^6
	$K_{L_1}^M$ [ML]	$10, 1$	10^3	10^3	—	10^{-1}	10^{-1}	10^{-2}
$C_2O_4^{2-}$	$K_{L_1}^M$	—	10^3	10^3	10^5	10^7	10^8	10^4
	$K_{L_1}^M$ [ML]	—	10	10	10^{-2}	1	1	10^{-4}
$NH_2CH_2COO^-$	$K_{L_1}^M$	—	10^2	10^2	10^{10}	10^5	10^9	10^4
	$K_{L_1}^M$ [ML]	—	1	1	10^3	10^{-2}	10	10^{-4}
$^-SCH_2COO^-$	$K_{L_1}^M$	—	10^4	10^4	—	—	—	10^4
	$K_{L_1}^M$ [ML]	—	10^2	10^2	—	—	—	10^{-4}
X-537	$K_{L_1}^M$	$10^2, 10^3$	10^3	10^4	—	—	10^4	10^4
	$K_{L_1}^M$ [ML]	$10, 10$	10	10^2	—	—	10^{-4}	10^{-4}

ligands such as cysteine seem most suitable for Zn(II) uptake. (5) To permit discrimination in the uptake of Ca(II) [versus Mg(II)], a specific ligand such as X-537, a polyether-carbonyl ionophore, may be required (see Section 9.1.2).

The above treatment is valid only when each individual element is present alone in the medium. In seawater or any real situation, a variety of elements are present in the vicinity of an organism or a cell, which, in turn, has a multitude of different receptor ligands. Therefore, one element may interfere with the uptake of another, or the presence of different receptor ligands may increase the selectivity of the uptake of cations. We consider the simplest such case: two elements M_1 and M_2 versus two receptor ligands L_1 and L_2. There are four reactions to take into account (where the state of complexation in the surrounding medium is, for convenience, ignored so that ML is now expressed simply as M):

$$M_1 + L_1 \rightleftharpoons M_1L_1 \qquad K_{L_1}(M_1) \qquad (8\text{-}4)$$

$$M_1 + L_2 \rightleftharpoons M_1L_2 \qquad K_{L_2}(M_1) \qquad (8\text{-}5)$$

$$M_2 + L_1 \rightleftharpoons M_2L_1 \qquad K_{L_1}(M_2) \qquad (8\text{-}6)$$

$$M_2 + L_2 \rightleftharpoons M_2L_2 \qquad K_{L_2}(M_2) \qquad (8\text{-}7)$$

If $[M_1]_{\text{total}} \ll [L_1]_{\text{total}}$ and $[M_2]_{\text{total}} \ll [L_2]_{\text{total}}$ and $K_{L_i}(M_i) > K_{L_j}(M_i)$, then M_1 would be specifically picked up by L_1 and M_2 by L_2; this is because no significant amount of M_1 would be left for binding with L_2 and only a trace of M_2 would be left to bind with L_1. In general, however, there is an interference between them. We define an apparent stability constant (after Frausto da Silva and Williams, 1976) as follows:

apparent $K'_{L_1}(M_1)$

$$= [M_1L_1]/([M_1]_{\text{total}} - [M_1L_1]) \times ([L_1]_{\text{total}} - [M_1L_1]) \qquad (8\text{-}8)$$

The material balances are:

$$[M_1]_{\text{total}} = [M_1] + [M_1L_1] + [M_1L_2]$$
$$= [M_1] + K_{L_1}(M_1)[M_1][L_1] + K_{L_2}(M_1)[M_1][L_2] \qquad (8\text{-}9)$$

$$[L_1]_{\text{total}} = [L_1] + [M_1L_1] + [M_2L_1]$$
$$= [L_1] + K_{L_1}(M_1)[M_1][L_1] + K_{L_1}(M_2)[M_2][L_1] \qquad (8\text{-}10)$$

Therefore:

$$\text{apparent } K'_{L_1}(M_1) = K_{L_1}(M_1)[M_1][L_1]/([M_1] + K_{L_2}(M_1)[M_1][L_2])/([L_1]$$
$$+ K_{L_1}(M_2)[M_2][L_1])$$
$$= K_{L_1}(M_1)/(1 + K_{L_2}(M_1)[L_2])/(1 + K_{L_1}(M_2)[M_2])$$
$$= K_{L_1}(M_1)/\alpha_{M_1}\alpha_{L_1} \qquad (8\text{-}11)$$

where

$$\alpha_{M_1} = 1 + K_{L_2}(M_1)[L_2] \tag{8-12}$$

$$\alpha_{L_1} = 1 + K_{L_1}(M_2)[M_2] \tag{8-13}$$

That is, the apparent stability constant becomes smaller than the true value by the factor α_{M_1} and α_{L_1}, which must always be larger than unity. In general,

$$\alpha_{M_1} = 1 + \sum_i \{\sum_j \mathscr{S}^j_{L_i}(M_1)[L_i]\} \tag{8-14}$$

$$\alpha_{L_1} = 1 + \sum_i \mathscr{S}^H_i[H^+]^i + \sum_i K_{L_1}(M_i)[M_i] \tag{8-15}$$

when there are n ligands $L_1, \ldots, L_n$ and m metal ions $M_1, \ldots, M_m$ and

$$\mathscr{S}^j_{L_i}(M_1) = \prod_j K^j_i(M_1); \qquad K^j_i(M_1): M_1(L_i)_{j-1} + L_i \rightleftharpoons M_1(L_i)_j \tag{8-16}$$

$$\mathscr{S}^H_i = \prod_i K^H_i; \qquad K^H_i: L_1(H^+)_{i-1} + H^+ \rightleftharpoons L_1(H^+)_i \tag{8-17}$$

Therefore, a meaningful measure of uptake potential would be $K_{L_1}{}'(M_1)[M_1]$ ($[M_1]$ in seawater, for example). The calculated log $K_{L_1}(M_1)[M_1]$ values at pH 8 and 25°C for a series of divalent metal ions [and Fe(III)] and several representative ligands (Frausto da Silva and Williams, 1976) indicate that Mg(II) is preferentially taken up by polyphosphates, Ca(II) by polycarboxylates, Fe(II) by phenanthroline, Fe(III) especially by oxamic acids, and Zn(II) by sulfide ligands (e.g., cysteine). In fact, it is known that many of the Ca(II)-binding proteins have the carboxylate groups of glutamate and aspartate residues situated at the Ca(II)-binding site. Similarly, oxamic acid is the basis of the typical scavengers of Fe(III) (Section 9.1.3). The binding sites of many Zn(II) enzymes include cysteine residues as well as histidine residues.

8.2.2 From Gut Lumen or Soil

Here we have a much more complicated situation, because the metal ions present very probably are complexed with a variety of different ligands other than water; they cannot simply be regarded as free ions. For example, Fe(III) could be present as Fe(oxalate)$_n$, Fe(amino acid)$_n$, Fe(hexose)$_n$, or Fe(porphyrin) in the intestine. Iron can also be in the Fe(II) state due to some reducing components in the foodstuff. In general, the uptake should then be expressed by:

$$ML + L_1 \text{ (receptor)} \rightleftharpoons ML_1 + L \qquad K_{L_1}(ML) \quad (8\text{-}18 = 8\text{-}1)$$

This equilibrium constant $K_{L_1}(ML)$ is equal to $K_{L_1}(M)/K_L(M)$ where

$$K_{L_1}(ML) : M(HOH)_n + L_1 \rightleftharpoons ML_1(HOH)_m \qquad (8\text{-}19)$$

$$K_L(ML) : M(HOH)_n + L \rightleftharpoons ML(HOH)_m \qquad (8\text{-}20)$$

Therefore, as expected, $K_{L_1}(M)$ must be greater than $K_L(M)$ in order for the uptake of M by L_1 to take place to a significant extent. A study (Forth, 1974) on the absorption of chelated Fe(III) by intestine showed that a chelated Fe(III) whose stability constant was lower than 10^{12} was transported into the mucosal cell, but the iron was not absorbed if the stability constant of the chelate was higher than 10^{12}. This result indicates that $K_{L_1}(M)$, the affinity of the receptor site on the mucosal membrane for Fe(III), is about 10^{12}–10^{13}. However, it seems that there is another mechanism for the uptake of iron on the mucosal membrane. This latter is an active transport system.

The fact is undeniable that the animal body or the plant root succeeds in absorbing necessary minerals selectively under normal conditions, provided there is an adequate supply present, despite this seemingly complicated situation. Then, the question becomes how this is accomplished. Some of the conclusions drawn in the previous section may continue to supply a partial explanation, but they do not seem to be sufficient for a complete understanding of this problem.

8.2.3 From Circulating System to Individual Cells

In passing from the gut lumen to the individual cells, a metal ion needs to go through at least three discriminating barriers. The first barrier is the mucosal membrane described above. The metal ion is then loaded onto a transfer protein in the circulating system and must finally be taken up by a cell. At each step the metal ion will be bound to an accepting molecule, and in each of these loading (and unloading) processes, some selectivity, strong or weak, will be exerted on the incoming metal ions to be processed. The source of this selectivity effect is, of course, the same thermodynamic one discussed in the previous two sections.

Let us illustrate this situation (see Fig. 8-3), assuming that two metal ions go through three barriers. In the gut lumen, the ratio of the amounts of M_1 to M_2 is ($[M_1]/[M_2]$). By the time these ions have gone through the first barrier, the ratio would become $(K_{L_1}(M_1)/K_{L_1}(M_2))([M_1]/[M_2])$. Let us assume that after this point the two metals compete for a common transfer molecule, L_2. Two molecules, M_1L_2 and M_2L_2, form and their

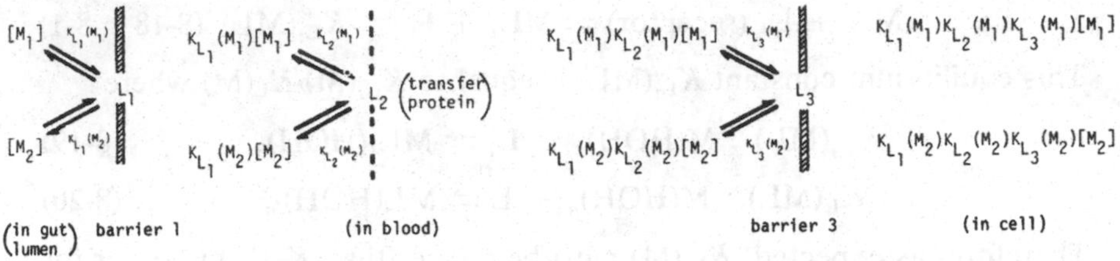

$$[M_1] \xrightleftharpoons{k_{L_1}(M_1)} K_{L_1}(M_1)[M_1] \xrightleftharpoons{k_{L_2}(M_1)} K_{L_1}(M_1)K_{L_2}(M_1)[M_1] \xrightleftharpoons{k_{L_3}(M_1)} K_{L_1}(M_1)K_{L_2}(M_1)K_{L_3}(M_1)[M_1]$$

$$[M_2] \xrightleftharpoons{k_{L_1}(M_2)} K_{L_1}(M_2)[M_2] \xrightleftharpoons{k_{L_2}(M_2)} K_{L_1}(M_2)K_{L_2}(M_2)[M_2] \xrightleftharpoons{k_{L_3}(M_2)} K_{L_1}(M_2)K_{L_2}(M_2)K_{L_3}(M_2)[M_2]$$

L_1 L_2 (transfer protein) L_3

$\begin{pmatrix} \text{in gut} \\ \text{lumen} \end{pmatrix}$ barrier 1   (in blood)   barrier 3   (in cell)

Figure 8-3. Selection through a multistep process.

concentration ratio will be $(K_{L_1}(M_1)K_{L_2}(M_1)/K_{L_1}(M_2)K_{L_2}(M_2))([M_1]/[M_2])$. They are then taken up by L_3 of the target cell. The ratio of the amounts taken up by this cell will now be $(\prod_i K_{L_i}(M_1)/\prod_i K_{L_i}(M_2))([M_1]/[M_2])$. If $K_{L_i}(M_1)/K_{L_i}(M_2)$ $(i = 1, 2, 3) = 10$ at each step, the selectivity for M_1 of the uptake at the first stage is only ten to one. However, by the time the final destination has been reached, the selectivity factor would have become a thousand to one. The fact is that a transfer protein usually has a very high selectivity for a given element. Even when this is not true, however, this sketch shows that the overall selectivity can become great when use is made of a multistep uptake process in which selective steps can overcome the shortcomings of less effective ones.

8.3 Kinetic Factors in Uptake of Metal Ions

In the target cell, a metal ion is incorporated into a specific functional unit; it could be, for example, protoporphyrin IX (heme group), chlorin (porphyrin of chlorophyll), or an apoenzyme. No mistake can be tolerated at this stage. For example, chlorin must incorporate Mg(II), not Zn(II) or any other metal, although it is known that Zn(II) can be incorporated into chlorin *in vitro*. This implies that, thermodynamically speaking, there is no very great difference between Mg(II)-chlorin and Zn(II)-chlorin; as a matter of fact, it has been shown that Zn(II) binds with chlorin more strongly than Mg(II) does. As outlined above, it might be that a high degree of metal ion selection could in such a case be achieved prior to the time the metal ion enters the target cell. However, since Zn(II) is itself an important and ubiquitous ion, it, too, may need to be present in the chloroplast, along with Mg(II) and other ions. What could then prevent Zn(II) from being incorporated into chlorin? Or, put another way, why is Mg(II) incorporated into chlorin preferentially?

Another situation of a different type is illustrated by Cu/Zn-superoxide dismutase. Cu(II) and Zn(II), 1 gram atom each, are bound in the superoxide dismutase. Each element must bind to its specific binding site; but these two binding sites happen to be very close to each other in this enzyme. How is the selectivity achieved?

Some solutions to these problems may be proposed. For example, the structure of the binding site might be such as to specify the binding ion to a large extent. Each metal ion has its own preferred structure for coordination. For example, typical preferred structures are: Mg(II), tetrahedral (T_d, four-coordinate); Ca(II), octahedral (O_h, six-coordinate); Mn(II), O_h; Fe(II, III), O_h; Zn(II), T_d; Cu(II), square-planar (four-coordinate); and Cu(I), T_d. Therefore, if the apoprotein has created a specific arrangement of binding ligands (T_d, O_h, or whatever), as well as a specific hole size, then the metal ion that prefers such a structure and hole size would bind to the site more easily than any other metal ion. This seems to be the case, for instance, with Cu/Zn-superoxide dismutase. The copper-binding site has a slightly distorted square-planar structure, whereas the zinc-binding site is a distorted tetrahedron (Richardson *et al.*, 1975). The structural effect in this case is again controlled by thermodynamic factors.

A second conceivable mechanism of selective uptake employs kinetic effects. The final uptake of a metal ion can be described by:

$$M_1 + A(acceptor) \longrightarrow M_1A \qquad (8\text{-}21)$$

That is, the formation of M_1A is regarded as irreversible in one way or another. It is also very likely that this final step is catalyzed by an enzyme. Such an enzyme, ferrochelatase (Dailey and Fleming, 1983), is known for the incorporation of Fe(II) into protoporphyrin IX, and it is assumed that one exists for the Mg(II) incorporation into chlorin. An enzyme of this kind would need to be fairly specific in its binding of a metal ion and in catalyzing the ion's insertion into a specific molecule (A in equation 8-21). It may be that it would also be capable of inserting wrong metal ions, though this would occur at a much slower rate. If the reaction is reversible, thermodynamic criteria would take control and a wrong chelate could eventually accumulate. Therefore, it is most important that reaction 8-21 be irreversible. If the formation rate constant of M_1A is large and the reverse reaction is very slow, then once M_1A is formed, it would be very difficult to replace M_1 by another metal ion M_2, even though M_2A may be thermodynamically favored over M_1A. Nonetheless, this approach cannot provide a perfect guarantee that M_2A will not be formed instead of M_1A in some cases. One way to prevent the reverse reaction of 8-21 (and the replacement of M_1 by M_2) is to remove the M_1A from the reaction site as soon as it is formed, so that M_2 will not have access to M_1A. For example, M_1A may be produced in the cytoplasm and then be rapidly incorporated into a membrane; M_2, the other metal ion that is present in the cytoplasm, will thus be prevented from coming into contact with M_1A. This mechanism may well apply to the case of chlorophyll.

One final interesting example is hemoglobin. Zn(II)-protoporphyrin has been extracted from the erythrocytes of lead-intoxicated or iron-deficient patients (Lamola and Yamane, 1974). Zinc is abundant in erythrocytes but normally is not incorporated wrongly into porphyrin. Lead is known to disrupt several steps in the heme synthesis, but must not have done so completely in this case, because Zn(II)-protoporphyrin was produced. The crucial step of incorporation of Fe(II) into hemoglobin is catalyzed by ferrochelatase, as was mentioned above, and the activity of this enzyme has been shown to be inhibited by lead. Zn(II) insertion, which is also catalyzed under normal conditions by the same enzyme, may well be able to occur nonenzymatically. This would explain the production of Zn(II)-porphyrin in the presence of lead. In the case of the iron-deficient patients, the ratio Fe(II)/Zn(II) is substantially lower than normal. If so, Zn(II) might then insert itself into the porphyrin to a significant degree, either through the catalytic influence of chelatase or by a nonenzymatic means.

8.4 Uptake of Anions—Thermodynamic Factors

8.4.1 Ionic Radius

The anions of biological interest are mostly oxyanions, except for a few simple ones like F^-, Cl^-, I^-, and OH^-. The ionic radii of relevant anions are given in Table 8-2. Note that the tetrahedral oxyanions with atoms from the second or lower rows of the periodic systems have similar radii, because the radius of an oxyanion is determined mainly by the oxygen atoms due to the small size of the central atom. The same can be said about CO_3^{2-}, NO_3^-, and BO_3^-. Protonation of an anion reduces its radius but not significantly in most cases. The largest change is seen in going from CO_3^{2-} (185 pm) to HCO_3^- (163 pm). It may be inferred from the data that common or similar uptake mechanisms or sites may exist for SO_4^{2-}, SeO_4^{2-}, and MoO_4^{2-}; all of them have the same charge and similar radii. For example, rabbit renal brush-border membrane has been shown to transport SO_3^{2-}, $S_2O_3^{2-}$, SeO_4^{2-}, MoO_4^{2-}, CrO_4^{2-}, WO_4^{2-} as well as SO_4^{2-} (E. G. Schneider et al., 1984). Then how can cells distinguish between, for example, SO_4^{2-} and MoO_4^{2-}?

Table 8-2. Ionic Radii of Anions[a]

Ion	Radius (pm)	Ion	Radius (pm)
F^-	136	SO_4^{2-}	230
OH^-	140	ClO_4^-	236
HCO_3^-	163	PO_4^{3-}	238
CO_3^{2-}	185	SiO_4^{4-}	240
NO_3^-	189	SeO_4^{2-}	243
BO_3^{3-}	191	MoO_4^{2-}	254
SH^-	195		
Br^-	195		
I^-	216		

[a] Adapted from Frausto da Silva and Williams (1976).

8.4.2 Thermodynamics of Anion Uptake

The uptake of an anion X^- is described by:

$$X^-(HOH)_n + L^+ \rightleftharpoons X^- — L^+ + n(HOH) \qquad (8\text{-}22)$$

In contrast to a cation, an anion requires a positively charged site, or at least a positively polarized dipole site. Surprisingly, there are only a few possible positive sites in biopolymers, in contrast to the wide variety of negatively charged sites. The possible protein sites are $—NH_3^+$ (lysine and N-terminal residues), imidazolium (histidine), and guanidium (arginine). A positive dipole may be provided by hydrogen bonds associated with $—OH$, $—NH$, and $—SH$; for example, $X^- \cdots H—N—$. However, the energy associated with a hydrogen bond alone does not seem to be sufficient to ensure secure binding. This means that at least one formal positive charge is definitely required for strong anion binding. The following treatment is largely based on Frausto da Silva and Williams (1976).

Equation 8-22 implies a stripping of associated water molecules from an anion upon its binding to the uptake site. If the surroundings of the uptake site are hydrophilic and hence contain water molecules, then the binding of X^- to L^+ would be hindered by the tendency of X^- to remain solvated by water. This suggests that a hydrophobic environment would favor the binding of an anion to a biomolecule, and that anion-binding sites would lie not at the surface of a protein but rather embedded in the protein.

The Gibbs free energy change for reaction 8-22 is given by ΔG:

$$\Delta G = \Delta H - T\,\Delta S = (\Delta H_p - \Delta H_h) - T(\Delta S_p - \Delta S_h) \qquad (8\text{-}23)$$

where $\Delta H_{h,p}$ and $\Delta S_{h,p}$ correspond to the following reactions:

$$X^- + L^+ \text{ (protein)} \rightleftharpoons X^- \!\!-\!\! L^+ \qquad \Delta H_p, \Delta S_p \qquad \text{(8-24)}$$

$$X^- + n\text{HOH} \rightleftharpoons X^-(\text{HOH})_n \qquad \Delta H_h, \Delta S_h \qquad \text{(8-25)}$$

The binding between X^- and L^+ is electrostatic (in contrast to the coordination bonding between a cation and a ligand), and the electric potential of an anion, which largely determines the electrostatic force, is rather small because of the comparatively large anionic radii. ΔH_p and ΔH_h may be approximated by:

$$\Delta H_p = -A_p z/D_p r_p, \qquad \Delta H_h = -A_h z/D_h r_h \qquad \text{(8-26)}$$

where D_p and D_h are the dielectric constants of the protein environment and water, respectively; r_p is the distance between X^- and L^+ in the protein, r_h is the distance between X^- and the center of the dipole of the associated water; z is the charge on the anion; and A_p and A_h represent constants. The hydration energy data given in Table 8-3 confirm the validity of formulas 8-26. Therefore,

$$\begin{aligned} \Delta G &= -A_p z/D_p r_p + A_h z/D_h r_h - T(\Delta S_p - \Delta S_h) \\ &= -[-z(A_p/D_p r_p + A_h/D_h r_h)] + T(-\Delta S_h + \Delta S_p) = -\Delta H' + T\Delta S' \end{aligned}$$

$$\text{(8-27)}$$

Table 8-3. Hydration Energy of Anions[a]

	ΔG_h (kcal/mole)	ΔH_h (kcal/mole)	ΔS_h (e.u.)	$z/r_h{}^{b}$
F^-	-102	-113	-36	4.52
Cl^-	-73	-80	-22	3.76
Br^-	-66	-72	-19	3.57
I^-	-59	-63	-13	3.32
OH^-	-104	-117	-43	4.44
ClO_4^-	-62	-68	-19	3.11
NO_2^-	-71	-79	-27	
NO_3^-	-63	-70	-22	3.65
CH_3COO^-	-87	-96	-31	
CO_3^{2-}	-298	-321	-76	7.40
SO_4^{2-}	-235	-254	-63	6.35
SeO_4^{2-}	-220	-239	-66	6.10

[a] Adapted from Frausto da Silva and Williams (1976).
[b] r_h = ionic radius + r_{H_2O} (0.085).

Both z and $A_p(A_h)$ can be taken to be positive; ΔS_h has a negative value. We can deduce a few general trends from this equation.

1. Smaller D_p, i.e., increased hydrophobicity of the protein environment, increases $\Delta H'$ and thus is advantageous for anion binding.
2. A small size of the anion (smaller r_h) tends to reduce $\Delta H'$. However, ΔS_h of a smaller anion will be larger in value (more negative) than in the case of a larger anion; therefore, the binding of a smaller anion is favored by the $T\Delta S'$ term and opposed by the $\Delta H'$ term.
3. In contrast, binding of a larger anion is favored by the $\Delta H'$ term but not helped by the $T\Delta S'$ term.

The binding strengths of anions to proteins, as deduced from swelling effects, is known as the "lyotropic series"; it decreases in the order:

$$ClO_4^- > NCS^- \sim I^- > ClO_3^- \sim N_3^- > NO_3^- \sim NCO^- > Br^- >$$
$$Cl^- > CH_3COO^- \sim F^- \sim OH^- > SO_4^{2-} > HPO_4^{2-}$$

The hydration energy of anions increases in an order that is almost exactly the reverse of this series; that is (from Table 8-3):

$$I^- < ClO_4^- < NO_3^- < Br^- < Cl^- < CH_3COO^- < F^- < OH^- < SO_4^{2-}$$

This sequence implies that the order of the lyotropic series is a function of the ease of stripping the associated water from an anion; in other words, the $A_h/D_h r_h$ is dominant in the expression 8-27. A study of the inhibitory effect of anions on acetoacetic decarboxylase produced the following series (Fridovich, 1963):

$$NCS^- > ClO_4^- > I^- > B(C_5H_6)_4^- > ClO_3^- > Br^- > Cl^- >$$
$$CCl_3COO^- > F^- > IO_3^-$$

This is essentially the same as the lyotropic series. Strongly hydrated anions such as SO_4^{2-} and HPO_4^{2-} apparently do not bind to this protein.

Strongly hydrated anions such as OH^-, SO_4^{2-}, and $P_2O_7^{2-}$ do, however, bind strongly to other proteins. For example, most kinases bind ATP (and hence $P_2O_7^{2-}$) strongly, carbonic anhydrase binds OH^-, α-chymotrypsin binds SO_4^{2-}, and ribonuclease binds SO_4^{2-} and PO_4^{3-} (AsO_4^{3-}). It seems that the anion-binding site in these proteins consists of a combination of several cations of the $-NH_3^+$ (lysine), guanidium, and imidazolium groups. Such multication systems may be able to overcome the troublesome water stripping of these anions; that is, the combined $A_p/D_p r_p$ terms can become comparable to, or may even exceed, the large $A_h/D_h r_h$ term in formula 8-27. In the case of carbonic anhydrase, the binding site of OH^- is a Zn(II) cation. In general, the metal cation in a metalloprotein or metalloenzyme can serve as an anion-binding site if the coordination sphere of the metal is partially open. For example, carbonic anhydrase binds, besides OH^-, various anions with binding strengths in the following order (Pocker and Stone, 1967, 1968): $I^- > Br^- > Cl^- > F^-$; $NCO^- > NCS^-$, $N_3^- \gg ClO_4^- > NO_3^- > CH_3COO^-$. If this binding is represented by $X^-(H_2O)_n + Zn(H_2O)_m \rightleftharpoons X^-Zn(H_2O)_{m-1} + (n+1)H_2O$ or the binding reaction occurs between an X^- and a simple $Zn(II)$ ion in water, then the binding strength series would be expected to be the reverse of the one observed. Thus, the binding of X^- to the $Zn(II)$ embedded in the protein seems still to be influenced very much by the hydrophobicity of the binding site environment. In other words, the stripping of water molecules from $X^-(H_2O)_n$ is still the most important factor in the binding of these anions, although the metal ion has clearly had an effect.

8.4.3 Other Factors in Anion Uptake

Some anions undergo oxidation or reduction before being incorporated into organic compounds. Other anions may condense with various biological organic compounds, e.g.:

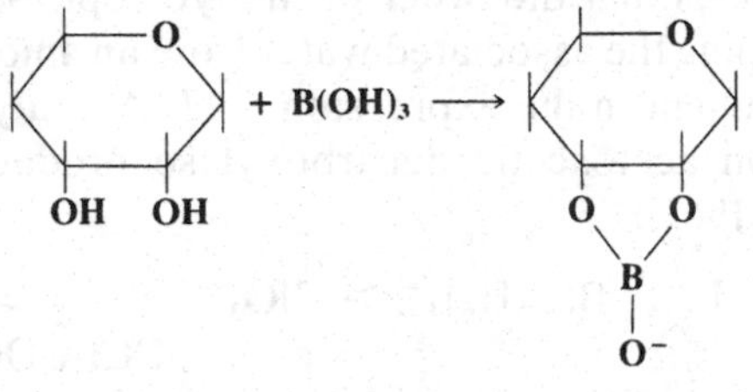

$$(8-28)$$

Phosphate would, of course, be incorporated into a large variety of phosphate esters such as ATP, AMP, and nucleic acids. Specific chemical reactions such as these (including redox reactions) would help improve the selectivity of anion uptake by cells, even though a great degree of selectivity may not be expected for simple thermodynamic anion binding, as was pointed out in the previous sections.

Specific proteins have been isolated that bind and carry sulfate or phosphate. On the other hand, there is some indication that sulfate and molybdate may be competitively absorbed in mammalian intestine. These cases will be discussed in detail in later chapters.

Ionophores, Channels, Transfer Proteins, and Storage Proteins

Cells must selectively absorb nutrients, including inorganic ones; indiscriminate absorption would seriously jeopardize their well-being. The general chemical principles of selective uptake were discussed in the previous chapter. In the present chapter we concentrate on some of the more concrete aspects of passive uptake including the role of ionophores (and scavengers), channels, (serum) transfer proteins, and storage proteins. The first two concepts are associated with the transport problems across the membrane, whereas the latter two involve specific proteins that bind inorganic elements in the cytoplasm and in blood serum or are used to store inorganic chemicals.

Other aspects of transmembranous translocation, including binding and carrier proteins, are related to active transport, and these will be discussed in the next chapter.

9.1 Ionophores

"Ionophore" literally means "ion-bearer" (from Greek, -$phoros$ = bearer); an ionophore can be defined for our purposes as a nonprotein compound that has a specific binding affinity and carrying property for a specific ion. Most ionophores act as ion carriers across a membrane. This function is not universally implied in the definition of an ionophore, however. Indeed, the physiological functions of some ionophores are not yet well understood. Ionophores are known to exist for such cations as H^+, K(I), Na(I), Mg(II), Ca(II), and Fe(III), but none are known for anions. Cations are soluble in water but not in nonpolar solvents. The interior of a membrane is made up mostly of the hydrocarbon chains of phospholipids and is hydrophobic. Thus, a phospholipid bilayer membrane is rather impermeable to most cations, whereas it is much more permeable to smaller anions and neutral molecules. A small organic anion such as acetate may also readily pass through a membrane either as the anion or, more likely, in its neutralized form. This difference in a membrane's permeability to cations and anions may partially explain the need for the existence of ionophores for cations and the fact that anion ionophores are yet to be discovered.

These basic considerations lead to the inference that ionophores involved in membrane transport would have hydrophobic character, at least at their surface, so that they will have a strong affinity for the membrane. Hydrophobic character is, in fact, a general property of known ionophores with the exception of siderophores. This observation in turn suggests strongly and supports the notion that ionophores play an important role in the transmembranous translocation of cations, thereby in most cases facilitating passive diffusion.

The cations for which ionophores are known are the ones all organisms require in large amounts or ones that are essential to all organisms but may exist in rather insoluble forms such as $Fe(OH)_3$. Thus, the ionophores for iron, which now are generally known as "siderophores," act as scavengers for the trace amounts of iron that are available. There may exist ionophores for other cations such as Zn(II), Mn(II), Co(II), and Cu(II), but, if so, they remain to be discovered.

9.1.1 Ionophores for Proton

Proton ionophores had not, until recently, been recognized as such, but had been instead classified in other ways. An example is the "uncoupler" involved in oxidative phosphorylation (Hinkle and McCarty, 1978). In the chemiosmotic theory of Mitchell (1977), the proton concentration gradient across a membrane is produced either by the electron transfer system in mitochondria or by photosystem I in chloroplasts, and it is supposed to drive the process of ATP formation. This ATP formation, coupled with the electron transfer system, is inhibited by, for example, dinitrophenol; thus, dinitrophenol was called an "uncoupler." The uncoupling is caused by the dissipation of the proton concentration gradient. H^+ binds with dinitrophenolate anion, forming a neutral molecule soluble in hydrophobic medium. The neutral molecule diffuses through the membrane along the proton concentration gradient, thus annihilating the proton gradient necessary for ATP formation. From this example it can be inferred that a typical proton ionophore is a weak organic acid whose main body is hydrophobic. A few exogenously used proton ionophores are shown in Fig. 9-1.

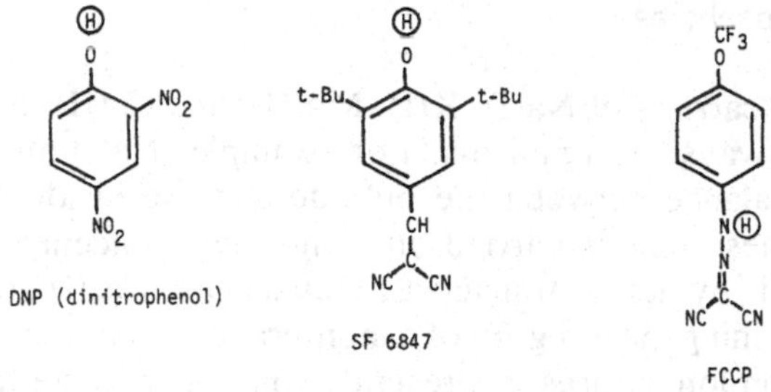

Figure 9-1. Proton ionophores.

It is quite conceivable that other endogenous as well as exogenous organic acids act as proton ionophores. For example, a group of cation ionophores, called "nigericins," has the general form XCOOH (see Section 9.1.2). When a nigericin binds a cation, it must give off H^+ from the carboxyl group, i.e.:

$$X—COOH + K(I) \rightarrow [K^I(X—COO^-)]^0 + H^+ \qquad (9\text{-}1)$$

$$X—COOH + Ca(II) \rightarrow [Ca^{II}(X—COO^-)]^+ + H^+ \qquad (9\text{-}2)$$

Thus, both cations and protons are carried, but in opposite directions. For instance, at the outer surface of the membrane a nigericin picks up K(I) as $[K(XCOO)]^0$, transports it along the concentration gradient across the membrane, and releases the K(I). It then picks up H^+ from the inner surface of the membrane and subsequently migrates back.

Three types of organic compounds involved in the oxidation–reduction catalysis invariably contain hydrogen atom(s) that they release in the form of H^+ at the same time that they donate electron(s). Thus:

$$NAD(P)H \rightarrow NAD(P)^+ + H^+ + 2e \qquad (9\text{-}3)$$

$$\text{dihydroubiquinone} \rightarrow \text{ubiquinone} + 2H^+ + 2e \qquad (9\text{-}4)$$

$$FADH_2 \ (FMNH_2) \rightarrow FAD \ (FMN) + 2H^+ + 2e \qquad (9\text{-}5)$$

These reactions are the very basis for the chemiosmotic theory, in which H^+ is assumed to be pumped out along with the electron transfer in which NADH, FAD, and ubiquinones are involved. Thus, in a sense these molecules, too, can be regarded as proton carriers. They are, however, not weak acids, but rather require an electron release or acceptance in the process of releasing or binding protons.

9.1.2 Cation Ionophores

These are carriers of Na(I), K(I), Mg(II), and Ca(II), cations that are vital in many ways to organisms. For example, Na(I) and K(I) control the osmotic balance between the outside and the inside of a cell. The transport of these ions is carried out, and their concentration gradients are maintained, by active transport mechanisms. Ca(II) seems to be involved in sustaining the integrity of a membrane, and it, too, is transported actively. Cation ionophores, if present, would jeopardize the "actively" established cation concentration gradients. This is the reason that most natural cation ionophores act as antibiotics; a cation ionophore acting on a microorganism could upset its cation metabolism and lead to its destruction.

Naturally occurring cation ionophores are further classified as macrotetrolides, depsipeptides, cyclic peptides, and carboxylates. Typical examples are listed in Table 9-1, along with their specificities for cations. The compounds included in the first three classes are ringlike, and contain either oxygen atoms and/or carbonyl groups that coordinate the cations, particularly alkali metal ions (see Fig. 9-2). The matching between the size of the hole created by the ringlike compound and the size of the cation determines the specificity. When coordinated to a cation, the ionophore exposes its hydrophobic side, which has an affinity for the hydrophobic interior of the lipid bilayer of a membrane. Thus, despite the

Table 9-1. Cation Ionophores

Ionophore	Specificity
Macrotetrolide	
Nonactin	K(I) > Rb(I) > Cs(I) > Na(I) > Ba(II)
Monactin	"
Dinactin	"
Depsipeptide	
Valinomycin	Rb(I) > K(I) > Cs(I) > Ba(II) > Na(I)
Eniatin B	K(I) (~Ba(II)) > Rb(I) > Na(I) > Cs(I)
Beauvericin	Ca(II) >> K(I) ~Rb(I) ~Cs(I) > Na(I)
Cyclic peptide	
Antamanide	Na(I) > K(I)
Carboxylate	
Nigericin	K(I) > Na(I)
Monesin	Na(I) > K(I)
X-537	Ba(II) > Sr(II) > Mg(II) > Ca(II) > Mn(II) > K(I) > Rb(I) > Na(I)
A23187	Ca(II) > Mg(II)

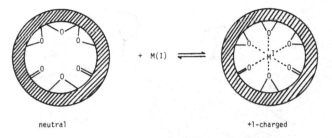

neutral   +1-charged

Figure 9-2. Cation ionophores of ring type.

positive charge of the resulting complex, its diffusion through the membrane is possible. The hydrophobic groups found in these ionophores are methyl, ethyl, propyl, and benzyl groups. Alkali metal ions do not have strong coordinating abilities; however, they prefer oxygen atoms as coordinating ligands (as compared to nitrogen, sulfur, and so forth), and multiple coordination such as that shown in Fig. 9-2 helps increase the binding stability.

Carboxylate ionophores (nigericins) have ether oxygen atoms, hydroxyl oxygen atoms, or, in a few cases, carbonyl oxygen atoms and amine nitrogen atoms as the coordinating ligands. In addition, as their name implies, they contain a carboxyl group. Some examples are shown in Fig. 9-3. When such an ionophore binds a cation, it loses H^+ from its COOH in exchange. The COO^- group thus created forms a hydrogen bond to the hydroxyl group at the other end of the molecule; thus, a ring is formed.

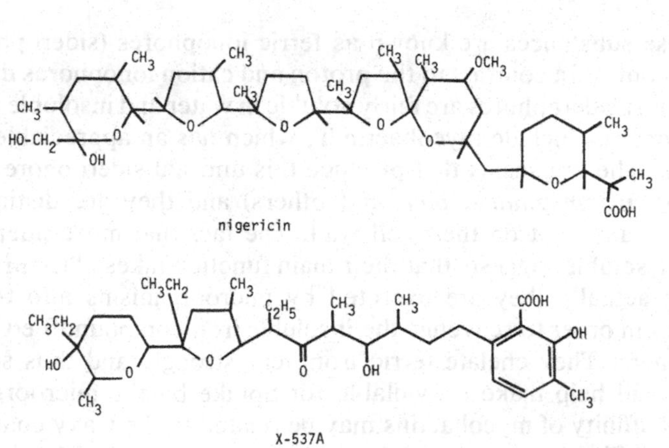

nigericin

X-537A

Figure 9-3. Cation ionophores of carboxylate type.

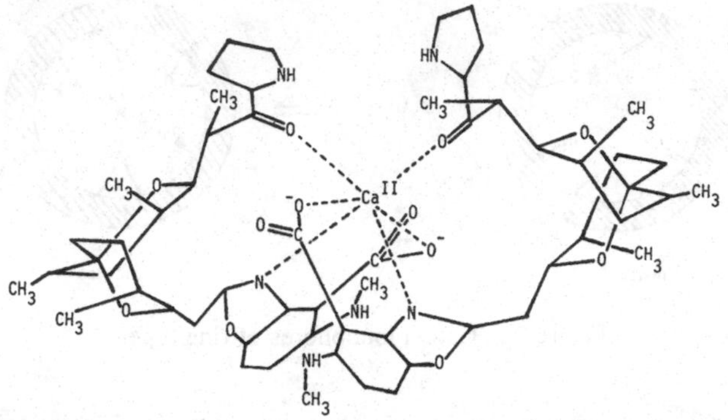

Figure 9-4. An example of cation ionophore (carboxylate type), A23187. (After Chaney *et al.*, 1976. *J. Antibiot.* 29:424.)

An example of a divalent carboxylate (A23187) complex is shown in Fig. 9-4 (Chaney *et al.*, 1976). As expected, 2 moles of A23187 react with 1 mole of a divalent cation and form a ring (cage)-like structure. As in the other groups of cation ionophores, the surface of the resulting cagelike structure is covered with hydrophobic residues. The cation specificity in this case seems to be determined by subtle matching between the size of the cation and the size of the preferred ringlike conformation of the nigericin.

9.1.3 Ionophores for Iron—Siderophores

These substances are known as ferric ionophores (siderophores) or siderochromes. In contrast to the proton and cation ionophores discussed above, most siderophores are fairly soluble in water and insoluble in lipids. The exceptions include mycobactin T, which has an appreciable affinity for lipids. The organisms that produce this unusual siderophore are *Mycobacterium* (*smegmatis, plei,* and others) and they are distinctive in having a waxy coat on their cell wall. The fact that most siderophores are water-soluble suggests that their main function takes place in aqueous medium; actually they are excreted by microorganisms into their surroundings in order to scavenge the insoluble iron-compound $Fe(OH)_3$ and its polymers. They chelate ferric iron very strongly and thus solubilize the iron and help make it available for uptake by the microorganisms. The lipid affinity of mycobactins may be related to the waxy coat of their producers. Clearly, mycobactin ionophores must be capable of diffusing through the waxy coat, and this necessitates lipid affinity.

Siderophores produced by microorganisms can be classified into a catechol (diphenol) type and a hydroxamate type, according to the co-ordinating groups employed (Hider, 1984).

9.1.3.1 Catechol-Type Siderophores

The prototype of the catechol siderophores is enterobactin (some-times called enterochelin). Enterobactin is produced by enteric bacteria such as *E. coli, Salmonella typhimurium,* and *Aerobacter aerogenes* when these bacteria are placed in low-iron media. Enterobactin can be regarded as the trimer of 2,3-dihydroxybenzoyl serine (see Fig. 9-5). The lower homologues of enterobactin that have been identified, such as dihydroxy-benzoyl serine, may simply be degradation products of enterobactin. As seen in Fig. 9-5, the two hydroxyl groups of the catechol residue chelate Fe(III). It was shown (Frost and Rosenberg, 1975; Rosenberg and Young, 1974) using ^{55}Fe and [^{14}C] enterobactin that the iron is taken up by the cell concomitantly with the ligand, presumably as an intact complex. However, as suggested above, since the complex is hydrophilic this trans-port cannot result from simple diffusion through the hydrophobic mem-brane. Instead, the Fe(III)-enterobactin complex is actively transported by a membrane protein system involving "fep" protein. After transport, the enterobactin moiety is rapidly hydrolyzed by an esterase in the cell, whereby the free iron is released.

In addition to the serine derivatives mentioned above, glycine and lysine derivatives have been isolated (Chimiak and Neilands, 1984). For example, 2,3-dihydroxybenzoyl glycine was isolated from *Bacillus meg-aterium,* and di(2,3-dihydroxybenzoyl) lysine from *Azotobacter vinelandii.*

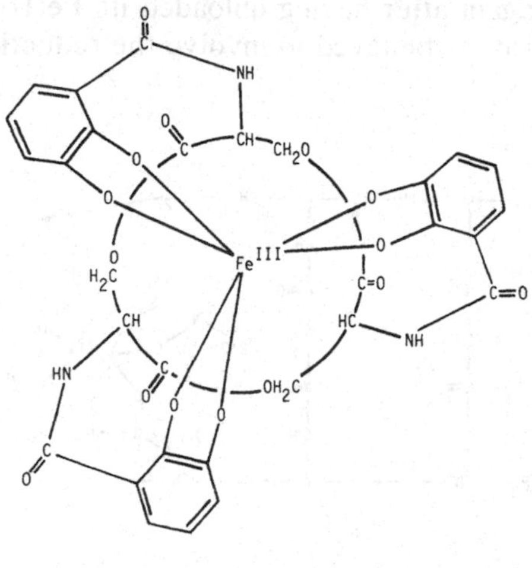

Figure 9-5. The structure of Fe(III)-en-terobactin. (After Raymond *et al.,* 1976. *J. Am. Chem. Soc.* 98:1763.)

9.1.3.2 Hydroxamate-Type Siderophores (Neilands, 1972, 1977; Emery, 1974)

Gram-positive *Bacillus* species produce both catechol- and hydroxamate-type siderophores. Hydroxamates are produced more commonly by molds and fungi, however. The most typical hydroxamate siderophores are ferrichromes, which are produced by *Aspergillus, Neurospora, Penicillium, Ustilago, Actinomyces, Streptomyces,* and others. Ferrichromes have the general formula shown in Fig. 9-6. The three hydroxamate groups chelate a ferric iron very strongly (the formation constant $= 10^{30}$). Other hydroxamate siderophores that have three coordinating hydroxamate groups include coprogen (from *Penicillium* and *Neurospora*), ferrioxamines (e.g., *Norcadia, Streptomyces, Actinomyces*), and desferal (*Streptomyces pilosus*). These siderophores form ferric complexes similar to those of ferrichromes.

The hydroxamate-type siderophores produced by *Aerobacter, Bacillus,* and *Mycobacterium* species contain as chelating groups hydroxyl or carboxyl oxygen atoms, or tertiary nitrogen in addition to two hydroxamate groups. For example, aerobactin (from *Aerobacter aerogenes*) has two carboxyl groups, schizokinen (from *Bacillus megaterium*) one carboxyl group and one alcoholic hydroxyl group, and mycobactin (from *Mycobacterium* species) a phenolic hydroxyl group and a tertiary nitrogen atom (Byers, 1974).

Ustilago sphaerogena was shown to rapidly take up ferrichrome when the siderophore was supplied after the organism was grown under iron-deficient conditions. This uptake showed all the attributes of active transport (Emery, 1974). Here, as in the case of the catechol-type siderophores, the ferrichrome was taken up as a whole, i.e., in the form of the ferric complex. The deferri-hydroxamate siderophore was found to be excreted again after having unloaded its Fe(III) in the cell. The unloading of the iron is believed to involve the reduction of Fe(III) to Fe(II).

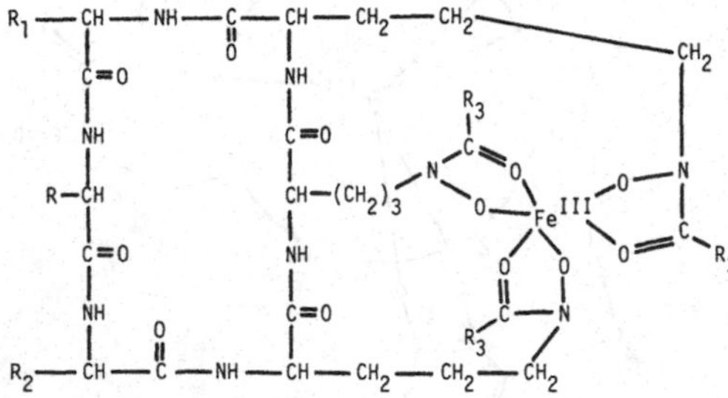

Figure 9-6. The general formula of ferrichrome (a hydroxamate siderophore).

There are a number of interesting problems associated with the hydroxamate type of siderophores, such as the relation between the conformational structure of a siderophore and its function (i.e., in ion transport or as an antibiotic) or activity, the phylogenetic relationships among various compounds, and so forth. See Neilands (1972) for the details of these problems. Receptors for these and other siderophores are reviewed also by Neilands (1982).

9.1.3.3 Other Siderophores

A number of microbial metabolites are known to contain hydroxamate groups; two examples are rhodotorulic acid and aspergillic acid. It is possible that these compounds, too, are involved in iron transport in one way or another.

Mutants of *E. coli* K12 and B/r that cannot utilize the enterobactin system were found to give a growth response to citrate under iron-poor conditions. This suggests the presence of a transport system for a ferric-citrate complex in these mutants; thus, citrate may function here as a siderophore (Frost and Rosenberg, 1975).

It has been reported that virtually every kind of cell is able to take up heme (Hutner, 1972; see also Granick and Sassa, 1978). Transport systems are presumed to exist for the heme group; a heme receptor on murine erythroleukemia cell has recently been characterized (Galbraith *et al.*, 1985). The addition of heme has been found to suppress porphyrin production in the cells of organisms from one of the most primitive bacteria, *Rhodopseudomonas spheroides,* to a mammal, mouse. These observations indicate that the heme can act as a siderophore, at least under certain conditions.

Most plants develop iron chlorosis in alkaline soil, where free available Fe(III) is insufficient. Some plants, especially barley, oats, and wheat, can, however, grow on such a soil. Their roots secrete very special iron-chelating polyamino acids (Sugiura and Nomoto, 1984). Typical examples are mugineic acid from barley and avenic acid from oats:

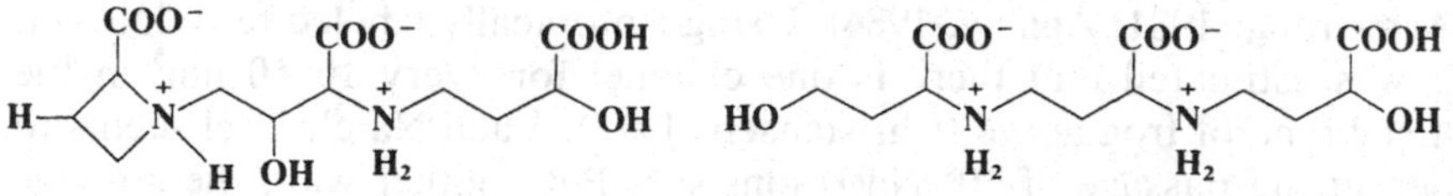

mugineic acid   avenic acid

These are in a sense similar to well-known chelating agents such as EDTA (ethylenediamine tetraacetate) and NTA (nitrilo triacetate); both O atoms (of COO^-) and amine N atoms of these compounds coordinate to a metal ion. Similar coordination patterns may be expected with these newly discovered polyamino acids. Fe(III)-mugineic acid complex exhibits a high-spin Fe(III) character in Mössbauer and EPR spectra, and a stability constant of about 10^{18} (Sugiura and Nomoto, 1984).

9.2 Channels

9.2.1 Channels for Cations

The unbalanced distribution of sodium and potassium ions between a cell and its environment is maintained by active transport mechanisms such as the sodium pump (see Chapter 10). The imbalance is responsible for electrically polarized states, particularly in nerve cells. This polarization amounts to approximately -70 mV in the resting state of many nerve cells. When a cell such as a nerve axon is electrically stimulated, the potential suddenly decreases, and can even reverse. This is called depolarization. In this way a stimulus is transmitted along the never fiber. The sudden change in potential is now believed to be caused mainly by a sudden inward movement of Na(I) ions. The lipid bilayer portion of the membrane is impermeable to Na(I) ions. To explain the observed behavior it is necessary to propose the sudden creation of a kind of pore through which Na(I) can move into the cell down the electrochemical potential gradient (i.e., passive transport). This pore is not thought to be a simple gap in the lipid bilayer, but is an integral membrane protein called the "Na channel." There also exists a similar K channel which is distinct from the Na channel. In the resting state, the gates of the channels are closed and they open only when stimulated. The opening of the Na channel is much faster than that of the K channel.

9.2.1.1 Na Channels

An interesting feature of the Na channel is that it is blocked selectively by tetrodotoxin and saxitoxin (Ulbricht, 1977; Ritchie and Rogart, 1977; Armstrong, 1981; Agnew, 1984). Using isotopically labeled tetrodotoxin, it was estimated that there is one channel for every 10–60 nm^2 in the membrane of frog nerve (Christensen, 1975). Each Na channel seems to permit the passage of 10^8 Na(I) ions/sec. Put another way, the passage time of a single Na(I) ion through the channel is only about 10^{-7} sec. The Na channel allows the passage of other cations in the following order of decreasing ease: $Na(I) > NH_3OH^+ > N_2H_5^+ > NH_4^+ >$ formamidinium

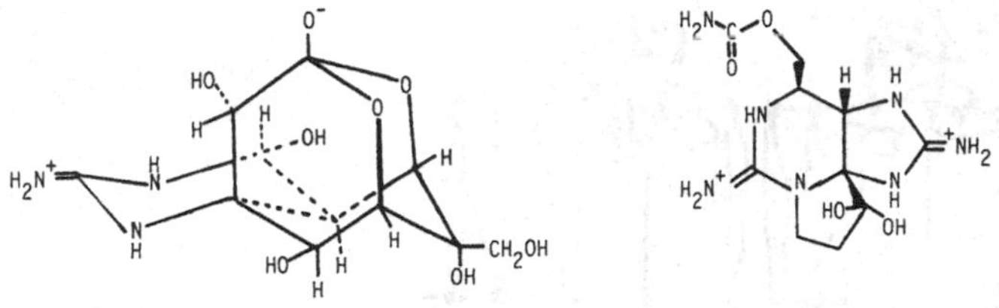

Figure 9-7. Neurotoxins: tetrodotoxin and saxitoxin.

> guanidinium ion > aminoguanidinium ion > dimethyl ammonium (Hille, 1971).

Tetrodotoxin is produced by a Japanese puff fish (fugu) as well as a type of salamander, and saxitoxin is found in the dinoflagellates that are considered to be responsible for the poisonous effect of the red tide (Martin and Martin, 1976). Both substances contain a guanidinium group (see Fig. 9-7), which binds with a component of the channel; the binding constants are large, 2×10^8 M^{-1} in the case of tetrodotoxin and about 1×10^9 M^{-1} in the case of saxitoxin (Ulbricht, 1977). Tetrodotoxin or saxitoxin is believed to bind to a receptor protein (Agnew, 1984) that is thought to be located at the entrance of the channel. The channel is hypothesized to have a constriction at the middle, which is called the selectivity filter. The filter is considered to be a 0.3×0.5 nm^2 pore formed by a ring of oxygen atoms, one group of the ring being an ionized carboxylate (Hille, 1971, 1975). The supposition that the filter is made of a ring of oxygen atoms is based partially on the fact that a ring of oxygen atoms of ether, hydroxyl, and carboxyl oxygen atoms is the chelating ligand in the cation ionophores. A detailed study of the behaviors of the Na channel, particularly its dependence on the precurrent, which is related to the so-called "gating" current, has led to the hypothesis that the Na channel is not a permanent structure but rather a functional aggregate consisting of three components: selectivity filter, gating molecules (hydrophilic proteins), and an inactivating gating substance (a small polypeptide with a net positive charge) (Rojas, 1977). The Na channel isolated from rat brain synaptosomes was also demonstrated to consist of $\alpha_1(\beta 1)_1(\beta 2)_1$; α subunit has a molecular weight of about 260,000, $\beta 1$ about 39,000, and $\beta 2$ 37,000 (total = 336,000) (Hartshorne and Catterall, 1984). $\beta 1$ subunit seems to be noncovalently attached to subunit whereas $\beta 2$ subunit is bound to subunit by a disulfide bond. A recent study (Noda et al., 1984), however, demonstrated that the Na channel from *Electrophorus*

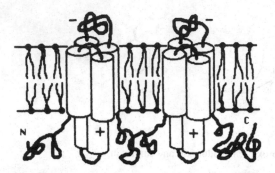

Figure 9-8. A model of the sodium channel. A cylinder represents an α-helical segment. (Modified from Noda *et al.*, 1984. *Nature* 312:121.)

electricus (an electric eel) is a single peptide of 1820 amino acid residues. The protein shows four repeated units, which are presumed to orient themselves in a pseudosymmetrical fashion in the membrane. Each homological unit turned out to contain a unique segment with clustered positively charged residues; this is believed to be involved in the gating structure. Figure 9-8 is a schematic model of this Na channel as deduced from the sequence. Presumably, the negatively charged sections sticking up act as the receptors for saxitoxin or tetrodotoxin.

9.2.1.2 K Channels

Several different types of K channels are present in membranes of nerve, muscle, and other groups of cells (Latore *et al.*, 1984). The K channel in the axon membranes in striated and cardiac muscle has a function of delayed rectifier; it participates in repolarizing the cell membrane during the late phase of the action potential. Ca(II)-activated K channels are also present in excitable and nonexcitable cells (Schwartz and Passow, 1983). This system seems to be activated by internal Ca(II) and external K(I); these synergistic actions are antagonized by internal Na(I). No molecular details of the channels are yet available.

9.2.1.3 Ca Channels

Ca channels play an important role in coupling membrane excitation to cellular responses such as secretion or contraction (Tsien, 1983). However, the molecular details of such channels are yet to be studied.

9.2.2 Channel-Forming Antibiotics

Several antibiotics (other than the cation ionophores mentioned earlier) have been shown to create channellike structures in artificial lipid

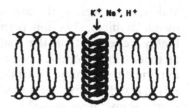

Figure 9-9. A model of a channel formed by gram-icidin A.

bilayers (Boheim *et al.*, 1977). Gramicidin A serves as a typical example. It is an oligopeptide consisting of 15 mostly hydrophobic amino acids arranged in linear fashion with alternating D and L configurations: OHC— NH — Val — Gly — Ala — Leu — Ala — Val — Val — Val — Trp — Leu—Trp—Leu—Trp—Leu—Trp—$CONHCH_2CH_2OH$. A lipid bilayer becomes permeable to small cations such as H^+, Na(I), and K(I) when it is doped with gramicidin A, apparently due to a channel formed by a dimer of the oligopeptide chains (Veatch and Blout, 1974; for a recent review see Anderson, 1984). The dimer may be a double-stranded helix as sketched in Fig. 9-9. It should be noted that the hydrophobic branches of the constituent amino acids extend outside of the channel, thus permitting the channel to become embedded in the lipid portion of the membrane. The helical dimer's central pore, lined by the carbonyl oxygens of the peptide backbone, has a diameter of 0.3–0.4 nm, as estimated from molecular models. About 10^7–10^8 K(I) ions per second pass through the channel when the membrane is immersed in 1 M KCl solution and the applied potential is 100 mV (Boheim *et al.*, 1977).

Another oligopeptide that forms a channellike structure in lipid bilayers is alamethicin (Boheim *et al.*, 1977). It has been isolated from the fungus *Trichoderma virides* and was demonstrated to increase the ion permeability in both biological membranes and artificial lipid bilayers. Alamethicin's peptide chain consists of 18 amino acid residues, 14 of which are hydrophobic. The main component of alamethicin has the following amino acid sequence (Martin and Williams, 1976):

R—Pro—Aib—Ala—Aib—Ala—Gln—Aib—Val—Aib—Gly—Leu—Aib—Pro—Val—
Aib—Aib—Aib—Glu—Gln (C-terminal)

$$\text{(phenyl)}\!-\!CH_2\!-\!CH\!-\!NH$$
$$\qquad\qquad\quad |$$
$$\qquad\qquad CH_2OH$$

In this sequence, the residue designated as Aib is α-aminoisobutyric acid, with all the others being common amino acids. The nature of the R group at the N-terminus is not known with certainty. Small cations such as Na(I)

and K(I), as well as the anion Cl⁻, can go through the channel induced
by alamethicin. The ion permeability of this channel was found to be
highly dependent on the voltage applied, unlike gramicidin A, whose
channel's ionic conductance is only weakly voltage-dependent. All of the
experimental findings, including the voltage-dependent behavior, can be
rationalized in terms of a model in which the pore is created in an oligo-
meric alamethicin complex. The pore diameter of the basic complex,
thought to contain an average of six monomer units, is too small to allow
conduction. A voltage-dependent addition of another unit, however, wid-
ens the pore, allowing conduction to occur (Boheim, 1974; Gordon and
Haydon, 1976). In this case, too, the pore of the channel is assumed to
be lined with carbonyl oxygens, and its diameter is estimated to be about
1 nm (Boheim *et al.*, 1977).

Other compounds that are suspected to form channellike structures
in biological as well as artificial membranes include monazomycin, an
antibiotic with the empirical formula $C_{62}H_{119}O_{20}N$, and a class of antifungal agents called polyenes (Gomez-Puyou and Gomez-Lojero, 1977). Two
such polyene antibiotics are shown in Fig. 9-10 (Martin, 1977). These
compounds, which contain several OH groups and a ringlike structure,
bear a striking similarity to some of the cation ionophores. The polyenes
have a special affinity for cholesterol derivatives. Bacterial membranes,

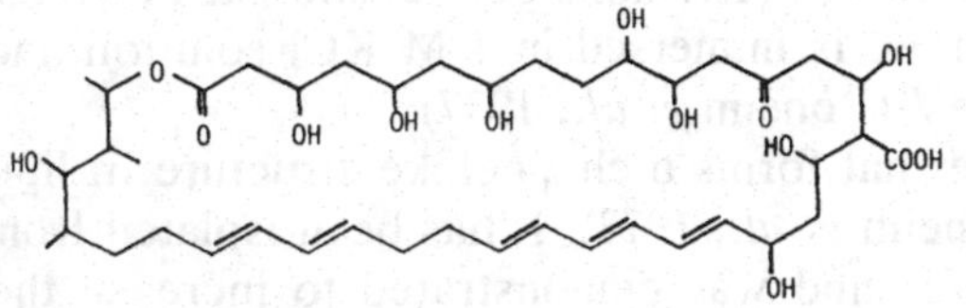

Nystatin (or Fungicidin, Polifungin)

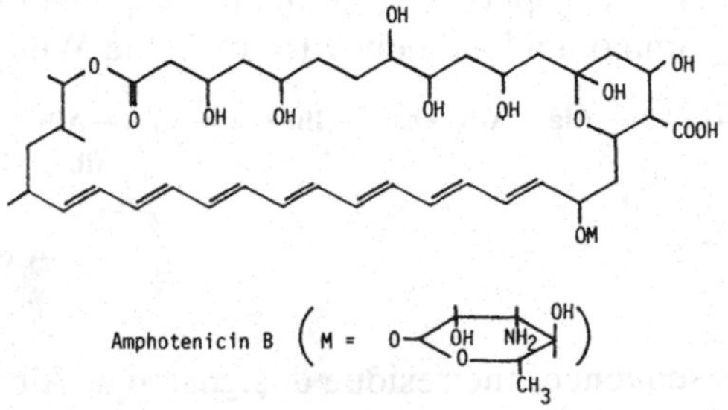

Amphotenicin B $\left(M = O\!-\!\langle\begin{array}{c}OH\\O\end{array}\begin{array}{c}OH\\NH_2\end{array}\rangle\!\!-\!\!\begin{array}{c}OH\end{array}\right)$

Figure 9-10. Polyene antifungal agents.

which contain no cholesterol units, are not subject to the disruptive effects of the polyene antibiotics. The cholesterol-containing membranes of eukaryotic cells, on the other hand, can be attacked; thus, the polyenes act as antifungal (but not antibacterial) agents.

9.3 (Serum) Transfer Proteins for Inorganic Compounds

The inorganic compounds taken up by the mucosal cells of the intestine are carried into the cells, pass through layers of mucosal cell serosal side membrane, cross the outer membrane of blood vessel cells, and finally enter the bloodstream. This part of the uptake process is not well understood but it may well be similar to uptake by the epithelial cells of the intestine. The absorbed inorganic elements and compounds have subsequently to be carried by the blood to the locations where they are required. For this transfer purpose, the blood contains specific proteins to bind specific elements. Examples of such proteins include transferrin [for Fe(III)], serum albumin [for Zn(II)], ceruloplasmin [for Cu(II)], and α-globulin [for Mn(II)]. It is possible that the metallic ions also are transported within the mucosal cells by proteins. At least one transferrin-like protein has been implicated in the mucosal cell transport of Fe(III). In the bloodstream, hemoglobin has been shown to bind CO_2, the end product of respiration; thus, hemoglobin acts as a carries of CO_2.

The pH of blood (human) is about neutral (pH 7.3). As a result, most metallic elements including Zn(II), Cu(II), Co(II), Mg(II), Ca(II), Na(I), and K(I) will exist in the bloodstream as free (i.e., aquo complex) ions; the important exception is Fe(III), which forms hydroxides at neutral pH and would thus precipitate out. This necessitates the presence of specific chelating proteins or some other device to keep Fe(III) in a soluble form. Two such proteins are transferrin, which carries Fe(III) ions, and ferritin, which stores them. Monovalent and divalent metallic ions could, in principle, be present as free ions in the blood, but they would then readily be excreted through the kidney. Protein binding prevents this loss, providing a second reason for the organisms' need for specific proteins.

9.3.1 Fe(III) Transfer Proteins—Transferrin

A number of monographs and reviews have been published on transferrin in recent years (Aisen, 1977; Aisen and Litowsky, 1980; Urushizaki et al., 1983). Transferrin constitutes 2–3% of the solid in blood serum. Similar substances are also found in milk and egg white; the former is called lactoferrin and the latter ovotransferrin or conalbumin. They are very similar to serum transferrin in every respect—molecular weight, spectral properties (EPR, optical), and Fe(III) affinity—but their physiological roles are still obscure. It is important to note that the transferrin molecule is not pure protein but contains a substantial amount of carbohydrate.

As its name suggests, the function of transferrin is definitely the transfer of iron to sites where iron is needed, particularly the erythroblasts and reticulocytes in the bone marrow. Transferrin has a molecular weight of about 77,000 and binds 2 moles of Fe(III) per mole. It can also bind Mn(III), Cu(II), and Cr(III) *in vitro,* but the binding and transfer of these ions is not regarded to be physiologically significant.

Transferrin has to load iron at the source and unload it in the erythroblasts. Here certain problems arise: at the time of loading the iron must be in the Fe(III) state; the carrier molecule should thus have an affinity for Fe(III), but it must also be able to unload the iron readily in the target cells. It has been shown in the previous section that Fe(III) ion has a strong specific affinity for the hydroxyl oxygen of certain aromatic compounds or hydroxamate. Therefore, it comes as no surprise that the Fe(III)-binding groups of transferrin include tyrosine residues: two tyrosine residues have been identified in the Fe(III)-binding site and another appears to be involved (Feeney *et al.,* 1983). It is an intriguing fact that transferrin requires 1 mole of CO_3^{2-} anion in order to bind 1 mole of Fe(III). This CO_3^{2-}, along with three tyrosine residues, would constitute four of the six required binding ligands for one Fe(III) ion (see D. J. Schneider *et al.,* 1984; Garratt *et al.,* 1986). The two other ligands are believed to be histidine residues (Feeney *et al.,* 1983). Altogether the protein has a binding constant of about 10^{23} M^{-1}. In the absence of the carbonate anion, the binding constant is very small; thus, the presence of CO_3^{2-} is an absolute requirement. CO_3^{2-} can be substituted by other similar anions such as $C_2O_4^{2-}$, at least for the purpose of binding iron, but this oxalate–Fe(III)–transferrin complex is much less efficient in releasing the iron to the reticulocytes or erythroblasts.

A high binding constant does not necessarily mean that a complex is inert; indeed, the iron in the transferrin complex is readily exchangeable with free Fe(III) ion in solution. However, the strong binding appears to be incompatible with the subsequent facile release of Fe(III) from the complex.

There are at least two possible chemical means to avoid this dilemma. One would be the reduction of Fe(III) to Fe(II), since Fe(II) is much less strongly bound to transferrin. Another way would be to make use of the fact that CO_3^{2-} is required to attain the high binding constant and a removal of CO_3^{2-} from the coordination site would drastically reduce the binding constant and probably Fe(III) would come off easily. No unequivocal evidence has been presented for the mechanism of the release of Fe(III) from transferrin in the target cell, but some relevant information is available. The known events occurring in the process of uptake of iron by erythroblasts from transferrin are summarized as follows (see Morgan, 1983; Octave *et al.*, 1983): (1) the complex, Fe(III)–transferrin–CO_3^{2-}, binds to the specific receptor site on the reticulocyte or erythroblast membrane (mature erythrocytes cannot take up iron); the receptor protein is considered to consist of one subunit of molecular weight 165,000 and two subunits of molecular weight 95,000; (2) the transferrin–receptor complex is endocytosed; (3) the endocytosed vesicle fuses with a lysosome; (4) iron is released from transferrin in this fused vesicle; and (5) apotransferrin is then exocytosed to be recirculated. The carbohydrates on the transferrin molecule are thought to be involved in its binding to the reticulocyte or erythroblast, but no unequivocal proof for this supposition has been obtained.

Our hypothesis for the mechanism of unloading Fe(III) from transferrin is illustrated in Fig. 9-11 and incorporates the suggestion made in the discussion above and also suggestions that bicarbonate is a prerequisite for iron release from transferrin (Aisen and Leibman, 1973; Egyed, 1973; Martinez-Medellin and Schulman, 1973). In this scheme, CO_3^{2-} is supposed to be protonated and then be dehydrated. There are a number of indications that the addition of proton is involved in the release of iron from transferrin (Morgan, 1983). CO_3^{2-}, having now become CO_2, can no longer hold the protein in such a conformation as to bring all the coordinating groups into a suitable proximity. Reduction of Fe(III) to Fe(II) might also play a role, but it is not a necessity. This proposed scheme is compatible with the fact that other anions such as oxalate can bind Fe(III) strongly, and the complex Fe(III)–transferrin–$C_2O_4^{2-}$ binds to the same receptor site on the cell membrane, but nonetheless fails to release Fe(III), for $C_2O_4^{2-}$ is much weaker a base than CO_3^{2-} is. It might also be mentioned that the lysosome content is much more acidic than the cytoplasm, thus providing the necessary protons.

It was mentioned earlier that 1 mole of transferrin binds 2 moles of Fe(III). Are these two binding sites equivalent? Nonequivalence of the two sites was first suggested by Fletcher and Huehns (1967, 1968) but is still not unequivocally established (Aisen *et al.*, 1978; Aisen and Litowsky, 1980). One site (referred to as C site, which is located near the C-terminus) is supposed to donate its iron to receptors on the erythroblasts or the reticulocytes, while the other (N site located near the N-terminus) is assumed to direct its iron to the receptors on iron-storage cells in the liver

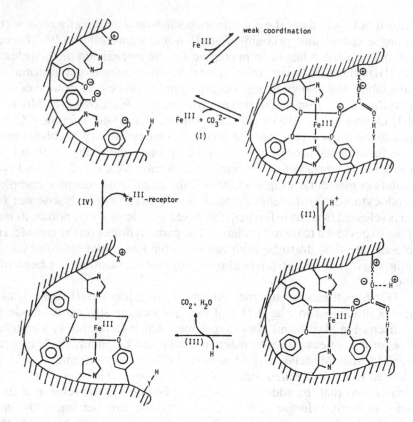

Figure 9-11. A proposed mechanism of loading and unloading of Fe(III) on transferrin.

(Fletcher and Huehns, 1967, 1968). It has been further hypothesized that the intestinal mucosal cells direct absorbed iron preferentially onto the N site of transferrin in the portal vein plasma to allow the newly absorbed iron to pass through the liver (which prefers C-site iron), whereby it can arrive safely at the bone marrow (Brown, 1975).

Iron, as it enters the mucosal cells, is believed to be in the Fe(II) state. The iron in the mucosal cell either binds to an intracellular transfer protein (Forth and Rummel, 1973) similar to transferrin or is incorporated into ferritin. In both cases, the iron would need to be in the Fe(III) state. As the iron is translocated across the serosal side membrane into the bloodstream, it is believed to be reduced to Fe(II). It must be oxidized again before binding to apotransferrin. Thus, the iron must undergo both oxidation and reduction while it is being taken up by the organisms. One important oxidation process for mucosal iron involves a copper enzyme, ceruloplasmin (Frieden, 1971). This enzyme catalyzes the oxidation of Fe(II) to Fe(III) by O_2. The enzyme, also known as ferroxidase, is considered to function also as a copper-transport protein in plasma. The reduction of Fe(III) will be discussed later in conjunction with the iron-storage proteins.

9.3.2 (Serum) Transfer Proteins for Other Metal Ions

Plasma (serum) contains transfer proteins other than transferrin, but they are not as well understood. Transferrin itself may also act as a carrier for Mn(III), since it binds Mn(III) as strongly as Fe(III). How the receptor cells distinguish Mn(III) from Fe(III) is not known. Transferrin may also carry Cr(III).

The most abundant protein in plasma (serum) is albumin; it is considered to act as a carrier for Cu(II) and/or Zn(II). The binding site for Cu(II) ion may be made up of nitrogen atoms of imidazoles and amines, whereas Zn(II) ion would bind more strongly with a cysteinyl sulfhydryl group. Ferroxidase (ceruloplasmin) is, as mentioned above, thought to be a carrier of copper.

Vitamin B_{12} is said to be predominantly in the methyl-B_{12} form in serum. Two serum proteins are known to bind B_{12} (Nexø and Olesen, 1982). Transcobalamin II is found in the β-globulin portion and thought to bind B_{12} before it enters the portal system, delivering it to the reticulocytes and the erythroblasts in the bone marrow. On binding cobalamin, transcobalamin II undergoes such a conformational change that increases its affinity for an acceptor on cell membranes. The transcobalamin II–cobalamin complex is attached to the acceptor and then internalized by endocytosis. The binding of the complex to the acceptor requires Ca(II), and the binding constant has been estimated to be about 5×10^9 M^{-1}. The other, haptocorrin (transcobalamin I), has a few different isoproteins. The plasma haptocorrin has a high content of sialic acid, whereas haptocorrin from leukocytes lacks sialic acid. The function of haptocorrin is not clear.

Certain phosphoproteins are considered to act as carriers for Ca(II) ion. On the other hand, no special carrier proteins appear to exist in serum for Mg(II), Na(I), or K(I). Thus, these cations are easily filtered out by the kidney and have to be reabsorbed by the adjoining tubules.

9.4 Calcium-Binding Proteins

There are a large number of proteins that bind Ca(II) specifically. The functions of these proteins are diverse. From the standpoint of structural and functional features, they may be grouped into several classes. Class I includes calmodulin, troponin-C, and parvalbumin. These are involved in the functioning of the intracellular Ca(II) as a second messenger (see Section 7.6), and have similar structures. The so-called vitamin D-dependent intestinal Ca-binding protein and S-100 may be included in this class from the structural standpoint. Calcium-binding proteins of class II contain a special amino acid, γ-carboxyglutamate; typical examples include prothrombin and osteocalcin (see Martin, 1984). The former is involved in the process of blood clotting and the latter is found in developing bones. A variety of proteins (class III) require Ca(II) for their proper functioning, mostly as a structural element. Examples of this class include α-amylase (from bacteria), staphylococcal nuclease, pancreas DNase, thermolysin, and some proteases. Enzymes of the next class (class IV) are activated by Ca(II), in conjunction with calmodulin in some cases. Phospholipase A_2 is activated by Ca(II); this seems to be exploited in transducing the effect of hormonal action, as phospholipase A_2 is involved in stimulus–response coupling (see Section 7.6). Ca/Mg-ATPase (Ca pump) also belongs to this class. Some of the other enzymes in this class, as discussed in Section 7.6, include cAMP-hydrolyzing phosphodiesterase and phosphorylase kinase. There are a number of proteins whose sole function seems to be binding Ca(II) (class V). A typical example is calsequestrin, which is found in the sarcoplasmic reticulum of muscle cells. This functions in Ca(II) storage.

The final class (VI) is a hodgepodge of proteins and protein derivatives. A phosphoprotein seems to function as a Ca(II) carrier in mammalian blood serum as mentioned earlier. Nerve cells reveal a number of phosphate-containing proteins; some of them are the so-called CBP-I and CBP-II from pig brain. Mitochondria are another Ca(II)-storage site, where phosphoproteins seem to bind Ca(II). Glycoproteins, with carboxylate, phosphate, or sulfate derivatives of carbohydrates, also bind Ca(II) specifically. They are involved in intercellular binding, nucleation matrix for calcification, and other functions, as discussed in Chapter 6. In addition, Ca(II) binds to nonprotein substances such as phospholipids, and carboxylate, phosphate, or sulfate derivatives of carbohydrates.

9.4.1 Some General Features of Calcium-Binding Proteins

Ca(II) has a relatively large ionic radius, about 100 pm, and no d electrons; hence, its coordination structure and coordination number are fairly flexible, from six to eight. Ca(II) shows a strong affinity toward oxygen ligands, especially negatively charged ones such as carboxylate and phosphate (see Section 8.2), but no strong affinity toward nitrogen ligands.

The variability of the coordination number and structure would allow proteins that bind Ca(II) to have a great degree of variation in their binding strength to Ca(II). This variation would be controlled by the number and kind of ligating groups and their spatial arrangement. This also implies that conformational changes on the protein part could modify the binding strength [to Ca(II)] significantly. The same could be said with regard to any coordinating cation (or anion for that matter), but seems to be especially true in the case of Ca(II).

The commonly found ligands for Ca(II) in regular proteins are carboxylate (either monodentate or bidentate) of aspartic acid and glutamic acid residues, OH group of serine, threonine, and tyrosine, and carbonyl group of the peptide bond. The combination of carboxylate, hydroxy, and carbonyl groups for Ca(II) binding is reminiscent of Ca(II) ionophores (Section 9.1). Some examples of Ca(II)-binding sites are given in Table 9-2. In all the cases listed, the coordination number about Ca(II) seems to be six.

Both carboxylate groups of γ-carboxyglutamate residue could be used, perhaps in the mode of malonate. Malonate (or γ-carboxyglutamate), being bidentate, binds Ca(II) more strongly than monodentate carboxylates.

There are essentially two different compartments for Ca(II) in multicellular organisms: extracellular fluid and intracellular fluid (cytosol = cytoplasm). Under typical conditions, the free Ca(II) concentration in the extracellular fluid, $[Ca(II)]_e$, is on the order of 10^{-3} M, and that in the resting cell, $[Ca(II)]_c$, is 10^{-8}–10^{-7} M (see Section 7.6). Accordingly, the extracellular Ca(II)-binding proteins would have binding constants on the order of 10^3–10^4 M^{-1}. In other words, $[Ca(II)]_e$ is controlled by these proteins. In a resting cell, Ca(II) remains bound to entities whose binding constants are on the order of 10^8–10^9 M^{-1}. A sudden rise of $[Ca(II)]_c$ triggers cellular responses such as muscle contraction and release of neurotransmitters (see Section 7.6). The [Ca(II)] fluctuation (from 10^{-8}–10^{-7} to 10^{-5} M) must then be controlled, i.e., it should not be allowed to rise much above 10^{-5} M for too long. Therefore, some kind of buffer is required that should have a binding constant on the order of 10^5–10^6 M^{-1}. The triggering proteins, by the way, should also have binding constants on the order of 10^5–10^6 M^{-1}.

The Ca(II)-binding constants required for Ca(II)-binding proteins thus range from 10^3 to 10^9 M^{-1}. A proper protein, i.e., with an appropriate binding constant, should then be used for a specific function. There exist

Table 9-2. Calcium-Binding Site in Some Proteins[a,b]

Protein	Sequence
Parvalbumin	
CD	Asp-51, Asp-53, Ser-55, Phe-57, Glu-59, Glu-62
EF	Asp-90, Asp-92, Asp-94, Lys-96, Glu-101, $1H_2O$
Troponin C	
(from rabbit	
skeletal muscle)	
I	Asp-27, Asp-29, Asp-33, Ser-35, Glu-38, $1H_2O$
II	Asp-63, Asp-65, Ser-67, Thr-69, Asp-71, Glu-74
III	Asp-103, Asn-105, Asp-107, Tyr-109, Asp-111, Glu-114
IV	Asp-139, Asn-141, Asp-143, Arg-145, Asp-147, Glu-150
Calmodulin	
I	Asp-20, Asp-22, Asp-24, Thr-26, Thr-28, Glu-31
II	Asp-56, Asp-58, Asn-60, Thr-62, Asp-64, Glu-67
III	Asp-93, Asp-95, Asn-97, Tyr-99, Ser-101, Glu-104
IV	Asn-129, Asp-131, Asp-133, Glu-135, Asn-137, Glu-140
S-100	Asp-61, Asp-63, Asp-65, Glu-67, Asp-69, Glu-72
Thermolysin	
a	Asp-57, Asp-59, Glu-61, $3H_2O$
b	Asp-138, Glu-177, Asp-185, Glu-187, Glu-190, $1H_2O$
c	Glu-177, Asn-183, Asp-185, Glu-187, $2H_2O$
d	Tyr-193, Thr-194, Ile-197, Asp-200, $2H_2O$
Staphylococcal	Asp-19, Asp-21, Asp-40, Thr-41, Glu-43, $1H_2O$
nuclease	
Phospholipase A_2	Tyr-28, Glu-30, Gly-32, Asp-49, $2H_2O$
Concanavalin A	Asp-10, Tyr-12, Asn-14, Asp-19, $2H_2O$

[a] Adapted from Levine, B. A., and Williams, R. J. P., 1982. Calcium binding to proteins and anion centers, in *Calcium and Cell Function* (W. Y. Cheung, ed.), Vol. II, Academic Press, New York, pp. 1–38.
[b] The amino acids that do not have OH nor COOH groups are involved in the Ca binding through their $C=O$ oxygen atom.

indeed Ca(II)-binding proteins, whose binding constants vary from 10^3 to 10^9 M^{-1}.

9.4.2 Examples

A few examples of Ca(II)-binding proteins that have been well characterized will be discussed briefly.

9.4.2.1 Parvalbumin

Parvalbumins are small (M_r 12,000) acidic proteins; they were initially isolated from lower vertebrate (fish and amphibian) muscle and subsequently found also in higher vertebrates (see Demaille, 1982). The structure of parvalbumin from carp was determined by X-ray crystallography (Kretsinger and Nockolds, 1973). There are two Ca-binding domains (loops); one loop (Asp-53 to Glu-62) is flanked by helical segments C and D, which are juxtaposed almost perpendicularly to each other, and the other (Asp-92 to Glu-101) surrounded by E and F helical segments perpendicularly disposed. This kind of Ca(II)-binding site flanked by two helical segments is, thus, often called "EF" hand or loop (Kretsinger and Nockolds, 1973). The ligating amino acids in parvalbumin are found in Table 9-2.

Parvalbumin has been shown to function as a relaxing (or buffering) factor in muscle contraction. Troponin-C immediately binds Ca(II) to its regulatory sites (see below) after a burst of Ca(II) from the sarcoplasmic reticulum. Excess Ca(II) is then captured by parvalbumin or Ca(II) bound to the regulatory sites of Tn-C slowly transfers to either the so-called Ca-Mg sites of Tn-C or to parvalbumin. The decrease of $[Ca(II)]_c$ by the sarcoplasmic Ca pump returns the cell to the relaxed state. The Ca(II)-binding constants of Tn-C are typically 10^5 M^{-1} at the regulatory sites and 10^8 M^{-1} at the Ca-Mg sites (Potter and Johnson, 1982). The former is specific for Ca(II) whereas the latter binds both Ca(II) and Mg(II), even though the Mg(II)-binding constants are not as large as those for Ca(II). This is because $[Mg(II)]_c$ is typically much higher (on the order of 10^{-3} M) than $[Ca(II)]_c$. Ca(II)- and Mg(II)-binding constants of parvalbumin are comparable to that of Tn-C regulatory sites, being 2×10^4–10^6 M^{-1} (depending on the species) and 3×10^4 to 6×10^4 M^{-1}, respectively (Demaille, 1982; Haiech et al., 1979). The question is then why parvalbumin would not prevent Ca(II) from binding to Tn-C. Under normal conditions, parvalbumin is saturated with Mg(II), whereas the regulatory sites of Tn-C are not. Ca(II) needs to displace Mg(II) in order to bind parvalbumin, which is relatively slow. This kinetic factor assures that Ca(II) binds first to Tn-C (Demaille, 1982).

9.4.2.2 Troponin-C

The function of Tn-C as the regulatory protein in muscle contraction is outlined in Sections 7.6 and 10.2. Recent X-ray studies (Herzberg and James, 1984; Sundaralingam et al., 1985) have revealed that Tn-C has a dumbbell shape; two spherical portions (lobes) are connected by a helical segment of 8–9 turns. The two Ca-binding sites in the N-terminal domain (lobe) of the molecule show a specificity for Ca(II) (with $K = 10^5$ M^{-1}), whereas the two binding sites in the other lobe (C-terminal domain) have a higher affinity for Ca(II) ($K = 10^7$ M^{-1}) but can bind Mg(II) as well. All the binding loops seem to have a structural feature similar to those found in parvalbumin (i.e., EF hand). From the binding characteristics, it can be inferred that the sites in the N-terminal domain control the regulatory function of Tn-C.

9.4.2.3 Calmodulin

The physiological functions of calmodulin are outlined in Section 7.6 and well documented (Cheung, 1980, 1982; Klee et al., 1980). It is a relatively small (M_r 16,800) acidic protein; some 30% of the amino acids are aspartic and glutamic acids. X-ray crystallography (Babu et al., 1985) has revealed a dumbbell shape like Tn-C. As in Tn-C, each lobe contains two Ca(II)-binding sites; sites 3 and 4 in the C-terminal lobe bind Ca(II) [and Mg(II), too] more strongly than those Ca(II)-specific sites 1 and 2 in the N-terminal domain (Haiech et al., 1981; Anderson et al., 1983; Klevit et al., 1983). The Ca(II)-binding groups of the former sites (3 and 4) have binding constants of about 10^7 M^{-1} and the latter, 10^6 M^{-1} (Demaille, 1982). Under resting conditions, sites 3 and 4 are likely to be saturated with Ca(II) and Mg(II). A binding of Ca(II) to sites 1 and 2 has been demonstrated to change the conformation of the protein, especially in the connecting helix region (Schutt, 1985). This connective region has been suggested to bind to a target protein such as phosphorylase kinase and phosphodiesterase.

9.4.2.4 Other Calcium-Binding Proteins

Vitamin D-dependent intestinal Ca-binding protein found in the intestinal mucosa has been shown by X-ray crystallography to have a structure analogous to that of parvalbumin (Szebenyi et al., 1981). The so-called S-100 obtained from mammalian brain has an amino acid sequence resembling those of parvalbumin and the intestinal Ca-binding protein, and hence is believed to assume a structure similar to that of parvalbumin. Calmodulin and Tn-C, having twice as high a molecular weight as parvalbumin, S-100, and the intestinal Ca-binding protein, are likely to have derived from the smaller protein by gene duplication (see MacManus et al., 1986).

Little structural information is available with regard to the other Ca-binding proteins, besides the information mentioned earlier.

9.5 Storage Proteins and Other Binding Proteins

9.5.1 Storage Proteins for Iron

Ferritin is the iron storage protein found in a wide variety of organisms, from fungi to higher plants and mammals. A similar protein, bacterioferritin, has also been found in some bacteria including *E. coli* and *Azotobacter vinelandii* (Yariv *et al.*, 1983). In fungi it is found in spores, thus acting as the source of iron for germination and further development. In mammals it acts as storehouse, source, and pool. Organisms use iron very economically; in humans only a small portion (about 1 mg) of the total load (3–4 g) is absorbed and lost daily. The blood plasma contains about 3 mg of iron at any one moment but as much as 35 mg of iron is being carried per day through the plasma. Most of this 35 mg is carried by transferrin on the way to the bone marrow. In view of these facts, it is apparent that somewhere there is a large iron pool. In fact, as much as 25% of total body iron is in such a pool, mostly bound to ferritin; the iron pool is located in the reticuloendothelial (or phagocytic) cells of the liver, spleen, bone marrow, and muscle. These cells degrade old red blood cells and store the iron from the degraded hemoglobin in erythrocytes (Lynch *et al.*, 1974).

Ferritin is quite unique in its binding of iron. In fact, it does not "bind" iron at all in the usual sense. It appears to simply surround an iron–hydroxide–phosphate micelle, so that the entire body—the covering protein (apoferritin) and the iron core—becomes soluble in the cytoplasm. In this way, 1 mole of subunit protein (M_r 18,500) can accommodate an enormous amount of iron, as much as 180 gram atoms of iron. In the ordinary form of binding utilizing a coordinate bond, it would be impossible for this rather small protein to accommodate so much iron. Ferritin is regarded as having a cagelike structure and to possess a catalytic activity for the autoxidation of Fe(II) to Fe(III). The composition of the iron core is approximately $FeO(PO_3H_2)–8Fe(OH)_3$ (Crichton, 1973). Therefore, iron as Fe(II) enters into the cage of ferritin and is at the same time oxidized to Fe(III). It then accumulates in the course of forming a micelle in the cage. A ferritin molecule usually consists of 24 subunits; the overall structure appears spherical (Crichton, 1984).

Release of iron from ferritin is accomplished by reducing the Fe(III) of the micelle in the cage. This reduction is effected by $FMNH_2$, $FADH_2$, or reduced riboflavin. The process itself is nonenzymatic, but the formation of $FMNH_2$ is catalyzed by an enzyme, ferritin reductase, in which the ultimate reductant is NADH (Frieden and Osaki, 1974). The incor-

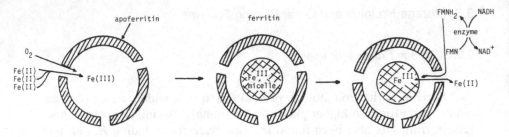

Figure 9-12. A model of the release of iron from ferritin.

poration of Fe(II) into apoferritin and the reductive release of iron from ferritin may thus be described by Fig. 9-12.

It has been shown that iron induces the synthesis of new apoferritin molecules (Harrison *et al.*, 1974). The induction is effected at the translational level of the protein synthesis, but its detailed mechanism is not known. When the organisms or tissues are overloaded with iron, both the biosynthetic apparatus for ferritin and the ability of ferritin to take up iron may become saturated. In such a case, another iron storage material, hemosiderin, may be formed (Harrison *et al.*, 1974). Hemosiderin is a granule with a high iron content: $29 \pm 8\%$ of its dry weight. Other components include phosphorus and a variety of organic constituents, including some proteins. Hemosiderin is considered to form as a result of the action of proteolytic enzymes on ferritin in lysosomes (Richter, 1983).

9.5.2 Storage and Other Binding Proteins for Other Elements

The liver is considered to act as a pool for a number of metal ions and compounds, including Zn(II), Cu(II), and vitamin B_{12}. Presumably it contains specific storage proteins for these materials. They should have specific binding characters for the respective bound substances, and at the same time should have the ability to release them readily as required. In the case of iron, its oxidation–reduction character and the drastic difference in coordinating ability between Fe(II) and Fe(III) are exploited by organisms to satisfy the conflicting demands. A similar trick might be utilized by the copper storage proteins. A small metallothionein-like protein is believed to play a role in the storage and transport of copper.

Oxidation–reduction processes cannot be applied to the case of Zn(II). A possible mechanism for strong binding and ready release of Zn(II) would involve a conformational change in the protein, somewhat similar to that observed with transferrin. A Zn(II)-binding protein was found in human milk (Eckhert *et al.*, 1977; Hurley *et al.*, 1977). A similar protein of low molecular weight was found in rat milk, but is reported to be absent in cow milk. This protein is considered to enhance the absorption of zinc in the neonate, which has not yet developed an efficient zinc absorption mechanism in its own intestine. These proteins may have some resemblance to a metallothionein involved in the zinc uptake in the intestinal mucosa. Metallothionein will be discussed in Chapter 14.

It is likely that the cobalamin storage protein in liver is similar to transcobalamin II, the serum storage protein referred to earlier. Transcobalamin II is believed to recognize the side chain of B_{12}. B_{12} would be degraded by the digestive enzymes in the gastrointestinal tract without some protection. Cobalt, as vitamin B_{12}, is absorbed mainly in the ileum, the last portion of the small intestine. The fundic mucosa of stomach secretes a protein with a molecular weight of about 40,000, called "intrinsic factor," which binds B_{12} and aids in its absorption by the ileum while at the same time protecting it from degradation. On binding cobalamin, the intrinsic factor molecule changes its conformation in such a way that its affinity for the ileal acceptor increases more than tenfold. As in the case of transcobalamin, this binding requires Ca(II). The intrinsic factor–cobalamin complex is then internalized by an energy-requiring mechanism (Nexø and Olesen, 1982).

There may exist specific storage proteins for manganese, chromium, molybdenum, and other elements. Xanthine oxidase or a similar molybdenum enzyme may itself act as a carrier and a storage for molybdenum.

Active Transport

10.1 Introduction

The relative amount of an element within a living cell is usually significantly different from that in its surroundings. For example, the sodium concentration inside a cell is typically only 2% of that in the environment; on the other hand, the potassium concentration is much higher inside a living cell than outside of it. A similar relation is found with respect to magnesium and calcium: the magnesium level is higher in a cell, whereas the free calcium concentration inside is very low compared to that outside, especially as compared to seawater.

The natural tendency is that an equal distribution be established between the cell and its environment. Although cell membranes are typically impermeable to cations, the impermeability is not perfect and, in the absence of some protective mechanism, cations can gradually diffuse in or out to equalize the concentrations. This natural tendency, movement down a concentration gradient, is what is known as "passive" diffusion. To maintain a differential distribution level of cations, cells must "pump" them in or out at the expense of some form of energy, usually the energy released by the hydrolysis of ATP (to ADP and P_i). This, then, is an "active" kind of transport. Other substances, including carbohydrates and amino acids as well as inorganic compounds, are transported actively by living cells.

The reverse process, i.e., passive transport down the concentration (electrochemical potential in general) gradient, can produce a certain kind of energy, which may be utilized to produce ATP from ADP and P_i. In fact, the mechanism to pump out sodium (and in potassium), which is often called the "sodium pump," is reversible to some extent, and has been shown to produce ATP under proper conditions. The calcium pump, which transports Ca(II) actively, can also be reversed to produce ATP.

However, by far the most important examples of ATP production are the so-called oxidative phosphorylation in mitochondria and the photosynthetic phosphorylation, both being caused by H^+ concentration gradient. This proton concentration gradient is a direct result of the electron transport processes in mitochondria and chloroplasts. In this chapter we shall examine the chemical bases of active transport processes and related effects.

10.2 General Features of Active Transport

The molecular mechanisms of active transport processes are still far from well understood. Some general features do, however, seem to be emerging out of the vast quantity of experimental data.

The main trick to be accomplished is the conversion of a chemical form of energy, that derived from ATP hydrolysis or similar reactions, into the mechanical work of transporting material. Another comparable instance of energy conversion is found in muscle contraction. This mechanical motion involves the conformational change of several subunits of the muscle protein, is triggered by Ca(II), and is supported energetically by ATP hydrolysis. Apparently the chain of events is essentially as follows. The binding (energy) of Ca(II) causes several subunits of the proteins in the thin filament to change their conformations. This change brings one of the components of the thin filament into contact with the head of a myosin molecule in the thick filament. This head is loaded with ATP. Its contact with actin of the thin filament then causes a conformational change in the thick filament myosin structure. This last change is manifested as a contraction of the muscle protein. What is to be noted here is the intimate relationship between the binding or release of an ionic species and a conformational change of a protein, as well as that between ATP hydrolysis (chemical energy) and a change in the conformation of the protein. The conformational change of a protein represents a mechanical movement, into which chemical energy has been directed.

How might the mechanical work of transporting material be accomplished with the use of chemical energy? As in muscle contraction, all active transport appears to be mediated by proteins. These proteins are called "transport" or "carrier" proteins, and should be distinguished from the channel-forming polypeptides described in the previous chapter. A molecule of the latter type merely defines a passageway for some entity, which then diffuses through the channel down a concentration gradient, whereas a protein of the former type must have an apparatus for converting chemical energy into a mechanical energy of some sort that can pull the entity through a membrane against a concentration gradient.

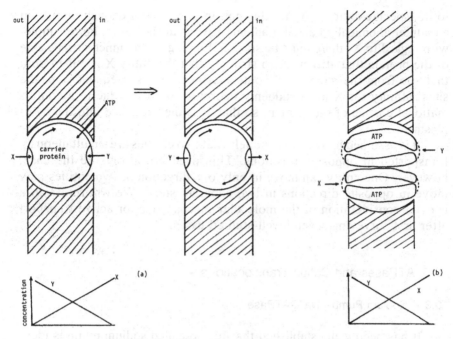

Figure 10-1. Models of active transport.

A number of molecular mechanisms have been proposed to explain the function of a transport protein; Fig. 10-1 illustrates two such mechanisms, which are intended here merely to point out some of the problems associated with active transport. In model (a), the binding of X at the outer surface of the protein and Y at the inner surface, along with ATP binding (and subsequent hydrolysis to ADP and P_i), somehow causes the protein to change its conformation and effectively rotate by about 180°. Now the X-binding site is exposed to the inner surface of the membrane and the Y site is at the outer surface. At this stage, the protein again changes its conformation so that its binding affinities for X and Y are significantly reduced and X and Y are released, despite their higher concentrations at the new locations. An important question to be addressed here is how feasible the turnaround of the transport protein is, as well as how fast it could happen. The answer would depend on the nature (fluidity) of the surrounding lipid bilayer of the membrane and on the interaction between the protein and the membrane. No detailed answer to the question seems to have yet appeared for a concrete case.

Model (b) might be called the "gated channel" model. Here the conformational change in the protein is realized as the opening of a gate to the specific entity X or Y. This notion alone is not enough to explain a counter chemical potential-gradient movement, however. For example, with regard to X, there must be specific binding on the inner surface side of the protein that attracts X so strongly that the entity X is pulled onto that site from the membrane's outer surface. In a subsequent step, the sites' affinity for X must suddenly decrease, probably due to a conformational change of the protein, so that X is then released into the cytoplasm.

It is assumed in these two models that two entities are simultaneously transported in opposite directions. This need not always be the case, however. One entity can move in only one direction or two entities may move in opposite directions in two separate steps. We will return to a general consideration of the molecular mechanisms for active transport after first surveying several well-studied cases.

10.3 ATPases and Cation Translocations

10.3.1 Sodium Pump—Na/K-ATPase

It has been well established that the so-called sodium pump is identical to a membrane enzyme, Na/K-ATPase. It requires Na(I) and K(I), as well as the usual Mg(II), for activity. The overall molecular weight of the purified enzyme seems to be about 250,000. It is a glycoprotein and contains two different subunits, one (α) with a molecular weight of about 95,000–100,000, and the other (β) with a molecular weight of 50,000–60,000. Evidently the complete enzyme consists of two units of α and one of β.

The α subunit is involved in the hydrolysis of ATP, and, as a result, it becomes phosphorylated. The site that is phosphorylated is an aspartate residue located in the sequence: Thr(or Ser)-Asp-Lys.

The Na(I)-binding site is located at the outer surface of the membrane. Its activity is inhibited by a cardiac glycoside, ouabain. The phosphorylation site on α is situated on the inner surface; this is not surprising, because ATP is supplied from the cytoplasmic side.

A partial digestion of subunit β was reported to produce a fragment that showed Na(I)-dependent ionophoric activity (Shamoo and Goldstein, 1977). Cyanogen bromide digestion of a purified Na/K-ATPase likewise yielded a fraction that showed K(I)-dependent ionophoric activity (Kuriki et al., 1976). Whether or not this fraction is a part of β is not clear (cf. Wilson, 1978).

The enzyme transports an average of 3 moles of Na(I) and 2 moles of K(I) per mole of ATP hydrolyzed. There is much evidence that the site to bind Na(I) is different from that for K(I) and, moreover, that these two ions are bound to different conformational states of the enzyme; the two conformational states are usually designated as E^1 and E^2.

The outer-surface and inner-surface Michaelis constants K_m for Na(I) and K(I) have been found to differ by a factor of about 10^2, as would be required by either of the models proposed in Fig. 10-1 (Wilson, 1978). That is to say, K_m at the outer surface for Na(I) is 100 times as high as K_m at the inner surface, which means that Na(I) is bound to the site on the inner surface 100 times as strongly as it is to the site on the outer surface. K_m for K(I) has been shown to have the opposite relationship with regard to the two surfaces.

The findings described above can be rationalized in the form of the following reaction model (Wilson, 1978). The first step is the binding of 3Na(I) and 1ATP to E^1. This is followed by hydrolysis of the ATP to form a phosphorylated enzyme and ADP. A conformational change of the protein results, producing E^2 ($E^2 \sim P$), which somehow allows the three Na(I) ions to move across the membrane. Then two K(I) ions bind to E^2 ($E^2 \sim P$) and $E^2 \sim P$ is hydrolyzed. This is accompanied by the conformational change of E^2 back to E^1 and the concomitant movement of the K(I) across the membrane. The existence of two distinct conformations has been demonstrated experimentally by a number of authors (see Gresalfi and Wallace, 1984). The whole process can now be expressed in the form of a series of chemical equilibria, where superscripts "i" and "o" refer to the "inside" and the "outside" of the cell, respectively:

$$E^1 + 3Na(I)^i + Mg(II) + ATP \xrightleftharpoons{K(Na)^i}$$

$$3Na-E^1(Mg) \sim P + ADP \xrightleftharpoons{\text{conformational change}}$$

$$3Na-E^2(Mg) \sim P \xrightleftharpoons{K(Na)^o} 3Na(I)^o + E^2(Mg) \sim P$$

$$E^2(Mg) \sim P + 2K(I)^o \xrightleftharpoons{K(K)^o} 2K-E^2(Mg) \sim P \xrightleftharpoons{\text{conformational change}}$$

$$2K-E^1(Mg) \sim P \xrightleftharpoons{+HOH, K(K)^i} E^1 + 2K(I)^i + P_i + Mg(II) \quad (10\text{-}1)$$

As indicated earlier, $K(Na)^i = 100K(Na)^o$ and $K(K)^o = 100K(K)^i$. Because $[K(I)^i] > [K(I)^o]$ and $[Na(I)^o] > [Na(I)^i]$, the entire sequence would occur in fact in the reverse direction (i.e., from the bottom to the top) were it not for the hydrolysis of ATP. That is:

$$\Delta G = \Delta G(\text{ATP hydrolysis})$$
$$+ \Delta G(\text{positive, due to the concentration gradient}) < 0 \quad (10\text{-}2)$$

where the large negative value of the first term is required to overcome the positive second term resulting from the given concentration gradient.

An interesting feature, which is also associated with the Ca pump to be discussed in the next section, is that this pump is known to be reversible: if it is provided with large enough concentration gradients [in terms of Na(I) and K(I)], the machine can produce ATP from ADP and P_i. In other words, when ΔG of the second term of equation 10-2 is very large, ΔG_{net} can become positive and the process occurs in the reverse direction, which includes ATP formation.

A conformational change such as that postulated here would be expected to be accompanied by fairly large values of ΔH and ΔS. The binding of both Mg(II) and P_i to the enzyme has been found to be highly exothermic, the ΔH values being -167 and -176 kJ/mole, respectively (Kuriki et al., 1976). Whether or not this binding of P_i represents the formation of $E^1 \sim P$ or $E^2 \sim P$ is not clear. The large negative value of ΔH would be counterbalanced by a large negative ΔS value, so that ΔG would not be of as large a magnitude as ΔH, though it could still be negative.

The results described still fall short of giving credence to any particular molecular mechanism. More detailed structural information on the enzyme itself must be uncovered before we will be in a position to understand the nature of conformational changes, how the binding of various entities bring them about, and how the hydrolysis energy of ATP is actually translated into the conformational changes of the protein. Readers are referred to Cantley (1981), Jorgensen et al. (1982), Repke (1982), and Glynn (1985) for more recent reviews of Na/K-ATPase.

10.3.2 Calcium Pump—Ca/Mg-ATPase (of Sarcoplasmic Reticulum)

Muscle contraction is triggered by a sudden release of Ca(II) from the sarcoplasmic reticulum. Soon afterward, Ca(II) has to be reabsorbed into the sarcoplasmic reticulum. This process is effected the by Ca pump in the sarcoplasmic reticulum membrane. As with the sodium pump, this pump has been identified as a Ca/Mg-ATPase.

Ca/Mg-ATPase consists of a single polypeptide (with a molecular weight of about 110,000) and a few proteolipid molecules (Wilson, 1978; Racker, 1977a). The polypeptide portion exhibits the ATPase activity, and some evidence suggests that the proteolipid is involved in the Ca(II) transport. In one study, for example, addition of the proteolipid to protein-free liposomes loaded with ^{45}Ca resulted in release of calcium ion into the medium, which was interpreted as showing that the proteolipid acts as an ionophore for Ca(II) (Racker, 1977a). In the same study, an ATPase preparation with a high proteolipid content was found to pump Ca(II) rapidly and efficiently, while the same ATPase preparation with a low proteolipid content did so only poorly (Racker, 1977a).

Others (Shamoo and Goldstein, 1977; Hermann and Shamoo, 1982), however, have assigned the ionophoric activity to a fragment (M_r 25,000) of the polypeptide. CH_3Hg^+, a cysteine-blocking agent, was found to inhibit the ATPase activity of the intact enzyme (i.e., protein of M_r 110,000) but not to inhibit the Ca(II)-ionophoric activity of either the intact enzyme or the fragment (M_r 25,000). $HgCl_2$ and $ZnCl_2$, however, do inhibit the ionophoric activity of both the enzyme and the fragment. These findings, in conjunction with the fact that the fragment contains little or no cysteine, suggest that Hg(II) [or Zn(II)], but not CH_3Hg^+, competes with Ca(II) for the Ca(II)-ionophoric site, a site that is not dependent on cysteine sulfhydryl groups.

Proteolytic degradation of this ATPase produces three fragments (including the one discussed above) (Green et al., 1977): fragment A_1 (M_r 33,000), A_2 (24,000), and B (55,000). It appears that fragment A_2 corresponds to the ionophoric fragment already described. Cysteine is a predominant constituent in the enzyme; the number of reactive cysteine residues is reported to be nine in A_1, two in A_2, and five in B. From kinetic studies on the reactivities of the cysteines, it was suggested that there is a clustering of thiols in the region of the hydrolytic site of the enzyme. Reaction of the thiols with N-ethylmaleimide abolished binding of ATP to the enzyme but left the binding of Ca(II) unchanged (Hasselbach, 1978). These observations indicate that the cysteine residues are somehow involved in the ATP binding but have little to do with the Ca(II) binding, a finding in accord with the conclusion reached on the basis of the effect of CH_3Hg^+. Recently, a monomeric solubilized protein has been demonstrated to bind Ca(II) and exhibit the ATPase activity (Martin et al., 1984).

The active site residue that is phosphorylated by ATP is an aspartate, just as in Na/K-ATPase. This aspartate residue is on fragment A_1 and shown in boldface below. The amino acid sequence of this fragment has been partially determined (Green et al., 1977) to be: (N-terminal) Ac— Met—Glu—Ala—Ala—Ala— His— Ser— Lys— Ser— Thr—...Thr —Leu—Gly—Cys—Thr—Ser—Val —Ile—Cys—Ser—**ASP**—Lys—Thr —Gly —Thr —Leu —Thr—Asn—Gln—Val—Cys—Met—Ser—Lys— ...Ala—Thr — Ile —Cys —Ala—Leu—Cys—Asx—Asx—Ser—Arg— Glu —Ala— Cys— Cys — Phe—...Ile—Ala—Arg—Asn — Tyr—Leu— Glu—Gly—COOH. It is seen from this sequence that cysteine residues are fairly amassed around the active site even in the primary structure. Another interesting feature, which the authors (Green et al., 1977) did not mention, is that there are a number of positively charged residues (Lys and Arg) in the vicinity of the active site. These might be involved in the binding of ATP, which is highly negatively charged.

The reaction mode of Ca/Mg-ATPase appears, judging from a number of studies, to be very analogous to that of Na/K-ATPase as shown in reaction scheme 10-1. Two moles of Ca(II) [and probably 1 mole of Mg(II), which moves in the opposite direction] are transported by the system at the expense of 1 mole of ATP. Thus:

$$E^1 + 2Ca(II)^o + ATP \xrightarrow{-ADP} 2Ca(II)-E^1 \sim P \rightleftharpoons 2Ca(II)-E^2 \sim P$$

$$\xrightarrow{-2Ca(II)} E^2 \sim P \xrightarrow{+Mg(II)'} Mg(II)-E^2 \sim P \rightleftharpoons Mg(II)^o + E^1 + P_i \quad (10\text{-}3)$$

Again, the superscripts refer to the outside (o) and inside (i) of the sarcoplasmic reticulum vesicle, and E^1 and E^2 represent two different conformational states of the enzyme. The formation of the phosphoenzyme $E^1 \sim P$ [E^1 plus Ca(II) and ATP] and $E^2 \sim P$ [from E^1 (or E^2) plus Mg(II) and P_i] has been well established (Racker, 1977a). As in the case of Na/K-ATPase, the binding of Mg(II) and/or P_i to the enzyme is found to be highly exothermic (Racker, 1977a). This energy can be utilized to change the conformation of the enzyme from E^1 to E^2 as represented by the backward reaction of the last step of reaction 10-3. The association constant of Ca(II) to E^1 has been estimated to be 5×10^6 M, whereas that of Ca(II) to E^2 is less than 5×10^2 (Hasselbach and Waas, 1982). This large difference is the basis of the loading and unloading of Ca(II). The entire process can be reversed by sufficiently raising the inside concentration of Ca(II), leading to a formation of ATP from ADP and P_i. This reversibility of the Ca pump has been well established (Hasselbach, 1978). Readers are referred to Inesi *et al.* (1982) for a more recent discussion of the mechanism of sarcoplasmic reticulum ATPase.

10.3.3 Mitochondrial H$^+$-ATPase (Proton Pump) and Chloroplast H$^+$-ATPase

10.3.3.1 Introduction

There are two separate but intimately related proton-pumping systems in mitochondria. One system is functionally separate from the ATPase–proton pump conglomerate; it is a mechanism of proton translocation coupled with electron transfer from NADH to O_2. Whether this is in fact a kind of proton pump (Ernster, 1977), or simply a closed loop of proton–electron combination as propounded by Mitchell (1977), has not been settled. The other system is the proton-driven ATP synthetic apparatus (Ernster, 1977; Mitchell, 1977; Boyer, 1977; Racker, 1977b; Slater, 1977). The connection between these two separate systems appears to be made by the movement of protons and the subsequently established proton electrochemical potential, although there are some indications and suggestions that the two systems are connected physically; not covalently, but in a way such that a conformational alteration in one system influences that in the other. I shall discuss these two systems as if they were separate, the ATP-synthesizing (or H^+-ATPase) part in this section and the proton pump coupled with the electron transfer system in a later section.

10.3.3.2 Mitochondrial H^+-ATPase

This ATPase is much more complex than the Na/K-ATPase and the Ca/Mg-ATPase discussed previously. A schematic representation of the entire mitochondrial ATPase is given in Fig. 10-2 (DePierre and Ernster, 1977). The system consists of three parts: the head part is called F_1 (fraction 1), and it is connected to the membrane-embedded proteolipid part via a connecting part, OSCP–F_{c2}. The portion of the enzyme left by removal of F_1 is often called F_0; i.e., F_0 consists of OSCP, F_{c2}, and the proteolipid. F_1 is believed to consist of different kinds of subunits, α, β, γ, δ, and ϵ; one F_1 molecule seems to have a subunit composition of $\alpha_3\beta_3\gamma\delta\epsilon$. See DePierre and Ernster (1977) and Pedersen (1982) for a summary of the physicochemical properties of these subunits.

What concerns us most here is how protons are translocated and how this process is associated with ATP formation (or degradation). The entire system can be isolated from the membrane by treatment with detergents under suitable conditions, and when the isolated holoprotein ($F_1 + F_0$) is incorporated into liposomes, it catalyzes an ATP-dependent proton translocation with the establishment of an electrochemical potential gradient. The F_1 portion can be released from submitochondrial particles or

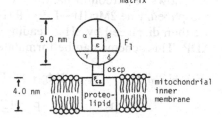

Figure 10-2. A schematic model of mitochondrial ATPase F_0–F_1. (After DePierre and Ernster, 1977. *Annu. Rev. Biochem.* 46:201.)

from the isolated ATPase by use of urea, NaBr, or cardiolipin. Solubilized F_1 has been found to have ATPase activity. Removal of the γ and δ subunits from F_1 led to a loss of coupling activity (i.e., F_1 could not be bound to F_0) but retention of ATPase activity. Removal of F_1 from the membrane holoenzyme left F_0, which did not show ATPase activity, but this removal made the membrane highly permeable to protons. Therefore, it may be concluded that F_1 binds ATP, ADP, and P_i, and possesses the ATPase activity, whereas F_0, especially the proteolipid portion, has a proton-conducting (translocating) activity.

When ATP formation is taking place, protons move down an electrochemical potential gradient; therefore, all that is required for the proton transfer system of the proteolipid is a kind of channel through which H^+ can travel freely. However, the same system (i.e., $F_1 + F_0$) can be made to function in the reverse direction, i.e., it can translocate H^+ against an electrochemical potential gradient at the expense of ATP. Hence, it seems that the proton-translocating system must be gated, just as are the ion transport systems in Na/K-ATPase and Ca/Mg-ATPase.

An influx of 2 moles of protons seems to be required to produce 1 mole of ATP from ADP and P_i. Several mechanisms have been proposed for ATP formation at F_1 with the accompanying proton influx. A mechanism proposed by Mitchell (1977) known as the direct mechanism can be formulated as:

$$PO_4^{3-}\text{—}F_1 + 2H^+ \rightleftharpoons [PO_3^{2-}]\text{—}F_1 + HOH$$

$$ADP + [PO_3^{2-}]\text{—}F_1 \rightleftharpoons ATP\text{—}F_1 \rightleftharpoons ATP + F_1 \qquad (10\text{-}4)$$

Here protons are hypothesized to react with bound P_i to form a bound reactive intermediate PO_3^{2-}. In a so-called indirect mechanism proposed by Boyer and Slater (Boyer, 1977; Slater, 1977), ATP formation is presumed to be spontaneous on F_1, and the proton flux is supposed to change the conformation of F_1 so that the tightly bound ATP is released from F_1. There is some experimental evidence that energy is required primarily for the release of bound ATP from F_1. Racker's (1977b) mechanism takes into consideration both the known facts for this case and the proposed mechanisms for Na/K-ATPase and Ca/Mg-ATPase. It can be described as follows: F_1^1 (conformation 1 of F_1) binds Mg(II) and P_i, forming a phosphoenzyme $2Mg(II)\text{—}F_1^2 \sim P$ (F_1^2 = conformation 2). The two Mg(II) are then displaced by $2H^+$, leading to $2H^+\text{—}F_1^2 \sim P$, which reacts with ADP. This scheme can be formulated as:

$$F_1^1 + 2Mg(II) + P_i \rightleftharpoons 2Mg(II)\text{—}F_1^2 \sim P \xrightarrow{+2H^+, \, -2Mg(II)} 2H^+\text{—}F_1^2 \sim P$$

$$2H^+\text{—}F_1^2 \sim P \xrightarrow{+ADP} 2H^+ + F_1^1 + ATP \qquad (10\text{-}5)$$

Scheme 10-5 is analogous to reaction schemes 10-1 and 10-3 previously proposed for Na/K-ATPase and Ca/Mg-ATPase.

10.3.3.3 Chloroplast H^+-ATPase

The ATPase in chloroplasts is represented as $CF_1 + F_0$; CF_1 is equivalent to the F_1 of mitochondria. Details of the structure, shape, and subunit composition of CF_1 have not been elucidated unequivocally, but these features are believed to be similar to those of F_1 of mitochondria.

10.3.4 A Summary Discussion of Cation-Translocating ATPases

Presuming that nature would employ analogous methods to carry out similar functions, we may hypothesize that the mechanisms of the three cation-translocating ATPases so far described are similar to one another, and that they may indeed be represented by reaction schemes 10-1, 10-3, and 10-5. What I try to do here is to explore possible molecular mechanistic models for this kind of process.

First, what could cause a conformational change in a protein, in particular an ATPase? In reaction schemes 10-1, 10-3, and 10-5, the binding of a cation and/or the formation of a phosphoprotein is supposed to be the cause. If phosphate binds to an aspartate residue, as in Ca/Mg-ATPase, what kind of effect might lead to the subsequent conformational change? For one thing, the binding of P_i to an aspartate residue would increase the negative electric charge at the site; that is:

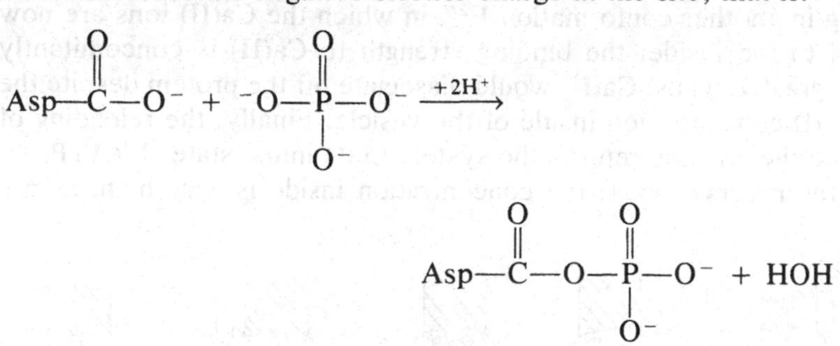

$$
\underset{\substack{\| \\ \text{Asp}-\text{C}-\text{O}^-}}{\overset{O}{}} + \underset{\substack{\| \\ {}^-\text{O}-\text{P}-\text{O}^- \\ | \\ \text{O}^-}}{\overset{O}{}} \xrightarrow{+2H^+}
$$

$$
\underset{\substack{\| \\ \text{Asp}-\text{C}-\text{O}-\text{P}-\text{O}^- \\ | \\ \text{O}^-}}{\overset{O \qquad O}{}} + \text{HOH} \qquad (10\text{-}6)
$$

This will in turn alter the electrostatic interaction between the reactive center and the nearby electrically charged amino acid residues. If the surrounding is dominated by negative charges, the addition of P_i would increase repulsive interactions and perhaps open up the phosphorylation site. If the surrounding is dominated by positive charges, on the contrary, the phosphorylation would close that particular cavity. This means that, in either case, displacements of amino acid residues around the phosphorylation site are to be expected; and such displacements are precisely what are needed to cause a conformational change of the whole protein molecule. A number of positively charged groups (Lys and Arg) are found around the active aspartate residue in Ca/Mg-ATPase, as seen in its amino acid sequence (see Section 10.3.2). Cysteine residues are also found in large number in the same region of the active site. In the normal pH range in which the ATPase operates, cysteine residues do not seem to be electrically charged. Thus, cysteine residues should not be significantly affected by the change in the electric charge at the aspartate. These cysteine residues could, however, be involved in the binding of ATP through hydrogen bonding.

The binding of a cation would cause a change opposite to that caused by the binding of a phosphate. Cation-binding sites would contain negatively charged groups such as the COO^- of glutamate and aspartate, as well as coordinative groups such as histidine imidazole and amine nitrogen. Binding of a cation tends to cause a contraction in protein structure.

An elaborate picture of one interpretation of the mode of action of Ca/Mg-ATPase is given in Fig. 10-3. Obviously, this picture represents mere conjecture, but it may serve as a useful model for additional studies. The enzyme is loaded with ATP and is in the conformational state E^1. The binding of two Ca(II) ions changes the conformation, and this somehow facilitates the phosphorylation of the enzyme by ATP. The compounded effect of the two events further alters the protein conformation (to E^2), exposing sites that would bind Ca(II) more strongly, causing the ions to migrate. The next step is the hydrolysis of $2Ca(II)-E^2 \sim P$, resulting in another conformation $E^{2'}$, in which the Ca(II) ions are now exposed to the inside; the binding strength to Ca(II) is concomitantly reduced greatly. Thus, Ca(II) would dissociate off the protein despite the high Ca(II) concentration inside of the vesicle. Finally, the reloading of ATP onto the enzyme returns the system to its initial state, $E^1(ATP)$.

In the reverse mode, the concentration inside is very high, as are

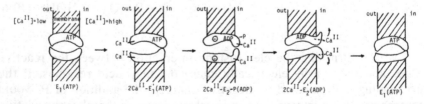

Figure 10-3. A model of Ca-ATPase (Ca pump).

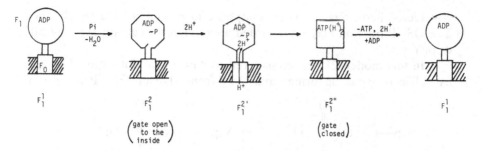

Figure 10-4. A model of H^+-ATPase (mitochondrial ATPase).

those of ADP and P_i (compared to that of ATP). The first step is then the binding of ADP (instead of ATP) to E^1; this would lead to the $E^{2'}$ conformation. $E^{2'}$ binds Ca(II) only weakly, but this binding could nonetheless be sufficient provided that the Ca(II) level is very high. $2Ca(II)—E^{2'}(ADP)$ then binds P_i, forming $2Ca(II)—E^{2'}(ADP) \sim P$. This step should be spontaneous, for the binding of P_i is fairly exothermic. ADP and P_i are now in close proximity, and the protein catalyzes the formation of ATP. As soon as ATP is formed, the conformation of the protein changes to E^1 and Ca(II) will be ejected to the outside. This last step would be spontaneous so long as the conformational change reduces the Ca(II) binding strength of the protein, for the Ca(II) concentration outside is low.

A similar (but not identical) mechanism has been put forward by Racker (1977a). An analogous model may easily be devised for Na/K-ATPase.

No reliable detailed model can as yet be offered for mitochondrial ATPase. Nonetheless, one possible model is suggested in Fig. 10-4 for the sake of discussion. F_1 binds ADP or ATP strongly in the conformation F_1^1. Addition of P_i [and probably Mg(II)] to $F_1^1(ADP)$ converts F_1^1 to the F_1^2 state, in which the protein is phosphorylated and the gate is now open to the entrance of protons. This gate could be located in the subunit ϵ (or γ or δ) of F_1. The entrance of $2H^+$ then brings about another conformation $F_1^{2'}$, in which the formation of ATP from the bound ADP and $F_1^2 \sim P$ takes place. This results in the creation of $ATP2H^+$ and in yet another conformation $F_1^{2''}$. This latter conformation is supposed to have an open structure and much less affinity for $ATP2H^+$. Thus, $ATP2H^+$ will be displaced by ADP, which exists in much greater abundance, and one reaction cycle is now complete. The reverse reaction, H^+ translocation at the expense of ATP, can proceed in just the opposite direction, except that the gate in the F_1^2 state is open to the outside. The proteolipid is considered here to be a proton ionophore or proton-conducting channel. The OSCP fraction may be regarded as serving to regulate the flow rate of protons.

In this model, $2H^+$ is considered to cause a conformational change as well as to assist the formation of ATP from ADP and $F_1 \sim P$ as follows:

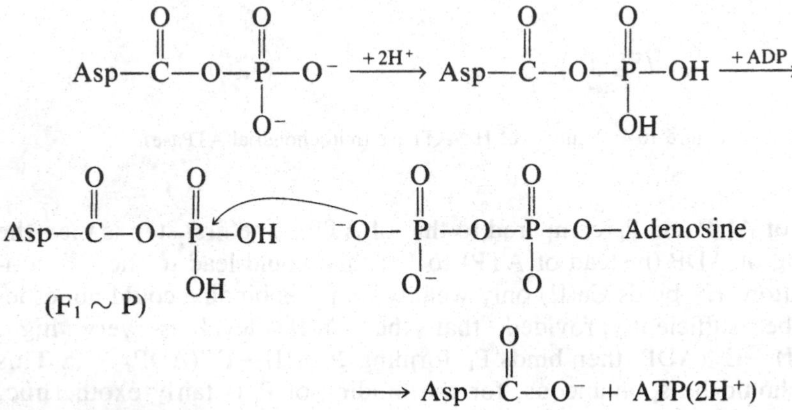

$$
\text{Asp}-\overset{\overset{\text{O}}{\|}}{\text{C}}-\text{O}-\overset{\overset{\text{O}}{\|}}{\underset{\underset{\text{O}^-}{|}}{\text{P}}}-\text{O}^- \xrightarrow{+2H^+} \text{Asp}-\overset{\overset{\text{O}}{\|}}{\text{C}}-\text{O}-\overset{\overset{\text{O}}{\|}}{\underset{\underset{\text{OH}}{|}}{\text{P}}}-\text{OH} \xrightarrow{+\text{ADP}}
$$

$$
\underset{(F_1 \sim P)}{\text{Asp}-\overset{\overset{\text{O}}{\|}}{\text{C}}-\text{O}-\overset{\overset{\text{O}}{\|}}{\underset{\underset{\text{OH}}{|}}{\text{P}}}-\text{OH}} \qquad {}^-\text{O}-\underset{\underset{\text{O}^-}{|}}{\overset{\overset{\text{O}}{\|}}{\text{P}}}-\text{O}-\underset{\underset{\text{O}}{|}}{\overset{\overset{\text{O}}{\|}}{\text{P}}}-\text{O}-\text{Adenosine}
$$

$$
\longrightarrow \text{Asp}-\overset{\overset{\text{O}}{\|}}{\text{C}}-\text{O}^- + \text{ATP}(2H^+) \qquad (10\text{-}7)
$$

The addition of proton to the phosphate oxygen would increase the electrophilicity of the phosphate phosphorus atom, thus rendering the attack of ADP onto this P atom easier as shown above. In a sense, this is a modified version of the so-called "direct" mechanism of Mitchell (Mitchell, 1977; Mitchell and Koppenol, 1982). Thus, the model (Fig. 10-4) may be regarded as a combination of the "direct" and the "indirect" (conformational) mechanisms.

10.4 Proton Pump Coupled with Electron Transfer Processes, and Other Proton Pumps

Two photoreaction centers in chloroplasts, photosystems PSI and PSII, are connected by an electron transfer system consisting of plastoquinones, cytochrome f, and plastocyanin. This system is also known to translocate protons and the proton electrochemical potential gradient thus created drives an ATPase system to produce ATP from ADP and P_i. The similarity between the chloroplast system and the mitochondrial system is rather striking and, perhaps, meaningful. The two systems are, after all, the most basic apparatuses for the whole contemporary biosphere, where oxygen is prevalent. Chloroplasts decompose water, with the use of energy from the sun, and produce electrons (and protons) and free oxygen. The oxygen is utilized in mitochondria to extract energy from the organic (and some inorganic) compounds produced by the chloroplasts. Structurally speaking, mitochondria and chloroplasts are more similar than they are different. Their structures are indicated diagrammatically in Fig. 10-5. Both have double membranes. The inner membrane of a mitochondrion is extensively invaginated, leading to structures called cristae. Electron microscopic observations have revealed numerous protrusions sticking out into the matrix from the inner membrane. These protrusions turned out to contain the H^+-ATPase. The functional unit in a chloroplast is a structure called the thylakoid, which is membrane-enclosed. Thylakoids are also studded with ATPase molecules. It is interesting to note that the thylakoid is formed as a sort of pinched-off crista, as shown in Fig. 10-5c. Therefore, a chloroplast can be regarded as a modified mitochondrion. The H^+-ATPase head faces the inside (matrix) in mitochondria, whereas that in chloroplasts faces outward. This contrast is reflected in the direction of the vectorial movement of protons in these two organelles. If mitochondria (loaded with NADH) are suspended in a medium under anaerobic conditions and oxygen is injected into the medium, the pH of the medium drops rapidly. This means that H^+ is translocated from the matrix side to the intermembrane space and then into the cytoplasm side (out of the mitochondrion). A similar experiment with a suspension of chloroplasts showed a movement of H^+ in the opposite direction, i.e., illumination of the suspension (instead of O_2 addition) increased the pH of the medium. This kind of experiment formed the basis for the formulation of the "chemiosmotic theory" of Mitchell (1977).

10.4.1 Mitochondrial System

The mitochondrial electron transfer system consists of NADH dehydrogenase (an alternate uses succinate dehydrogenase), ubiquinones,

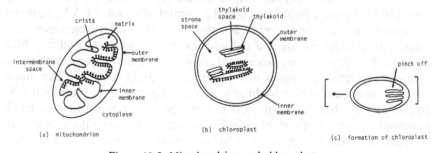

(a) mitochondrion

(b) chloroplast

(c) formation of chloroplast

Figure 10-5. Mitochondrion and chloroplast.

cytochromes (b, c_1, and c), and cytochrome c oxidase. NADH dehydrogenase contains, as prosthetic groups, FMN and iron–sulfur groups, whereas succinate dehydrogenase contains FAD and iron–sulfur groups. Cytochrome c oxidase contains cytochromes a and a_3 and two copper atoms per molecule. The iron–sulfur proteins and cytochromes are supposed to be specifically located, either embedded in the membrane, at its surface, or spanned across the entire membrane (DePierre and Ernster, 1977). As a pair of electrons are transferred through this system from NADH to O_2, protons are translocated from the matrix side to the cytoplasmic side. According to Mitchell (1977) (see also Hinkle and McCarty, 1978), protons are carried across the membrane, along with electrons, by both FMN and ubiquinone. Thus, in each case two protons and two electrons are released. That is:

$$NADH + H^+ \xrightarrow{\text{FMN}} NAD^+$$

$$FMNH_2 \xrightarrow[\text{surface of membrane}]{\text{carried to the other}} FMN + 2e + 2H^+ \qquad (10\text{-}8)$$

$$Q + 2H^+ + 2e \rightarrow QH_2 \xrightarrow[\text{surface of membrane}]{\text{carried to the other}} Q + 2H^+ + 2e \qquad (10\text{-}9)$$

In this model, a double loop involving ubiquinone (Q) is postulated. That FMN (or FAD) and quinone can act as proton ionophores was discussed in Section 9.2.1. Three pairs of H^+ ($6H^+$ altogether) are translocated across the membrane according to this model, i.e., the ratio H^+/site here is 2.

Other studies, however, have shown the ratio H^+/site to be 3–4 (Chance, 1977). This result cannot easily be accommodated by Mitchell's hypothesis. Chance (1977) described the electron transfer system as composed of four pools. Pool 1 has a mid potential of about -0.3 V, consists of iron–sulfur units, and can be identified as NADH dehydrogenase. Pool 2 consists of ubiquinones and cytochrome b, the mid potential of which is about 0 V. Pool 3 contains cytochromes c_1, c, and a along with some iron–sulfur units, and has a mid potential of about $+0.3$ V. The major component of pool 4 is cytochrome a_3 and its potential is about $+0.4$ V. In this model, there are gaps between the pools; each of the gaps has a potential difference of roughly 0.3 V (except that between pools 3 and 4). This gap in potential is postulated to be converted into an electrochemical potential difference across the membrane by the proton translocation system. This poses a question, however, of how H^+ is to be translocated if not the ionophore transport postulated in the chemiosmotic theory.

The existence of a H^+/site ratio greater than 2 suggests a pumping mechanism. No details about the mechanism of this pumping action are yet known. Cytochrome c oxidase has been shown to have proton-translocating activity (Wilkström, 1977, 1982). The proton translocation is apparently linked to a conformational change of the protein, with an associated spin state transition in the heme structure. The stoichiometry of this proton translocation is $2H^+/2e$ net. $(1/2)O_2$ consumes $2H^+$ per two electrons. Therefore, if the O_2 reduction site (H_2O formation site) is on the inside, $2H^+$ would have been translocated, whereas $4H^+$ would have been moved (from the inside to the outside) if the H_2O formation site is on the outer surface of the membrane. There is still a controversy over the H^+-pumping function of cytochrome c oxidase (e.g., see Papa, 1982; Hatefi et al., 1985).

A suggestion has been made that the basic mechanism of this proton translocation resembles the Bohr effect (pH dependence) of hemoglobin (Chance, 1977). There are, in fact, several indications that the reduction potential of some of the electron transfer units is dependent on pH. The Bohr effect can be described by:

$$
\begin{array}{ccc}
\text{Hb(hemoglobin)} & \rightleftharpoons & \text{HbO}_2 \\[4pt]
K_a(\text{Hb}) \Big\updownarrow + H^+ & & + H^+ \Big\updownarrow K_a(\text{HbO}_2) \\[4pt]
H^+\,\text{Hb} & \rightleftharpoons & H^+\text{HbO}_2
\end{array}
\qquad (10\text{-}10)
$$

In this reaction scheme, $K_a(\text{HbO}_2) > K_a(\text{Hb})$; that is, Hb binds H^+ more strongly than HbO_2. This is considered to be a result of a conformational change in hemoglobin upon binding O_2. In other words, when hemoglobin binds O_2, it changes its conformation (as is well known in the phenomenon called "heme–heme interaction"; see Section 4.1.4) and, as a consequence, protons become more easily dissociable from the hemoglobin molecule. Likewise, when a component in the electron transfer system changes its oxidation state, it also can change its conformation. At least a part of the energy released from the change in the reduction potential might be utilized to cause such a conformational change. The alteration in conformation may cause H^+ released to the outside from particular amino acid residues in the protein, and the protons lost would later be replenished from inside.

10.4.2 Chloroplast System

The electron transfer system in chloroplasts connects PSI with PSII, and the passage of a pair of electrons through this system is believed to produce two ATP molecules. Accordingly, it has been postulated that four protons are translocated across the membrane (of the thylakoid). Several models have been put forward for the mechanism of proton translocation in chloroplasts (Hauska and Trebst, 1977). One of the mechanisms invokes a looping (single or double) of plastoquinone, just as in the chemiosmotic theory of the mitochondrial electron transfer system.

10.4.3 Bacteriorhodopsin

Halobacterium halobium is a peculiar bacterium in that it grows best in 4.3 M NaCl solution (Stoeckenius, 1976; Stoeckenius and Bogomolni, 1982), a medium in which most organisms cannot live. It does not metabolize carbohydrates anaerobically, and it usually requires the addition of amino acids or peptides to its nutrient solution. It can live anaerobically, however, provided it is supplied with illumination. Its membrane can be fractionated into several components. One of them, the purple fraction, has been found to be a proton pump energized by illumination (Stoeckenius, 1976; Stoeckenius and Bogomolni, 1982). The purple fraction contains retinal bound to a protein; this complex is known as "bacteriorhodopsin" (Hess *et al.*, 1982). A similar retinal–protein complex, rhodopsin, is found in the pigment of the eyes of animals (Zurer, 1983). Retinal, an aldehyde derivative of carotene, is bound to an NH_2 group of the protein by a Schiff base type of conjugation in the case of bacteriorhodopsin; that is:

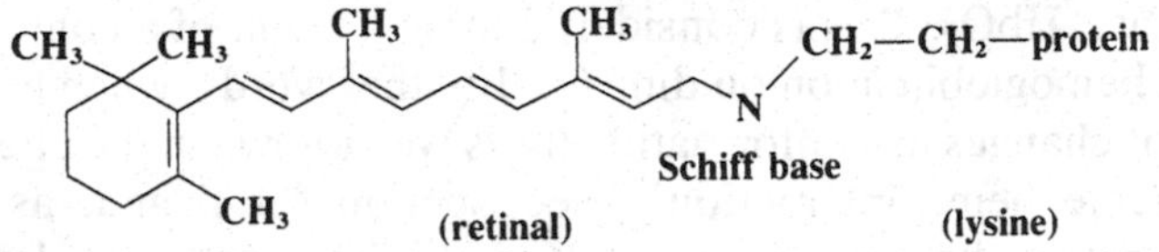

(retinal)   Schiff base   (lysine)

In the case of rhodopsin, the bound retinal is normally in the state in which all of the C=C bonds except that at C=11 are in the *trans* form. Illumination changes it to another form in which the double bond at C=11 is *trans* (i.e., all-*trans*). This change causes the scission of the retinal moiety from the protein (called "opsin").

The intact bacteriorhodopsin has an absorption maximum at 570 nm.

Light causes a bleaching of the purple color and the absorption shifts to 412 nm (Stoeckenius, 1976). This change is reversible, i.e., the new state reverts in the dark to the original one with a peak at 570 nm. Retinal itself absorbs at 370 nm in the free state. The retinal part in rhodopsin, particularly the Schiff base portion, seems to be wrapped up by hydrophobic residues of the proteins.

The spectral change (570 to 412 nm) occurs rapidly, with a $t_{1/2}$ of less than 10 nsec after a light flash. Following this spectral change a proton extrusion occurs, which is much slower than the spectral change and is not dependent on light. This creates a proton concentration gradient (an electrochemical potential gradient), which drives ATP formation by an ATPase located at the inner surface of the bacterial membrane.

10.5 Other Modes of Active Transport

In *H. halobium,* a light-induced proton electrochemical potential gradient is also utilized to translocate Na(I) out of the cell (Lanyi, 1977). As protons go into the cell, Na(I) ions come out; this kind of transport is called "antiport." Active transports of the antiport type, energized by proton movement, are known for K(I), Ca(II), and some amino acids in *E. coli*. The extra protons extruded (i.e., three or four protons as compared to the two protons needed to produce one ATP) seem to be utilized in the case of the mitochondrial electron transfer system as a means of antiporting ATP out of the structure. In some systems, a Na(I) concentration gradient is used to antiport amino acids and other nutrients.

Our interest here is how antiporting system works in molecular terms. It is likely that the cation [H^+ or Na(I)] producing the concentration gradient binds to the transport protein in the membrane. This results in a change in the conformation of the transport protein, which is accompanied by a transport of the countersubstance across the membrane. The whole process can be regarded as a combination of the sort of protein conformational change caused by H^+ binding [or Na(I) binding] with a movement of a cation of other substance in the direction opposite to that of H^+ movement. The details of these mechanisms remain to be elucidated, however.

10.6 Active Transport of Other Cationic Elements

The concentrations of essential inorganic elements in cells are usually higher than in the surroundings, especially in the hydrosphere. A few exceptions to this general rule include Na(I) and Ca(II). Both of these cations are usually less concentrated in the cell than in the surrounding. Na(I) and Ca(II), therefore, tend to migrate into the cells. The cells must use some of their metabolic energy to keep pumping these unwanted ions out. Na/K-ATPase and Ca/Mg-ATPase provide means to do this.

Other essential cations, including Mg(II), Fe(II, III), Zn(II), Mn(II), and Co(II), must constantly enter the cells, often against concentration gradients and through cell membranes that are impermeable to these cations.

Chapters 11 and 12 will treat in detail the ways in which organisms deal with the uptake of the essential cations. Absorption mechanisms have been shown to be of active type by demonstrating the dependence on temperature, and by identifying metabolic inhibitors such as CN^-, ouabain, and dinitrophenol. Certain systems have rigorous specificities for cations, but others (indeed, the majority) seem to have less rigorous specificities. Cations with similar characteristics (e.g., size, charge) can usually compete for a receptor site, though many details of these processes remain to be investigated.

10.7 Active Transports of Anions

Cells also have specific transport proteins for anions such as chloride, sulfate, and phosphate. Many oxyanions have common characteristics, however; for example, SO_4^{2-}, SeO_4^{2-}, and MoO_4^{2-} are all similar to each other (see Section 8.4). Therefore, it comes as no surprise that the sulfate transport system seems to be able to translocate SeO_4^{2-} or MoO_4^{2-} as well.

The so-called band 3 is an integral membrane protein that catalyzes the exchange of anions across the red cell membrane (Falke et al., 1984a,b). Its proper function seems to transport HCO_3^-, which is a waste product, from the cell. It exchanges Cl^- with HCO_3^-, and has an affinity for F^- and I^- in addition to Cl^-.

It is now recognized that there are a whole host of Cl^--transport systems in a variety of organisms (Gerencser, 1984). For example, a Cl^--ATPase that electrogenically transports Cl^- has been identified in marine species of Acetabularia (an alga) (Gradman, 1984). Other organisms or tissues for which active Cl^--transport systems have been identified include plants, intestines of crustaceans, insects, and molluscs, and renal tubules of invertebrates (Gerencser, 1984).

The membrane of *Halobacterium* species contains a halorhodopsin, in addition to the bacteriorhodopsin mentioned earlier (Schobert and Lanyi, 1982; Spudich and Spudich, 1985). Halorhodopsin-containing vesicles, when irradiated by light, were shown to transport Cl^- inward against the electrical and concentration gradients; this inward transport was accompanied by a passive movement of cations such as H^+, Na(I), and K(I) (Schobert and Lanyi, 1982).

Biological Aspects I— Metabolism of the Elements

Metabolism of Elements by Bacteria, Fungi, Algae, and Plants

11

From a structural standpoint, single-celled bacteria, algae, and fungi can all be regarded as being essentially the same. They have cell walls with cytoplasmic membrane underneath them. The cell walls are made of a variety of compounds, including cellulose, chitin, and peptidoglycans. The walls seem to have no specific physiological functions with regard to the absorption of inorganic (or, for that matter, organic) substances except in the case of the outer cell wall (or outer membrane) of gram-negative bacteria. The specific absorptions are instead carried out by the cytoplasmic membrane. The cell wall can, however, adsorb cationic metallic elements, since it is composed of compounds with negatively charged groups (e.g., phosphate, uronate) or, in certain organisms, containing coordinating groups.

The requirements for inorganic elements differ from one class of organism to another, especially as a function of their trophic status, e.g., photoautotrophic, chemoautotrophic, heterotrophic. All organisms seem to have developed mechanisms to absorb the elements they need, mechanisms that, in most cases, involve active transport.

Multicellular plants, particularly terrestrial ones, have specific organs (roots, rhizoids) for nutrient absorption. Vascular plants have translocating apparatuses such as xylem and phloem for the long-distance movement of nutrients. Nonvascular plants apparently depend on intercellular diffusion for the transport of nutrients. Lichen and moss seem to have an especially high capability for adsorbing cations.

Excretion of inorganic substances from these organisms is not well understood. Since the biological functions of essential elements were discussed in Part II, they will not be reviewed in this chapter or hereafter; instead, emphasis will be placed on various aspects of uptake, transport, and excretion entailed in the metabolisms of inorganic elements.

11.1 Bacteria

A convenient classification of eubacteria is the one according to gram-stain behavior. The classification as to gram-positive and gram-negative has been found to reflect the difference in the cell wall compositions and is thus appropriate for a discussion of uptake of inorganic elements.

11.1.1 Gram-Positive Bacteria

11.1.1.1 Iron (Neilands, 1974a; Hider, 1984)

Except for a few classes, most bacteria (most organisms for that matter) live in an environment in which pH is higher than 3 or 4, so that Fe(III) exists as the insoluble $Fe(OH)_3$ or its polymers. Exceptions include *Thiobacillus* species (gram-negative). *T. ferroxidans* is found in acid drainage waters of iron and bituminous coal mines; in this acidic water, iron exists as soluble $FeSO_4$. Therefore, this organism would not face the usual difficulty of obtaining iron from its environment. In fact, it uses the oxidation of Fe(II) (as well as sulfur compounds) to Fe(III) as its source of energy; it is a chemoautotrophic organism.

Ordinary bacteria, however, face a problem in securing iron, an element that is essential to all but a few species. A solution to this problem was the invention of siderophores. The chemistry of siderophores was dealt with in Section 9.1.3. A typical gram-positive bacterium of *Bacillus* family produces siderophores of two types: the catechol type and the hydroxamate type. Gram-negative bacteria such as *E. coli* and *Salmonella typhimurium* produce only the catechol type, although other gram-negative bacteria (e.g., *Aerobacter aerogenes*) produce a hydroxamate, aerobactin, along with a catechol-type siderophore. Hydroxamates are produced mainly by higher organisms such as fungi and molds.

Bacillus megaterium produces and excretes schizokinen when grown in media low in iron. This is a siderophore comprising two hydroxamate groups, one carboxylate group, and one alcoholic group. [Fe(III) (schizokinen)] is absorbed by an active transport system (Arceneaux *et al.*, 1973). Once inside, the iron of the complex is thought to be reduced to Fe(II) so as to be released. Fe(II) then is assumed to be incorporated into a number of proteins and enzymes. The freed schizokinen is then excreted. The control point of the iron uptake may be at the stage of ligand synthesis, absorption of the chelate complex, and/or excretion of the chelating agent.

There are a few organisms that do not require iron, or that lack typical iron-containing proteins or enzymes (Neilands, 1974b). Gram-positive anaerobic or facultative Lactobacillaceae species are an example; these are found in milk, and they ferment carbohydrates such as glucose and lactose with the formation of lactic acid. Many species of *Streptococcus* are said to be devoid of the usual heme-enzyme catalase, and instead contain a nonheme hydrogen peroxide-decomposing enzyme. A Mn-containing catalase has recently been isolated from *Lactobacillus plantarum* (Kono and Fridovich, 1983a). *Streptococcus faecalis* lacks cytochromes (Neilands, 1974b); this is understandable, because it (and other Lactobacillaceae) depends only on glycolysis for energy production and does not respire. Most anaerobic heterotrophic bacteria dependent solely on fermentation also do not require cytochromes.

Some strains of *Lactobacillus* species have been found to be entirely devoid of heme compounds. Their growth is independent of environmental sources of iron. This indicates further that they may not need any iron-containing enzymes or proteins. For example, the ribonucleotide reductase of *L. leichmanii* does not depend on an iron-containing B2 protein, unlike most other organisms (including mammals). Instead, a cobalt-containing B_{12} coenzyme is utilized (see Section 7.2).

11.1.1.2 Other Elements (Weinberg, 1977)

Mg(II) transport by *Bacillus subtilis* and *Staphylococcus aureus* has been found to be active and roughly follows a Michaelis–Menten formula (Jaspar and Silver, 1977). The K_m value [for Mg(II)] in *B. subtilis* is about 250 μM, which is somewhat larger than the K_m in *S. aureus* (= 70 μM) and K_m values in gram-negative bacteria (see below). This Mg-transport system in *B. subtilis* also shows an affinity for Mn(II) [K_m(Mn(II)) = 500 μM]. Mg(II) accumulation is inhibited by Co(II) in *B. subtilis* and by Zn(II), Mn(II), and Co(II) in *S. aureus* (Jaspar and Silver, 1977). Whether these cations are themselves transported by the same system as that for Mg(II) is not clear. It is highly likely that these cations, being similar to Mg(II) in charge and size, may competitively inhibit the binding of Mg(II) to the acceptor site, even though the organism may have a separate transport system for each individual cation [see below for Mn(II)]. It is interesting to note that Mn(II) or nonradioactive Mg(II) [but not Co(II) or Zn(II)] promotes the release of previously accumulated ^{28}Mg(II) from *B. subtilis* and *S. aureus* (Jaspar and Silver, 1977). This fact indicates that the Mg(II)-transport system works both ways (inward and outward) and that carrying of a cation is obligatory: Mn(II) from outside is carried in, and the system then has to transport the internal ^{28}Mg out. Co(II) was shown to inhibit the loss of ^{28}Mg from *B. subtilis* but induced it from *E.*

coli (Jaspar and Silver, 1977). This fact may reflect differences in the cell wall–membrane structures between the two organisms.

B. subtilis and *B. megaterium* (as well as *E. coli*) were observed to accumulate ^{45}Ca at low temperature (below 5°C), or upon exposure of the cells to energy poisons or uncouplers (Silver, 1977). These observations suggest active transport systems, which normally work to pump out Ca(II) in these organisms. In fact, everted membranes (inside-out) of *B. megaterium* take up Ca(II) with a high affinity (K_m = 9 μM) (Silver, 1977). An alkaline external pH is required for Ca(II) accumulation with the everted membrane; this indicates a Ca(II)/H$^+$ antiport mechanism in which the H$^+$ chemical potential gradient between the inside and the outside energizes the Ca(II) transport.

Bacillus species form heat-resistant endospores; the cortex of the endospore contains a high level of dipicolinic acid (DPA). DPA is synthesized in the mother cell compartment and rapidly moves into the spore. Ca(II) is also taken up into the spore and seems to form a complex with DPA. This complex can constitute as much as 20% of the dry weight of the spore and is considered to be responsible for the thermostability of the endospore.

Mn(II) activates a great variety of enzymes (see Chapter 5). The spore formation in *Bacillus* species specifically requires Mn; spores were shown not to form in the absence of Mn(II) (Silver and Jaspar, 1977). Mn(II) was found specifically to activate phosphoglycerate phosphomutase in spore-forming cells of *B. subtilis* (Oh and Freese, 1976); no other cation tested was effective, including Mg(II), Co(II), Ca(II), and Zn(II). Mn(II) has also been shown to be required for the synthesis of a small cyclic peptide, mycobacilin (which perhaps works as a cation channel; see Chapter 9), in *B. subtilis* and a secondary metabolite called bacitracin in *B. licheniformis* (Silver and Jaspar, 1977). These secondary metabolites may be involved in some way in the transport of Mn(II).

B. subtilis was shown to have an active transport system specific for Mn(II) with K_m = 1 μM; this transport is not affected by the presence of a 10^4-fold excess of Mg(II) or Ca(II) (Silver and Jaspar, 1977). An interesting finding is that a small molecule (M_r ~ 1500) that stimulates the Mn(II) uptake was synthesized under Mn(II)-deprived conditions. This compound is heat-stable and water-soluble, and was shown not to be bacitracin.

S. aureus (and *Rhodopseudomonas capsulata*) was also found to have a highly specific active transport system for Mn(II) (Silver and Jaspar, 1977). Cd(II) is a specific inhibitor of Mn(II) accumulation in a Cd-sensitive strain of *S. aureus* (Silver and Jaspar, 1977). A Cd-resistant strain (in which the resistance gene is carried on a plasmid) does not accumulate Cd(II), nor does it bind Cd(II) (Silver and Jaspar, 1977).

These observations on cation transport in bacteria suggest that the organisms have developed elaborate and specific mechanisms to take up required elements.

11.1.2 Gram-Negative Bacteria

11.1.2.1 Iron (Neilands, 1974a, 1977; Hider, 1984)

Siderophores such as enterobactin and ferrichrome protect E. coli and Salmonella typhimurium from certain antibiotics, bacteriocins, and phages. This suggests that there are receptor sites on the outer cell wall (outer membrane) common to siderophores and to other factors such as antibiotics, bacteriocins, and phages. This observation was utilized to determine the nature of the receptor sites of siderophores (Neilands, 1977), because the receptor sites for the other factors were already known. The receptor site in E. coli for ferrichrome was identified to be the one controlled by a gene named tonA; this receptor site also binds albomycin (another hydroxamate, an antibiotic), colicin M, phage T1, T5, and ∅80 (Neilands, 1977). The site controlled by the gene tonB is the receptor for Fe(III)-enterobactin (the proper siderophore produced by E. coli) as well as colicin B. More recent developments on siderophore receptor proteins have been reviewed by Neilands (1982). The transport (across the cytoplasmic membrane) protein for Fe(III)-enterobactin is synthesized in E. coli by the fep gene (Frost and Rosenberg, 1975). A metabolic scheme proposed for Fe(III)-enterobactin places the control locus on the enterobactin gene, i.e., excess Fe(II) released within the cell represses the enterobactin gene and stops the transcription of the gene and hence the synthesis of the siderophore. See Section 9.1.3 for other types of siderophores.

11.1.2.2 Other Elements (Weinberg, 1977)

Two distinct Mg(II)-transport systems have been identified in E. coli; both systems have $K_m = 30 \ \mu M$ (Jaspar and Silver, 1977). One is not entirely specific for Mg(II) but shows affinities for Co(II), Mn(II), Ni(II), and Zn(II), as well (Jaspar and Silver, 1977). The other is very specific for Mg(II) and is not inhibited by Co(II). Rhodopseudomonas capsulata (a photoautotrophic bacterium) also was shown to have a Mg(II)-transport system ($K_m = 55\ \mu M$), which is competitively inhibited by Co(II), Mn(II), and Fe(II) (Jaspar and Silver, 1977).

There seems to be no great difference between gram-positive and gram-negative bacteria as far as Ca(II) transport is concerned, except that gram-negative bacteria do not form endospores. In other words, an active transport mechanism to pump out Ca(II) (similar to the one found in gram-positive bacteria) has been shown to exist in *E. coli, R. capsulata,* and other gram-negative bacteria (Silver, 1977).

One transport system in *E. coli* has a very high affinity for Mn(II), the K_m 0.2 μM (Silver and Jaspar, 1977). This transport system is highly specific for Mn(II) and is not affected by the presence of a 10^5-fold excess of Mg(II) and Ca(II). It does not transport Ni(II), Cu(II), and Zn(II), but does seem to be inhibited competitively by Fe(II) or Co(II). These observations indicate that the acceptor site is designed to fit a transition metal cation of a certain size, and that it prefers an octahedral structure. That this transport system is active has been shown by its temperature dependence and the inhibitive effects on it of CN^-, DNP, and *m*-chlorophenylcarbonylcyanide hydrazone (Silver and Jaspar, 1977).

A sulfate transport protein was isolated from a strain of *Salmonella typhimurium* (Kotyk and Janacek, 1975). The transport system accumulates sulfate or thiosulfate against a concentration gradient. The isolated protein binds 1 mole of sulfate per mole of the protein, which has a molecular weight of 32,000 (Kotyk and Janacek, 1975). It was estimated that there are about 10^4 molecules of this protein per cell (Kotyk and Janacek, 1975). Molybdenum toxicity in *S. typhimurium* was relieved competitively by SO_4^{2-} or $S_2O_3^{2-}$; mutants lacking a SO_4^{2-}-transport system also failed to transport molybdate (MoO_4^{2-}) (Hutner, 1972). Therefore, it is likely that the same transport system serves both MoO_4^{2-} and SO_4^{2-}.

A phosphate-binding protein was isolated from the periplasmic space (the region between the cytoplasmic membrane and the cell wall) of *E. coli;* K_m was determined to be about 0.7 μM (Kotyk and Janacek, 1975). One mole of the protein binds 1 mole of phosphate and there are about 2×10^4 molecules of the protein per cell (Kotyk and Janacek, 1975).

11.2 Fungi (Yeast, Mold, Mushroom)

11.2.1 Iron

These heterotrophic organisms are saprophytes and are in need of some mechanisms for capturing trace elements, especially iron, from the environment. A number of fungal species excrete siderophores (Neilands, 1972; Hider, 1984). As mentioned earlier, these siderophores are derivatives of hydroxamate. Examples include ferrichromes from species of *Aspergillus, Neurospora, Penicillium, Ustilago, Actinomyces,* and *Streptomyces;* coprogen from *Neurospora* and *Penicillium;* and ferrioxamines from *Norcadia, Actinomyces,* and *Streptomyces.* It is now assumed that all members of *Ascomycetes, Basidiomycetes,* and *Fungi Imperfecti* produce siderophores.

The mechanisms of uptake of Fe(III)-siderophore complexes by these organisms are yet to be studied in detail; however, they are very likely to be similar to those in bacteria. Ferritin, first recognized as an iron-storing protein in animals, has also been isolated from some fungi, including *Phycomyces blakesleanus* (David, 1974). It is found to be associated with lipid droplets in mycelia, sporangiophores, and spores. Iron stored in the sporangiophore is utilized in spore formation, and the iron in spores is utilized in spore germination. It has also been shown that, as with ferritin in animals, apoferritin is inducible by the presence of iron (David, 1974).

11.2.2 Other Elements

The Mg(II)-transport system in *Saccharomyces cerevisiae* (baker's yeast) was found to be similar to the one in *E. coli* (Jaspar and Silver, 1977) (see Section 11.1.2.2). The accumulation of Mg(II) in *Penicillium chrysogenum* was shown to be effected by active transport and require the presence of phosphate, as in *S. cerevisiae.* Mg(II) in fungi is bound mostly to ribosomes or exists as a polymeric magnesium orthophosphate (Jaspar and Silver, 1977). The latter material has been detected in *P. chrysogenum, Aspergillus niger, A. oryzae, Phycomyces blakesleanus, Endomyces magnusii,* and others. The formation and breakdown [and subsequent release of Mg(II)] or such complexes is considered to be one of the mechanisms used to regulate the cellular level of Mg(II) in these organisms. This mechanism bears some resemblance to the iron-level-controlling mechanism using ferritin, although the latter involves the synthesis of a protein. Magnesium is localized as free Mg(II) aquo species in vacuoles of *E. magnusii, S. cerevisiae,* and *Saccharomyces carlsbergensis* (Jaspar and Silver, 1977).

Acyla (a mold) develops from a single-cell spore. Ca(II) was found to be essential at all stages of its development (Silver, 1977). In general, Ca(II) can be assumed to be essential to the development of any cellular organism. The accumulation of Ca(II) by *Acyla* is not energy-dependent

(K_m = 40 μM) (Silver, 1977). It is thus transported along the concentration gradient; pumping Ca(II) out of the cell is probably an active process, however, as seen earlier in the case of bacteria.

Zn(II) is another element essential to all organisms. For example, Zn(II) was found to stimulate development of the following fungi from filamentous to the yeast form: *Histoplasma capsulatum, Ustilago sphaerogena, Mucor rouxii,* and *Candida albicans* (Failla, 1977). The uptake of Zn(II) by *Candida utilis* was shown to occur only during the lag phase and the latter half of the growth phase (David, 1974). The maximum intracellular Zn(II) level attained by *C. utilis* during the lag phase is about 40 μmole per milligram of cell. This level is independent of the Zn(II) concentration in the growth medium (David, 1974), indicating the presence of a controlling mechanism for Zn(II) uptake.

Phosphate and arsenate compete for the same uptake sites in *S. cerevisiae; K_m* is about 4 $\times$ 10^{-4} M for both (Kotyk and Janacek, 1975). The same was also observed to be true in *Rhodotorula rubra* (Failla, 1977). This fact (i.e., competition of PO_4^{3-} and AsO_4^{3-} for the same uptake site) can be regarded as being fairly universal in biological systems, because of their chemical similarity. The phosphate uptake by these organisms was shown to be energy-dependent (Failla, 1977).

The uptake of sulfate by *S. cerevisiae* and *Neurospora crassa* was also shown to be energy-dependent (Kotyk and Janacek, 1975). It was revealed, as expected, that SO_4^{2-}, $S_2O_3^{2-}$, SeO_4^{2-}, and MoO_4^{2-} share a common transport system (also energy-dependent) in the mycelia of *Penicillium* and *Aspergillus* species (Failla, 1977).

11.3 Algae

Healy (1973) has provided a comprehensive review on inorganic nutrients in algae; the description presented here is largely based on his review.

Algae are autotrophs and have specific requirements for CO_2 and inorganic nitrogen compounds such as NH_3 and NO_3^- in addition to the other common inorganic nutrients such as S, P, Mg, Zn, Fe, Mn, Na, K, Cl^-, and others. Silicon is specifically required by one group of algae, the diatoms (see Section 6.3.1). Copper is also required universally by algae and plants.

Carbon dioxide is present as CO_3^{2-} (or HCO_3^-) in the marine and other aqueous media in which most algae exist. The presence of carbonic anhydrase has been established in a variety of algae. It is assumed that this enzyme dehydrates HCO_3^- (to CO_2 + OH^-), and that the CO_2 thus formed, being a neutral molecule, can easily go through the algal cell membrane.

Many algae can utilize NH_3, NO_3^-, or NO_2^- as their nitrogen source; in addition, some algae can use organic nitrogen compounds such as urea, purines, and amino acids. Nitrate reductase is found in most plants including algae and fungi; it is a molybdenum enzyme. Nitrite reductase, found in plants and *Ankistrodesmus braunii* (alga), requires iron and manganese. *Chlorella vulgaris* grown in an NH_3-containing medium did not begin uptake of NO_3^- until 2 hr after the nitrogen source was changed to NO_3^-. This suggests that NH_3 represses the synthesis of nitrate reductase. Similar observations were reported with *Anabaena cylindria* and *Scenedesmus* species.

Sulfate is commonly used by algae. $S_2O_3^{2-}$ can support the growth of *Chlorella pyrenoidosa*. SO_3^{2-} was found to be utilized by two cyanobacteria, and SO_3^{2-} or $S_2O_3^{2-}$ was found to be effective for the growth of *Porphyridim cruentum*. S-containing amino acids themselves were shown to support the growth of *Chlorella* species, *Anacystis nidulans*, and *Anabaena variabilis*. The uptake of SO_4^{2-} in algae is energy-dependent just as it is in bacteria and fungi, and it is partially independent of the subsequent reduction of SO_4^{2-}. This implies that the intracellular level of SO_4^{2-}, and not that of sulfur-containing organic compounds, controls the uptake. In vacuolated algae, SO_4^{2-} is most likely stored in the vacuole; this is consistent with the fact that the uptake and the reduction of SO_4^{2-} were found to be completely independent of each other in *Chora australis* (a vacuolated alga).

Phosphate uptake by algae is energy-dependent and is stimulated by light. The latter phenomenon is probably caused by the general increase in activity of the cell upon exposure to light. As in the case of bacteria and fungi, the phosphate-transport system of algae is capable of transporting arsenate. In *Chlorella pyrenoidosa*, phosphate is accumulated into polyphosphate and metabolic phosphate pools such as phospholipid, sugar phosphate, and nucleic acids. When external phosphate becomes scarce, the polyphosphate is degraded to satisfy metabolic needs. Phosphate deficiency induces alkaline phosphatase activity on the cell surface of several diatoms, chrysophytes, cyanobacteria, some green algae, and dinoflagellates (as well as bacteria and fungi). This enzyme degrades organic phosphates available in the surrounding fluids so that they release inorganic phosphate, which is then taken up. Acid phosphatase was observed to be induced by a phosphate deficiency at the surface of *Euglena gracilis*.

The only form of silicon available for uptake by diatoms is H_4SiO_4. As expected, germanium, in the form of H_4GeO_4, inhibits the uptake of silicon in such diatoms as *Navicula pelliculosa* and *Cylindrotheca fusi-*

formis. That the Si uptake by diatoms is energy-dependent has been inferred from the usual experimental observations that the uptake is inhibited by DNP and other respiratory inhibitors (see Simpson and Volcani, 1981).

It is interesting to note that no alga has been found to produce and secrete siderophores, despite the fact that the production of siderophores is so common in bacteria and fungi. It may be that the iron requirement of algae is less severe than that of the other organisms. A second possible explanation is that the environment in which algae thrive, i.e., marine and other hydrospheres, is not iron deficient to the same extent as other environments inhabited by other types of organisms. A third possible reason is that even if their environment did lack sufficient supply of iron, it would not be economical for algae to produce and secrete siderophores in order to capture the scant iron available. This is true simply because the siderophores, being water-soluble, would be dispersed and lost to the environment (hydrosphere). Therefore, iron uptake in algae is considered to be effected by active transport mechanisms. Ferritin-like bodies have been observed in some algae, but they probably serve as a means of iron storage.

Magnesium and manganese are two elements specifically required for photosynthesis in algae as well as other (green) plants. The Mg(II) transport by *Euglena gracilis* was found to be energy-dependent and inhibited by Mn(II) and Co(II) (Jaspar and Silver, 1977). The uptake of Mn(II) by Mn-deficient algae seems to be dependent on phosphorylation; this suggests an active nature for the Mn transport.

Copper is another specific element required by the algal photosynthetic apparatus. Copper is also an important component of a key mitochondrial enzyme, cytochrome oxidase. It is required only in trace amounts, however. The uptake of copper by *Chlorella vulgaris* was shown to be faster in dead than in living cells. This suggests the existence of a transport mechanism that normally serves to pump copper out or at least keep it from coming in.

Zinc uptake by *Chlorella fusca* was shown to be characterized by two phases. One is a rapid initial phase, which represents a reversible binding of Zn(II) to anionic sites in the cell wall. The second, a slower and prolonged phase, shows the character of an energy-dependent transport. This reported behavior (i.e., in two distinct phases) is certainly not unique for Zn(II) in this organism; indeed, it can be regarded as characteristic of cation uptake by any cell. A Zn deficiency in *E. gracilis* leads to an increase in the content of Fe and Mn. Evidently the organism tries to incorporate any similar available cation in an attempt to replace the needed Zn.

The uptake of sodium, potassium, and chloride are dependent on each other. It appears that two separate mechanisms are present for the transport of these ions in algae. One is a Na/K-ATPase similar to the one discussed in Chapter 10. The other is an active Cl^- uptake that is accompanied by Na and K influx.

11.4 Lichen and Moss

A lichen is a composite of a fungus and an alga; the fungus forms a dense web of threads (hyphae) within which cells of the alga grow. Moss is a typical nonvascular plant; some mosses, however, have well-developed rhizoids, supporting and conducting tissues. These tissues are lacking in peat moss (*Sphagnum*).

The metabolisms of inorganic elements in these organisms are not well understood. An interesting feature of the organisms is that they take up metallic elements to unusually high levels and thus are used as indicator organisms for environmental pollution, particularly that due to heavy metals. For example, radioactive nuclei such as Sr-90 and Cs-137, presumably derived from nuclear explosions, have been detected in lichens growing as far north as the tundra region of Finland.

Laboratory experiments on the lichen *Cladonia stellaris* and *C. rangiformis* showed that: (1) Cs-137 was more mobile than Sr-90 in dead lichen thalli, and (2) with living lichen, only limited translocation of Cs-137 occurred when the tip or base of the thallus was dipped in solution, whereas Sr-90 became evenly distributed within the thallus (Brown, 1976). These results are interpreted to indicate that in the living lichen Cs(I) is incorporated into the cell cytoplasm by an active process (such as Na/K pump) and that Sr(II), being similar to Ca(II), is not so readily incorporated into the cell but is reversibly bound to ionized groups in the cell wall (Healy, 1973).

A laboratory experiment on heavy metal uptake by *C. rangiformis* grown on a lead slag heap showed the following results (Healy, 1973) (the units are micromoles of the element per gram of lichen). When metal ions are added individually at 1.9 nM, the resulting lichen contents are: Pb(II), 37.6 > Cu(II), 31.4 > Zn(II), 25.9 > Ni(II), 25.2 > Co(II), 23.5. When the metal ions are instead given as a mixture at 1.0 nM each, different results are obtained: Pb(II), 26.3 > Cu(II), 11.3 > Ni(II), 5.1 > Zn(II), 4.6 > Co(II) 4.1. The order Zn(II) < Cu(II) > Ni(II) > Co(II) is the so-called Irving–Williams series, representing, in general, the relative complexing ability of the transition metal ions. The fact that this particular strain of lichen absorbs Pb(II) at abnormally high levels was first taken to indicate that it has adapted to its Pb-rich environment, developing a mechanism to fix Pb(II) and to tolerate it. However, this hypothesis turned out to be wrong, for the population of the same lichen obtained from a coal slag heap that did not contain a high level of Pb was found to show a very similar pattern in its relative uptake capacity for the five metal ions (Healy, 1973). Furthermore, it was also shown that even dead *C. rangiformis* specimens absorb these cations to a similar extent (Healy, 1973). Therefore, it was concluded that the absorption of cations by *C. rangiformis* is simply a reversible binding to anion sites. It was also reported that the site of the passive cation uptake is extracellular, probably consisting of some of the cell wall components (Healy, 1973). One such component is a substance that contains carboxyl groups. The contribution of these uronic acid-based pectins to the binding of cations has been estimated to be about 16% (of the total cation-binding capacity) in *C. stellaris*, 37% in *Umbilicaria deusta*, and 82% in *Parmelia saxatilis* (Healy, 1973). It was suggested that the polyuronate could be the major binding site of cations in the case of mosses. Other cell wall constituents of fungal portions of a lichen are also likely to be involved in cation binding, including chitin, proteins, and phosphate groups of phospholipids. The cation binding by chitin, polyuronate, and similar compounds will be reviewed in the next section.

11.5 Binding of Metal Ions to Polysaccharides and Derivatives

The hypothesis was noted above (and will be referred to hereafter, as well) that metal cations are reversibly adsorbed on the cell wall surfaces prior to, and independent of, active uptake. The cell wall materials involved are diverse. They include chitin in fungi and crustaceans, and polyuronic acids (uronic acid = carboxylate derivative of monosaccharide) in many algae. There are also other anionic polysaccharides such as the so-called mucopolysaccharides that could potentially adsorb cations. These include chondroitin sulfate and ascophyllan, which contain both COOH and SO_3H groups, and fucoidan and dextran sulfate. The latter contain only the SO_3H group.

The adsorption of cations onto these materials has not really been studied systematically, but some of the results reported are summarized in Table 11-1. It is well known that pectate, a component of the primary cell wall of the plant cell, depends on Ca(II) for its adhesive effect in bridging adjacent cells. Table 11-1 indicates that transition metal cations can bind to polyuronates including pectate, as well as, or even more strongly than, Mg(II) and Ca(II). Chitosan, having one amine group per glucose unit, strongly binds metal cations. Chitin has one acetylated amine group ($-NHCOCH_3$) per glucose unit, which can be regarded to bind a metal cation through the acetyl carbonyl oxygen. This binding is not expected to be as strong as that of an amine group, however.

In general, cation binding to an anionic polymer is rather complicated. The extent of binding depends not only on the strength of the metal-to-ligand bond but also on the conformation of the polymer, the distribution pattern of negative charges, and so forth. Therefore, it is not possible to present a generalized picture of trends in cation binding to the anionic polymer.

11.6 Plants

It has long been known that green plants require, in addition to CO_2 and H_2O, inorganic elements such as N, P, S, K, Mg, Ca, Mn, Fe, Cu, Zn, Mo, Cl, and B. The mechanisms of the uptakes of these elements are, however, poorly understood. The following is a summary of the most important information currently available.

11.6.1 Absorption from Roots

Findings are summarized in a review by Pitman (1976). Our discussion in this section is based largely on this source.

The absorptions of K(I), Na(I), and Cl^- have been studied in the roots of many plants, especially barley, maize, *Phaseolus,* and oat. That the transport systems for these ions are energy-dependent and hence active has been established mainly by the inhibitory effects on these systems of DNP, carbonylcyanide *p*-trifluoromethoxyphenylhydrazone, CN^-, N_3^-, and ouabain. K(I)-specific ATPases have been isolated from oat roots and other plants (Nissen, 1974), but whether this is the only mechanism available has not been settled.

Similar inhibitory effects are utilized to characterize the transport systems of other cations and anions. For example, the uptake of Ca(II) by maize roots was reduced by 91% upon addition of 50 μM DNP; the inhibition of Mn(II) transport in barley roots was complete at levels as low as 10 μM DNP. The uptake of Mg(II) is also thought to be active, and it was found to be dependent on the presence of Ca(II). The content and the rate of uptake of Mg(II) by barley and maize roots were greater

Table 11-1. Metal Cation Binding to Polysaccharides and Derivatives

	Mg(II)	Ca(II)	Sr(II)	Ba(II)	Mn(II)	Fe(III)	Co(II)	Ni(II)	Cu(II)	Zn(II)	Cd(II)	Ag(I)	Hg(II)	Pb(II)
					Relative K values[a]									
Polysaccharide with —COOH														
Polyglucuronate	1	40	150	130	—	—	7	—	320					
Polymannuronate	1	1.8	1.2	10	—	—	2	—	22					
Alginate[b]	1	1.6	1.4	13	—	—	1.6	—	18					
Pectate	1	7.0	9.6	10.1	—	—	5	—	224					
Hyaluronate	1	0.9	—	—	—	—	—	—	2.7					
Polysaccharide with —COOH and —SO₃H														
Chondroitin sulfate	1	1.1	1.0	1.3	—	—	—	—	2.6					
Ascophyllan	1	1.1	1.3	2.1	—	—	1	—	7.7					
Polysaccharide with —SO₃H														
Fucoidan	1	1.3	1.6	2.9	—	—	1.3	—	2.0					
Dextran sulfate	1	1.1	2.0	7.0	—	—	1.3	—	1.0					

Metal ion adsorbed (mmoles metal/g substrate)[c]

Chitosan[a]	—	—	1.44	1.18	2.47	3.15	3.12	3.70	2.78	3.26	5.6	3.97
Bark[e]	—	—	0.10	0.43	0.21	0.22	0.35	0.63	0.23	0.73	0.62	0.74

Relative binding strength[f]

Chitin[g]

Ni(II) > Zn(II) > Co(II) > Cu(II) > Pb(II) > Fe(III) ≒ Mn(II) > Mg(II) > Ca(II)

Metal ion adsorbed (mmoles metal/g wool)[c]

Wool (keratin)	0.13	—	—	—	0.32	0.04	0.04	0.33	0.52	0.30	1.12	2.25	0.58

[a] A. Huag and O. Smidsrød, 1970. *Acta Chim. Scand.* 24:843.
[b] 92% mannuronate.
[c] M. S. Masri and M. Friedman, 1974. in *Protein–Metal Interactions* (M. Friedman, ed.). Plenum Press, p. 551.
[d] Polyglucosamine).
[e] Lignin and humic acid.
[f] T. Yoshinari and V. Subramanian, 1977. in *Environmental Biogeochemistry* (J. O. Nriagu, ed.), Ann Arbor Science Publ., p. 541.
[g] Poly(N-acetylglucosamine), from lobster shell.

from a solution containing only Mg(II) than from a solution containing both Mg(II) and Ca(II). It was suggested that Ca(II) may have a noncompetitive effect on a site other than that involved in the Mg(II) transport.

Two different types of Mn(II)-transport systems have been recognized: one is independent of Ca(II) and the other is affected by it. Ca(II) was shown to inhibit Mn(II) uptake by the roots of *Trifolium* and *Medicago* species, and the degree of this inhibition was found to be pH-dependent. At lower pH the inhibition of Mn uptake by Ca(II) was more complete. Zn(II) uptake also appears to be active and is inhibited strongly by Ca(II).

The uptake of iron by plant roots seems to be active, but its study is complicated by the presence of an oxidation–reduction process. In aerated media at neutral pH, iron most certainly exists as Fe(III) and is probably chelated by ligands. Such ligands include citric acid and special polyamino acids such as mugineic acid and avenic acid isolated from roots of barley and oat (see Section 9.1.3.3). It has been repeatedly shown that Fe(III) is reduced to Fe(II) on the surface of a root. The mechanism of this reduction is, however, obscure. As discussed earlier, the uptake of Fe(III) by bacteria and fungi involves siderophores and does not require the reduction of Fe(III) at the cell surface. The absorption of iron into intestinal mucosa, on the other hand, is believed to involve reduction of Fe(III) to Fe(II) prior to the uptake. An iron-storage system similar to ferritin has been observed in plant chloroplasts (David, 1974; Hyde *et al.*, 1963).

The uptake of SO_4^{2-} was also shown to be active in the roots of *Helinthus*. It was suggested that there are two different kinds of sites for phosphate uptake: one for high concentration of phosphate and the other for low concentration. The latter was shown to be affected by Ca(II), but the former is not. At least 90% of the phosphate entering the roots is converted into organic phosphate. In a number of species, however, it has been shown that phosphate is stored as polyphosphate during part of the year and only used later during a growing period, just as was found to be the case in *Chlorella pyrenoidosa* (see above).

Nitrate uptake and metabolism involves reduction, as was previously discussed in the case of algae. It has been suggested that the NO_3^--transport system is inducible by NO_3^-. Needless to say, NH_3 can replace NO_3^- as the nitrogen source in plants.

The uptakes of borate (H_3BO_3) and silicate (H_4SiO_4) appear to be passive in contrast to all of the ionic species discussed above. This has been inferred from results showing that the total uptake of silica is roughly proportional to the rate of transpiration in intact plants, indicating that silicate somehow is absorbed along with water.

11.6.2 Absorption by Other Tissues

The uptakes of Cu(II), Zn(II), and Mn(II) by sugarcane (*Saccharum officinarum*) leaf tissue were shown to conform to a Michaelis–Menten formulation (Bowen, 1969; Clarkson, 1974). The rate of Mn(II) uptake was unaffected by the presence of Zn(II) or Cu(II). Conversely, neither the uptake of Cu(II) nor that of Zn(II) was affected by Mn(II). Nonetheless, a marked competition was observed between Cu(II) and Zn(II) uptake. These results indicate that a transport system carries both Cu(II) and Zn(II), but that it is entirely independent of another transport system that carries Mn(II). It was also found that 0.1 nM solutions of Na(I), Li(I), Cs(I), Ag(I), Co(II), Cr(II) Al(III), Mg(II), Ba(II), and NH_4^+ were completely without effect on the uptake of Mn(II), Cu(II), and Zn(II) by these tissues.

11.6.3 Long-Distance Transport in Plants

Iron is, as discussed above, supposed to be reduced to Fe(II) at the surface of a root prior to absorption. It was shown, however, that Fe(III)-citrate was the form actually translocated in the xylem of the mature root systems of tomato and soybean (Tiffin, 1967; Anderson, 1975). Fe(III)-citrate was suggested to be the major form present in other xylem transport systems, too. Available data (Tiffin, 1967) indicate that Ca(II), Mg(II), Mn(II), Co(II), and Zn(II) are all translocated as simple hydrated cations in the xylem of mature tomato roots (Anderson, 1975). It was found that silicon in xylem sap is entirely in the form of H_4SiO_4 (Barber and Shone, 1966).

Metabolism of Elements in Mammals and Vertebrates

12

12.1 Metabolism of Iron

A remarkable feature of iron metabolism is that there is no specific excretory mechanism for iron, or if there is one (as yet to be discovered), the excretion rate is very low. Iron will, however, be lost through hemorrhage and other incidents. Organisms use iron very economically. It is essential and required in substantial quantity, but rather hard to obtain. Thus, organisms cannot afford to lose it easily and hence have developed extensive mechanisms to utilize their iron over and over again. Two-thirds of all the iron in mammals is in the form of hemoglobin. The daily iron turnover caused by destruction of red blood cells is equivalent to about 2% of that in all the hemoglobin. However, this iron is reutilized to produce more hemoglobin. The average half-life of hemoglobin iron has been estimated to be about 8 years in an adult human (Finch and Loden, 1959). This is enormously long, compared to, for example, the half-life of calcium in bone, which is also considered to be "long" but estimated to be less than 9 months.

Because of lack of an excretory mechanism, the iron level in the body is controlled by its absorption rate and various storage mechanisms. Ferritin and hemosiderin are utilized for storage purposes, as discussed in Section 9.3.1.

An outline of iron metabolism in mammals is shown in Fig. 12-1. Iron in the diet is absorbed mainly from the upper part (duodenum) of the small intestine, but also to a lesser extent from the middle (jejunum) and lower parts (ileum). The iron absorbed by the epithelial cells (mucosal cells) of the intestine passes to the serosal side and is then loaded onto transferrin, the plasma iron-carrier protein (see Section 9.3.1). Transferrin later unloads iron onto the bone marrow cells, where hemoglobin is produced. Smaller portions of the absorbed iron may be taken up by liver

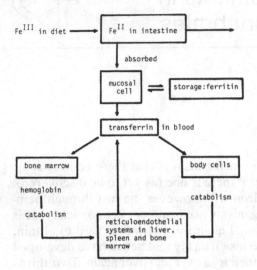

Figure 12-1. The metabolism of iron in humans.

and other tissues that produce various iron-containing enzymes. As red blood cells age, they are phagocytosed by macrophages of the reticulo-endothelial systems of the liver, spleen, and bone marrow. The main reaction involved in the catabolism of hemoglobin in the reticuloendo-thelial systems is a series of oxygenations. Hemoglobin is converted to bilirubin (the main component of the bile) with the release of iron. This iron is then incorporated into ferritin and either stored or utilized as needed to produce hemoglobin and other iron-containing enzymes and proteins.

12.1.1 Absorption of Iron

This topic was extensively reviewed by Forth and Rummel (1973), Worwood (1974), and more recently by Narins (1980). There are three compartments to be considered in the transport of iron from the intestinal lumen to the blood capillaries, as shown in Fig. 12-2. These are separated

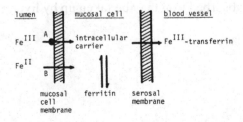

Figure 12-2. The iron absorption in the mammalian intestine.

by three barriers, the lumen side membrane of the mucosal cells and the membranes between the mucosal cells and the blood vessels. In the course of studies on dose-dependency, it was observed that in low dosage ranges (i.e., the physiological range) the transfer of iron across mucosal cells is determined mainly by specific binding sites, which have a limited capacity, whereas in the higher dose range (i.e., the toxic range) nonspecific binding sites with nearly unlimited capacity determine the absorption rate. In terms of time, it could be shown that there are two phases in iron absorption: an initial rapid phase and a subsequent slow phase. The rate of absorption of iron also depends greatly on the existing iron load of the body. In iron-deficient organisms, the initial rapid absorption becomes predominant. The capacity of the iron-transfer system is limited, however, especially in cases of iron deficiency. Taken together, these observations may be interpreted to indicate that there are two basic mechanisms of iron transfer (absorption) in the intestine: one is a specific carrier-mediated and (perhaps) energy-dependent mechanism, and the other is a passive type of absorption. The level of protein(s) in the former absorption system is known to be controlled genetically.

12.1.1.1 Carrier-Mediated Transport at Intestinal Lumen–Mucosal Cell Border

This active transport system can be regarded as being responsible for the observed initial rapid-phase absorption. As noted above, it has a limited, but variable, capacity. At relatively high body iron load the system shows low iron-binding capacity, whereas in iron deficiency it shows a high binding capacity. The observed difference seems to apply not only to the quantity of the iron being bound but also to the binding strength. For example, a normal mucosal cell (iron loaded) cannot take up iron from Fe(III)-conalbumin ($K = 10^{23}$ M^{-1}), or Fe(III) (8-hydroxyquinoline-5-sulfonate) (Forth and Rummel, 1973). This finding implies that the mucosal iron transport system under normal conditions does not have a very high binding strength for iron and hence cannot release iron from highly stable iron complexes such as these. However, under iron-deficient conditions the mucosal cells can utilize both complexes as iron sources (Forth and Rummel, 1973). It is not very clear whether the carrier systems employed under normal and deficient conditions are indeed identical or whether they are different. It is at least conceivable that iron deficiency somehow influences the conformation of the normal transport system in such a way that the system shows a higher affinity for iron. On the other hand, it may simply be that iron deficiency induces the deployment of another transport system that, by nature, has a higher iron affinity.

There is some evidence that the extrusion of iron from the mucosal cells into the blood (rather than the uptake of iron from the intestinal lumen into mucosal cells) is the rate-limiting step in iron absorption, and that the transport system from the mucosa into the blood becomes saturated at doses in the range of 5–10 μg in normal rats and 50–100 μg in iron-deficient rats (Forth and Rummel, 1973). Thus, an iron deficiency creates a low iron level in the mucosal cells as a result of the fact that the transfer of iron from the mucosal cells to the blood becomes more rapid.

The binding constant for iron on the acceptor site of the transmucosal iron-transport system under normal conditions seems to be in the range of 10^{12}–10^{13} M^{-1}. Chelated iron complexes whose stability constants are lower than 10^{12} M^{-1} are taken up by the system but other chelated iron complexes whose stability constants are higher than 10^{13} M^{-1} cannot be taken up (Forth, 1974). Iron, however, seems to be reduced to Fe(II) before being transported by this system; this has been demonstrated using rabbit intestinal brush-border membranes (Marx and Aisen, 1981) and mouse intestinal brush-border membranous vesicles (Muir $et\ al.$, 1984). The reduction mechanism of Fe(III) entities to Fe(II) is not clear.

The absorption of iron, i.e., Fe(II) [as well as Co(II) and Mn(II)], in rat duodenal segment has been shown to be approximated by the Michaelis–Menten equation, with the constants V_{max} (μmoles per 30 min) and K_m determined to be: Fe, $V_{max} = 1008 \pm 94$, $K_m = (1.6 \pm 0.4) \times 10^{-3}$ M; Co, $V_{max} = 1221 \pm 77$, $K_m = (2.6 \pm 0.4) \times 10^{-3}$ M; Mn, $V_{max} = 713 \pm 103$, $K_m = (4.5 \pm 1.5) \times 10^{-3}$ M (Thomson $et\ al.$, 1971a,b). At least in the case of iron, K_m can be regarded as representing the value of k_2/k_1 since, as noted above, the corresponding value of $k_{-1}/k_1 = 10^{-12}$–10^{-13}.

As reflected in the V_{max} and K_m values, transition metals such as cobalt and manganese could compete with iron for the iron-transport system (Worwood, 1974). It was shown that iron transfer across intestinal mucosa is inhibited by an equimolar concentration of cobalt in rats with deficient, normal, or high iron loads. However, uptake of cobalt occurred only in iron-deficient rats (Worwood, 1974). Manganese absorption from the duodenum (in rats) was also increased under conditions of iron deficiency but it could be inhibited by the presence of iron (5 nM) in the perfusion medium. Nickel, zinc, and copper were also shown to be absorbed $in\ vitro$ by isolated segments of the small intestine of rats. The ratios of the transfer rates in iron-deficient versus normal rats are 11 for Fe, 4 for Ni(II), 3 for Mn(II) and Co(II), 2 for Zn(II), and 1 for Cu(II) (Forth and Rummel, 1973). Except for Cu(II), the transfer rate increases for all cations studied when the organism becomes iron-deficient. However, it is clear that the specificity of the transport system is highest for iron, as would be expected. The fact that divalent cations such as Co(II), Mn(II), Ni(II), and Zn(II) compete with iron for the transport system also indicates that the form of iron that is transported is very likely to be Fe(II). Several lines of evidence indicate that iron transport in iron-deficient animals is energy-dependent but that it can be maintained by anaerobic glycolysis (Forth and Rummel, 1973).

12.1.1.2 Intracellular Iron Carriers in Mucosal Cells

After the brush-border and nuclear fractions, the cytoplasmic (particle-free) fraction of mucosal cells was found to contain the highest percentage of ^{59}Fe taken up during a 10-min exposure (Forth and Rummel, 1973). This ^{59}Fe in the cytoplasmic fraction disappeared within 1 hr in iron-deficient rats, whereas its level remained unchanged in normal rats (Forth and Rummel, 1973). Apparently, the cytoplasmic fraction of the ^{59}Fe under conditions of iron deficiency is bound with a protein and is readily exchangeable. In normal cases, however, the iron is bound with another protein from which the release of iron is fairly slow. It is very likely that ferritin is the main iron store in the iron-sufficient state and that it also serves to store excess iron and thus protect against iron toxicity.

Several iron-binding proteins have been identified in mucosal cells of rats. One glycoprotein from the mucosa of rats binds iron with a high affinity (Forth and Rummel, 1973). Other authors have found a transferrin-like iron-binding protein to be present. This protein was shown to play a significant role in determining the rate of iron transfer across the cytoplasm of mucosal cells in the duodenum section of the intestine (Forth and Rummel, 1973). Findings corroborating the involvement of a protein in iron absorption include the following: (1) iron absorption is impaired in the presence of inhibitors of protein synthesis, such as cycloheximide and tetracycline (Forth and Rummel, 1973); and (2) the ability of intestinal iron absorption (in mice) to adapt to an increased demand is genetically controlled. It should be noted, however, that the question of whether the inhibited or genetically controlled protein is really the intracellular transferrin-like carrier protein, or whether it might instead be related to the membranous transport system, has not been unequivocally delineated.

12.1.1.3 Passive Transport

Passive forms of transport may represent the slower phase of iron absorption in the intestine. They seem to exist in the upper intestinal portion but become significant only in the middle and lower portions. The following observations indicated that iron complexes bypass the specific transport system discussed above, and that the mechanism by which they are ultimately absorbed involves a passive form of transport. First, the characteristic differences in ^{59}Fe transfer and tissue content between jejunal segments of normal and iron-deficient rats are abolished if iron is administered in the form of a stable complex such as [Fe^{III}(nicotine hydroxamate)] (Forth and Rummel, 1973). This observation is in accord with the hypothesis that the iron complex is not taken up by the specific transport system, since its ability to function depends greatly on the state of the iron load. Furthermore, two similar complexes, one electrically neutral and the other ionic, are absorbed very differently. [Fe^{III}(8-hydroxyquinoline)$_3$]^0 was absorbed much faster than was a free iron, but [Fe^{III}(8-hydroxyquinoline-5-sulfonate)$_3$]$^{3-}$ was absorbed much more slowly than free iron (Forth and Rummel, 1973). The former complex is neutral and lipid-soluble, whereas the latter is anionic and lipid-insoluble. Therefore, the former could be expected to penetrate the lipid membrane rather readily.

Heme is one of the iron chelates of low molecular weight known to occur within mucosal cells under physiological conditions. Several lines of evidence support the hypothesis that heme is taken up as such by mucosal cells (Forth and Rummel, 1973). However, the iron from heme subsequently appears in the nonheme fraction of the mucosal cells. This implies that absorbed heme is destroyed within the cells. The heme-splitting factor here is believed to be xanthine oxidase, a sharp contrast to the process employed in the reticuloendothelial system.

12.1.2 Transport in Blood Vessels

The iron-carrying protein in blood is transferrin. The nature of transferrin and the mechanism of iron loading and unloading are detailed in Section 9.3.1. Here we need to consider only one facet of the process, the role of ferroxidase, a copper-containing enzyme also known as ceruloplasmin. This is the major copper-containing protein in plasma. It had long been known that one symptom of copper deficiency is anemia. The missing link between the symptom, anemia, and the cause, copper deficiency, turned out to be ceruloplasmin. In order for iron to be incorporated into transferrin, it must be in the Fe(III) state. As discussed earlier, however, the major iron-uptake mechanism absorbs iron in the Fe(II) state. It can be partially oxidized within the mucosal cell, but the majority of it remains as Fe(II) when it emerges from the serosal side into the bloodstream. Then it must be oxidized by ferroxidase before combining with apotransferrin.

12.1.3 Hemoglobin Production

Red blood cells are produced mainly in the bone marrow in normal adult humans. The biosynthesis of hemoglobin has been treated extensively in a monograph (Bunn et al., 1977) and is summarized in Fig. 12-3.

12.1.4 Catabolism of Hemoglobin

Aged red blood cells (erythrocytes) are phagocytosed by the reticuloendothelial system of phagocytic cells that line the blood capillaries of the liver, splenic sinusoids, lymph nodes, bone marrow, adrenal and pituitary gland (Lynch et al., 1974). The mechanism of how phagocytic cells recognize the aged erythrocytes is believed to be as follows (Lynch et al., 1974). Normal erythrocytes are biconcave, presumably to enhance their surface area relative to volume. Erythrocytes have an average di-

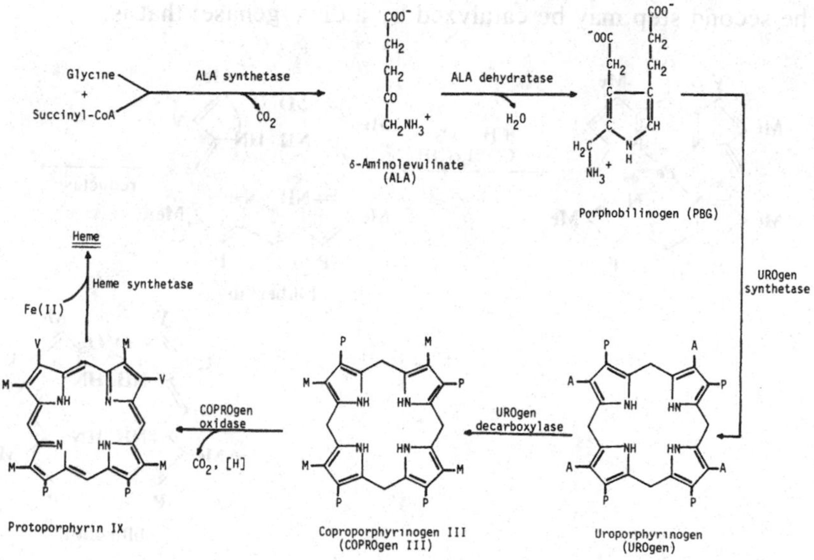

Figure 12-3. The biosynthesis of hemoglobin.

ameter of about 8 μm and have to negotiate channels (capillaries) as narrow as 3 μm wide; therefore, they must be flexible. Once erythrocytes become stiff, perhaps due to oxidative damages (see Chapter 13) in their membranes, they become trapped and undergo phagocytosis by the reticuloendothelial system. The ability of an erythrocyte to alter its shape declines with age. The average lifetime of erythrocytes is about 120 days in man.

In phagocytosis, red blood cell proteins and lipids are destroyed by hydrolytic enzymes in lysosomes. Heme thus released then undergoes oxygenative degradation. One enzyme involved is an oxygenase dependent on cytochrome P-450 (see Chapter 4 for a detailed discussion of P-450, i.e., the first step is believed to be):

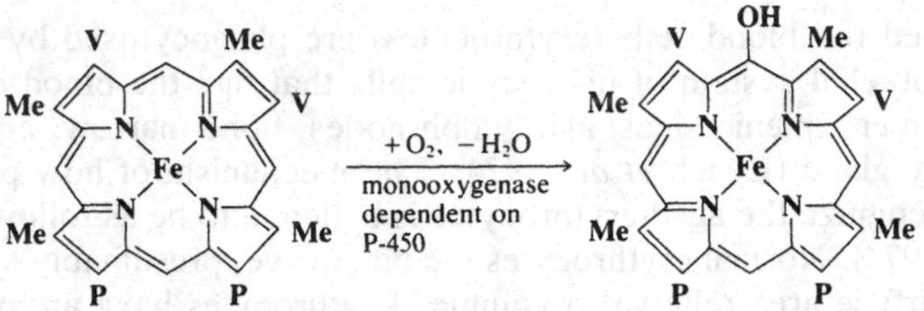

(12-1)

The second step may be catalyzed by a dioxygenase; that is:

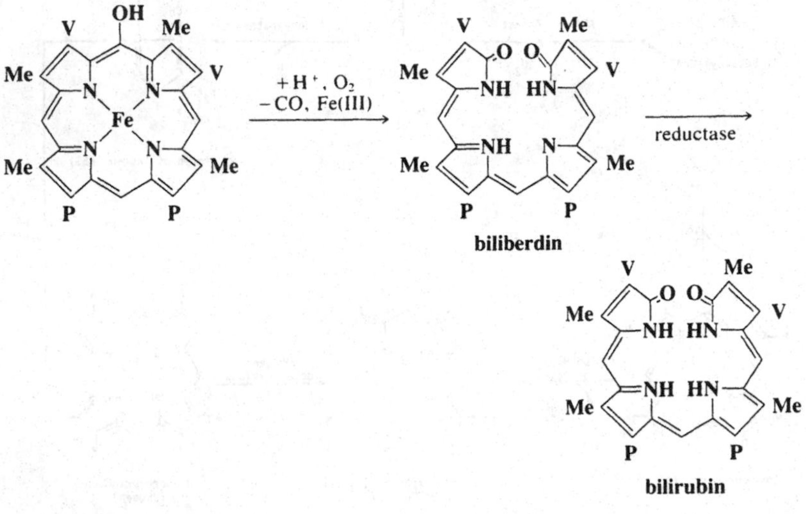

biliberdin

bilirubin

(12-2)

The released iron is stored as ferritin or hemosiderin (see Section 9.3.1).

The heme oxidative reaction was found to be inducible in liver and kidney by certain endogenous compounds as well as by particular trace metals (Kappas and Maines, 1976). Cadmium and cobalt, for example, enhanced the rate of heme oxidation in liver 5–10 times over its normal rate. A more effective inducer is tin. A single dose of stannous chloride increased the rate of heme oxidation in rat kidneys by 20–30 times. Concomitant with this striking increase in heme oxidation, the microsomal content of P-450 in kidney was decreased by an average 50%. It was also found that administration of $SnCl_2$ produced an approximately 35% decrease in the kidney tissue content of mitochondrial cytochromes a, b, c, and c_1. The suggestion was made that this heme oxidation-enhancing effect may represent an aspect of the toxicity of tin.

12.2 Metabolism of Copper

12.2.1 Distribution and Functions

In general, the highest concentration of copper is found in liver, followed by brain, lung, kidney, and ovary. The mean copper concentrations in these and other organs in the adult human (ppm, wet basis) are: liver 14.7, brain 5.6, lung 2.2, kidney 2.1, ovary 1.2, testis 0.8, lymph nodes 0.8, and muscle 0.7 (Underwood, 1977). Similar distributions have been found in other mammals. Unusually high levels of copper occur in the pigmented part of the eye: as high as 105 ppm (dry basis) in the iris and 88 ppm in the choroid (the intermediate layer of the eyeball wall) of the eye of freshwater trout (Underwood, 1977).

Liver is the center of metabolism of copper as well as other elements. Liver produces ceruloplasmin, the major copper-containing protein in the plasma. Another important enzyme found ubiquitously, especially in erythrocytes, is superoxide dismutase (Section 4.2), which used to be called erythrocuprein, cerebrocuprein, or hepatocuprein. Erythrocuprein is produced in erythroblasts. Yet another ubiquitous enzyme containing copper is cytochrome c oxidase, which is found in all mitochondria. A copper enzyme that may be peculiar to the nervous system is dopamine-β-hydroxylase, involved in the metabolism of neurotransmitters (catecholamines) such as dopamine and norepinephrine. The main factor in the pigment of eyes is melanin, which is produced from tyrosine through the action of a copper-containing enzyme, tyrosinase. The presence and function of these proteins and enzymes make it possible to rationalize the copper distribution described above.

One specific effect of copper deficiency is a group of skeletal abnormalities observed in rabbits, chicks, pigs, dogs, and foals (Underwood, 1977). It is caused by poor mineralization and is responsive to copper supplementation (Underwood, 1977). Amine or lysyl oxidase, which is dependent on copper (Shieh and Yasunobu, 1976), seems to be involved in the cross-linking of collagen fiber, so that a copper deficiency would naturally lead to diminished stability and strength of bone collagen (Underwood, 1977). There are also some observations indicating that copper is somehow involved in keratin formation. The normal growth of hair and wool (mostly consisting of keratin) ceases in copper-deficient rats, rabbits, cattle, dogs, and sheep (Underwood, 1977). Copper toxicity will be dealt with in Chapter 14.

12.2.2 Absorption and Transport of Copper

Copper absorption occurs in the stomach and small intestine. It appears that two different copper-absorption mechanisms exist. One is energy-dependent and the other is protein-mediated (Evans, 1973).

Many observations support the conclusion that amino acids facilitate the intestinal absorption of copper (Evans, 1973). Amino acid transport across the intestinal mucosa is energy-dependent and hence it is thought that copper is transported as a copper–amino acid complex via the energy-dependent amino acid transport system (Evans, 1973).

There is also an indication that copper absorption is, in part, protein-mediated. The protein involved in this mechanism seems to be identical to the so-called metallothionein or copper chelatin (see Section 14.3.5.2 for a further discussion of metallothionein and chelatin). Zn(II) was demonstrated to interfere with copper absorption in the duodenum of rats; this is in accord with the hypothesis that the protein involved is a metallothionein. In fact, the largest copper-containing fraction in the cytoplasm of mucosal cells was shown to be a protein of approximately 10,000 molecular weight and which has all the characteristics of metallothionein. The presence of metallothionein (or chelatin) in the cytoplasm, however, has nothing directly to do with the absorption mechanism. Instead, it controls the free copper level in the cytoplasm by sequestering the absorbed copper and hence regulates excessive copper toxicity. As expected, Cd(II), Hg(II), and Ag(I), as well as Zn(II), can affect copper absorption (Evans, 1973).

Copper in the blood is found mainly in erythrocyte superoxide dismutase, albumin, and ceruloplasmin or it is bound with amino acids, or as an unidentified copper complex in erythrocytes. After oral administration of ^{64}Cu (in humans and rats), the ^{64}Cu is found to be associated mainly with the albumin fraction in the portal blood; it disappears rapidly from the portal blood, there occurring a concomitant increase of ^{64}Cu level in the liver (Evans, 1973). These results clearly indicate that albumin is the transport protein of copper, at least in the portal vein. An interesting observation in this connection is the fact that dog serum albumin does not bind copper (Appleton and Sarker, 1971) and that dogs are extremely susceptible to copper toxicity (Goresky et al., 1968).

Because copper in the circulating blood is mostly bound as ceruloplasmin or present in erythrocytes, very little copper permeates the glomeruli and thus urinary excretion of copper is minimal under normal conditions (Evans, 1973).

12.2.3 Copper in Liver

Liver plays the central role in copper metabolism; it sequesters and stores copper, converts it into a bile component, and produces some of the copper proteins, including ceruloplasmin. Among various fractions of liver cells, the large granular fraction and the nuclear fraction contain most of the total copper, in each case approximately 20% of the total present (Evans, 1973). The large granular fraction contains mitochondria and lysosomes. The lysosomes have been found to play a vital role in copper metabolism in the liver. Lysosomes sequester excess cellular copper and also produce the bile, which is the main excretory vehicle for copper. The role and state of copper in the nuclear fraction are not known. The predominant copper fraction dissolved in the cytoplasm of the liver cell is associated with metallothionein. The microsomal fraction, which comes mostly from endoplasmic reticulum, contains a minor amount of copper (10% of the total). The endoplasmic reticulum provides the protein synthetic apparatus; the copper in this fraction, therefore, is mostly contained in newly synthesized copper-containing enzymes and proteins.

12.3 Metabolism of Zinc

12.3.1 Distribution and Functions

Zinc is ubiquitous and an essential component in a large number of enzymes (see Chapter 5). Among the metallic elements, it is second only

to iron in the total amount contained in the human body. About 20% of
the total body zinc is found in the skin. An appreciable portion of the
total zinc is also distributed in hair, fur, nails, bones, and dentin. The
highest zinc concentration in normal tissues occurs in the choroid of the
eyes; levels in the choroids of dog, fox, and marten have been estimated
to be 14,600, 69,000, and 91,000 ppm (dry basis), respectively (Underwood, 1977). The typical zinc levels in normal tissues of human bodies
are 510 ppm (dry basis) in prostate, 275 in kidney, 270 in muscle, 165 in
heart, 145 in pancreas, 105 in spleen, 85 in testis, 75 in lung, and 70 in
brain (Underwood, 1977). Other mammals show similar zinc distribution
patterns. The high zinc level in the prostate is notable. Other studies give
the following values (dry basis): human prostate gland 859 ppm, human
semen 910–2930 ppm, human spermatozoa 1900 ppm (Underwood, 1977).

Seventy-five to eighty-eight percent of the total zinc of normal human
blood is contained in erythrocytes, 12–22% in the plasma, and 3% in
leukocytes (Underwood, 1977). Most of the zinc in erythrocytes is present
in carbonic anhydrase. About two-thirds of the plasma zinc is loosely
bound with albumin and the rest is firmly bound with an α_2-macroglobulin
(Underwood, 1977).

Profound changes occur in the zinc level in the blood plasma and
other tissues under various disease states and under stress conditions.
Depressed zinc levels have been reported in patients with atherosclerosis,
cirrhosis, pernicious anemia, malignant tumors, chronic and acute infec-
tions, tuberculosis, leprosy, and even acute injury (Underwood, 1977;
Prasad and Oberleas, 1976).

As outlined in Chapter 5, the enzymatic zinc activity is very diverse.
Some of the more prominent and physiologically important Zn enzymes
include DNA polymerase, RNA polymerase, carbonic anhydrase, alkaline
phosphatase, and the dehydrogenases, which are involved in the catabolism of carbohydrates, lipids, alcohols, and other compounds. As suggested by the high level of zinc in the prostate, zinc deficiency is known
to retard development of male reproductive organs. At least a part of the
reason for this effect is related to the fact that zinc is the active catalytic
element in DNA polymerase and RNA polymerase. There are numerous
reports that indicate this fact both directly and indirectly (Underwood,
1977; Kirchgessner et al., 1976).

Skeletal abnormalities are a conspicuous feature of zinc deficiency
in birds. The physiological mechanism of this effect is not clear, but a
part of it may be attributable to defects in alkaline phosphatase, which is
supposed to provide the bone formation apparatus with its required phosphate (see Section 6.3.4). Zinc is also believed to be involved in keratin-

ization, the formation of keratin tissues (e.g., wool, hair, skin). The physiology of this effect of zinc is not well understood.

12.3.2 Metabolism of Zinc

Pancreas secretes into the duodenum a ligand with which Zn(II) seems
to combine. The nature of the compound is not known, except that it has
a rather low molecular weight. Its purpose seems to be to facilitate the
intestinal absorption of Zn(II). The discovery (Hurley *et al.*, 1977; Eckert
et al., 1977) of a Zn(II)-binding ligand in human milk and rat milk is
interesting in this connection. The compound from human milk has a low
molecular weight; similar Zn(II)-binding compounds were identified in
both milk and intestinal mucosa of rats. The one found in the intestine
was present in rats 16 days of age or older, but was absent in rats from
birth to 16 days. The authors (Hurley *et al.*, 1977; Eckert *et al.*, 1977)
have proposed the hypothesis that the Zn(II)-binding ligand of maternal
milk enhances zinc transport in the neonatal period before the development of the intestinal mechanisms for zinc absorption. Paradoxically, no
similar ligand was found in cow milk.

Complexed zinc is transported into the mucosal cells by an unknown
mechanism with which Ca(II) seems to interfere. Zn(II) is then transferred
to the binding sites on the basolateral plasma membrane (Evans, 1976).
Albumin in the plasma interacts with the basolateral plasma membrane
and removes Zn(II) from the receptor sites (Kirchgessner *et al.*, 1976).
There is some indication that metallothionein may be involved in the
regulation of Zn(II) absorption in the intestine (Evans, 1976).

Zn(II) carried in the plasma is incorporated at differing rates into
different tissues. The most rapid accumulation and turnover occur in the
pancreas, liver, kidney, and dorsolateral prostate (Underwood, 1977).
Liver is the major organ involved in zinc metabolism (as true for copper
and other trace elements). Zn(II) injection induces a *de novo* synthesis
of metallothioneins, which sequester Zn(II) (Underwood, 1977). Metallothionein is regarded as the Zn(II) storage protein in liver.

Zinc is excreted largely by way of the feces (Underwood, 1977).
Pancreatic juice also carries some zinc and a part of it, too, will be lost
in the feces. Only a minor portion is excreted in the bile. Urinary excretion
of zinc also is not very high (0.3–0.5 mg/day in normal human adults)
(Underwood, 1977). Excretion of zinc by way of sweat, on the other hand,
is known to be substantial.

12.4 Metabolism of Other Trace Elements—Cobalt, Manganese, Molybdenum, and Others

12.4.1 Cobalt

12.4.1.1 Functions

The only biocompound definitely known to contain cobalt is vitamin B_{12} (cyanocobalamin) and its derivatives. Although Co(II) can substitute for Zn(II) in many of the zinc enzymes, cobalt is not regarded as the physiological component of these enzymes.

A characteristic feature of cobalt-deficient ruminants (e.g., cattle, sheep) is loss of appetite, emaciation, and listlessness (Underwood, 1977). The main source of energy in ruminants is not glucose (which is the energy source in humans and other mammals) but acetic and propionic acids produced by fermentation (by bacteria) in the rumen. Methylmalonyl-CoA mutase, which catalyzes the conversion of methylmalonyl-CoA to succinate, is known to be B_{12}-dependent and its activity is severely depressed in B_{12} deficiency. Propionic acid must first be bound to CoA, resulting in propionyl-CoA, which is then converted into methylmalonyl-CoA. Succinate formed from methylmalonyl-CoA by the action of this enzyme then enters the TCA cycle. Therefore, a reduction of this enzymatic activity due to B_{12} deficiency would result in propionic acid being nonutilizable, and hence in an accumulation of it and methylmalonic acid. This is believed to be the cause of the reduced appetite (Underwood, 1977).

Methionine synthetase, which also requires B_{12}, catalyzes the following reaction:

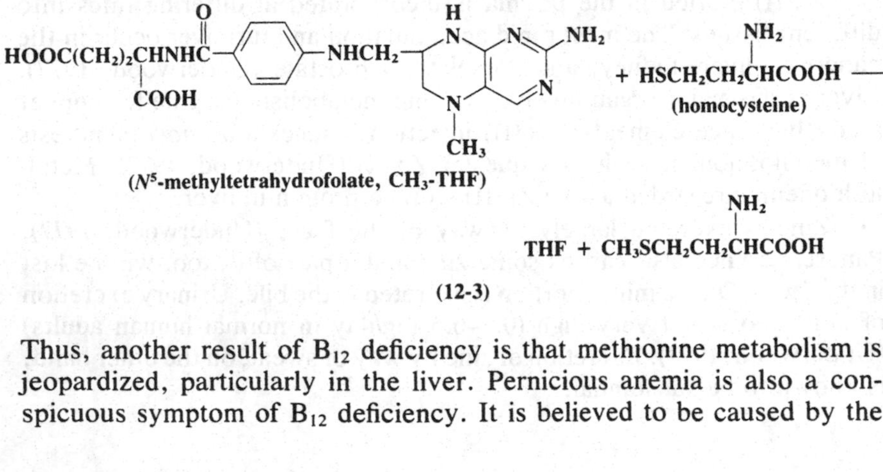

$$HOOC(CH_2)_2\underset{COOH}{CHNHC} \overset{O}{\parallel} -\text{⟨ ⟩}- NHCH_2 - [N^5\text{-methyltetrahydrofolate, } CH_3\text{-THF}] + HSCH_2CH_2\underset{NH_2}{CHCOOH} \longrightarrow$$

(homocysteine)

$$THF + CH_3SCH_2CH_2\underset{NH_2}{CHCOOH}$$

(12-3)

Thus, another result of B_{12} deficiency is that methionine metabolism is jeopardized, particularly in the liver. Pernicious anemia is also a conspicuous symptom of B_{12} deficiency. It is believed to be caused by the lack of the so-called "intrinsic factor" secreted in the stomach; this substance is essential for the intestinal absorption of B_{12} (see Section 9.5.2).

B_{12} deficiency in man can be a cause of the development of macrocytic (megaloblastic) anemia, although the mechanism through which B_{12} is involved in the process of red blood cell formation is not clear. One suggestion is that B_{12} deficiency indirectly prevents the conversion of deoxyuridylate to thymidylate, a process that requires folic acid. B_{12} is apparently essential to the formation of the active form of folic acid in a process believed to be related to reaction 12-3. Since thymidylate is to be incorporated into DNA, a deficiency of B_{12} would stop the production of DNA, thus prohibiting mitosis. This could then result in larger (megaloblastic) cells.

12.4.1.2 Metabolism

There seem to be two different transport systems for cobalt in the intestine. One is designed for Co(II)-inorganic compounds, whereas the other is designed for B_{12}. The former is not specific for Co(II); its proper function is regarded to be the transport of Fe(II). As mentioned earlier, there is a competition between the absorption of Fe(II) and Co(II) or Mn(II). For example, the absorption of ^{59}Fe from jejunal loops of iron-deficient rats is reduced by two-thirds in the presence of a 10-fold excess of Co(II). A further 100-fold excess of Co(II) suppresses the ^{59}Fe absorption nearly completely. A 10-fold excess of Fe(II), on the other hand, decreases the ^{58}Co absorption to about 50% and a 100-fold excess to less than 20% of the value measured in a control (Underwood, 1977).

The liver is the storage organ for B_{12} and its derivatives. The rate of loss of B_{12} from the liver is very slow; its biological half-life is estimated to be about 400 days (Harper, 1971). The major route of excretion of cobalt in man is in the urine; minor amounts of cobalt are lost in feces, sweat, and hair.

12.4.2 Manganese

12.4.2.1 Distribution and Functions

A relatively small amount (12–20 mg/70-kg man) of manganese is distributed widely in the body; liver, pancreas, and kidney show the highest levels. Manganese tends to be concentrated in mitochondria more than in any other organelle.

Manganese is known to be essential for the normal growth of bones, at least in mice, rats, rabbits, and guinea pigs. This is supposed to be related to a possible involvement of manganese in the synthesis of the organic matrix components (particularly chondroitin sulfate) of the cartilage. The critical sites of manganese function in chondroitin sulfate synthesis have been assigned to two enzymes both of which are dependent on manganese (Leach *et al.*, 1969). One enzyme polymerizes UDP-*N*-acetylgalactosamine and UDP-glucuronic acid to form the polysaccharide. The other enzyme is supposed to incorporate galactose from UDP-galactose into a galactose–galactose–xylose trisaccharide; the latter serves as the linkage between polysaccharides and their associated protein. It is inferred from these findings that Mn(II) is also widely involved in carbohydrate metabolism.

Manganese has been shown also to play a role in the biosynthesis of cholesterol. In the overall process of cholesterol formation from acetyl-CoA, two steps (Olson, 1965) between acetate and mevalonate and one additional enzyme, farnesyl pyrophosphate synthetase, are Mn(II)-dependent (see the reaction in Fig. 12-4). In female rats, minor manganese deficiency resulted in birth of some viable young, though all showed ataxia. Severe deficiency led to stillbirth (or short life of young) and sterility. Severely Mn-deficient male rats and rabbits exhibited sterility, associated with testicular degeneration. The precise locus or mode of action of manganese in preventing these reproductive defects has not been established. DNA polymerase, which is particularly active in male gonads, requires not only Zn(II) but also Mn(II) or Mg(II) as a cofactor. Cholesterol, the derivatives of which are sex hormones, may be the main linking factor; depressed production of cholesterol in Mn deficiency may lead to limited synthesis of steroid hormones.

12.4.2.2 Metabolism

The intestinal absorption of orally administered manganese is fairly low (3–4%). As mentioned earlier, Mn(II) can compete with Fe(II) for the iron-transport site and it seems, therefore, that Mn(II) is primarily absorbed via the iron-transport system. Whether the intestinal tract has a specific Mn(II)-transport system in addition to the one shared by Fe(II) and Co(II) is not clear.

Manganese as Mn(III) seems to bind tightly to transferrin in the blood; transferrin is a carrier for Mn(III). Again, liver is the metabolic center for manganese. Manganese is excreted in the bile and hence in the feces. The urinary excretion of manganese is virtually absent. This could be due to the fact that Mn(III), like iron, is tightly bound with a protein, transferrin, in plasma.

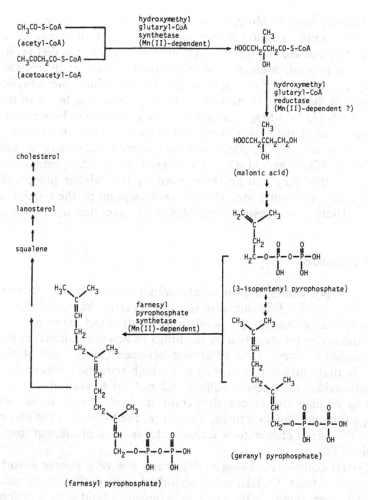

Figure 12-4. The participation of Mn(II) in steroid formation (lipid metabolism).

12.4.3 Molybdenum

Enzymes dependent on molybdenum are discussed in Sections 3.3 and 7.1. Important Mo enzymes in mammals include xanthine oxidase, aldehyde oxidase, and sulfite oxidase.

Molybdenum occurs in low concentration in all tissues and body fluids; the highest concentrations are found in liver and kidney, followed by spleen, lung, brain, and muscle.

Molybdenum, as MoO_4^{2-}, is readily and rapidly absorbed. A characteristic feature of molybdenum metabolism is its dependence on the sulfate status of the organism. High sulfate levels in the diet reduce the absorption of molybdenum, and high sulfate body levels increase the urinary excretion of molybdenum and hence reduce the molybdenum retention in the body. These effects are explainable in terms of the similarity of MoO_4^{2-} to SO_4^{2-}, and hence the competition between them in transport. The plausibility of such an effect is obvious in the case of uptake by the mucosal cells. The effect on the excretory process may be explained as follows. SO_4^{2-} and MoO_4^{2-} are filtered out by the glomeruli (of the kidney) so that they can be reabsorbed by the tubular lining cells. At higher SO_4^{2-} concentration, MoO_4^{2-} reabsorption by the tubular cells is more restricted; as a result, more MoO_4^{2-} is excreted in the urine.

12.4.4 Chromium

The major chemical form of chromium that is readily absorbable is chromate, CrO_4^{2-}. It is similar in size and electric charge to SO_4^{2-}; thus, it is conceivable that CrO_4^{2-} is taken up by the SO_4^{2-}-transport system. CrO_3 fumes can be absorbed by the lungs as has been shown in the case of industrial workers. One difference between CrO_4^{2-} and MoO_4^{2-} or SO_4^{2-} is that chromate (or CrO_3) has a high reduction potential, i.e., it is highly oxidative, whereas sulfate and molybdate are not. CrO_4^{2-} thus tends to oxidize substances that come in contact with it, resulting in damage to biological structures. Therefore, the toxic effect of chromium, especially that of chromate or dichromate, is more prominent than is the beneficial effect.

Cr(III) complexes, being highly inert, are very poorly absorbed in the intestinal tract. Cr(III), either absorbed or as derived from reduction of Cr(VI) compounds in the cells, is believed to bind with transferrin. It is also bound to albumin.

Chromium administered as the glucose tolerance factor (see Section 7.3.1) was found to accumulate mainly in the liver, followed by the uterus, kidney, and bone (Baetjer, 1974; Underwood, 1977). Lung, heart, intestine, spleen, pancreas and brain acquired about 50% of the chromium concentration found in the liver; considerably lower concentrations were found in muscle, ovaries, and aorta (Baetjer, 1974; Underwood, 1977).

Almost half of the total cellular chromium in the liver was shown to be concentrated in the nuclei (Baetjer, 1974). It is noteworthy that the testis accumulates chromium very rapidly after the injection of a tracer dose (Baetjer, 1974). These observations suggest a structural role of chromium in DNA and/or RNA.

The urinary pathway is the major route of excretion for ingested chromium (Baetjer, 1974). Nearly all the chromium in urine is present in the form of low-molecular-weight complexes (Baetjer, 1974). Protein-bound chromium, a minor component in the blood, is excreted slowly.

12.4.5 Nickel

The essentiality of nickel in animals has been well established in certain organisms; see Sections 3.2.5 and 5.3.3 for a discussion of the essentiality of nickel in plants and bacteria. There are, however, many reports indicating that it is also essential in animals. This can be inferred particularly from two sets of data (Nielsen et al., 1974; Sunderman, 1975). One set indicates that there are homeostatic mechanisms to control the nickel level in the body. In another set of reports, it has been shown that inadequate amounts of nickel in the diets of experimental animals result in an impairment of metabolic processes, whereas administration of nickel would correct the impairment.

Nickel deficiency in chicks and rats results in a decrease in oxygen uptake by liver homogenates in the presence of α-glycerophosphate, as well as an increase in liver total lipids and a decrease in liver lipid phosphorus (Nielsen et al., 1974). Ultrastructural abnormalities were also observed in liver cells, including dilation of cisterns of the rough endoplasmic reticulum and swelling of the mitochondria (Nielsen et al., 1974). It was also observed that a nickel deficiency led to a significant loss (15–19%) of pups in rats. Ni(II) has been suggested to have a specific role in the preservation of the compact structure of ribosomes (Nielsen et al., 1974). Ni(II) is also known to have a stabilizing effect on DNA (and RNA) structure (Nielsen et al., 1974; Sunderman, 1975).

Nickel is poorly absorbed in the gastrointestinal tract. Therefore, most nickel ingested orally is subsequently excreted in the feces. Absorbed nickel is excreted in the urine and sweat.

Nickel in the blood is present in three forms (Sunderman, 1975): an ultrafilterable form, as nickel bound to albumin, and in a nickel– metalloprotein complex termed "nickeloplasmin". The association constants of Ni(II) with albumins from different sources are $K_1 = 5 \times 10^4 \text{ M}^{-1}$ (dog), 1.6×10^5 (pig), 4×10^5 (rat), 6×10^5 (human), and 6×10^5 (rabbit). The Ni(II)-binding serum protein nickeloplasmin was found to be an α-macroglobulin with a molecular weight of approximately 7×10^5; its physiological significance is unknown.

Several enzymes have been shown in vitro to be activated by Ni(II); examples include arginase, enolase, phosphoglucomutase, acetylcoenzyme A synthetase, pyridoxal phosphokinase, thiamine kinase, and pyruvate oxidase. There is a possibility that some of these activating effects may have genuine significance *in vivo* as well. Nickel is an essential component of urease.

12.4.6 Vanadium

The intestinal absorption of vanadium compounds is poorly understood. In the case of industrial exposure to V_2O_5 (or V_2O_3), the main entrance is the respiratory system.

When trace amounts of [48]V were injected intravenously in rats, the liver, kidney, spleen, and testis accumulated it during the first 4 hr and retained most of the radioactivity for up to 96 hr, by which time most other major organs had retained only 14–84% of that present after the first 10 min. After 96 hr, 46% of the [48]V had been excreted in the urine and 9% in the feces (Hopkins and Tilton, 1966). Radioactivity in the liver subcellular supernatant fraction decreased from 57% to 10% (of the total liver content of [48]V) by 96 hr, whereas that in the mitochondria and in nuclear fractions increased from 14% to 40% (Hopkins and Tilton, 1966).

When [48]V was injected subcutaneously into young rats, the highest uptake was seen to be in the dentin and bone. This may be related to the fact that VO_4^{3-} has some similarity to PO_4^{3-}, which is one of the components of bone and dentin. The observation that [48]V was incorporated and retained in the mitochondrial fraction of liver may well have an analogous explanation.

12.5 Metabolism of Major Inorganic Elements

This topic is treated in detail in most textbooks of physiology, especially those dealing with physiological chemistry.

12.5.1 Sodium and Potassium

The major physiological functions of sodium and potassium are the maintenance of osmotic balance and excitability of nerve cells, though they also show activating effects on a number of enzymes. One important enzyme that is involved in both osmotic balance and excitation is an ATPase activated by sodium and potassium. Its major function is to pump out Na(I) from the cells and to transport K(I) into the cells. (See Chapter 10 for a discussion of this topic.) The body levels of these cations are controlled by the kidney. The Na(I) concentration is always high in extracellular fluids and blood plasma, hence some is lost daily through the urine despite the reabsorption in the kidney tubules. This loss is necessary for physiological purposes. Sodium is also lost in sweat. Potassium loss is not as high as that of sodium, largely because potassium is more concentrated in cells than in extracellular fluids.

12.5.2 Magnesium

A normal human adult contains about 21 g of magnesium. About 70% of the total is present in bones, perhaps as a substitute for calcium in the complex calcium salts of the bone. The remainder (30%) is in the soft tissues and the body fluids; Mg(II) is more concentrated within the cells than outside of the cells.

Mg(II) functions as an activator for a great variety of enzymes (see Chapter 5), and hence is distributed universally among all tissues and organs.

About 44% of ^{28}Mg orally administered was shown to be absorbed in the intestine at a normal diet level of 10 nmoles/day (Underwood, 1977). At a lower diet level, about 1 nmole/day, 76% of ^{28}Mg was absorbed, whereas only 24% was absorbed at a high diet level (24 nmoles/day) (Underwood, 1977). In the first 48 hr after administration, only about 10% of the ^{28}Mg absorbed was excreted in the urine (Underwood, 1977). This indicates that the reabsorption of Mg(II) by the renal tubules is effective.

12.5.3 Calcium

About 1 kg of calcium is present in the body of a normal human adult of 70 kg; almost all of it is in the bones and teeth. Traces of calcium are present in all tissues and body fluids. It is of great importance to physiological processes such as muscle contraction and nerve stimulus transmission. (See Section 7.6 for a discussion of intracellular Ca function.)

The Ca(II) absorption from the intestinal tract depends on many factors, including pH and the presence of phosphate, phytate, oxalate, free fatty acids, proteins, and vitamin D. Phytate, oxalate, fatty acid, and phosphate all tend to form insoluble Ca(II) compounds and this results in a lowered absorption of Ca(II). Proteins increase the Ca(II) absorption perhaps because they stabilize Ca(II), protecting it from forming insoluble compounds. Vitamin D is presumed to cause the synthesis of a specific binding protein. Ca(II)-binding proteins have been found in the intestinal mucosa of chicks, rats, dogs, cows, monkeys, and humans. Whether the Ca(II)-binding proteins provide a transport mechanism across the mucosal membrane is not clear. [See Section 9.4 for a review of Ca(II)-binding proteins.]

Most (70–90%) of the calcium leaving the body is excreted in the feces. This represents almost entirely unabsorbed dietary calcium. The body level of calcium is balanced through the bones and is influenced by parathyroid hormone. A low level of calcium in extracellular fluids promotes increased secretion of parathyroid hormone. The hormone, in turn, acts on the bones, increasing their rate of release of calcium and phosphate until the level of calcium returns to normal.

Biological Aspects II—Toxicity of and Defense against the Elements

Oxygen—Its Toxicity and Defense against It

13

Life originated in an oxygen-free atmosphere. The first organisms are thus considered to have been strictly anaerobic; they could not live in the presence of oxygen. As the atmosphere became oxygenic, aerobic organisms evolved. This occurred because aerobic respiration provided organisms with mechanisms to make more effecient use of nutrients. Most of the organisms on the present earth are aerobes. Thus, the strong oxidizing power of oxygen is beneficial, but it can also be deleterious to organisms. The very reason that allows oxygen to serve as an efficient ultimate electron acceptor in respiration, its strong oxidizing effect, poses a threat to the survival of organisms. The organisms on the present earth live rather precarious lives in the presence of oxygen, and this life has been made possible by the development of elaborate defensive mechanisms to act against oxygen toxicity. We will here survey both the toxic effects of oxygen and the major defense mechanisms.

13.1 Toxicity of Oxygen and Its Derivatives

O_2 is a spin-triplet in its ground state ($^3\Sigma_g$), and the first excited state is the so-called singlet oxygen ($^1\Delta_g$). Important derivatives of oxygen include superoxide (O_2^-), hydrogen peroxide (HOOH), hydroperoxide (ROOH), and hydroxyl radical (HO). All of these species have different reactivities; the only common property is their strong oxidizing power. The most stable oxidation state of oxygen is that of oxide (O^{2-}). Supposing that the ultimate product from oxygen and its derivatives is water, we can express the oxidizing power of the various forms in terms of their reduction potentials. The pertinent reduction potential values are given in Fig. 3-3 and Table 4-1. However, the various chemical reactivities of these species are more relevant with respect to the toxicity of oxygen.

13.1.1 Dioxygen in the Ground State

O_2 is a biradical ($^3\Sigma_g$), but it is rather inert in its reactivity toward molecules in a singlet state. This is due to its having rather high pairing energy. Being a biradical, it is, however, very reactive toward molecules with unpaired electrons. One of the two unpaired electrons in O_2 molecules couples readily with the unpaired electron in another entity; that is:

$$O_2 + \cdot R \rightarrow ROO\cdot \qquad (13\text{-}1)$$

$R\cdot$ can be an alkyl radical or a transition metal ion with unpaired electrons, such as Co(II) or Fe(II) (in high spin). The reactivity of O_2^- in the form of $ROO\cdot$ is rather restricted in some of the O_2 complexes of transition metals but is quite high otherwise. For example, an alkyl peroxyl radical $ROO\cdot$ reacts readily with another alkyl entity, forming a new radical:

$$ROO\cdot + R'H \rightarrow ROOH + R'\cdot \qquad (13\text{-}2)$$

$R'\cdot$ then further reacts with O_2 forming $R'OO\cdot$ and so on. Thus, a free radical chain reaction can result. A reaction like 13-2 is called a propagation reaction; the chain reaction is thus initiated by $R\cdot$. An alkyl radical $R\cdot$ can be formed in many ways. Examples include:

$$RH \xrightarrow{\text{(irradiation)}} R\cdot + H\cdot \qquad (13\text{-}3)$$

$$RH + X\cdot \text{ (radical)} \rightarrow R\cdot + HX \qquad (13\text{-}4)$$

Radiation, particularly of a short wavelength, can thus initiate a chain reaction involving organic molecules. Other radicals can result from reaction 13-4 by intervention of species like O_2^-, $HO\cdot$, $\cdot CCl_3$, and so forth. To complicate matters further, hydroperoxides (ROOH) can also decompose to produce a radical, particularly in the presence of a transition metal ion; thus:

$$ROOH + M(+n) \rightarrow RO\cdot + OH^- + M(+n+1) \qquad (13\text{-}5)$$

$$ROOH + M(+n+1) \rightarrow ROO\cdot + H^+ + M(+n) \qquad (13\text{-}6)$$

Steps like these make up what is sometimes called the "Haber–Weiss" mechanism, which was supposed to explain the promoting effect of transition metal ions on the air-oxidation of lipids.

The crux of all of these reactions is the abstraction of a hydrogen atom from a hydrocarbon, and the main site of reaction in a cell seems to be the membranes, whose hydrocarbon chains of the phospholipids are vulnerable. The reactivity of a hydrogen atom toward abstraction depends on the nature of the carbon atom to which the hydrogen is bound. Generally speaking, the reactivity increases in the order: primary

RCH_2—H < secondary RR'CH—H < tertiary RR'R"C—H. The reactivity is especially high if the carbon atom is adjacent to an electron-withdrawing group or a double bond. For example, a position like that shown below is considered to be readily attacked (Mead, 1976):

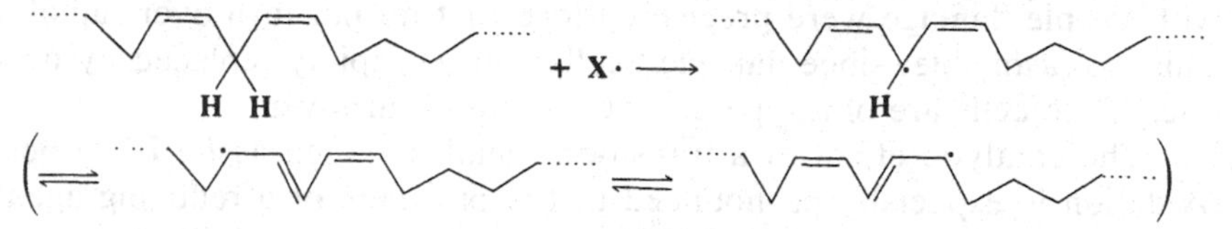

$$+ \ X \cdot \longrightarrow$$

(13-7)

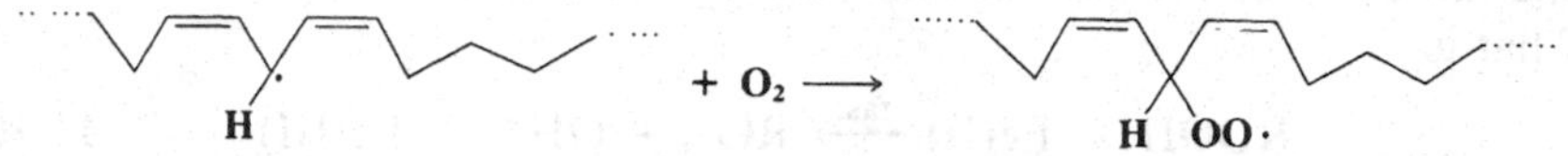

$$+ \ O_2 \longrightarrow$$

(13-8)

$$+ \ R'H \longrightarrow$$

$$+ \ R' \cdot$$

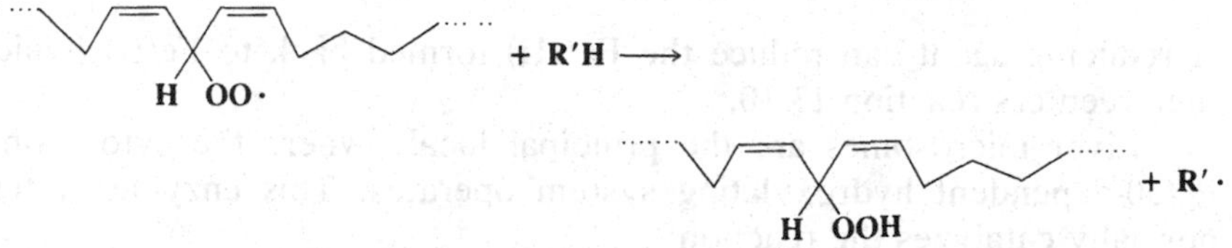

(13-9)

The product, a hydroperoxide, can decompose further, causing the molecule to be split into shorter chains with the result that the integrity of the membrane structure may be disrupted.

The foregoing example illustrates what is thought to be one of the main damaging effects of oxygen on membranes, but certainly it is not the only possibility; many other types of reactions can be envisioned as well. Nonetheless, its importance warrants further consideration of its likely implications. Reaction 13-9 requires that the initial radical center be able to come close to another hydrocarbon chain R'H. This may take place more readily in a relatively flexible membrane. The flexibility of a membrane in turn depends partially on its unsaturated hydrocarbon content. Membranes comprised mainly of saturated hydrocarbon chains are expected to be much less susceptible to peroxidation for two reasons: they are less readily attacked by a free radical, and they are less flexible. Since the unsaturated hydrocarbon content differs from one membrane to another, so does the susceptibility to peroxidation or damaging oxidative effects.

In an investigation of *in vitro* peroxidation of biological substructures in the presence of ascorbic acid, iron, and oxygen, the cell microsomal fraction (endoplasmic reticulum) was found to be the most susceptible, followed by mitochondria and muscle (Mead, 1976). The same study showed that tissue homogenates from brain, liver, and kidney were much more susceptible to peroxidation than those of testis and intestine. The tissues with visible damage were precisely those that do not turn over rapidly. This is reasonable, since damaged cells can be rapidly replaced by new ones if the cells are of a type undergoing rapid turnover.

The catalytic effect of a transition metal ion such as Fe(II) in peroxidation is especially pronounced in the presence of a reducing agent such as ascorbic acid (see above). This is considered to be due to lessening the difficulty of the second reaction of the "Haber–Weiss" mechanism. That is:

$$ROOH + Fe(II) \xrightarrow{fast} RO\cdot + OH^- + Fe(III) \qquad (13\text{-}10)$$

$$ROOH + Fe(III) \xrightarrow{slow} ROO\cdot + H^+ + Fe(II) \qquad (13\text{-}11)$$

A reducing agent can reduce the Fe(III) formed back to Fe(II), which then reenters reaction 13-10.

Liver microsomes are the principal locale where the cytochrome P-450-dependent hydroxylating system operates. This enzyme system normally catalyzes the reaction:

$$RH + O_2 + 2H^+ + 2e \rightarrow ROH + HOH \qquad (13\text{-}12)$$

which normally would not contribute to a free radical chain process. If the microsomal membrane is disrupted, however, abnormalities might occur, including the production of $\cdot O_2^-$ and of $\cdot CCl_3$ from CCl_4. $\cdot O_2^-$ is considered to be one of the intermediates in the P-450-dependent enzyme system (see Chapter 4). The same enzyme system sequentially transfers two electrons from NAD(P)H to the reacting center. CCl_4, if it were present, could gain access to this electron transfer system embedded in the membrane due to its lipid solubility and could then divert one of the electrons to form a radical in the following way:

$$CCl_4 + e \rightarrow Cl^- + \cdot CCl_3 \qquad (13\text{-}13)$$

Both $\cdot O_2^-$ and $\cdot CCl_3$, being radicals, can initiate peroxidation of the lipid membrane, thus aggravating the damaging effect. This is considered to be the main reason for the toxicity of CCl_4 (Recknagel, 1967).

13.1.2 Superoxide Radical, Hydroxyl Radical, and Hydrogen Peroxide

Superoxide radical can be formed by a number of enzyme systems, especially one containing flavin. Flavin residues such as those in FAD and FMN can exist in any one of three oxidation states: fully oxidized (quinonoid), semireduced (semiquinone), and fully reduced (hydroquinone), as shown in 4-76. That is, oxidation of $FADH_2$ (or $FMNH_2$) by oxygen can take place in two separate steps:

$$FADH_2(FMNH_2) + O_2 \rightarrow FADH\ (FMNH) + HOO \cdot$$

$$FADH\ (FMNH) + HOO \cdot \rightarrow FAD\ (FMN) + HOOH \qquad (13\text{-}14)$$

Thus, enzymes dependent on FAD (FMN) can produce $HOO \cdot$ as well as HOOH either as by-products or as a main product. An example is xanthine oxidase, which produces $HOO \cdot$ as a by-product. Another interesting case is NADPH oxidase; its main function seems to be production of $\cdot O_2{}^-$, which is utilized to kill bacteria in leukocytes (see below). It is very likely that this enzyme, like the similar NADH dehydrogenase, contains flavin residue.

As seen above, hydrogen peroxide is another possible product of flavin-dependent enzymes; it is also one of the main products of reactions catalyzed by a copper enzyme, amine oxidase, which is believed to be free of flavin.

Hydroxyl radical forms as a result of one-electron reduction of hydrogen peroxide:

$$HOOH + e + H^+ \rightarrow HOH + \cdot OH \qquad (13\text{-}15)$$

Any kind of electron source can be employed, and the reaction can be viewed as related to a Haber–Weiss mechanism (13-5). If the electron source is $\cdot O_2{}^-$ (as $HOO \cdot$), then:

$$HOOH + HOO \cdot \rightarrow HOH + O_2 + \cdot OH \qquad (13\text{-}16)$$

The oxygen molecule produced here can be either in the ground state or in an excited state (see below).

If radicals like $\cdot O_2{}^-$ (or $HOO \cdot$) and $HO \cdot$ are once produced, they can initiate the free radical chain peroxidation of lipids in membranes.

In recent years the notion that $O_2{}^-$ is a very reactive agent and is one of the culprits for the toxicity of oxygen has become questioned, for $\cdot O_2{}^-$ has turned out to be less reactive toward cellular components than originally thought (see Cohen and Greenwald, 1983; Bielski, 1983; Borg and Schaich, 1983). It must be noted, however, that $HOO \cdot$ is generally 10^3–10^4 times more reactive than $O_2{}^-$ and may still be considered to play a significant role in oxygen toxicity (Youngman, 1984). Since the pK_a of $HOO \cdot$ is about 4.7 (Bielski, 1983), the significant portion of superoxide species could exist in the form of $HOO \cdot$ under physiological conditions. In general, the $\cdot OH$ free radical is now believed to be the agent actually causing the damage; however, it is very effective at this only in the immediate vicinity where it (the ·OH radical) is produced.

The possibility has been pointed out (Wolff *et al.*, 1986) that a free radical such as those discussed above may damage a protein, rendering it inactive or more susceptible to enzymatic proteolysis.

13.1.3 Singlet Oxygen

Recent evidence points to the possible formation of singlet oxygen ($^1\Delta_g$) in some biological systems (Foote, 1976; Krinsky, 1977). $^1\Delta_g$ is the first excited state of oxygen molecule and lies 22 kcal (92 kJ) above the ground state. The second excited state is also a singlet, $^1\Sigma_g$, and is 37 kcal (155 kJ) above the ground state. The latter singlet, being higher in energy, seems likely to have much less significance in biological systems and our discussion will be confined to the $^1\Delta_g$ state.

$O_2(^1\Delta_g)$ has a reactivity quite different from that of $O_2(^3\Sigma_g)$. Some of the characteristic reactivities of this singlet oxygen are worth exploring further. For example:

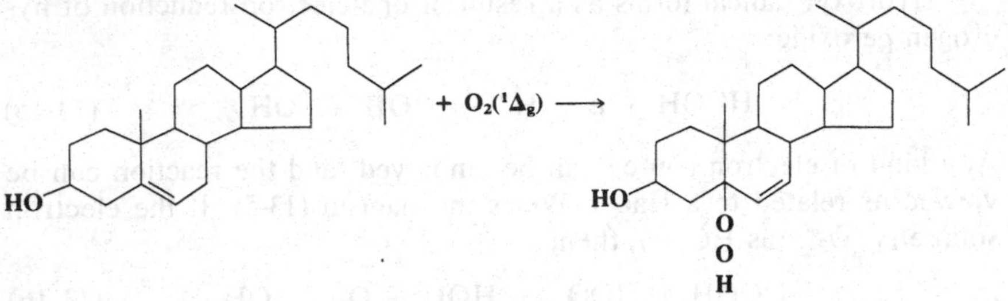

(13-17)

A free radical peroxidation of the same compound would yield a mixture of several hydroperoxides, but reactions with singlet oxygen are in general quite specific. Other examples of reactivity of $O_2(^1\Delta_g)$ would include:

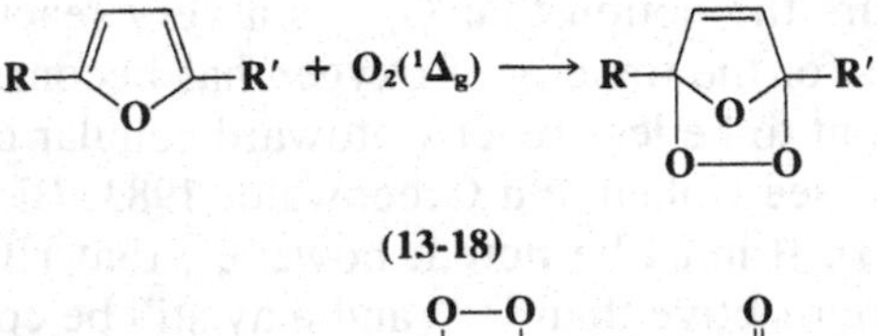

(13-18)

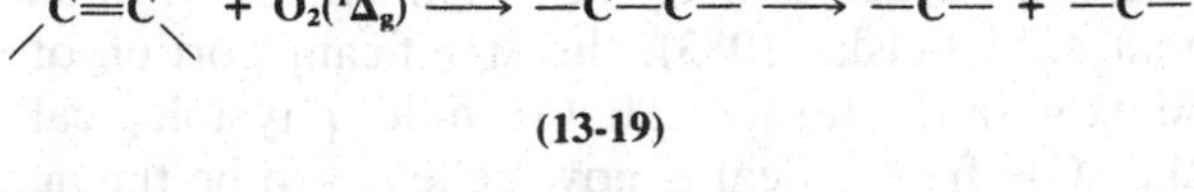

(13-19)

It is clear, therefore, that $O_2(^1\Delta_g)$ would have a damaging effect on the unsaturated hydrocarbon chains of membranes; the effect is similar to that of $O_2(^3\Sigma_g), \cdot O_2^-$, and others, in terms of the ultimate results, even though the mechanisms are quite different.

When $O_2(^1\Delta_g)$ returns to the ground state, it emits energy in the form of light in two ways:

$$O_2(^1\Delta_g) \rightarrow O_2(^3\Delta_g) + h\nu \qquad (\lambda = 1269 \text{ nm})$$
$$2O_2(^1\Delta_g) \rightarrow 2O_2(^3\Delta_g) + h\nu \qquad (\lambda = 634 \text{ nm}) \qquad (13\text{-}20)$$

This emission of light, particularly that at 634 nm, is utilized to establish the involvement of $O_2(^1\Delta_g)$. For example, a luminescence at $\lambda = 620$ and 580 nm was discovered in a system: O_2–NADPH–Fe(III)–microsome from rat liver (Nakano $et\ al.$, 1975). This emission was comparable to that of $O_2(^1\Delta_g)$ chemically produced from a mixture of NaOCl and HOOH. Another method of detecting the involvement of $O_2(^1\Delta_g)$ is to observe the effect of the addition of singlet oxygen quenchers. Carotenoids, amines, and tocopherols are utilized as the quenchers.

Singlet oxygen may form in a number of ways. Its formation from the reaction of $\cdot O_2^-$ with HOOH (13-16) and from the reaction of HOOH with OCl^- have been mentioned. Other possible reactions leading to the formation of singlet oxygen are (Foote, 1977; Krinsky, 1977):

$$OO^- + OO^- + 2H^+ \rightarrow HOOH + O_2(^1\Delta_g) \qquad (13\text{-}21)$$
$$HOOH + HOO^- \rightarrow HOH + OH^- + O_2(^1\Delta_g) \qquad (13\text{-}22)$$
$$OO^- + HO\cdot \rightarrow OH^- + O_2(^1\Delta_g) \qquad (13\text{-}23)$$

Unlike the nonenzymatic reaction 13-21, the enzyme superoxide dismutase seems to produce the ground state O_2. Reaction 13-22 is the disproportionation of HOOH; again the enzymatic decomposition of HOOH does not, in general, appear to form singlet oxygen, though a reaction of HOOH catalyzed by myeloperoxidase was reported to yield $O_2(^1\Delta_g)$ (see below). The third reaction listed (13-23) is the oxidation of $\cdot O_2^-$ by the strongly oxidizing hydroxyl radical.

13.1.4 Killing of Microorganisms by Phagocytic Cells

Phagocytic cells play an important role in defense against invading microorganisms (Roos, 1977). There are a number of different types of phagocytic cells, such as leukocytes and macrophages (endothelial reticulum system). Phagocytic leukocytes can move toward foreign microorganisms, somehow recognize them as foreign, and then ingest (phago-

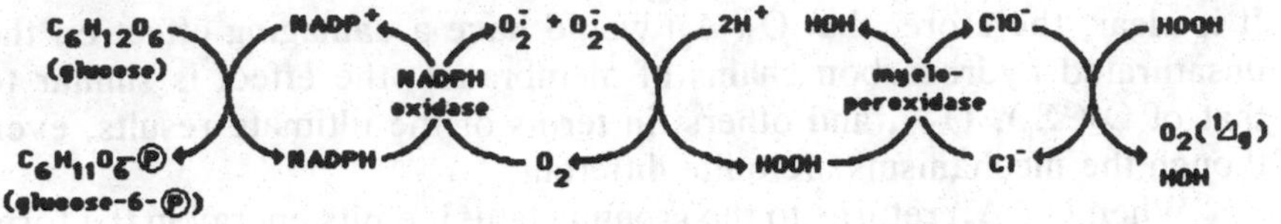

Figure 13-1. The formation of O_2^-, HOOH, and $O_2(^1\Delta_g)$ in macrophage.

cytose) and digest them. There are two kinds of phagocytic leukocytes: polymorphonuclear leukocytes and mononuclear phagocytes. The attacked microorganisms are wrapped in vesicles called phagosomes with which lysosomes (vesicles containing hydrolytic enzymes) merge. The pH of the fused phagolysosome suddenly drops, and, thus, acid as well as hydrolytic enzymes (e.g., proteinases, lipases) are freed to destroy the microorganismic cells. However, this is not the only mechanism involved.

As soon as a leukocyte binds to a microorganism, a sudden rise in the production of HOOH from O_2 is observed. Also observed are the activation of NADPH oxidase (Allen *et al.*, 1974), O_2^- production (Babior *et al.*, 1973), and emission due to singlet oxygen (Allen *et al.*, 1974). It is, therefore, apparent that the phagocytic leukocytes exploit the toxicity of oxygen and its derivatives to kill the invading microorganisms. It has been suggested that, under these conditions, NADPH is produced by the hexose monophosphate shunt using glucose as hydrogen donor. A proposed mechanism (Allen, 1975) to explain the whole process of the production of various oxygen derivatives is summarized in Fig. 13-1. The highly oxidizing agents $\cdot O_2^-$, HOOH, $\cdot$OH (which can be derived from the previous two species), and $O_2(^1\Delta_g)$ thus produced are assumed to attack the microorganisms, especially by peroxidizing their membranes.

13.1.5 Peroxidation, Aging, Carcinogenicity, and Inflammation

Two pigments, known as lipofuscin (Miquel *et al.*, 1977) and ceroid (fluorescent), have been observed to accumulate with age, especially in those tissues that show peroxidation to the greatest extent. One hypothesis asserts that peroxidized polyunsaturated fats (from membranes) decompose into malonaldehyde and other compounds, and that the malonaldehyde forms a conjugated Schiff base with some amine compounds (Miquel *et al.*, 1977; Reddy *et al.*, 1973). A model of such a Schiff base was shown to fluoresce at 430–470 nm, a wavelength of emission similar to that observed from the aging pigments.

The compounds called prostaglandins, all of which are derived from the unsaturated fatty acid arachidonic acid, have been attracting attention because they show many varied and important physiological functions (for a recent review see Gale and Egan, 1984). Two recently discovered intermediates of the decomposition of the endoperoxide PGH_2 that is derived from arachidonic acid through PGG_2 (see equation 13-24) are thromboxane (designated as TXA_2) and prostacyclin (PGI_2):

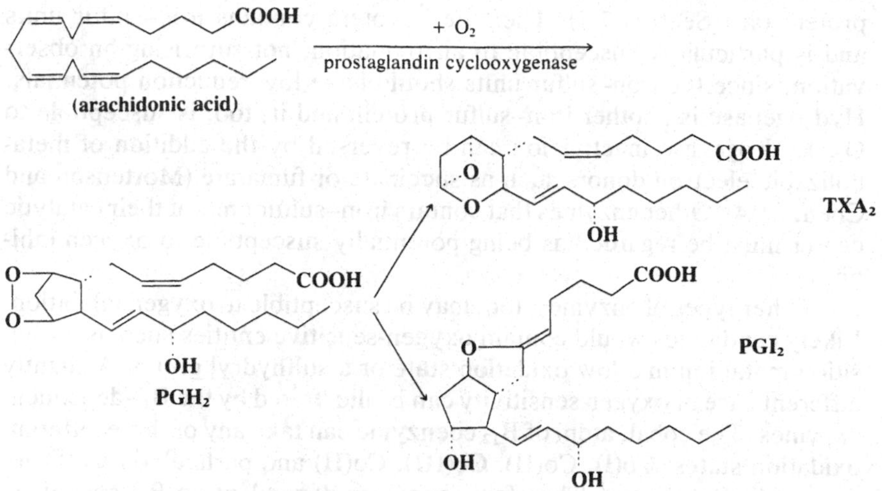

$$+ O_2 \xrightarrow{\text{prostaglandin cyclooxygenase}}$$

(arachidonic acid)

TXA₂

PGH₂

PGI₂

(13-24)

These two compounds, whose lifetimes *in vivo* are short, have contrasting effects on blood vessels: TXA_2 enhances platelet aggregation and blood vessel contraction, whereas PGI_2 inhibits platelet aggregation and enhances blood vessel dilation. An interesting observation relevant to our discussion here is that the peroxide derivatives of membrane lipids inhibit the biosynthesis of PGI_2. Decreasing the normal supply of PGI_2 would lead to platelet aggregation and the contraction of blood vessels (hypertension) and it might thus be a cause of thrombosis and atherosclerosis, typical diseases of the aged.

When phagocytic leukocytes are stimulated by capturing a microorganism, the superoxide radical as well as HOOH is produced at a heightened rate and some portion of it is excreted. The attack of this free radical on surrounding tissues is considered to be a partial cause of inflammation (McCord, 1974; Levy, 1976). The mutagenicity of superoxide radical has been discussed in a recent report (Moody and Hassan, 1982).

13.1.6 Inhibition of Enzymes by Oxygen and Its Derivatives

A number of enzymes with low reduction potentials may be inhibited by oxygen, because the catalytic center of such an enzyme may be oxidized by oxygen and thus inactivated. Two such examples are nitrogenase and hydrogenase.

Nitrogenase is extremely sensitive to oxygen (Robson and Postgate, 1980). Nitrogenase consists of two proteins, "Mo–Fe" protein and "Fe" protein (see Section 7.1). The "Fe" protein contains iron–sulfur units and is particularly susceptible to air oxidation, not surprising an observation, since the iron–sulfur units should have low reduction potentials. Hydrogenase is another iron–sulfur protein and it, too, is susceptible to O_2. Hydrogenase inactivation can be reversed by the addition of metabolizable electron donors such as succinate or fumarate (Mortenson and Chen, 1974). Other enzymes that contain iron–sulfur units at their catalytic center must be regarded as being potentially susceptible to oxygen inhibition.

Other types of enzymes, too, may be susceptible to oxygen inhibition. Likely candidates would contain oxygen-sensitive entities such as a transition metal ion in a low oxidation state or a sulfhydryl group. A slightly different case of oxygen sensitivity can be illustrated by the B_{12}-dependent enzymes. The cobalt atom of B_{12} coenzyme can take any of three different oxidation states: Co(I), Co(II), Co(III). Co(II) and particularly Co(I) are susceptible to oxygen. Therefore, enzymes dependent on B_{12} coenzyme in its low oxidation state may be inhibited by oxygen.

Enzymes dependent on the air-sensitive Fe(II) ion constitute another group that is likely to be susceptible to disruption by oxygen. Examples include aconitase (iron–sulfur unit) and enzymes similar to it, and the so-called protein B2 of ribonucleotide reductase. *In vitro* activation of aconitase requires the copresence of a reducing agent such as cysteine along with Fe(II). This fact implies that the Fe(II) added *in vitro* may readily be oxidized by air.

13.2 Defense Mechanisms against Oxygen Toxicity

As seen above, the culprits that damage cells, particularly membranes, are ground-state oxygen ($^3\Sigma_g$), singlet oxygen ($^1\Delta_g$), free radicals (including $\cdot O_2{}^-$ and $\cdot OH$), hydrogen peroxide, and hydroperoxide. The protection of cells from these agents requires the presence of quenchers (of the excited state $^1\Delta_g$) and radical inhibitors or other devises to destroy free radicals and peroxides. Many organisms have been shown to be fairly well equipped with these devices.

13.2.1 Superoxide Dismutase

The major sources of $\cdot O_2^-$ were described above. Another possible source that has not been mentioned may be hemoglobin. Hemoglobin binds O_2 in the form of $Fe(III)—O_2^-$. Normally the bound O_2^- would not come apart as such, but under certain special conditions it may dissociate as $\cdot O_2^-$. In addition, there are enzymes that activate O_2 and oxygenate substrates (see Chapter 4). These enzymes, too, may yield $\cdot O_2^-$ as a by-product under unusual conditions.

In any event, the possible sources of $\cdot O_2^-$ are widespread, as is superoxide dismutase, which catalyzes the following reaction:

$$OO^- \cdot + OO^- \cdot + 2H^+ \rightarrow HOOH + O_2 \qquad (13\text{-}25)$$

That is, it converts $\cdot O_2^-$ into the less toxic HOOH and O_2. The chemistry of superoxide dismutase is discussed in Chapter 4. This enzyme is now recognized to be more widespread among organisms than catalase (Fridovich, 1976), presumably because $\cdot O_2^-$ is much more reactive than HOOH and there is more chance of $\cdot O_2^-$ formation than HOOH formation. Superoxide dismutase is found even in anaerobic bacteria such as *Clostridium, Chlorobium, Chromatium,* and *Desulfovibrio* (Hall, 1977). It is found also in cyanobacteria, aerobic bacteria, yeast, plants, and animals. In mammals it is found especially in brain, liver, and erythrocytes, the places where O_2 is consumed briskly. Superoxide dismutase found in prokaryotes and the mitochondria of eukaryotic cells contains either iron or manganese at its catalytic center, whereas that found in the cytoplasm of eukaryotic cells contains copper and zinc. The significance of this difference in relation to the evolution of life was mentioned in Chapter 1.

The notion that superoxide dismutase was created for the purpose of dealing with oxygen toxicity is in accord with the fact that the enzyme can be induced by oxygen and its level decreases in the absence of oxygen (Fridovich, 1976). For example, an increase in the enzyme level was observed to be induced by oxygen in *Streptococcus faecalis, Saccharomyces cerevisiae, E. coli,* and rat lungs (Fridovich, 1975). K_{12}–C_{600} strains of *E. coli*, when grown under anaerobic conditions, contained less than 20% of the amount of enzyme found in the parental strain (Fridovich,

1976). The induction of the enzyme by 2% oxygen in anaerobically grown *Bacteroides* species has also been observed (Gregory *et al.*, 1983).

13.2.2 Catalase and Peroxidases

Catalase catalyzes the following reaction:

$$2HOOH \rightarrow 2HOH + O_2 \qquad (13\text{-}26)$$

Peroxidase from horseradish causes decomposition of HOOH or ROOH in the following manner:

$$ROOH + AH_2 \rightarrow ROH + HOH + A \qquad (13\text{-}27)$$

Both enzymes have the ability to change ROOH or HOOH into less toxic substances. Catalase is much less widely distributed than superoxide dismutase. It has not, for example, been detected in most facultative and strictly anaerobic microorganisms, though a Mn-containing catalase-like protein has recently been discovered in some *Lactobacillus* species (Kono and Fridovich, 1983a,b). Respiring microbes that lack catalase are known to secrete HOOH. In humans the catalase content is high in liver, kidney, and blood but very low in brain, testis, thyroid, and other tissues (Fridovich, 1976). Catalase in the blood is supposed to scavenge HOOH from the latter tissues. This implies that cells in these tissues are much more tolerant to HOOH than $\cdot O_2^-$. The fact that the brain contains a high level of superoxide dismutase has already been noted.

The function of horseradish peroxidase and similar peroxidases is not well understood. The chemistry of catalase and peroxidases was discussed in Chapter 4.

Glutathione peroxidase is an entirely different kind of enzyme. It catalyzes the following reaction:

$$HOOH + 2GSH \rightarrow 2HOH + GSSG$$

$$ROOH + 2GSH \rightarrow 2ROH + GSSG \qquad (13\text{-}28)$$

It is widely distributed in mammalian cells and is effective at low concentrations of HOOH, lower than the effective concentration for catalase. As indicated above, it can act on lipid hydroperoxides as well as HOOH. Ducks are reported to lack catalase but they instead have glutathione peroxidase to counteract accumulation of HOOH (Fridovich, 1976). One of the physiological functions of glutathione peroxidase is believed to be the decomposition of the lipid hydroperoxides from membranes, thus reducing the toxicity of the hydroperoxides (Diplock, 1976; see also Flohe,

1982). The enzyme contains selenium as an essential element; the chcm-istry of selenium including its role in glutathione peroxidase is discussed in Section 3.6 and the essentiality and toxicity of selenium are dealt with in Chapter 14.

13.2.3 Radical Inhibitors, Antioxidants, Singlet Oxygen Quenchers

Radical reaction inhibitors of various types are well known in the realm of organic chemistry. A basic requirement for such a compound is its ability to form a stable, unreactive radical; that is:

$$R\cdot + I—H \text{ (inhibitor)} \rightarrow RH + I\cdot \text{ (stable)} \qquad (13\text{-}29)$$

A radical of the type $I\cdot$ thus formed would not react further, and a free radical chain reaction would be terminated when this reaction takes place. The most commonly used inhibitors are the derivatives of p- or o-hydro-quinone. Hydrogen abstraction from such a structure leads to a semi-quinone, which has a substantial stability and is particularly unreactive in such a compound whose reaction center is sterically hindered. Phenol derivatives, if the hydroxy radical formed is well blocked (e.g., by 2,6-butyl groups), can also function as radical inhibitors.

Vitamin E shows a protective effect on membranes, and it is believed to function as a biological antioxidant (Chiu *et al.*, 1982; Burton *et al.*, 1983). One component of vitamin E, α-tocopherol, has the following struc-ture:

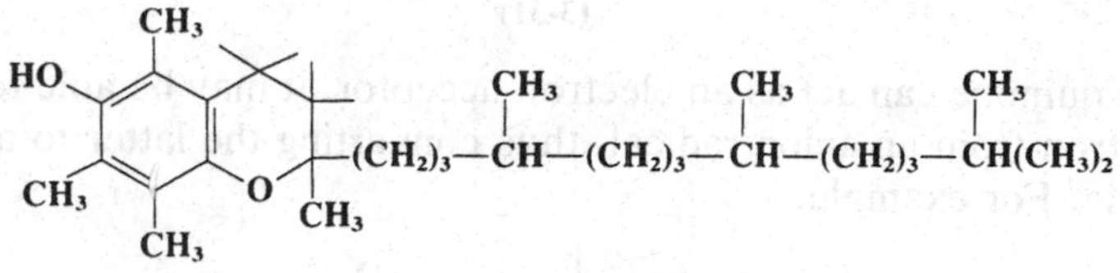

α-tocopherol (TH$_2$)

The reactions involved are usually described to be (Chiu *et al.*, 1982):

$$ROO\cdot + TH_2 \rightarrow ROOH + TH$$

$$ROO\cdot + TH \rightarrow ROOH + T \qquad (13\text{-}30)$$

The first product of hydrogen abstraction, TH, is a phenoxy radical, and the second product, T, is considered to be a quinone-type compound.

In addition to its role as an antioxidant, vitamin E is considered to play a role in maintaining the structural integrity of membranes (Diplock, 1976, 1983). There are many observations to indicate that the functional integrity of some membrane-bound enzymes is impaired by vitamin E deficiency. A model building study suggested that the stabilization of membranes by vitamin E may be due to specific interactions between the phytyl side chain of α-tocopherol and the chains of membrane-bound polyunsaturated fatty acids. If the effect of a vitamin E deficiency is a disruption of the integral structure of membranes, it would still be relevant to the oxidation problem, since disordered structure renders a membrane more susceptible to a free radical peroxidation. This indirect effect is particularly evident in those membranes that are most susceptible to disruption in structure due to the absence of vitamin E, i.e., those that contain large amounts of polyunsaturated fatty acids. Needless to say, these are also the membranes that are especially susceptible to free radical peroxidation.

A derivative of p-phenylenediamine might function as a radical inhibitor analogous to hydroquinones. For example, N,N'-diphenyl-p-phenylenediamine has been shown to have anti-inflammatory activity (Levy, 1976). It could be expected to participate in the following kind of reaction:

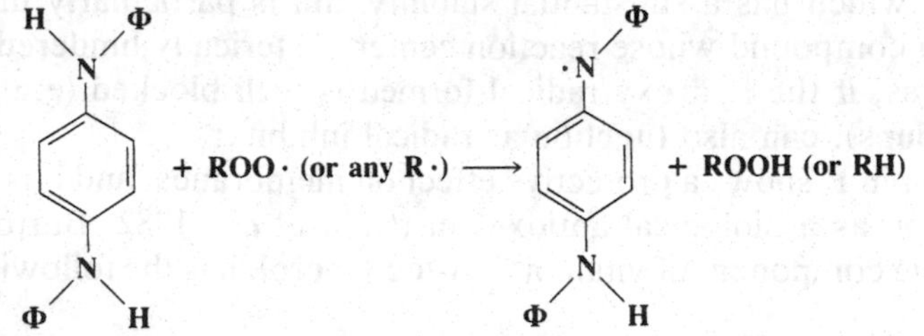

$$+ ROO \cdot \text{ (or any } R \cdot) \longrightarrow \quad + ROOH \text{ (or RH)}$$

(13-31)

Since a quinone can act as an electron acceptor, it may be able to accept an electron from an anion radical, thus converting the latter to a neutral molecule. For example:

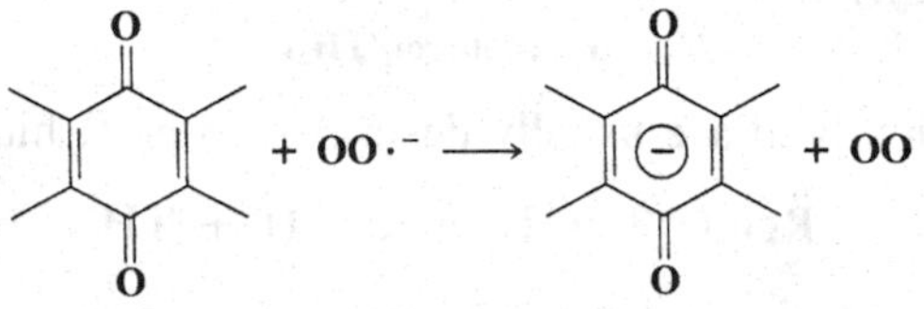

$$+ OO \cdot^- \longrightarrow \quad + OO$$

(13-32)

Thus, quinones could serve as quenchers of the toxic $\cdot O_2^-$ radical.

Whether these compounds actually act in this way under physiological conditions is not known. Other potential physiological antioxidants include sulfhydryl compounds and perhaps similar compounds of selenium. The hydrogen atom of a sulfhydryl group can easily be abstracted, resulting in a disulfide:

$$RSH + R \cdot \rightarrow RS \cdot + RH, \ 2RS \cdot \rightarrow RSSR \qquad (13\text{-}33)$$

Finally it should be noted that the quenching of $O_2(^1\Delta_g)$ cannot be viewed as an ordinary chemical reaction; it is, rather, an energy transfer from O_2 to the quenching molecule. Carotenoids are an example of physiological compounds known to be effective in quenching $O_2(^1\Delta_g)$. β-Carotene has been shown also to be an effective radical-trapping antioxidant at lower oxygen pressure (Burton and Ingold, 1984).

13.2.4 Protection of Nitrogenase from Oxygen

The protection of nitrogenase from destruction or blocking by oxygen presents a special problem due to similarity of the substrate, nitrogen (N_2), to oxygen (O_2). They are similar in size and have similar solubilities in water and lipids, so that it would be rather difficult to exclude O_2 from N_2-binding sites by mechanical means.

Nitrogen fixation is carried out by aerobic organisms as well as anaerobic organisms. No problem arises in anaerobes; they live and fix nitrogen only under anaerobic conditions. Moreover, most facultative organisms can fix nitrogen only under anerobic conditions. Thus, it is only aerobic nitrogen-fixing organisms that somehow have to cope with the potential inhibiting effect of oxygen.

The simplest means of protection is the employment of an effective respiratory system so as to maintain the free oxygen level in a cell at a fairly low level (Postgate, 1974; Robson and Postgate, 1980). An advanced aerobe such as *Azotobacter* uses this mechanism. Nitrogen fixation in *Azotobacter* is most efficient when the oxygen supply is kept within limits by respiration; this protection can be overcome, however, by exposing the organism to a pure oxygen atmosphere. *Mycobacterium flavum,* considered to be less evolved than *Azotobacter,* is very sensitive to oxygen when fixing nitrogen, and it is normally unable to grow without a supply of some fixed nitrogen compounds.

Azotobacter species seem to have yet another protective device. *A. vinelandii*, when fixing nitrogen, was shown to develop an intracellular network of subcellular membranes (Robson and Postgate, 1980); such a network is absent from an ammonia-grown population. It has been postulated that these subcellular membranous structures contain both nitrogenase and an oxygen-protective device of some sort. The subcellular particles have been shown to contain cytochromes and other proteins and to be oxygen-tolerant.

Cyanobacteria, another group of nitrogen-fixing organisms, have an additional problem. They themselves produce free oxygen in their photosynthetic apparatus. Many nitrogen-fixing cyanobacteria form special cells called "heterocysts," which are thought to be the specific site of nitrogen fixation. They are connected to the adjacent cells through pores. It is believed that the heterocysts contain some unique oxygen-excluding mechanism (Robson and Postgate, 1980). Some other cyanobacteria fix nitrogen but do not form heterocysts. The majority of these organisms, however, fix nitrogen only at low oxygen pressures at low levels of illumination, where the photosynthetic production of oxygen is minimal. This implies that they are not very well equipped with oxygen-defending or -excluding facilities. Some cyanobacteria, including *Plactonema,* contain superoxide dismutase. This might be somehow involved in the protection mechanism; it could also serve to protect the photosynthetic apparatus from damage by oxygen.

Rhizobium, the nitrogen-fixing symbiont of legume nodules, is known to contain leghemoglobin, an oxygen-binding protein. This protein is important physiologically in the nitrogen fixation process of legume nodules, even though it is not an essential component of the bacteroid's nitrogen-fixing system. The leghemoglobin, however, is said to be substantially unoxygenated in functional nodules. These facts taken together suggest that the nodules possess presumably effective oxygen-excluding mechanisms, but that leghemoglobin is present to function as a supplementary mechanism to take care of the last trace of free oxygen. It is interesting to note (even though it does not suggest anything concerning the nature of defense mechanisms against oxygen) that the host plant supplies the gene for leghemoglobin and the bacterium contains the synthetic apparatus for the protein. This must be seen as a result of the long evolutionary history of the symbiotic relationship between the two organisms.

Toxicity of Heavy Metals 14

The term "heavy metal" is rather vague. For our purpose we take it to encompass cadmium, lead, mercury, and copper, but we will also include zinc and silver. Of these elements, copper and zinc are essential to life, but they can have deleterious effects when present in excess. A comprehensive review on toxicity of these and other elements has been published (Friberg *et al.*, 1979), and the toxic effects of metals on reproductive and developmental processes have recently been reviewed (Clarkson *et al.*, 1983).

As the uptake and excretory mechanisms of these elements differ from one organism to another, their metabolism also varies. The mechanism of their toxic effects does not vary, however, although the toxicity symptoms may. The modes of action can be subdivided into a few basic groups:

1. Blocking essential functional groups of biomolecules, including enzymes and polynucleotides
2. Displacing essential metal ions from biomolecules and other biological functional units
3. Modifying the active conformation of biomolecules, especially enzymes and perhaps polynucleotides
4. Disrupting the integrity of biomembranes
5. Modifying some other biologically active agents, e.g., potentiating the endotoxin produced by bacteria

Many organisms have developed means to combat some of the adverse effects of heavy metals.

I will first discuss the chemical bases of the toxicity of these elements, then investigate some physiological consequences, and finally deal with the mechanisms by which organisms cope with toxic metals.

14.1 Chemical Bases of Toxicity of Heavy Metals

The five toxicity mechanisms listed above are all based on the strongly coordinating abilities of these metallic ions. The physiological activities of biological functional molecules—proteins, enzymes, polynucleotides, and others—often depend on metal-coordinating groups such as amine (of amino acids and nucleotides), imine (of histidine and nucleotides), sulfhydryl (of cysteine), and carboxylate groups. Consequently, a heavy metallic ion may displace an existing essential element bound to an active site, or simply bind itself to the active site, resulting in an inhibition of the biomolecule's activity.

Replacement of an essential element by these metallic elements may not necessarily be of the coordination type. It can instead involve replacement of one element by another in an ionic crystal. An example could be cadmium incorporation into bone, which consists mostly of calcium phosphate. Cd(II) appears to be able to take the place of Ca(II) in the crystals. Replacement of this type is considered to take place readily between two metallic ions of similar size and charge. The ionic radius of Ca(II) is 114 pm; cations of similar size and charge include Cd(II) (106 pm), Hg(II) (110 pm), and Pb(II) (133 pm). The favored coordinating structure of Cd(II) is octahedral and hence similar to that of Ca(II); Hg(II) prefers a tetrahedral structure. These characteristics of Cd(II) make it a good competitor for Ca(II).

Let us now take a look at the coordinating abilities of these elements, since coordination bonding is more often a cause of toxicity than is crystal substitution. Some representative data of the stability constants for relevant elements are given in Table 14-1. Hg(II) is remarkable in its ability to bind very strongly with all of the ligand types represented here, especially sulfhydryl or sulfide ligands. This is, indeed, the basis of the toxic effects of Hg(II) and its derivatives. With regard to strong binding with sulfur-containing ligands, Hg(II) is followed by Cu(II) and Ag(I). Cd(II) and Pb(II) also bind strongly with sulfur-containing ligands but much less so than Hg(II), Cu(II), and Ag(I). Hg(II) binds strongly also with amine-type ligands, being followed by Cu(II) and then Ni(II) in this respect.

Zn(II) has been shown to be the essential element in a number of enzymes; as seen in Table 14-1, Zn(II) can easily be replaced (in the thermodynamic sense) by Cu(II), Cd(II), Hg(II), and Pb(II). Metalloenzymes in which the essential Zn(II) ion has been replaced by Cu(II), Cd(II), Hg(II), or Pb(II) are almost invariably inactive. In general, only Co(II)-substituted carboxypeptidase A and a few other cobalt-substituted zinc enzymes have enzymatic activity. There are, of course, always a few exceptions; e.g., Cd(II)- or Hg(II)-carboxypeptidase A has been shown to retain substantial esterase activity, though not peptidase activity.

Table 14-1. Binding Constants[a] of Cations with Ligands

Ligand (coordinating group)	Mg^{II}	Mn^{II}	Fe^{II}	Co^{II}	Ni^{II}	Cu^{II}	Zn^{II}	Hg^{II}	Pb^{II}	Ag^{I}	(log K_{Cd})
Oxalic acid (COO$^-$)	0.30	2.0	15	15	60	—	14	—	—	3×10^{-4}	(3.52)
OH$^-$ (M(OH)$_2$)	1×10^{-3}	5×10^{-2}	14	45	1×10^4	9×10^4	9×10^2	4×10^{11}	22	—	(13.95)
Ethylenediamine (2NH$_2$)	3×10^{-5}	5×10^{-3}	5×10^{-2}	2.2	1×10^2	1×10^5	1.7	6×10^8	—	—	(5.55)
HSCH$_2$CH$_2$NH$_2$ (SH and NH$_2$)	—	—	—	3×10^{-2}	0.31	—	0.18	—	0.34	—	(10.97)
HSCH$_2$CH$_2$N(CH$_2$COOH)$_2$	4×10^{-13}	4×10^{-8}	1×10^{-5}	9×10^{-3}	1×10^{-3}	—	0.16	0.28	2.0	—	(16.72)
Glutathione (NH$_2$, SH)	—	—	—	—	—	—	6×10^{-3}	6×10^{3}[b]	1.3	—	(10.50)
S^{2-} (MS)	—	1×10^{-16}	3×10^{-8}	—	—	3×10^7	3×10^{-6}	1×10^{24}	0.10	8×10^{6}[b]	(28)

[a] Relative K values referred to $K_{Cd} = 1$. The actual values of log K_{Cd} are given in the last column.
[b] Estimate; otherwise the sources are Stability Constants and B. Charlot, Qualitative et les Reactions en Solution, Masson (1957)

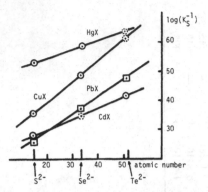

Figure 14-1. The formation constants ($K_s{}^{-1}$, K_s = solubility product) of MX, where M = Cu(II), Cd(II), Pb(II), and Hg(II), and X = S^{2-}, Se^{2-}, and Te^{2-}.

Figure 14-1 provides an alternative way of looking at the relative binding strengths of Cd(II), Pb(II), Cu(II), and Hg(II) to sulfide, selenide, and telluride. With all these anions, the order of binding strength is again Hg(II) > Cu(II) > Pb(II) > Cd(II). The binding constant becomes larger as S^{2-} is successively replaced by Se^{2-} and Te^{2-}; this fact has some significance in relation to the effectiveness of selenium in detoxifying the heavy metals, as discussed later.

Metallic elements must first succeed in getting to the target organ or structure before they can exhibit their effects positive or negative. The absorption or uptake and the transport of essential elements including copper and zinc are discussed in other chapters (8–12). Here we need only focus on transmembrane movement of heavy metallic elements. There are, of course, certain types of effects that result from interaction at a cell surface, but these are not our immediate concern. The membrane, be it cytoplasmic or otherwise, represents a formidable barrier to any cationic species that must pass through it. One fairly general rule in this regard is that the more lipophilic and less charged compounds can more readily penetrate a membrane. This is because phospholipids are the major components of the biomembranes and their tail portions are long hydro-carbon chains that constitute the interior of the membrane. Here again mercury distinguishes itself in having unique properties that are highly relevant. Hg(II) can readily form organometallic derivatives such as CH_3Hg^+, $(CH_3)_2Hg$, $(C_6H_5)Hg(OCOCH_3)$, and so forth. CH_3Hg^+ [as a salt such as CH_3HgCl, $CH_3Hg(OCOCH_3)$] is very stable in aqueous media, though the phenyl derivative is less stable. These organomercury com-pounds are soluble in lipid because of their organic portion and also because of the covalent character of the bond between mercury and car-bon. Thus, organomercury compounds can readily penetrate biomem-branes. Furthermore, the conversion of inorganic mercuric ion Hg(II) to

CH_3Hg^+ and $(CH_3)_2Hg$ can occur readily in the natural environment. Methane bacteria and other microorganisms have been found to accomplish the conversion, making use of the methyl derivative of vitamin B_{12}. The reaction can occur even without a biological catalyst, i.e.,

$$Hg^{II} + CH_3—Co^{III}—B_{12} + HOH \rightarrow$$
$$CH_3—Hg^+ + HOH—Co^{III}—B_{12} \quad (14\text{-}1)$$

The reaction proceeds as shown due to the electrophilicity of Hg(II) and the strong covalency of the resulting Hg—C bond. Metallic mercury itself is rather inert but even it has been shown (Holm and Cox, 1975) to be oxidized to Hg(II) by aerobic bacteria, and the Hg(II) thus formed can in turn be methylated. Methylcobalamin has also been implicated in the methylation of other metal ions such as Pb(II, IV), Sn(II, IV), Pd(II), and Pt(II) (Ridley et al., 1977). Ridley et al. (1977) proposed that methylation of these metal ions by methylcobalamin may be of a free radical type:

$$(Fe^{III} + Sn^{II} \rightarrow) Sn^{III} + CH_3—Co^{III}—B_{12}$$
$$\rightarrow CH_3—Sn^{IV} + Co^{II}—B_{12} \quad (14\text{-}2)$$

They also suggested that there was a relationship between the reduction potential of a metallic element and the type of methylation reaction it can be expected to undergo (Ridley et al., 1977). It appears more reasonable, however, to correlate the metal's electrophilicity and the covalency of the metal–carbon bond with the readiness of reaction 14-1. Reactions of type 14-2 might be important in the case of good radical acceptors. Other types of methylation reactions are also known to exist, including one involving S-adenosylmethionine.

The mechanisms of methylation aside, it is important to look into the reactivities and the covalencies of the metal–carbon bonds of organic derivatives with other heavy metals. Alkyl derivatives of Zn(II) and Cd(II) exist, but they are very reactive and decompose in aqueous media. No stable alkyl derivative is known for Cu(II) though Cu(I) can form some organometallic compounds. The only organometallic lead compound of interest is $(C_2H_5)_4Pb$, which is used as an antiknock gasoline additive. It is completely soluble in lipids and can thus be readily accumulated, particularly in nerve tissues where it is believed to be converted to the toxic form $(C_2H_5)_3Pb^+$.

The other half of the general rule mentioned above relates to the electric charge of the compound to be transported. A stable electrically neutral compound, even if somewhat hydrophilic, is much more readily transported across biomembranes or lipid bilayers than is an ionized compound. For example, cis-$Pt(NH_3)_2Cl_2$, which is now utilized as an anticancer drug (Rosenberg, 1980; Roberts, 1981; Lippard, 1982), can be more readily absorbed than, e.g., $cis\text{-}Pt(NH_3)_2(H_2O)_2^{2+}$.

14.2 Metabolism and Physiological Consequences of Toxic Effects of Heavy Metals

14.2.1 In Algae and Plants

Many phytoplankton (single-celled algae) are known to accumulate metallic elements (Rice et al., 1973). The concentration (or enrichment) factors, defined as the ratio of the concentration (ppm) in the cell to that in the surrounding seawater, are normally on the order of 10^4. This enrichment factor of course differs from one metallic element to another and from one organism to another. It also depends on many other factors such as the pH of the medium, and whether or not other elements are present. A summary of the enrichment factors observed for Zn and Pb in three organisms (alga, moss, and angiosperm) is given in Table 14-2. It should be noted that a moss, in this case Hygrophypnum ochraceum, concentrates heavy metals enormously, as discussed in Chapter 11, and that much higher enrichment factors are obtained in a clean environment. The latter fact suggests that there are mutual interactions among coexisting elements or interactions between the cations and various chelating organic compounds that may be present in, for example, a polluted stream (Wilson, 1972). These interactions may restrict the uptake of the elements by the organisms.

Since the concentrations of heavy metals in seawater range from low, 0.003 ppb, to high, 30 ppb [0.02–0.35 ppb for Pb, 2.0–30 ppb for Zn, 0.025–0.25 for Cd, 0.003–0.36 ppb for Hg (Rice et al., 1973)], the corresponding concentrations in the marine phytoplankton cell could range from 0.03 to 300 ppm (assuming a 10^4 enrichment factor). It is believed that the magnitude of concentration factors is largely a consequence of surface adsorption or ion exchange processes (Rice et al., 1973). For example, $^{65}Zn(II)$ uptake by an alga was shown to be approximated by a Langmuir isotherm, characteristic of surface adsorption. Cell concentrations of up to 100 ppm may be readily accounted for by adsorption onto the cell surface (see the note on p. 391 of this chapter).

Laboratory experiments on the effects of heavy metals on phytoplankton have mainly been concerned with inhibitory effects on growth and on photosynthesis ($^{14}CO_2$ uptake) (Rice et al., 1973). Some of the reported results are summarized in Table 14-3. The growth inhibitory effect of Cu(II) and Pb(II) is markedly different in marine plankton from

Table 14-2. The Accumulation[a] of Zinc and Lead by Plants from a Moderately Polluted Stream and an Unpolluted Stream[b]

River	Site	Concentration in water (mg/liter)			*Lamanea fluviatilis*		*Hygrophypnum ochraceum*		*Mimulus guttans*	
		Ca	Zn	Pb	Zn	Pb	Zn	Pb	Zn	Pb
Polluted	A	27.0	0.18	0.072	—	—	—	—	940	1,430
	B	29.3	0.15	0.063	10,000	13,500	45,000	79,000	—	—
Unpolluted		16.5	0.02	0.01	150,000	16,000	120,000	110,000	27,000	13,000

[a] The values are given in terms of enrichment factor (concentration in organism/concentration in environment).
[b] From B. A. Whitton and P. J. Say, in *River Ecology* (B. A. Whitton, ed.), Blackwell, pp. 286–311 (1975).

Table 14-3. Effects of Heavy Metals on Algae

Phytoplankton	Cu^{II}	Pb^{II}	Cd^{II}	Hg^{II}	Ref.[a]
Concn. in medium (ppm):	0.1	0.1	0.1	<0.1, inorg.	
Pheacodatylum tricornutum (marine)[b]	−	−	−	+	1
Chaetoceras galvestonensis (marine)[b]	−	−	−	+	1
Cyclotella nana (marine)[b]	−	−	−	+	1
Chlorella (freshwater)[b]	+	+	−	+	1
Nitschia delicatissma (marine)[c]		diphenyl Hg: + at 10 ppb; phenyl Hg(OAc): + at 1 ppb; CH_3Hg^+: + at 1 ppb			2
Marine phytoplankton[c]		$HgCl_2$: + at 5 ppb; CH_3Hg^+: + at 0.05 ppb			3

[a] 1, P. J. Nanna and C. Patouillet, 1972, *Report of NRL Progress*, Vol. 1; 2, R. C. Harris, D. B. White, and R. B. McFarlane, 1970, *Science* 170:736; 3, G. A. Knauer and J. H. Martin, 1972, *Limnol. Oceanogr.* 17:868.
[b] Significant inhibitory effect on cell growth (+); no significant effect (−).
[c] Significant inhibitory effect on $^{14}CO_2$ uptake (+).

that in freshwater species. This may be a reflection of a difference in cell surface structure. If we assume the same order of magnitude for the concentration factor (i.e., 10^4 for the uptake of these metals by the organisms at a laboratory environmental level of 0.1 ppm), then the level of the metals in the organisms would be quite high, as high as 10^3 ppm. It is likely that metals at this high level would completely cover the cell surface through adsorption with significant amounts left over to be incorporated into the cells. Concentration factors as high as 10^4 may not be operational under these experimental conditions, however, because such a high factor is considered to be obtainable only when metal ion concentrations in the medium are rather low so that the cell surface adsorption sites are not saturated; see the data from an unpolluted area in Table 14-2. At any rate, at metal ion concentration as high as 0.1 ppm, a significant fraction of the adsorption sites on the cell surface are covered with metal ions and a substantial amount may enter the body.

Inorganic mercuric compounds are less effective in inhibiting $^{14}CO_2$ uptake than organic mercury compounds. One major cause of this difference is the lipophilicity and consequent ease of uptake of organo-mercury compounds. CH_3Hg^+ or RHg^+ in general (R = alkyl or phenyl group) is strongly inhibitive toward enzymes, whereas R_2Hg is much less so. R_2Hg would not be expected to bind to a target site, whether it be a sulfhydryl group or otherwise, because mercury here is saturated in terms of coordination or bonding capacity. Such compounds may have to be partially decomposed to RHg^+ before they ever show any inhibitory effect.

The physiological (or biochemical) mechanism of the inhibitory effects of the heavy metals has not been well studied. One interesting observation is that K(I) loss from the cell (yeast cells or red blood cells) is one of the first detectable cellular responses to heavy metal effects (Passow et al., 1961). A similar result was noted in the effect of Cu(II) on Chlorella vulgaris; K(I) efflux was proportional to Cu(II) binding. That the binding of Cu(II) to the membrane changed the membrane permeability was shown by counting cells permeable to a dye, methylene blue. Even when the loss of cellular K(I) was as high as 86%, however, about half of the cells were shown still to be able to grow. This suggests that a change in membrane permeability is not the sole reason for the inhibitory effect of Cu(II). When C. vulgaris was allowed to absorb copper anaerobically, both the subsequent respiration (aerobic) and photosynthesis (and, hence, growth) were inhibited severely (McBrien and Hassal, 1967). The inhibition was not observed to occur under similar aerobic conditions. It has been suggested that Cu(I), formed as a result of the reaction of Cu(II) with SH groups, may bind at SH groups in catalytically active sites (Passow *et al.*, 1961). One enzyme to which this may apply and which is subject to the inhibitory action of copper is β-fructofuranosidase (invertase).

Another interesting observation with *C. pyrenoidosa* is that at a Cu(II) level of 100 ppb, nearly total growth inhibition was observed at a population of 1.8×10^{10} cells/liter but that no effect was obtained at 1.6×10^{11} cells/liter (Kamp-Nielsen, 1971). A rough estimate (see the note on p. 391) would show that 1×10^{10} cells/liter could be a critical cell concentration in this case. The copper concentration (100 ppb) is equivalent to about 10^{18} atoms/liter. If one assumes that a cell with a radius of 10 μm has an adsorptive site-concentration of 1 per 100 Å^2, then a cell could adsorb as many as 10^7–10^8 atoms of Cu(II), and all the adsorptive sites on the cells would be bound with copper when the cell concentration is 10^{10} cells/liter. When the cell concentration is 10^{11} cells/liter, only 10% of all the adsorptive sites on the cell surface would be occupied by Cu(II). If one assumes that the cells could tolerate this level of Cu(II) on their surface, they would survive (see also Levitt, 1972; Whitton and Say, 1975).

14.2.2 In Invertebrates and Fish

Invertebrates usually go through several molting stages; as a general rule, the younger stages, especially the larval stage, are the more sensitive to the toxic effects of heavy metals. Again it is found that sensitivity to heavy metals varies greatly from one organism to another due to differences in structure, uptake mechanism(s), and other factors.

Molluscs are known to be among the most sensitive of organisms; they are regarded as indicator organisms for environmental pollution, especially that due to heavy metals. They are also known to concentrate metallic elements very effectively, the concentration (enrichment) factors in clams and scallops being on the average 3×10^3 for Zn(II) and Cu(II), 2×10^6 for Cd(II), 3×10^3–10^5 for Pb, and 5–10×10^4 for Hg(II). Oysters have a great ability to concentrate Zn (enrichment factor $= 10^5$), Cu (10^4), and Cd (3×10^5). These enrichment factor values are about equal to, or even ten times as high as those in algae. Adsorption onto the cell surface is the primary cause. It has been noted that the digestive and respiratory systems of molluscs are relatively simple and that their mucous layers are directly exposed to any seawater taken in so that these layers are in the position to readily adsorb cations from the seawater. The mucous layer seems to consist mainly of acidic polysaccharides such as polyuronic acids and polyglucose sulfate, substances that are known to adsorb cations (see Section 11.5). Oysters exposed to toxic amounts of copper showed a dissociation and regressive change of digestive diverticula and desqua-mation and necrosis of the stomach epithelia. Food particles trapped on the mucous layer in the food groove of the mollusc are transported to the mouth and then to the stomach. Any heavy metal adsorbed on the mucous would thus readily be carried into the gastrointestinal tract where it could produce the observed deleterious effects.

In general, the observed concentration (enrichment) factors in the natural environment lie in the range of about 10^4 for phytoplankton, 10^2–10^3 for seaweeds, 10^3–10^4 (up to 10^5) for molluscs, and 10^2 for fish (Carroll et al., 1970). These values appear to reflect the different uptake and excretory mechanisms of the different groups of organisms. As discussed above, adsorption of the cell surface is the main mechanism of uptake in algae and molluscs, and these organisms do not have effective excretory mechanisms for the adsorbed cations. Seaweeds, being macroscopic in size, have a low surface/body weight ratio compared to phytoplankton; thus, even if seaweed surfaces do adsorb heavy metals, the levels of the metals may not rise very high. Fish usually take up heavy metals through the food chain and they do have some excretory mechanisms.

The excretion of heavy metals from organisms is usually described by a first-order rate law, so that the excretion rate can be expressed as a half-time, $t_{1/2}$. These values have been measured for several species; for example, $t_{1/2}$ for ^{210}Pb excretion from the soft parts of a mussel (Mytilus edulis) was found to be about 300 days (Mietinen, 1975). This is quite slow, and suggests either a very strong, immobile binding of Pb(II) to the soft tissues and/or a poor excretion mechanism. $t_{1/2}$ for ^{210}Pb from the shell of the mussel was immeasurably long. ^{210}Pb $t_{1/2}$ values for other invertebrates include 170 days in Mesidotae entomon and 50 days in Hermatoe species (Mietinen, 1975).

$t_{1/2}$ data for mercury compounds are summarized by Mietinen (1975). Usually the excretion of mercury compounds shows two exponential com-ponents: one is fast with $t_{1/2} = 1$–10 days, while the other is slow with $t_{1/2} = 200$–1200 days. The excretion rates of inorganic mercury com-pounds are invariably higher than those of methylmercury CH_3Hg^+. CH_3Hg^+ is more readily absorbed into cell bodies and more tightly bound to biomolecules than are inorganic mercury compounds. CH_3Hg^+ ab-sorbed by rainbow trout (Salmo gairdneri) directly from aquarium water through the gills was eliminated faster ($t_{1/2} = 296$ days) than CH_3Hg^+ administered orally ($t_{1/2} = 326$ days). Other $t_{1/2}$ data for mercury com-pounds in various invertebrates and fish are as follows:

$t_{1/2}$ in mussel: 23 days for inorganic Hg compounds; 100–400 days for CH_3Hg^+

in flounder (*Pleuronectas flesus*): 700 ± 50 days for CH_3Hg^+
in pike (*Esox lucius*): 640 ± 120 days for CH_3Hg^+
in eel (*Anguilla vulgaris*): 1030 ± 70 days for CH_3Hg^+

The elimination of ^{109}Cd by rainbow trout was found to follow a power function rather than the usual exponential one (Mietinen, 1975). The bulk of ^{109}Cd administered was rapidly eliminated so that after 42 days only about 1% of the original dose was retained in the body. The excretion of the remaining cadmium was very slow (Mietinen, 1975).

The main pathways for metal incorporation in fish under normal conditions involve the gills and the mouth. These organisms absorb oxygen through their gills. The gills are always exposed to water and seem to be the prime target of attack by heavy metals. The gastrointestinal tract's contribution may be distinctly different in marine organisms from that in freshwater species. This is because of the different rate of water drinking in these two groups of organisms as a result of osmotic pressure regulation; marine fish drink water continuously whereas freshwater fish do not drink. Treatment of starved brown bullheads with nonlethal levels of Zn(II) for 96 hr led to zinc accumulation particularly in the gills and the gut, with lower levels found in the liver, kidney, skin, muscle, bone, and spleen (Whitton and Say, 1975). It is an interesting observation that fish killed by a high level of Pb(II) or Zn(II) were usually found to have high amounts of mucus around their gills (Whitton and Say, 1975). This could be a defense mechanism against heavy metals; mucus, as noted before, can adsorb (chelate) metal ions.

A crab (*Maia sequinado*) was found to accumulate Hg(II) in its gills, esophagus, proventriculus, and antennary glands (Corner, 1959). In developing echinoderm eggs, silver (Ag(I)) was shown to be about 80 times as toxic as Zn(II), 20 times as toxic as Cu(II), and 10 times as toxic as Hg(II) (Wilber, 1969). This could be due to the smaller charge of Ag(I) allowing it to penetrate the cell membrane more readily than can the divalent cations.

14.2.3 Copper in Mammals (Including Humans)

14.2.3.1 Chronic Toxicity (Hill, 1977)

In general, copper poisoning is characterized by a gradual accumulation of copper in the liver over a long period of time without any outward symptoms of toxicity being evident. This period is followed by a sudden onset of hemolysis, hemoglobinemia, and hemoglobinuria coincident with the release of copper from the liver. The level and the period of tolerable accumulation of copper in the liver depend on the species, the general health condition of the subject, and the levels of some other metals such as molybdenum.

These features of chronic copper toxicity indicate the presence in the liver of a copper-sequestering mechanism, but one with a limited capacity. Beyond this limit, excess copper seems to trigger the activation of lysosomal enzymes or to cause lesions of the lysosomes. It is very likely that the small protein called chelatin (or metallothionein, as discussed in Section 14.3.5.2) is involved in the sequestering of copper.

14.2.3.2 Wilson's Disease (Venugopal and Luckey, 1978)

This rare inheritable disease is caused by an unusual accumulation of copper, especially in erythrocytes, kidney, liver, and brain. Excess deposition of copper is due to excessive absorption and poor excretion (due to biliary obstruction, although urinary excretion is usually increased), or the inability to store excess copper in ceruloplasmin. The clinical symptoms include tremor, ascites, psychosis, and slurring of speech; pathologically the copper accumulation causes hepatic cirrhosis, necrosis and sclerosis of the corpus striatum, and trauma in the brain, which leads to death. Treatment with chelating agents such as penicillamine can reduce the tissue copper levels.

14.2.4 Cadmium in Mammals

14.2.4.1 Metabolism (Mietinen, 1975)

Ingested ^{115}Cd was eliminated from human males to the extent of about 75% within 3–5 days and 94% within 10–15 days. This elimination occurred primarily via feces, probably in the form of bile. Only a minor portion was excreted in urine. This rapid elimination was followed by a very slow process, the half-time of which was estimated to be 15–20 years. Cd(II) circulates in blood primarily bound to red blood cells, but it is considered also to be bound to albumin and other proteins. When bound rather unspecifically either to albumin or to red blood cells, Cd(II) can be sequestered by the liver, in which it induces the synthesis of metallothionein. With continued exposure, however, cadmium is accumulated increasingly in the kidney. Most of the renal cadmium is concentrated in the proximal renal tubules.

14.2.4.2 Toxicity

Characteristic symptoms of cadmium toxicity are nephrotic tubular atrophy (and its consequences—e.g., proteinuria), testicular necrosis, osteomalacia, and hypertension (Nordberg, 1976a; Goyer, 1975; Piscator, 1976).

A high dose (1 mg/kg) of Cd(II) supplied by injection injures the stem cells of the testis (in rats) and causes severe testicular necrosis (Goyer, 1975). This is considered to be caused by vascular lesions. At lower doses, Cd(II) induces metabolic lesions at the level of spermatogonia and spermatocytes (Lee and Dixon, 1973). The hemorrhagic necrosis (due to vascular lesion) and the effect of Cd(II) on spermatocytes were found to be prevented by simultaneous or prior administration of Zn(II). The male gonad is one of the organs that accumulate Zn(II), probably due mainly to the fact that Zn(II) is required for the enzyme DNA polymerase. DNA polymerase must be especially abundant in spermatogonia (precursor of spermatozoa), where DNA molecules are duplicated constantly. The protective effect of Zn(II) administered prior to Cd(II) is believed to be a result of zinc-stimulated formation of metallothionein, which can, in turn, sequester Cd(II) administered later. The protective effect provided by simultaneous administration of zinc can be regarded as due to thermodynamic competition between zinc and cadmium for various active sites. The difference between spermatogonia and spermatocytes with regard to the protective effect of zinc may be due to the fact that chromosome duplication is actively carried out in the spermatogonia whereas no chromosome duplication occurs in the primary spermatocytes. If Cd(II) exerts an inhibitive effect on DNA polymerase, or if Cd(II)-substituted DNA polymerase is inactive, then the consequence would be serious in spermatogonia and probably not reversible by the administration of Zn(II) whereas the primary spermatocytes would hardly be affected. This hypothesis is in accord with the observation (Lee and Dixon, 1973) that thymidine incorporation into DNA was irreversibly suppressed for as long as 55 days after cadmium administration.

The chronic effect of cadmium is most manifest in the kidney. It appears that most cadmium sequestered in the proximal tubular lining cells is not excreted until it reaches a critical level (Goyer, 1975). It is hypothesized that the metallothionein-bound cadmium is sequestered by the tubular cells and that cadmium inhibits the enzymatic catabolism of proteins. However, when the cadmium content exceeds a certain critical cellular level, cellular necrosis (perhaps caused by lesions of lysosomal membranes) occurs, which is accompanied by renal tubular dysfunction, proteinuria, and an increased urinary excretion of cadmium (Goyer, 1975).

Osteomalacia is one of the characteristic symptoms of the "itai-itai disease" believed to be caused by prolonged ingestion of cadmium. Its main characteristic feature is the development of decalcified and fragile bone. It is possible that Cd(II) directly replaces Ca(II) in the bone although this cannot be regarded to be the major factor involved. Osetomalacia is now considered to be a secondary effect of a malfunctioning of the kidney, which regulates the calcium and phosphorus balance in the body (Nordberg, 1976a). Cadmium is also known to interfere with the intestinal absorption of calcium. A low-calcium diet enhances cadmium absorption.

Animal experiments (on rats) have shown that hypertension can be induced by oral or parenteral administration of cadmium at low levels, although the effect is less pronounced than was once thought (Perry, 1976).

14.2.4.3 Other Effects

An interesting finding is that Cd(II) specifically binds to α_1-antitrypsin in human blood (Chowdhury and Louria, 1976). No other metal ion was found to do so, including Pb(II), Hg(II), Fe(II), Ni(II), and Zn(II). It was suggested that prolonged cadmium exposure leads to severe α_1-antitrypsin deficiency and hence pulmonary emphysema.

Endotoxin comprises the haptenic portion of the cell walls of Enterobacteriaceae and thus is believed to consist of lipopolysaccharides. Endotoxins produce several of the effects of gastrointestinal infection, including nausea, diarrhea, and intestinal hemorrhage. Cadmium and lead have been found to increase enormously the effectiveness of endotoxins when administered to experimental animals at trace level (Cook et al., 1975). For example, a combination of a Cd(II) salt and the endotoxin from Salmonella enteritidis produced a greater mortality rate than the endotoxin alone. It was estimated that Cd(II) increased the effectiveness of the endotoxin by 12,500-fold (Cook et al., 1975). Endotoxin is known to suppress phagocytic activity but an alteration of the phagocytic activity by the metals is not considered to be the primary factor involved because an altered hepatic parenchymal cell function and defective carbohydrate metabolism in the hepatocytes were also noted (Cook et al., 1975). It is possible that the metal inactivates some enzymes in the phagocytic or other cells, enzymes that are involved in the destruction of the lipopolysaccharides (endotoxin). Another possibility is that the metals disrupt the integrity of the cell membrane, rendering it more susceptible to attack by endotoxins or more permeable to them.

Enzymes that are known to be inactivated by Cd(II) include zinc enzymes such as alcohol dehydrogenase, carbonic anhydrase, dipeptidase and carboxypeptidase, DNA and RNA polymerases, as well as other enzymes such as amylase, glutamic oxaloacetate transaminase, lipase, and tryptophan oxygenase (Vallee and Ullmer, 1972).

14.2.5 Lead in Mammals

14.2.5.1 Metabolism (Mietinen, 1975; Goyer, 1975)

The gastrointestinal absorption of lead is not very high, 1–16%. In the blood, lead is bound mainly by the red blood cells. Lead passes into the brain barrier rather readily (more readily than cadmium), especially in the form of triethyl lead $(C_2H_5)_3Pb^+$, which can be produced in the liver from tetraethyl lead. Lead is reabsorbed in the tubular section of the kidney, and thus, urinary excretion of lead is very low under normal conditions. In the case of lead poisoning, most of the lead in the proximal tubular lining cells is found in inclusion bodies in the nuclear fraction (see Section 14.3.5.3). In an adult human, as much as 90% of the total body lead is found in the bone. This represents a fixed pool of lead, from which the excretion rate is very low, $t_{1/2}$ being estimated to be as long as 90 years.

14.2.5.2 Toxicity—Clinical Symptoms (Nordberg, 1976b; Goyer and Moore, 1974)

The major toxic effects of lead appear in three systems: the hematological (hemoglobin synthesizing), central nervous, and renal. Lead exposure may give rise to microcytic anemia and shortened life span of circulating erythrocytes. The anemia is caused by the disruption of hemoglobin synthesis in bone marrow cells. The enzymes most affected by lead are δ-aminolevulinic acid dehydratase (ALA-D), which catalyzes the formation of porphobilinogen from δ-aminolevulinic acid, and ferrochelatase (FC or heme synthetase), which incorporates an iron atom into protoporphyrin IX. ALA-D is believed to be a zinc enzyme (Cheh and Neilands, 1976). The mechanism of the inhibition of this enzyme by lead is not clear. Other steps in hemoglobin synthesis are also known to be subject to lead poisoning. For example, high lead levels were implicated in defective synthesis of α- and β-globin chains of hemoglobin (White and Harvey, 1972), and δ-aminolevulinic acid synthetase was shown to be inhibited by a high level of lead (Diesel and Falk, 1956).

The presence of lead is known to shorten the life span of red blood cells but not severely (to about 100 days from the normal 120 days); details of the cause are not known.

The typical symptoms of encephalopathy caused by lead are ataxia, coma, and convulsion. Lead has been found to concentrate in the cortical gray matter and the basal ganglia of the brain though the level in the whole brain does not become especially high. One observation (Silbergeld *et al.*, 1974) is interesting in this context: 8×10^{-5} M Pb(II) affects the contraction force of the phrenic nerve-stimulated hemidiaphragm in rats, and this effect can be counteracted by 10^{-9} M Ca(II).

Acute exposure to relatively high levels of lead compounds leads to functional and morphological changes in proximal renal tubular cells. The clinical effects include aminoaciduria, glucosuria, and hyperphosphaturia. Observed morphological changes include nonspecific degenerative changes, mitochondrial swelling, and the presence of nuclear inclusion bodies. Lead in the inclusion bodies was found to be mobilized by the administration of the chelating agent EDTA.

Long-term chronic exposure to lead results in a slightly different renal dysfunction. For example, the tubular dysfunction is manifested by a defect in uric acid secretion, often resulting in hyperuricemia and sometimes in gout.

14.2.5.3 Other Effects at the Cellular Level (Goyer and Moore, 1974)

There are several observations that suggest some adverse effects of lead on chromosomes and protein synthesis (Goyer and Moore, 1974). An injection of lead was observed to enhance DNA synthesis and proliferation of renal tubular cells within 2 days. On the other hand, $Pb(OCHCH_3)_2$ (1% in diet) caused gaplike aberrations in chromatids in a leukocyte culture from mice. A similar aberration was observed in industrial workers engaged in lead processing. Lead has also been found to cause a deaggregation of polysomes (protein synthesis apparatus), depolymerization of RNA, and inhibition of leucine incorporation into RNA.

Lead toxicity is always associated with alteration in the ultrastructure and function of mitochondria in hepatic cells, renal tubular cells, and reticulocytes in the bone marrow. Lead is known to be concentrated in mitochondria and most of the enzymes involved in heme synthesis are located in the mitochondria of bone marrow cells.

The potentiation of endotoxin by Pb(II) was mentioned earlier. In addition to those cited above, enzymes that are known to be inhibited by lead include acetylcholine esterase (whose inhibition may be involved in the encephalopathic effect of lead), acid and alkaline phosphatase, ATPases, carbonic anhydrase, fructose-1,6-diphosphatase, and glutamate dehydrogenase (Vallee and Ullmer, 1972).

14.2.6 Mercury in Mammals

14.2.6.1 Metabolism (Mietinen, 1975; Goyer, 1975)

Inorganic Hg(II) compounds, either free or bound with proteins, are not absorbed very effectively (at most 15%) from the gastrointestinal tract, whereas monomethylmercury CH_3Hg^+ is absorbed almost completely (up to 95%) (Mietinen, 1975). The difference is due to the fact that membranes are poorly permeable to Hg(II) in general, but are fairly permeable to CH_3Hg^+. Based on measurements with ^{203}Hg, $t_{1/2}$ of typical inorganic Hg(II) compounds in man is 42 ± 3 days, whereas that of CH_3Hg^+ is 70–76 days (Mietinen, 1975).

Inorganic Hg(II) is mostly found in plasma in the blood, but CH_3Hg^+ is bound to the red blood cells (and to hemoglobin as well). It appears that mercury is concentrated in the liver lysosomes (Goyer, 1975). About 60–75% of the total mercury in rat kidney was found to be bound to metallothionein (Goyer, 1975). Hg(II) induces the synthesis of metallothionein in kidneys but not in liver. That metallic mercury Hg(0) can be oxidized to Hg(II) in biological systems was mentioned earlier.

14.2.6.2 Toxicity (Nordberg, 1976c)

Water-soluble inorganic Hg(II) salts affect the kidney, whereas the typical toxic symptoms of CH_3Hg^+ are manifested in the central nervous system. This difference between inorganic Hg(II) and CH_3Hg^+ again reflects difference in their solubilities in water and lipids. The metabolism and toxicity of phenyl mercury compound C_6H_5HgX appear to be intermediate between those of inorganic Hg(II) and of alkyl mercury compounds. This is because $C_6H_5Hg^+$ is relatively unstable in aqueous medium and decomposes rather rapidly to inorganic Hg(II). $C_6H_5Hg^+$ is as soluble as CH_3Hg^+ in lipids.

The symptoms of severe CH_3Hg^+ intoxication include numbness and tingling of the lips, mouth, hands, and feet, dysarthria, ataxia, concentric constriction of the visual field, blurred vision, blindness, deafness, and an impaired level of consciousness (Nordberg, 1976c). Infants who were exposed prenatally to CH_3Hg^+ showed signs of cerebral palsy, convulsions, and, in some cases, blindness, while their mothers seemed to be without toxic mercury effects (Nordberg, 1976c). This fact suggests an easy transfer of methylmercury across the placenta (Miller and Shaikh, 1983). Morphological findings include severe changes in the cerebral cortex, particularly of the visual cortex; a similar topographical localization was observed in the form of subcortical accumulation of methylmercury.

Despite the dramatic toxic effects of mercury, it has been difficult to identify the specific chemical causes of the physiological effects. One reason is that Hg(II) or CH_3Hg^+ is a potent and indiscriminate inhibitor of hundreds of enzymes or proteins as well as of membrane activities. Especially those enzymes whose active sites contain cysteine sulfhydryl groups appear to be most vulnerable. A few examples will suffice to illustrate the point. CH_3Hg^+ was shown to be a potent inhibitor of membrane adenyl cyclase in rat liver (Storm and Gunsalus, 1974). Since adenyl cyclase is a pivotal enzyme in a large variety of biofunctions (including mitosis and hormone actions), this finding indicates that CH_3Hg^+ could have effects on a very wide range of biological phenomena. Inorganic as well as organic mercury compounds were found to reduce (in rabbit ileum) the mucosal entry of sugars and amino acids to 80–90% of control levels within several minutes of its administration at 1 mM (Stirling, 1975). Inhibition of red blood cell membrane Na/K-ATPase was demonstrated in industrial workers exposed to mercury (Nordberg, 1976c).

14.3 Adaptation–Avoidance–Defense–Resistance–Tolerance

14.3.1 General Comments

Because of the devastating effects of the heavy metals and their rather ubiquitous presence (albeit normally at low levels), organisms have developed a variety of protective mechanisms. I will review some of these briefly. The mechanisms can be classified variously as tolerance, avoidance, adaptation, resistance, and defense. A few comments are in order concerning the use of these terms.

Tolerance implies the inherent presence in normal organisms of mechanisms that somehow allow the subjects to tolerate the presence of heavy metals in their bodies to a certain extent. Beyond that threshold, toxic effects would become manifest. *Avoidance* is a response in which the organism avoids taking up the dangerous metals, either because it has mechanisms to reject them or because it simply stays physically removed from any locale where heavy metals are found. *Adaptation* implies a genetic physiological change that is induced by repeated exposure to heavy metals (or any other dangerous situation for that matter) so that the organism becomes "resistant" to them or finds ways to "tolerate" or "avoid" them. *Defense* is a short-term effect elicited by exposure to heavy metals. It may or may not involve a genetic alteration.

While these definitions may seem reasonably straightforward, in practice the definitions are not terribly helpful. Often it is difficult to classify a mechanism unambiguously, and little may be gained by doing so. Therefore, I will here use *defense* as a general term encompassing most of the above.

The actual biochemical and physiological mechanisms employed in heavy metal defense are the principal subject of interest and relevance here. Important mechanisms have been summarized by Antonovics *et al.* (1971), especially with regard to tolerance in plants. Table 14-4 is my version, modified from theirs. A few brief comments follow regarding each of the mechanisms listed.

1. The cell wall is rather inert with respect to the cellular activity. Adsorbing materials such as mucopolysaccharides and glycoproteins can form a natural part of the cell wall or they may be produced intracellularly and secreted in response to the presence of heavy metals.
2. Siderophores (see Section 9.1.3) are an example; they are utilized to chelate Fe(III) so that it can be incorporated. Secretion of substances to fix heavy metals could be envisioned, but has not been observed.
3. This is equivalent to selective uptake of specific cations. In fact, no organism is known to possess a specific uptake mechanism for

Table 14-4. Defense Mechanisms against Heavy Metals

Mechanism	Note; example
A. Extracellular but caused intracellularly	
1. Cell wall, adsorbing material	Nonspecific; mucus on the gill of fish, bacterial cell wall
2. Secretion of chelating substances	Can be specific
B. Intracellular and internal	
3. Differential uptake of ions	No transport system for heavy metals
4. Removal by pumping out	None is known for heavy metals
5. Removal by deposition in some locale where heavy metals are better tolerated	Nuclear inclusion bodies; cell wall, hair, etc.
6. Removal by converting a heavy metal into an innocuous form	Metallothionein, CuS, HgSe, etc.
7. Removal by converting a heavy metal into a more readily excretable form	Hg(II)–(CH$_3$)$_2$Hg, etc.
8. Increased concentration of enzyme that is inhibited	
9. Alternate metabolic pathway bypassing inhibited sites	

heavy metals. The problem here is that the typical uptake systems for essential metals are not very specific and can result in the inadvertent absorption of heavy metals or that certain heavy metal compounds may rather readily penetrate a membrane. To the extent that this occurs, organisms are compelled to utilize the other mechanisms listed below.

4. No mechanism is known for pumping out heavy metals, except that Cd(II) is said to be excreted by Cd(II)-resistant cells of *Staphylococcus aureus* (Silver, 1984; Tynecka *et al.*, 1981).

5. A particular intracellular or extracellular vesicle or structure, one that may not be crucial for the normal functioning of the cell body, is utilized to sequester and store the toxic metals, i.e., compartmentation. Examples of such vesicles and structures include cell walls, lysosomes, vacuoles, hair, and nucleus (nuclear inclusion body). Even milk might function as a medium in which to deposit a material unwanted by the mother. PCB accumulation in milk is an example.

6. The intensity of the toxic effect of a metal depends greatly on the chemical form in which it occurs. For example, Hg(0) is much less toxic than Hg(II), and $(CH_3)_2Hg$ as such is not particularly toxic (the toxic form is CH_3Hg^+). Other examples of chemical detoxification include the chelation of heavy metals by metallothionein, and the conversion of heavy metals into sulfides or selenides, which are very insoluble and thus would not be able to exert any toxic effect.

7. Unfortunately, most heavy metals are quite injurious to the kidney; therefore, this mechanism would probably not provide most animals with a defense against heavy metals.

8, 9. These are possibilities, but it has not been shown that they are actually employed in defense against heavy metals. One reason for this may be that the effect of a heavy metal can be very wide-ranging and rather indiscriminate; therefore, an adaptation to such an agent by mechanism 8 or 9 would be unlikely to be effective.

14.3.2 Defense in Bacteria and Fungi

14.3.2.1 In Bacteria

The uptake of PbI_2, $PbBr_2$, or PbClBr by *Micrococcus luteus* and *Azotobacter* species has been studied (Tornabene and Edwards, 1972).

More than 99% of the total lead taken up was found to reside in the cell wall and cell membrane fractions.

In many bacteria, resistance to toxic heavy metals or metalloids such as mercury, cadmium, arsenate (or arsenite), lead, cobalt, nickel, and zinc is known to be associated with a plasmid (Myenell, 1972; Novick and Roch, 1968; Silver, 1984). The so-called R-plasmid in $E. coli$ carries a resistance factor gene to mercury, cobalt, and nickel, and another plasmid-carrying penicillinase gene in $Staphylococcus aureus$ mediates resistance to mercury, cadmium, arsenate (arsenite), lead, and zinc. The frequency of heavy metal-resistant strains was studied with regard to arsenic, cadmium, mercury, and lead using several hundred strains each of $E. coli$, $Klebsiella pneumoniae$, $Pseudomonas aeruginosa$, and $S. aureus$ isolated from hospital patients (Nakahara et al., 1977). The frequency of occurrence of resistant strains was found to be quite high (higher than that of drug-resistant strains); observed resistance frequencies (%) among the strains studied were: 36–75% for mercury, 93–98% for cadmium, 49–99% for arsenic, and 92% for lead.

Mercury detoxification mechanisms employed by bacteria include the formation of $Hg(0)$, or $(CH_3)_2Hg$, both of which are rather innocuous, volatile, and readily excretable, and the conversion of $Hg(II)$ to HgS. An example is the reduction of $Hg(II)$ to $Hg(0)$ in $E. coli$ (Summers and Sugarman, 1974; Fox and Walsh, 1982). The $Hg(II)$-reducing activity in the cytoplasm was found to be dependent on glucose-6-phosphate, glucose-6-phosphate dehydrogenase, and 2-mercaptoethanol. Treatment of the active fraction with sulfhydryl-blocking agents such as N-ethyl-maleimide and phenyl mercuric acetate, cyanide, or m-chloro-carbonyl cyanide phenylhydrazone inhibited the mercury-reducing activity. This result suggests the presence of SH group in the $Hg(II)$-reducing system, which may be dependent on the respiration. The system has been shown to be a flavoprotein with NADPH as the electron donor (Fox and Walsh, 1982). $Au(III)$ can also be reduced to $Au(0)$ by the isolated (cell-free) $Hg(II)$-reducing system (Summers and Sugarman, 1974).

The cadmium resistance in $S. aureus$ is carried by penicillinase plasmid (Silver, 1984). Strains sensitive to penicillin incorporated $CdCl_2$ into the cells, whereas resistant strains did not (Kondo et al., 1974). The incorporation of cadmium by the sensitive strains was temperature-dependent and did not occur at 4°C; it could be counteracted by $CaCl_2$ (Kondo et al., 1974). These results suggest that $Cd(II)$ is carried into the cell by a transport system designed for $Ca(II)$. A gene on the plasmid, designated as $CadA$, has been shown to encode an outward transport system of $Cd(II)$ in the resistant cells (Tynecka et al., 1981).

14.3.2.2 In Fungi

The spraying of fungicides to control pathological fungi is common. One of the oldest fungicides, Bordeaux mixture, contains copper, whereas the more modern fungicides contain various heavy metal compounds, one of the most effective of which is methyl mercury chloride (CH_3HgCl) or phenyl mercury acetate [$CH_3Hg(OCOCH_3)$].

It was observed a century ago that a mold grew in a solution containing 2% copper. Later it was also observed that a different mold, *Penicillium glaucum,* could be made to adapt to such heavy metals as copper, zinc, cadmium, mercury, manganese, cobalt, nickel, lead, and thallium. Ashida (1965) summarized similar findings regarding the adaptability of various fungal species to heavy metals. I cite a few examples from his list in Table 14-5. A careful study and appraisal of these results has suggested that copper-resistant mutants actually arise from the originally sensitive cells after exposure to a copper-containing medium. Here we are dealing with a whole population, not with the response of the individual cells. This comment applies also to work with bacteria. We must bear it in mind, when evaluating these findings in relation to the situation for mammals, for example. When a small number of sensitive cells are mixed with many resistant cells and the resulting population is cultured in the absence of copper, the proportion of the nonresistant cells may increase with time so that the resistance of the whole culture may be rapidly lowered. This does not necessarily mean that the acquired

Table 14-5. Adaptation of Fungi to Metal Toxicant[a]

	Cu[II]	Hg[II]	Organo-mercurial	Other metals[b]
Botrytis nicerea	+ +	−	+	
Gibberella zeae		+	+	
Hypochnus centrifugus		+ *		
Penicillium glaucum	+ *	+		Zn, Cd, Mn, Co, Ni, Pb, Tl
Penicillium notatum	+ +	+ +	−	
Piricularia oryzae	+ +(+ *)	+ +(+ *)	+	
Saccharomyces cerevisiae	+	+		Cd, Co, Ni, Mn, Ag, As, Cr
Scleotinia frusticola	+ *	+ *	−	
Stemphylium sarcinaeforme	+ *	+	−	

[a] +, successful adaptation (in culture); −, no successful adaptation; + +, stable resistance, + *, reversible resistance; + +(+ *), either + + or + * depending on strains.
[b] Successful adaptation to these metals.

resistance has been lost; it may be simply due to the fact that the resistant strain somehow is handicapped under normal conditions.

One copper-detoxifying (or resistance) mechanism found in many fungal species is the excessive production of H_2S, which combines with copper to form insoluble (and hence, inactive) CuS or Cu_2S. Thus, certain strains of yeast acquire a brown color when cultured in the presence of copper; the color is due to CuS/Cu_2S. Microscopic studies have shown the CuS/Cu_2S to be located mainly in and around the cell wall. This is not the only defense mechanism found in fungi. Yeast strains are known that are resistant to copper but do not produce significant amounts of H_2S. In these cells, the atomic ratio of total copper to total sulfur was found to be about 3:1 (Ashida, 1965). It has been found that some fungal species produce metallothioneins that bind copper as well as zinc, cadmium, and other metal ions (Prinz and Weser, 1975). The ratio of cysteine residues to metal ion is usually 3:1 in metallothionein.

In contrast to the case of copper, cadmium-resistant fungal cells do not accumulate much cadmium and do not depend on sulfur, but rather on nitrogenous compounds. The details are not known but the difference in the response between the two metals is clear-cut just as was shown to be the case of cadmium and mercury in bacterial cells.

14.3.3 Defense in Plants

Heavy metal tolerance in plants was extensively reviewed by Antonovics et al. (1971). This survey showed that the existence of metal-tolerant species is a common phenomenon among plants. There is no evidence, however, that a species has constitutional tolerance to heavy metals. Therefore, tolerance must be acquired through evolution. The survey also showed that many species found on copper-bearing soils also occur on soils containing nickel, suggesting some kind of common defense mechanism against copper and nickel; this is chemically plausible.

It was shown, in accord with the suggestion made above for fungi, that zinc-tolerant Anthoxanthum odoratum is competitively inferior to nontolerant races when grown on noncontaminated soil. A similar competitive inferiority of tolerant races compared to nontolerant ones was also observed for Agrostis tenuis and Plantago lanceolata.

Laboratory studies have shown that most selection occurs at the seedling stage; nontolerant seed in contaminated soils can germinate, but it fails to root and the seedlings die before they can produce many leaves.

There are two different types of behavior observed in plant uptake of heavy metals. One is exemplified by the uptake of zinc. The quantity of zinc found in plants is related to the level of zinc in the soil or the culture solution, often in a fairly linear fashion. The other mode is exemplified by copper. Copper uptake in plant shoots and tops stays low and constant at low levels of copper in the soil (or other medium) but beyond a certain threshold level in the medium, this "resistance" breaks down. It is noteworthy that copper content in plants rarely varies outside the range of 5–15 ppm whereas the zinc content may vary between 20 and 10,000 ppm. Plant roots can absorb far more copper, however, reaching copper levels as high as 3250 ppm (dry basis) compared to a maximum 40 ppm in the shoots. The uptake behavior of lead and nickel shows patterns similar to that observed for copper. The fact that there is a threshold level in copper tolerance, and also that roots can take up copper to a high level, may be interpreted as indicating that a complexing mechanism for copper exists in the roots, as a result of which copper is prevented from reaching the upper parts of the plants until the root system itself is severely damaged.

Apparently, changes in sulfur metabolism are not important in the defense mechanism employed by higher plants, unlike the case of fungi. Many studies have indicated that, in tolerant strains, zinc and copper are concentrated in the cell walls, thereby preventing them from entering more sensitive regions within the cells. Pectate fractions from cell walls of tolerant races were shown to contain five to six times as much zinc as the corresponding fractions from nontolerant races of *Agrostis tenuis* and *A. stolonifera*. See Section 11.5 for a discussion of the cation binding by mucopolysaccharides and similar compounds. The distribution of zinc in copper-tolerant plants was found to be similar to its distribution in nontolerant plants. This observation indicates that the tolerance mechanisms for copper and zinc are different. Zinc was shown also to be taken up in some plants by the soluble RNA fraction and/or by the vacuole.

As metallothionein has been confirmed to exist in fungi, it is very likely that similar chelating proteins may also be involved in the defense mechanism employed by plants. There has been no definite report in confirmation of this hypothesis, however.

14.3.4 Defense in Invertebrates and Fish

Some races of *Nereis diversicolor* (a clam worm) were found to thrive in an estuary where high concentrations of heavy metals were recorded in the sediment (Bryan, 1974). In contrast to the case of plants described above, the levels of copper found in the tolerant worms were roughly proportional to those in the sediments, whereas the zinc levels in the worms seemed to be controlled (i.e., 100–300 ppm in worms versus 100–3000 ppm in the sediment). The amount of lead in the worms was roughly proportional to the level in the sediment, as in the case of copper. Copper-tolerant worms were found to absorb copper more rapidly than non-tolerant ones, and to deposit copper mainly in the epidermis and parts of the nephridia. The zinc-tolerant worms on the other hand were about 30% less permeable to zinc than were normal ones.

It has been mentioned (Section 14.2) that the secretion of mucus especially at the gill surface, may function as a heavy metal defense mechanism in fish.

14.3.5 Defense in Mammals

14.3.5.1 General Comments

It is unlikely that organisms would, in general, possess specific uptake and transport mechanisms for strictly toxic elements; they do, however, have such mechanisms for essential elements, copper and zinc. Heavy metals, being present as cations and having typical physicochemical characteristics associated with cations (in terms of size, charge, and so forth), can enter the cells via the uptake and transport systems designed for essential metal cations. Organometallic compounds, exemplified by CH_3Hg^+, can passively penetrate cell membranes because of their affinity for lipids. These considerations suggest that it would be difficult to control specifically the absorption of potentially toxic heavy metals, except by the use, for example, of such unspecific "curtains" as the mucus layer.

Mammals, therefore, have developed mechanisms to control the levels of heavy metals specifically at the postabsorption stage. Two important organs involved in detoxification of (defense against) heavy metals (or most other toxicants, for that matter) are the liver and kidney. Substances absorbed in the human gastrointestinal tract are carried by the portal vein first to the liver, which absorbs most nutrients while sequestering drugs and many trace elements. The liver is equipped with a variety of detoxifying devices including the microsomal cytochrome P-450-dependent enzyme system, which metabolizes drugs. One of the most prominent mechanisms used to detoxify heavy metals in the liver depends on the production of metallothionein.

The kidney is another pivotal organ, one through which water-soluble metabolites are excreted. The kidney is continuously exposed to heavy metals and other toxic substances and hence is also equipped with detoxifying mechanisms. These mechanisms again include the formation of metallothionein as well as the formation of nuclear bodies.

Hair, consisting of the sulfur-rich protein keratin, can bind heavy metals to high levels; indeed, the metal contents in hair are often utilized as an indicator of metal toxication. It is possible that the pigment melanin is also involved in the binding of heavy metals. Sequestering of heavy metals by hair may thus be regarded as a defense mechanism. As mentioned earlier, bone constitutes a sink for lead, thus keeping lead away from more sensitive tissues.

14.3.5.2 Metallothionein (Kojima and Kagi, 1978; Foulkes, 1982)

In 1957 Margoshes and Vallee reported that horse kidney cortex contained a small protein rich in cadmium. It was also found to contain unusually high levels of sulfur, thus the name metallothionein. Similar proteins of low molecular weight (6000–10,000) have subsequently been identified in a wide variety of organisms, most mammals but also mold and yeast (Kojima and Kagi, 1978). The organisms for which metallothioneins have been reported include human, horse, calf, pig, seal, rabbit, rat, mouse, chicken, molluscs, *Neurospora*, and copper rock fish (for a list of marine organisms for which metallothioneins have been demonstrated, see George, 1982). Metallothioneins have been shown to contain not only cadmium but also copper, zinc, and mercury. Even more interesting is the finding that metallothionein production can be induced by exposure of the organisms to these metals as the metal salts. The fact that metallothioneins are inducible by toxic metals strongly indicates that they are involved in detoxification or metabolism of these metals. Ag(I) has also been shown to induce the synthesis of a similar protein in some organisms. It has been reported recently (Ikebuchi et al., 1986) that Pb(II) injected in the form of lead acetate induced a metallothionein-like protein (M_r 6900) containing Pb in the liver of rats; it contains 28% cysteine.

The observed metal content of metallothionein is usually 7 gram atoms per mole of the protein based on a molecular weight of 6800 (Kojima and Kagi, 1978). The metal composition in naturally occurring metallothionein does, however, show some variation from one tissue to another. For example, cadmium and zinc are often contained in about equal amounts in the protein from kidney, whereas zinc is usually the principal element in the protein from liver. This fact suggests that a metallothionein is indeed a controlling factor in the metabolism of the essential element zinc in the liver. More recent studies (Briggs and Armitage, 1982; Winge and Miklosky, 1982) indicate that calf liver cadmium metallothionein contains two seemingly dissimilar domains: one contains a Cd_3 cluster and the other a Cd_4 cluster. In the case of Zn/Cu metallothionein (Briggs and Armitage, 1982), copper seems to bind to a Cu_3 cluster site and zinc to a Zn_4 cluster

10   20

N-Acetyl-Met-Asp-Pro-Asn-Cys-Ser-Cys-Ala-Ala-Gly-Asp-Ser-Cys-Thr-Cys-Ala-Gly-Ser-Cys-Lys-

Cys-Lys-Glu-Cys-Lys-Cys-Thr-Ser-Cys-Lys-Lys-Ser-Cys-Cys-Ser-Cys-Cys-Pro-Val-Gly-

Cys-Ala-Lys-Cys-Ala-Glu-Gly-Cys-Ile-Cys-Lys-Gly-Ala-Ser-Asp-Lys-Cys-Cys-Ser-Cys-Ala

Figure 14-2. The amino acid sequence of a metallothionein. (From Kojima and Kagi, 1978. *Trends Biochem. Sci.* 3:90.)

site. Metallothioneins have also been isolated from the intestine, pancreas, spleen, and placenta of rats.

Actinomycin D (an inhibitor of the transcription DNA → mRNA) inhibited zinc uptake as well as the synthesis of metallothionein in liver (Richards and Cousins, 1975). This suggests that metallothionein synthesis is controlled at the transcriptional level (Richards and Cousins, 1975). It may involve a derepression of the metallothionein gene operon by zinc(II) or other heavy metals.

In most cases, metallothioneins are found to be of several distinct types with similar chain lengths of slightly different amino acid compositions. The amino acid sequence has been determined for metallothioneins from several different sources (Kojima *et al.*, 1976). Each of these proteins was found to have 19 (or 20) cysteine residues (out of a total 60 or 61 amino acid residues), and they were always at exactly the same relative positions. Figure 14-2 shows the amino acid sequence of human liver metallothionein-1. The cysteine residues are believed to bind such heavy metals such as Zn(II), Cd(II), Cu(II), and Hg(II).

A protein induced by $CuCl_2$ in rat liver is significantly different from other metallothioneins: it has a molecular weight of approximately 8000 and contains 14.6% cysteine and 6 gram atoms of copper per mole (Premakumar *et al.*, 1975). Rapid anaerobic addition of reduced apoprotein to sodium acetate buffer solution containing Cu(II) leads to an apparent reconstitution of the holoprotein. Because of the apparently significant differences, the authors suggest that this protein be called "copper chelatin." The rather low cysteine content implies that residues other than cysteine, probably histidine and/or lysine, contribute significantly to the binding of copper. Copper chelatin is also believed to be involved in the normal metabolism of copper.

14.3.5.3 Lead Inclusion Body (Goyer and Moore, 1974)

Nuclear inclusion bodies were first reported in 1936 in hepatic parenchymal and tubular lining cells of children dying of acute lead encephalopathy. Similar inclusion bodies have subsequently been reported in pigs, rabbits, chickens, and rats, particularly in the kidneys of the animals. It is of significance that a similar inclusion body has been observed in one type of plant, moss (Skaar et al., 1973). The nuclear inclusion bodies isolated from the kidneys of poisoned rats contained an average of 57 ppm (fresh basis) of lead, whereas the lead content averaged 0.8 ppm in the whole kidneys of the poisoned rats and only 0.009 ppm in the whole kidneys of the control rats (Goyer *et al.*, 1970).

As the name indicates, a nuclear inclusion body is a dense granular structure usually localized in cell nuclei but independent of the nucleoli. It has a dense central core and an outer fibrillar zone. The amino acid composition and solubility characteristics of inclusion body protein resemble those of the content of acidic fraction of proteins in normal nuclei. The protein has a high content of aspartic acid, glutamic acid, glycine, cysteine, and tryptophan. This composition is reminiscent of calcium-binding proteins. Inclusion bodies contain lesser amounts of calcium, iron, zinc, copper, and cadmium. The lead in an inclusion body is readily removable *in vitro* by metal-chelating agents such as EDTA.

It has been suggested that the formation of inclusion bodies serves as a protective mechanism. In the course of excretion by transtubular flow in renal tubular lining cells, a portion of the lead enters the nucleus, where it is sequestered by proteins and becomes nondiffusible. This mechanism thus succeeds in maintaining a relatively low lead level in the cytoplasm, thereby reducing the toxic effects of lead on sensitive cellular sites such as mitochondria.

14.3.5.4 Other Compartmentation Mechanisms (George, 1982)

Some intracellular as well as extracellular structures of compartments could sequester toxic elements and keep them from reaching susceptible parts of the cells. The nuclear inclusion body mentioned above is but an example of such compartmentation. Plant cell walls and animal hair could act as extracellular compartments to sequester toxic metals, as mentioned earlier. Calcium, together with either carbonate or phosphate, often forms intracellular (as well as extracellular) granules, which could take in such metals as Mn, Mg, Zn, and Pb. Bones and shells could act as such compartmentation agents in similar fashion.

Lysosomes are known to sequester Ag, Au, Cr, Cu, Fe, Hg, Ni, and Pb (George, 1982). The accumulation of hemosiderin in lysosomes as a degradation product of ferritin is a well-characterized example. High concentrations of copper have been reported in large vacuoles in hepatopancreas of *Crangon* species, *Assellus* species, *Austropotanobius* species (crayfish), and some *Sepia* species (cuttlefish) (George, 1982). These represent degradation products of hemocyanin and seem to act as a storage of copper required for resynthesis of hemocyanin (George, 1982).

Another granule associated with lysosomes isolated from the kidney of *Mytilus edulis* (mussel) contained high concentrations of Fe, Zn, Cd, Cu, Hg, and Pb. It was determined to be made up of insoluble lipopigment, lipofuscin (see Section 13.1.5 for a description of lipofuscin) (George, 1982).

14.3.5.5 Heavy Metal–Selenium Antagonism

Selenium compounds, particularly selenite, were first recognized as having protective properties against cadmium toxicity in testis necrosis, and were subsequently shown to inhibit cadmium-induced destruction of nonovulating ovaries and placenta, cadmium-induced lethal syndrome during pregnancy, and the teratogenic effects of cadmium (Parizek, 1976). On the other hand, injection of selenite has been shown to increase the level of cadmium in testis. Simultaneous oral administration of cadmium and selenium (selenite) led to an increased selenium intake and prevented cadmium-induced elevation of blood pressure. These results indicate that cadmium is not excluded by selenium from the target tissues, but rather that it is fixed into some innocuous form as a result of the presence of selenium. The increase of cadmium level in the testis as a result of selenium administration is associated with a two- or threefold reduction of cadmium in a protein fraction (probably metallothionein) and with a simultaneous increase in cadmium binding to a high-molecular-weight protein containing selenium (Chen *et al.*, 1974). A protein with a molecular weight of approximately 30,000 was isolated and shown to accumulate cadmium (Chen *et al.*, 1972). The binding of cadmium to high-molecular-weight proteins (presumably containing selenium) was also observed in the kidney and liver (Chen *et al.*, 1975).

Selenite also protects against mercury toxicity (Parizek, 1976). As in the case of cadmium, the administration of selenium along with Hg(II) leads to markedly increased levels of both selenium and Hg(II) in blood plasma. Tuna and bonito, being very active, tend to accumulate mercury to a high level. However, the fish themselves are healthy. It has been found that tuna contains a high level of selenium with the mole ratio Hg/Se being close to 1:1 (Ganther *et al.*, 1972). Tuna meat was shown to protect quail from the toxic effects of methylmercury (Ganther *et al.*, 1972). Subsequently, a linear relationship (mole ratio 1:1) was reported for the mercury and selenium contents in seals and other marine mammals (Koeman *et al.*, 1973) as well as mercury-mine workers (Kosta *et al.*, 1975).

These observations taken together suggest the formation of a Hg-Se compound of mole ratio 1:1 as the mechanism of detoxification. The simplest compound one could envisage would be HgSe, which is known to be highly insoluble and stable. Laboratory experiments have shown that the renal cortices, mesenteric lymph nodes, and livers of rats that received doses of $HgCl_2$ and Na_2SeO_3 were either black or dark gray (Groth et al., 1976). Microscopic studies further revealed black particles in the macrophages in the mesenteric lymph nodes, spleen, lungs, liver, and kidneys, and intranuclear inclusion bodies in the renal tubular cells (Groth et al., 1976). A neutron activation analysis showed 301 ppm of Hg and 152 ppm of Se in the liver of rats that were given mercury and selenium. This corresponds to a nominal mole ratio of Hg/Se = 1.5:1.9. It should be noted that this does not represent the composition of the black particles, but rather the average of the total liver contents (Groth et al., 1976).

Selenate was found to prevent the lethal toxic effects of thallium compounds. As with cadmium and mercury, selenite increased the Tl content in liver, kidney, and bone (Parizek, 1976).

NOTE: The So-Called ppm

1. Binding to Biomolecules

The amount of a trace element found in biological material is often expressed in units of ppm. This is normally calculated as grams of the element per 10^6 g of material, though it literally refers to parts per 10^6 parts. What we are interested in here is the extent of binding of a trace element that such ppm values indicate. Let us take the metallic elements Ca—Zn, Cd, and Hg as examples of binding elements, and proteins, polysaccharides, and polynucleotides as the binding matrix in a biological system. The corresponding atomic weights of the metallic elements are 40 (Ca), 65 (Zn), 112 (Cd), and 200 (Hg). We will approximate these values by 50 (for Ca—Zn), 100 (Cd), and 200 (Hg) for the sake of convenience of calculation. Proteins consist of amino acids, which have an average molecular weight of about 120. We assume it to be 100 for simplicity. Polysaccharides are composed of glucose (or generally, hexoses), glucuronic acid (a hexuronic acid), and/or sulfate esters of glucose. These units have molecular weights ranging from 160 to 250; we take 200 as the average molecular weight. The average molecular weight of nucleotide monomers is about 300.

Metallic element content in biological materials is often expressed using one of two different bases: either "dry" basis or the "fresh" basis. On the average, the water content of most biomaterials is about 80%. Therefore, the content of an element reported on a dry basis will be approximately five times that in the same material if calculated on a fresh basis. In the following discussion, we use values corresponding to a dry basis, where 1 ppm is equivalent to about 0.2 ppm expressed on a fresh basis.

Now let us consider the example of 1 ppm of a metallic element whose atomic weight is 50 contained in a largely protein sample. Then:

$$1 \text{ ppm} = 1 \text{ g element}/10^6 \text{ g sample} = (1/50) \text{ mole}/10^6 \text{ g sample}$$
$$= 1 \text{ mole}/5 \times 10^7 \text{ g} = 1 \text{ mole}/5 \times 10^5 \text{ moles of amino acid units}$$
$$= 1 \text{ mole of the element}/10^3 \text{ moles of protein*}$$
$$100 \text{ ppm} = 1 \text{ mole of the element}/5 \times 10^3 \text{ moles of amino acid units}$$
$$= 1 \text{ mole of the element}/10 \text{ moles of protein*}$$

The values with asterisks were obtained on the assumption that the average number of amino acid residues in a protein is 5×10^2 or equivalently, the average molecular weight of proteins is 5×10^4.

At 1 ppm, one metallic atom is bound to one out of 1000 protein molecules. If the sample's content of protein is only 50% (the other components being lipids, carbohydrates, and so on) and the metallic element is preferentially bound to proteins, then one out of 500 protein molecules is bound with the element.

A single cell of *E. coli* is considered to contain 2000 to 3000 different kinds of proteins, with a total of 10^6 individual protein molecules. On the average, there are only about 400 molecules present for each kind of protein. At 1 ppm, 2×10^3 molecules of protein (out of 10^6) in one cell incorporate the metallic element since, as indicated, one protein molecule in 500 is involved, assuming also that one atom of the element binds to one protein molecule. If the element is Cd or Hg, then the number of protein molecules bound with the element will be 10^3 (with Cd) or 500 (with Hg). This extent of binding of the metallic element could seriously disrupt the function of an entire cell, particularly if those proteins bound with the metal function as catalysts or play some similar role. A similar situation would be created by the intrusion of metallic elements in other cells including human body cells. On the other hand, the element may be readily tolerated if it is bound to a noncatalytic protein such as hair, or perhaps to some muscle protein.

Table 14-6 gives estimates of the extent of binding of metal atoms to various important cell components at a level of 1 ppm based on considerations like those above. Even at 1 ppm, the extent of metal binding is surprisingly large if it is bound exclusively to RNA or DNA. This is

Table 14-6. Magnitude of Metal Binding at 1 ppm Dry Basis in the Whole Cell[a]

Element (atomic weight)	Proteins	Carbohydrates	Polynucleotides
Ca–Zn (50)	(a) 1 mole/10^3 moles	1 mole/2.5×10^5 units of hexose	1 mole/1.7×10^5 units of nucleotide
	(b) 1 mole/5×10^2 moles	1 mole/2.5×10^4 units of hexose	1 mole/3.3×10^4 units of nucleotide
	(e) 2×10^3 molecules of proteins bind the element in a cell		(c) 1 mole/10 moles RNA (ribosome)
			(d) 15×10^2 moles/mole DNA
Cd (100)	(a) 1 mole/2×10^3 moles	1 mole/5×10^5 units of hexose	1 mole/3.3×10^5 units of nucleotide
	(b) 1 mole/10^3 moles	1 mole/5×10^4 units of hexose	1 mole/7×10^4 units of nucleotide
	(e) 10^3 molecules of proteins bind the element in a cell		(c) 1 mole/20 moles RNA
			(d) 800 moles/mole DNA
Hg (200)	(e) 500 molecules of proteins bind the element in a cell		(c) 1 mole/40 moles RNA
			(d) 400 moles/mole DNA

[a] Assumptions of calculation are: (a) metal binds to a sample that is exclusively protein, carbohydrate, or polynucleotide. The average molecular weight of the proteins was assumed to be 50,000. (b) Metal binds to the said fraction of a cell. The content of the fraction of a dried cell sample is protein 50%, carbohydrate 10%, and polynucleotide 20%. These values are for *E. coli*, but the composition in other cells may not be much different from this. (c) Ribosomal RNA is the major component of polynucleotides and metal binds exclusively with this RNA whose average number of nucleotide units is 3000. (d) DNA is about one-fifth of the total polynucleotide weight, and metal binds exclusively with DNA whose molecular weight is 3×10^9. (e) A cell contains approximately 10^6 molecules of proteins.

important because RNA is crucial to the synthesis of proteins, and DNA is the genetic material, disruption of which may cause the development of a tumor in the cell. Carbohydrates serve as energy sources and also have structural functions, and they may tolerate fairly extensive binding of metallic elements. Carbohydrates, however, usually have low affinity for metallic elements. The binding of metallic elements to carbohydrates of special types, such as mucopolysaccharides, can, however, be of significance in the metal defense mechanisms of plants and some invertebrates.

2. Adsorption onto the Cell Surface

In actual situations it is more likely that most metal cations are adsorbed on cell surfaces and only a minor portion actually enter the cells.

Assume that a cell is spherical with a radius r and density d. Then the weight of the cell w is $(4/3)d\pi r^3$ and its surface area is $4\pi r^2$. If the cell surface has one adsorption site every s units of area, then this cell could adsorb as many as $4\pi r^2/s$ molecules (atoms) of the metal. The maximum ppm attainable due to the adsorption of a metal element (of atomic weight m) on the cell surface would then be given by

$$\text{max ppm} = (4\pi r^2/s) \times 10^6/(4/3)\pi r^3 d = (3m/Nd)(1/sr) \times 10^6$$

where N is Avogadro's number. For the three typical m values considered in the preceding discussion, the corresponding maximum surface adsorption ppm values are given by the following formulas assuming $d = 1$ g/cm^3:

> for $m = 50$ (Ca–Zn), max ppm $= 10^{-15}/4sr$
> for $m = 100$ (Cd), max ppm $= 10^{-15}/2sr$
> for $m = 200$ (Hg), max ppm $= 10^{-15}/sr$

If we assume rather arbitrarily the values $s = 100$ Å$^2 = 10^{-14}$ cm^2 or 1000 Å$^2 = 10^{-13}$ cm^2, we obtain:

> (a) $s = 100$ Å^2, max ppm $= 10^{-1}/nr$
> (b) $s = 1000$ Å^2, max ppm $= 10^{-2}/nr$

where $n = 4$, 2, and 1 for Ca–Zn, Cd, and Hg, respectively. The size of a typical bacterial cell is on the order of $r = 10^{-4}$ cm; consequently in this case maximum possible ppm data for adsorbed metals would be:

> (a) 250 ppm (for $m = 50$), 500 ($m = 100$), 1000 ($m = 200$)
> (b) 25 ppm (for $m = 50$), 50 ($m = 100$), 100 ($m = 200$)

Whether an s value of 100 Å^2 or 1000 Å^2 is closer to reality is not known; it would certainly depend on the species involved as well as on the metal cation to be adsorbed. However, it may still be concluded that metal contents as high as 100 ppm can readily be accounted for solely by adsorption on cell surfaces, at least in the case of bacteria.

A typical eukaryotic cell is about 10 to 100 times as large as a typical bacterial cell, i.e., $r = 10^{-3}$–10^{-2} cm. Accordingly, the maximum ppm values attainable by cell surface adsorption here would be:

(a) 2.5–25 ppm (for $m = 50$), 5–50 ($m = 100$), 10–100 ($m = 200$)
(b) 0.25–2.5 ppm (for $m = 50$), 0.5–5 ($m = 100$), 1–10 ($m = 200$)

That is, as much as 10 ppm may be accounted for by surface adsorption in the case of singular eukaryotic cells. However, this kind of simplistic calculation cannot be applied to multicellular organisms, because of the complicated intercellular interactions.

Toxicity of Other Elements   15

The characteristic features of several elements will be reviewed in this chapter. Each of these elements has its own unique properties, the differences among them being much greater than those found among the heavy metals treated in the previous chapter. The elements to be discussed are selenium, arsenic, beryllium, aluminum, vanadium, chromium, manganese, and nickel. These are some of the more important toxic elements, and represent, altogether, wide enough a variety to allow us to delineate some general principles of the toxic effects of elements and the biological defense mechanisms against them. For a comprehensive review of the toxicities of all these and other elements, the reader is referred to several monographs and reviews (Frieden, 1985a; Luckey and Venugopal, 1977; Goyer and Mehlman, 1977; Friberg *et al.*, 1979).

15.1 Selenium

Selenium is one of the elements that have been and are being studied most intensively and extensively in recent years. There exist several reviews and monographs (Shamberger, 1983; Stadtman, 1974; Diplock, 1976; Ganther, 1974; Rosenfeld and Beath, 1964). The description in this section is based largely on the review by Diplock (1976).

15.1.1 Characteristic Features of the Chemistry and Metabolism of Selenium

This topic has already been treated in Section 3.6, particularly with regard to its oxidation–reduction, where it was shown that selenium can be expected to be much more reactive (in a thermodynamic sense) than sulfur. Moreover, H_2Se ($pK_a = 3.8$) is a considerably stronger acid than

397

H_2S (pK_a = 7.0). Accordingly, the pK_a of selenocysteine (pK_a = 5.24) is much lower than that of cysteine (pK_a = 8.25).

One of the characteristic reactions of selenium compounds is reported to be:

$$4GSH \text{ (glutathione)} + H_2SeO_3 \rightarrow GSSeSG + GSSG + 3HOH \quad (15\text{-}1)$$

This reaction occurs readily without the need of an enzyme. GSSeSG is subsequently reduced further by glutathione reductase as follows:

$$GSSeSG + NADPH + H^+ \xrightarrow{\text{glutathione reductase}}$$
$$GSH + GSSeH + NADP^+ \quad (15\text{-}2)$$

$$GSSeH + NADPH + H^+ \xrightarrow{\text{reductase}} GSH + HSeH + NADP^+ \quad (15\text{-}3)$$

This is believed to be a normal metabolic pathway in liver. The end product, H_2Se, is extremely toxic. The normal way in which H_2Se is detoxified is its conversion to a dimethyl or trimethyl derivative; that is:

$$HSeH + 2S\text{-adenosylmethionine} \xrightarrow{\text{methyl transferase}} (CH_3)_2Se, (CH_3)_3Se^+$$
$$(15\text{-}4)$$

$(CH_3)_2Se$, being volatile, is exhaled through the lungs, and $(CH_3)_3Se^+$ is the major chemical form released in the urinary excretory system.

H_2Se can also be incorporated into other organic compounds; e.g.,

$$HOCH_2CH(NH_2)COOH + HSeH \xrightarrow{-HOH}$$
$$HSeCH_2CH(NH_2)COOH \quad (15\text{-}5)$$

Selenomethionine can arise in an analogous manner to S-methionine:

$$CH_3-THF + HSeCH_2CH_2CH(NH_2)COOH \rightarrow$$
$$THF + CH_3SeCH_2CH_2CH(NH_2)COOH \quad (15\text{-}6)$$

15.1.2 Selenium Deficiency in Animals

It has been established that selenium at low levels is essential in various animals. Selenium deficiency results in characteristic diseases; examples include liver necrosis in rats and pigs, exudative diathesis (vascular lesions) in chicks, necrosis of heart, liver, kidney, skeletal muscle, pancreas, and testis in mice, white muscle disease in young lambs and calves, myopathy in turkeys and pigs, and dystrophy in sheep and cattle. Many hypotheses have been proposed to explain these selenium deficiency diseases, and it appears that a lack of selenium is manifested in two basic ways. One involves glutathione peroxidase, and the other involves a protein(s) containing acid-labile Se^{2-}.

15.1.2.1 Pathological States Related to Lack of Glutathione Peroxidase

Selenium deficiency leads to a reduced level of glutathione peroxidase. As discussed in Section 13.2.2, glutathione peroxidase protects biomembranes from peroxidative damage by helping to decompose lipid hydroperoxides.

In exudative diathesis the permeability of the peripheral capillaries is thought to be increased, with the result that fluid is exuded into the extracapillary tissue space causing the plasma to become depleted of proteins, particularly albumin. When vitamin E is absent, the integrity of the cell membranes is disrupted and the membranes become rather vulnerable to peroxidation. If glutathione peroxidase is present to a sufficient extent in the vicinity of a damaged membrane, lipid hydroperoxide will be disposed of and no further damage will take place. In the absence of glutathione peroxidase, however, the lipid hydroperoxide would be discharged into the lumen of the capillaries and aggravate the peroxidative damage to the capillary membranes. This, in turn, would lead to exudative diathesis.

Necrosis in liver, kidney, and other organs and tissues is considered to be caused by the absence of glutathione peroxidase with consequences similar to those outlined above. It seems probable that the normally functioning liver microsomal system (smooth endoplasmic reticulum) produces hydroperoxides regularly. This may be related to its function in the monooxidative metabolism of drugs and other substances, which is dependent on cytochrome P-450 (see Chapter 4).

15.1.2.2 Pathological States Related to Lack of Se^{2-}-Dependent Proteins

Selenium has been shown to be an integral part of the smooth endoplasmic reticulum; it exists there as acid-labile selenide. Here the function of selenium appears to be different from that associated with glutathione peroxidase. In fact, the smooth endoplasmic reticulum itself contains little or no glutathione peroxidase, but it has been shown that a selenium-dependent nonheme iron protein exists in the microsomal fraction.

The main characteristics of the pathology of white muscle disease are muscular degeneration and calcium accumulation. One clinical finding related to white muscle disease is a heightened level of plasma enzymes such as lactate dehydrogenase, malate dehydrogenase, glutamate-oxalo-acetate transaminase, creatine phosphokinase, and fructose dephospha-tase. The high plasma enzyme concentrations are presumed to be due to leaking of tissue enzymes into the plasma and may be explained in terms of a lack of glutathione peroxidase.

Comparative studies of ^{75}Se distribution in normal lambs and in lambs in the early stage of white muscle disease showed that in normal lambs, selenium was bound with a few different proteins in the kidney and with three different proteins in the plasma, liver, and pancreas. Four different selenium-binding proteins were found in the heart and semitendinous mus-cle of normal lambs, whereas only three selenium-binding proteins were found in these same tissues taken from lambs with the disease. The protein missing in the diseased lambs has a molecular weight of about 10,000 and is presumed to be involved in the prevention of white muscle disease. There is some indication that this protein contains 1 mole of heme (which is similar to cytochrome b_5) and 1 gram atom of selenium per mole. The structure of this interesting protein and the mechanism by which it pre-vents white muscle disease are not known. It has been suggested, how-ever, that the acid-labile Se^{2-} in this protein may be functioning as a component of an iron–sulfur unit with the selenium atom substituting for one of the sulfur atoms. Similar protein-bound Se was also observed in mitochondria.

15.1.3 Toxicity of and Defense against Selenium

15.1.3.1 In Plants (Ganther, 1974; Rosenfeld and Beath, 1964; Gunn, 1976)

Plants can be divided into three groups according to their abilities to accumulate selenium. Primary selenium accumulator plants contain se-lenium at as high as 3000 ppm. Examples of this group include many species of *Astragalus* and some species of *Machaeranthera, Haplopap-pus,* and *Stanleya.* Secondary selenium-absorber plants normally contain less than a few hundred ppm of selenium. Included in this group are many species of *Aster,* and some species of *Atriplex, Castelleja, Grindelia, Gutierrezia, Machaeranthera,* and *Mentzella.* Many other plants are non-accumulators, including weeds, crop plants, grains, and grasses; they rarely contain more than 30 ppm of selenium.

The selenium in primary selenium-accumulator plants is largely water-soluble and occurs in compounds of low molecular weight. Most of the selenium in the secondary selenium-absorber plants has been found to be SeO_4^{2-}; small amounts are present in organic forms.

Some of the organoselenium compounds isolated from plants warrant a close look. Se-methyl selenocysteine is the predominant soluble form of selenium in the primary accumulator plants. The biosynthesis of Se-methyl selenocysteine from selenite has been demonstrated in species of *Astragalus, Oonopsis condensata,* and *Stanleya pinnata.* This suggests that Se-methyl selenocysteine is the major metabolic product in the accumulator plants and that it is not toxic. Another compound found in *Astragalus* species is selenocystathionine, $[HOOCCH(NH_2)CH_2]_2Se$. Its formation from SeO_3^{2-} was demonstrated in *S. pinnata,* and it was identified as the cytotoxic compound in seed of *Lecythis ollaria.* Se-methyl selenomethionine is the predominant soluble organoselenium compound (biosynthesized from SeO_3^{2-}) in several of the species of *Astragalus* that do not accumulate selenium. Se-methionine, an analogue of methionine, would be incorporated into proteins. This would produce proteins with inappropriate characteristics. Thus, the compound could be toxic. Se-methyl selenocysteine, however, does not have a sulfur counterpart among the naturally occurring amino acids. Thus, it would not be expected to be incorporated into proteins. It is likely that the ability of the metabolic system to incorporate selenium into Se-methyl selenocysteine serves as a defense mechanism in the accumulator plants.

In nonaccumulator plants, soluble selenium compounds are injurious to seed germination and growth, producing dwarfism. A characteristic symptom of SeO_4^{2-} toxicity in cereal crops is a snow-white chlorosis. This may indicate that the SeO_4^{2-} destroys chloroplasts, resulting in a loss of pigments.

15.1.3.2 In Animals (Diplock, 1976; Gunn, 1976)

Acute selenium toxicity manifests itself in the liver, resulting in liver cirrhosis, which leads to death in a very short time. Two characteristic symptoms of chronic selenium toxicity are "blind staggers" and "alkali disease." The former occurs in cattle and sheep that graze on pastures containing moderate amounts of seleniferous plants such as *Astragalus* and *Stanleya.* Blind stagger disease is characterized by central nervous disorders. Alkali disease is seen in horses, cattle, sheep, and pigs that consume plants or grain containing protein-bound selenium; the clinical features of the disease include growth retardation, emaciation, deformities, shedding of hoofs, loss of hair, and arthritic disorders of the joints.

In experimental studies with laboratory animals, it was shown that

SeO_3^{2-} has a striking activity in increasing vascular permeability. Accordingly, it has been observed histologically that vascular manifestations (hemorrhage and parenchymatous degeneration and congestion) are very prevalent in acute selenium poisoning. These consequences could be attributable to reactions of the following type, similar to reaction 15-1:

$$4RSH + H_2SeO_3 \rightarrow RSSeSR + RSSR + 3HOH \qquad (15-7)$$

Since RSH molecules such as cysteine residues in proteins are integral parts of membranes, this reaction could result in a change of membrane conformation and a disruption of membrane integrity. This may be the basis for vascular permeability change and hence hemorrhage and other effects. The lesions characteristic of "alkali disease" are believed to be a result of chronic progressive degeneration caused similarly.

H_2Se gas is one of the most toxic substances known; it is absorbed through the respiratory system and is considered to enter the blood vascular system readily. It is probable that ingested HSe^- or Se^{2-}, being capable of binding with metallic cations, exerts inhibitory effects on a number of metalloenzymes and proteins.

Selenium compounds, especially Na_2SeO_3 and selenomethionine, have been shown to produce deformity in chick embryos (teratogenicity). The details of the mechanism of action are unknown, though the finding comes as no surprise in view of the wide-ranging effects of selenium compounds on membranes and other vital bioconstituents.

As mentioned earlier, the main means of detoxification of selenium in animals is its conversion to $(CH_3)_2Se$ and/or $(CH_3)_3Se^+$.

15.1.3.3 Interactions with Other Elements (Diplock, 1976; Gunn, 1976)

Interactions of selenium with cadmium and mercury are discussed in Section 14.3.5.5. It has been found that 5 ppm of As as Na_2AsO_3 in drinking water completely alleviated the liver damage in rats caused by 15 ppm of dietary selenium in the form of seleniferous wheat. It was also shown that subacute doses of arsenic decrease the exhalation of volatile selenium compounds while increasing the excretion of selenium through the gastrointestinal tract. The latter result is due to an enhanced biliary excretion of selenium caused by arsenic. Arsenite is more effective than arsenate in stimulating this biliary secretion. The mechanism of arsenic's action has not been delineated, but one possible explanation is as follows. Arsenic, both As(III) and As(V), is known to form a soluble sulfide complex ion:

$$2HAsO_2 + 3HSH \rightarrow As_2S_3 + 4HOH$$

$$As_2S_3 + S^{2-} \rightarrow 2AsS_2^- \qquad (15-8)$$

Similar reactions can be expected to occur between arsenite and selenide; perhaps $AsSe_2^-$ (or similar complex) may be even more stable than the corresponding sulfur complex. This may be the form excreted in the bile. Other elements capable of forming such complexes include Sb(III, V), Sn(IV), Ce(IV), V(V), Mo(VI), and W(VI).

15.2 Arsenic

15.2.1 Characteristic Features of the Chemistry of Arsenic

The important oxidation states of arsenic are $+V$, $+III$, and $-III$, with the typical inorganic species being arsenate AsO_4^{3-} (As(V)), arsenite AsO_2^- (As(III)), and arsine AsH_3 (As($-III$)). Arsenate is very similar to phosphate in its chemical properties. Whereas PO_2^- is rather obscure, AsO_2^- is well known and plays an important part in the metabolism of arsenic compounds. AsH_3 is analogous to PH_3 as well as NH_3.
Some relevant reduction potentials are as follows:

$$As^0/AsH_3 \qquad E_0 = -0.60 \text{ V (pH 0)} \qquad (15\text{-}9)$$

$$HAsO_2/As^0 \qquad E_0 = +0.25 \text{ V (pH 0)} \qquad (15\text{-}10)$$

$$H_3AsO_4 + 2H^+ + 2e \rightarrow H_3AsO_3 + HOH \quad E_0 = +0.56 \text{ V (pH 0)} \quad (15\text{-}11)$$

The last reaction, the reduction of As(V) to As(III), is very slow under normal conditions in the absence of a catalyst. The potential values cited are to be compared to those of the corresponding phosphorus systems: $P^0/PH_3 = -0.06$ V, $PO_3^{3-}/P^0 = -0.35$ V, $PO_4^{3-}/PO_3^{3-} = -0.50$ V (pH 0). The reduction of PO_4^{3-} or PO_3^{3-} is very difficult (in the thermodynamic sense) and hence the only important oxidation state of P is $+V$ (i.e., PO_4^{3-}), whereas both AsO_2^- ($+III$) and AsO_4^{3-} ($+V$) are readily accessible through oxidation–reduction; this explains the importance of the two oxidation states in metabolism.

In addition to these inorganic compounds, organoarsenicals must be brought into the picture when the toxicity of arsenic is considered, because of their widespread use in pesticides and other purposes. Examples include mono- or di-sodium methanearsonate (pesticide, herbicide) and 4-nitrophenyl arsenic acid (food preservative). Other interesting naturally occurring organoarsenicals include dimethylarsine $(CH_3)_2AsH$, trimethylarsine $(CH_3)_3As$, and methyl arsenic acid $CH_3H_2AsO_3$. These are considered to be metabolites in microorganisms such as fungi and *Methano-*

bacterium. The reaction sequence involving them has been revealed to be:

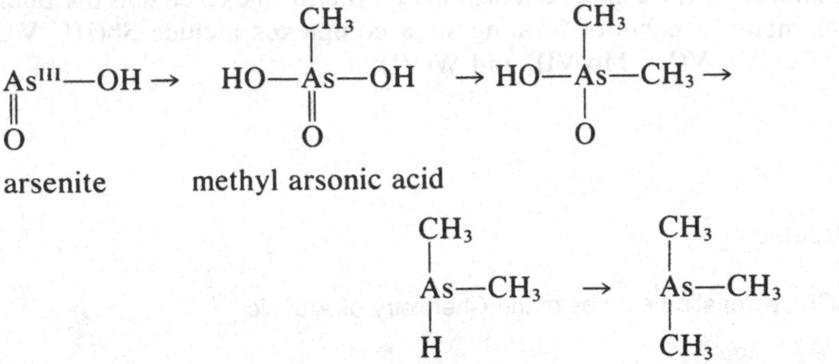

$$As^{III}{-}OH \rightarrow \quad HO{-}\underset{\underset{O}{\|}}{\overset{\overset{CH_3}{|}}{As}}{-}OH \quad \rightarrow HO{-}\underset{\underset{O}{\|}}{\overset{\overset{CH_3}{|}}{As}}{-}CH_3 \rightarrow$$

arsenite   methyl arsonic acid

$$\underset{\underset{H}{|}}{\overset{\overset{CH_3}{|}}{As}}{-}CH_3 \quad \rightarrow \quad \underset{\underset{CH_3}{|}}{\overset{\overset{CH_3}{|}}{As}}{-}CH_3 \qquad (15\text{-}12)$$

dimethylarsine trimethylarsine

These reactions seem to be carried out by *S*-adenosylmethionine as the source of CH_3^+ operating in conjunction with a reductase system (McBride *et al.*, 1978). Methylation of inorganic arsenic compounds has also been demonstrated in animals and humans (Fowler *et al.*, 1979).

15.2.2 Toxicity of and Defense against Arsenic

15.2.2.1 In Bacteria, Plants, and Invertebrates

Douglas fir trees grown in areas with high soil levels of arsenic accumulate the element in their bark and needles (Fowler, 1977). The trees themselves are apparently healthy. A strong competition was observed in the uptake of arsenic between the available phosphorus and the arsenic in the soil (Fowler, 1977). This fact can be interpreted as indicating that AsO_4^{3-}, being analogous to PO_4^{3-}, is taken up by the phosphate transport system in the roots, and that deposition of the arsenic in the bark or needles may serve to remove the toxic substance from sensitive regions to other insensitive regions of the tree. Some other plants exhibit similar behavior.

Many species of fish and shellfish, particularly crustaceans, are known to accumulate arsenic (Fowler, 1977). It is conceivable that, here also, AsO_4^{3-} is taken up along with PO_4^{3-}. Arsenic was found to be present in some marine organisms as a highly stable organoarsenical; the conversion of AsO_4^{3-} to AsO_2^- to this stable organoarsenical may serve as a detoxification mechanism.

A mechanism of defense against arsenate in bacteria is a reduced accumulation of arsenate by induced resistant cells (Silver, 1984). This seems to be brought about by a plasmid-governed active efflux of arsenate (Silver, 1984). This energy-dependent transport of arsenate has been speculated to be an ATPase system (Silver, 1984). This system must be highly specific for arsenate; otherwise the cell could lose the essential PO_4^{3-}.

15.2.2.2 In Animals—Cellular Mechanisms of Toxicity

AsO_4^{3-} and AsO_2^{-} are readily absorbed from the gastrointestinal tract, and are largely excreted in the urine (Fowler, 1977). Most of the administered arsenic is distributed to the liver and kidney. AsO_4^{3-} was reported to be reduced to AsO_2^{-} in the kidney and the excreted urinary arsenic was found to consist largely of methylated metabolites of arsenic, such as $(CH_3)_4As^{+}$ (Fowler, 1977). The toxicity behavior is known to depend heavily on the chemical form of the arsenic present.

Arsenate is known to be an uncoupler of mitochondrial oxidative phosphorylation. This is readily understandable, because AsO_4^{3-} competes with PO_4^{3-} for binding to ADP. It is accumulated by mitochondria, probably for the same reason. It can also replace PO_4^{3-} in DNA (and RNA); this may be the basic cause of the teratogenic effects of arsenic (Fowler, 1977). Likewise, AsO_4^{3-} inhibits normal DNA repair processes; this may explain its carcinogenicity.

Arsenite or its organic derivatives, possessing the $O{=}As^{III}$ unit, act as a sulfhydryl-trapping agent; that is:

$$As^{III}{=}O + 2RSH \xrightarrow{-HOH} As\begin{matrix} \diagup SR \\ \diagdown SR \end{matrix} \qquad (15\text{-}13)$$

For example, the pyruvate oxidase system is especially sensitive to As(III), because $O{=}As^{III}$ reacts with lipoic acid in the enzymatic system, blocking the essential SH groups. Arsenite inhibits the respiration of rat liver mitochondria; the respiratory system contains a number of sulfhydryl enzymes and proteins that are potentially subject to reaction with and inhibition by arsenite. Arsenite was also found to stimulate ATPase activity.

Organoarsenicals, such as derivatives of arsenic acid, are more potent inhibitors of cellular respiratory enzymes and serum cholinesterase than is arsenite, partially because of their lipid solubility. Other enzymes inhibited by organoarsenicals include pyruvate oxidase, decarboxylase, α-ketoglutarate oxidase, and malate oxidase.

15.3 Beryllium

15.3.1 Chemistry and Metabolism

The chemistry of beryllium is exclusively that of Be(II). Because of its extremely small ionic radius (r = 41 pm), the charge-to-radius ratio (which is related to the acidity of a cation; see Chapter 2) is very high: z/r = 95.4 nm^{-1}. This is much higher than that of Mg(II) (z/r = 74.8). Because of this high z/r value, Be(II) is readily hydrolyzed to form Be(OH)$_2$. Be(II) is stable only at a pH below 5; at neutral pH or basic pH, Be(II) precipitates. At a pH above 12, Be(II) redissolves, forming BeO$_2{}^{2-}$. Be(II) with its high z/r value is a so-called "hard acid" and prefers to bind with oxygen ligands, doing so with much more avidity than does Mg(II). The affinity of Be(II) for nitrogen ligands is weak, much weaker than that of Mg(II).

Be(II) salts are poorly absorbed by the gastrointestinal system of animals (Venugopal and Luckey, 1977; Reeves, 1979). This is largely due to the formation of the insoluble Be(OH)$_2$. Once absorbed from the gastrointestinal tract, through the lungs (inhaled beryllium compounds), or upon being administered intravenously, Be(II) distributes to bone, kidney, liver, and other tissues (Venugopal and Luckey, 1977). The main excretion path of orally ingested beryllium is, of course, through the feces, given the poor absorption characteristics noted above. Any Be(II) that is absorbed from the digestive tract, or that given parenterally, is excreted very slowly, if at all; only a minor portion is found in the urine, and fecal excretion persists over a long period of time. The excretion of inhaled beryllium in humans is mainly urinary and it, too, is very slow (Venugopal and Luckey, 1977).

15.3.2 Toxicity

Symptoms of acute beryllium toxicity in animals (including humans) include skin ulcers and inflammation of the lungs, nose, and the mucous membranes of trachea and bronchi (Tepper, 1972; Venugopal and Luckey, 1977; Reeves, 1979). Other effects are edema, hemorrhage with mild edema of the brain, and inflammation of the liver. Some of the symptoms of chronic beryllium toxicity in animals caused by exposure to airborne beryllium are granulomatous lung disease, enlargement of the liver and spleen, pulmonary dysfunction, heart enlargement, cyanosis, malignant tumors, cellular infiltration in the interstices of various organs and tissues, and calcific inclusions in cells and tissues (Venugopal and Luckey, 1977).

The pathogenesis of these toxic effects of beryllium is still unclear. A basic mechanism may, however, be the substitution of Be(II) for Mg(II) in a number of enzymatic systems. Mg(II) ion is an activator for a great variety of enzymes. Because of its higher z/r ratio, Be(II) may bind to such enzymes more strongly than Mg(II), hence disrupting the enzymatic activities. Enzymes that have been shown to be inhibited by Be(II) *in vitro* include alkaline phosphatase, pyrophosphatase, phosphoglucomutase, ATPases, amylase, enolase, RNA polymerases, and chymotrypsin (Venugopal and Luckey, 1977). Of particular significance is the disruption of DNA synthesis (Witshi, 1970). Be(II) tends to accumulate specifically in nuclei and to inhibit the incorporation of $[2\text{-}^{14}C]$thymidine into DNA. DNA polymerase is a zinc enzyme that requires Mg(II) for activity. Perhaps this is the basis for the inhibitory effect of Be(II) on the mitotic activity of some cultures (Bassler, 1965) and also for the carcinogenicity of beryllium (Reeves and Vorwald, 1967).

15.4 Aluminum

15.4.1 Chemistry and Metabolism

The aqueous chemistry of Al(III) is similar to that of Be(II); here, too, the charge-to-radius ratio $z/r = 134$ nm^{-1} is quite high. Al(III) is stable only below pH 4, but precipitates as Al(OH)$_3$ at pH 4–10; above pH 10 it redissolves as AlO$_2{}^-$. It forms a very insoluble phosphate salt, AlPO$_4$.

Because of the formation of insoluble Al(OH)$_3$ or AlPO$_4$, the gastrointestinal absorption of Al(III) in animals is very poor. Following absorption, aluminum is distributed mainly in the bones, liver, testis, kidney, and brain as well as in other soft tissues, though at much lower levels. The excretion of aluminum in mammals is mainly fecal, representing both unabsorbed aluminum and the aluminum excreted in bile into the gastrointestinal tract. Only small amounts of aluminum are excreted in the urine under normal conditions. Excess aluminum in blood, however, seems to be excreted efficiently by the urinary system.

15.4.2 Toxicity in Animals

Aluminum compounds are generally inert toxicologically, largely because of their poor absorption. They are, of course, more toxic when injected parenterally. Some of the acute toxic effects of aluminum are gastrointestinal disturbance, skin lesions, and nervous afflictions. One cause of aluminum toxicity is the formation of the insoluble $AlPO_4$; this results in a lowered phosphate absorption from the digestive tract, and hence in growth retardation. Absorbed Al(III) can affect phosphate metabolism in tissues through the same mechanism, i.e., the formation of $AlPO_4$. For example, it was observed that the incorporation of ^{32}P into phospholipid, RNA, and DNA was decreased in the liver, kidneys, and spleen of rats following chronic and acute intoxication by $AlCl_3$. ATP levels also decreased concomitant with an increase in ADP and AMP levels. These effects can be understood in terms of the removal of PO_4^{3-} in the form of $AlPO_4$.

High concentrations of aluminum in the brain were reported in aluminum ball mill operators, in patients undergoing kidney dialysis (who were treated with large quantities of aluminum hydroxide phosphate-binding gels), and in those with Alzheimer's disease (a kind of senile dementia) (Venugopal and Luckey, 1977; Mayor *et al.*, 1977). These subjects all exhibited a progressive encephalopathy characterized by severe motor and behavioral abnormalities (Mayor *et al.*, 1977). In experimental studies of $AlCl_3$-induced encephalopathy in rats, it was shown that selective impairment of short-term memory and associated learning occurs first, whereas the brain later suffers neurofibrillary degeneration (Venugopal and Luckey, 1977). An interesting observation on rats is that parathyroid hormone increases gastrointestinal aluminum absorption and alters its tissue distribution (Mayor *et al.*, 1977). Administration of parathyroid hormone significantly increased (up to five times, compared to controls) aluminum levels in bones, muscle, kidney, and especially, the gray matter of brain, particularly under hyperparathyroid conditions. The mechanism of these toxic effects of aluminum is yet to be studied.

15.4.3 Toxicity in Organisms Other Than Animals

Aluminum toxicity data for aquatic organisms were extensively tabulated by Burrows (1977). Toxic effects on algae were observed in the form of growth retardation and death of cells. The sensitivity to aluminum differs greatly from one species to another. It has been noted that desmids, which are more resistant to aluminum than other green algae, proliferate in oligotrophic waters and thus it may be speculated that they evolved under circumstances that favored the ability to tolerate high concentrations of metals freshly leached from adjacent rocks and soils. The tolerance mechanism itself very probably involves differentiated cell wall structure (Burrows, 1977).

Toxic effects of aluminum are known for a variety of bacteria, resulting in some unusual observations. Aluminum at 10 μg/liter altered the course of fermentation in *Aerobacter aerogenes* and stimulated the production of certain B vitamins. At levels below 32 mg/liter, aluminum limited the production of flagella in *Bacillus megaterium*. Aluminum toxicity has also been observed in a number of protozoa, invertebrates, and vertebrates (Burrows, 1977).

The characteristic feature of aluminum toxicity in fish is gill hyperplasia (a swollen, congested condition), and the ultimate cause of death is usually anoxia caused by the damage to the gill. This effect seems to be exerted most severely by aluminate (AlO_2^-). Such congestion of the gill was also observed as one of the toxic effects of heavy metals (Section 14.2.2). In the latter case, the congestion seems to be caused by increased secretion of a mucous material that is able to bind cations. The mechanism of the congestion of the gill caused by the anion AlO_2^- is not clear.

15.5 Transition Metals

15.5.1 Vanadium (Byerrum, 1974)

The chemical basis of the biological function of vanadium and the metabolism of vanadium are discussed in Sections 7.4 and 12.4.6.

Rats exposed to V_2O_3 fumes showed the symptoms of supurative bronchitis, septic bronchopneumonia, pulmonary emphysema, and moderate interstitial pulmonary sclerosis. VCl_3 administered to rats and rabbits caused protein and fatty dystrophies of cells in the liver, kidney, and myocardium, partial necrosis of the tissues of some organs, and reduction in the RNA and DNA content of cells of the liver, kidney, myocardium, stomach, intestine, and lung.

15.5.2 Chromium (Baetjer, 1974)

The chemical bases of the biological functions of chromium and its metabolism are discussed in Sections 7.3 and 12.4.4. It was mentioned in Section 7.3 that chromium at low concentrations is beneficial to many plant species. At higher concentrations, chromium in the same plants is toxic, however. For example, Cr(VI) in nutrient solution at 5–10 ppm produced iron chlorosis in oat plants. At 15–50 ppm it retarded the growth of oat plants, as well as tomato, potato, and others. Tobacco leaves grown on a soil with high chromium content showed no apparent toxic signs at chromium levels of 14 ppm (dry weight basis) but toxic effects were visible at 18–34 ppm. In the roots, concentrations as high as 175 ppm were harmless, but toxicities were evident at 375–400 ppm. This latter fact suggests the presence of some tolerating mechanism in the roots. Moss (*Hypnum cupressiforme*) near a ferroalloy plant was found to contain as much as 12,000 ppm of chromium without sign of damage.

In experimental animals and humans exposed to chromate fumes, the typical symptoms are ulceration and perforation of the nasal system, and respiratory cancer due to the strong oxidizing effect of Cr(VI) compounds. On the other hand, the toxicity of Cr(III) compounds is very low. Their poor absorbability from the gastrointestinal tract is the main reason. $CrCl_3$ administered orally causes vomiting and diarrhea. Cr(VI) compounds are rather readily absorbed because of the similarity between CrO_4^{2-} and SO_4^{2-} and they cause kidney lesions.

15.5.3 Manganese (Lieben, 1973)

Several manganese-dependent enzymes are discussed in Chapters 4 and 6, and the essentiality and metabolism of manganese are dealt with in Chapter 11 and Section 12.4.2.

Manganese is required universally, especially in the photosynthetic apparatus (PSII) of plants. Normal manganese levels in plant tops range from 20 to 3000 ppm, most often being 50–500 ppm, and the values are heavily dependent on the species. Manganese toxicity in plants is characterized by marginal chlorosis, which in turn is associated with localized manganese accumulation. The toxic level of manganese varies widely and also depends on the species. Manganese toxicity can, in general, be reduced by increasing the concentrations of other cations such as iron, zinc, calcium, and magnesium.

Two major manganese toxicity defense mechanisms used by plants are decreased uptake or trapping the excess in roots or other parts. For example, the tolerance against excess manganese by *Trifolium subterraneum* is attributed to both mechanisms. Azalea plants have been reported to deposit excess manganese in their woody parts.

Manganese is among the least toxic of the trace elements with respect to animals. It has been suggested that there is a connection between the metabolic modifications observed in Parkinson's disease and chronic manganese poisoning. The concentration of dopamine (one of the neurotransmitter catecholamines) in the brain of monkeys to which manganese was injected subcutaneously was found to be significantly lower than that in the controls. A correlation was observed to exist between the extent of the reduction of the dopamine content and the degree of toxicity, as determined by observed neurologic abnormalities. No reduction was seen in the levels of other catecholamines such as epinephrine and serotonin. This lower dopamine content is analogous to the situation found in Parkinsonism.

15.5.4 Nickel (Sunderman, 1975)

The metabolism of nickel is dealt with in Section 12.4.5. Inorganic Ni(II) salts are rather well tolerated by animals when given orally. This is mainly due to their poor absorption by the gastrointestinal tract. However, when administered intravenously or subcutaneously, nickel salts are highly toxic. For example, a single intravenous dose of 10–20 mg/kg (of $NiCl_2$) was found to be lethal to dogs.

Nickel carbonyl $Ni(CO)_4$ is highly toxic when inhaled, being about 100 times as toxic as CO. The target tissue of $Ni(CO)_4$ toxicity is the pulmonary parenchyma, a result being edema with focal hemorrhage and hyperplasia of the bronchiolar epithelium and of the alveolar lining cells. The nuclei of the affected cells become enlarged and are found to contain numerous dense granules, and atypical mitoses are frequent. Death usually occurs within 3–5 days.

Nickel is a typical carcinogen. Readers are referred to two articles (Loeb and Zakour, 1980; Loeb and Mildvan, 1981) on the effects of metal ions on the fidelity of enzymatic DNA or RNA polymerization reactions as a possible cause of carcinogenicity by metal ions (see Section 5.3.1.3).

References

Abbound, M. M., Sim, W. J., Loeb, L. A. and Mildvan, A. S., 1978. Apparent suicidal inactivation of DNA polymerase by adenosine-2′,3′-riboepoxide 5′-triphosphate, *J. Biol. Chem.* 253:3415–3421.

Abeles, R. H., and Dolphin, D., 1976. The vitamin B_{12} coenzyme, *Acc. Chem. Res.* 9: 114–120.

Abramowicz, D. A., and Dismukes, G. C., 1984. Manganese proteins isolated from spinach thylakoid membranes and their role in oxygen evolution, I and II, *Biochim. Biophys. Acta* 765:309–317,318–328.

Adams, M. J., Buehner, M., Chandrasekhar, K., Ford, G. C., Hackert, M. L., Liljas, A., Rossman, M. G., Smiley, I. E., Allison, W. S., Everse, J., Kaplan, N. O., and Taylor, S. S., 1973. Structure–function relationship in lactate dehydrogenase, *Proc. Natl. Acad. Sci. USA* 70:1968–1792.

Addadi, L., and Weiner, S., 1985. Interactions between acidic proteins and crystals: Stereochemical requirements in biomineralization, *Proc. Natl. Acad. Sci. USA* 82:4110–4114.

Adman, E. T., and Jensen, L. H., 1981. Structural features of azurin at 2.7 Å resolution, *Isr. J. Chem.* 21:8–12.

Agnew, W. S., 1984. Voltage-regulated sodium channel molecules, *Annu. Rev. Physiol.* 46: 517–530.

Aisen, P., 1977. Transport of iron by transferrin, in *Bioinorganic Chemistry-II* (K. Raymond, ed.), American Chemical Society, Washington, D.C., pp. 104–124.

Aisen, P., and Leibman, A., 1973. Role of the anion-binding site of transferrin in its interaction with the reticulocyte, *Biochim. Biophys. Acta* 304:797–804.

Aisen, P., and Listowsky, I., 1980. Iron transport and storage proteins, *Annu. Rev. Biochem.* 49:357–393.

Aisen, P., Leibman, A., and Zweier, J., 1978. Stoichiometric and site characteristics of the binding of iron to human transferrin, *J. Biol. Chem.* 253:1930–1937.

Alberts, B., Bray, D., Lewis, J., Raff, M., Roberts, K., and Watson, J. D., 1983. *Molecular Biology of the Cell*, Garland, New York, Chapter 13.

Allen, R. C., 1975. Role of pH in the chemiluminescent response of the myeloperoxide–halide–hydrogen peroxide antimicrobial system, *Biochem. Biophys. Res. Commun.* 63:684–691.

Allen, R. C., Yevich, S. J., Orth, R. S., and Steele, R. H., 1974. Superoxide anion and singlet oxygen, their role in the microbicidal activity of the polymorphonuclear leukocyte, *Biochem. Biophys. Res. Commun.* 60:909–917.

413

Almassy, R. J., Janson, C. A., Hamlin, R., Xuong, N.-H., and Eisenberg, D., 1986. Novel subunit–subunit interactions in the structure of glutamine synthetase, *Nature* 323:304–309.

Ameyama, M., Nonobe, M., Shinagawa, E., Matsushita, K., Takimoto, K., and Adachi, O., 1986. Purification and characterization of the quinoprotein D-glucose dehydrogenase apoenzyme from *Escherichia coli*, *Agric. Biol. Chem.*, 50:49–57.

Amsler, P. E., and Sigel, H., 1976. Hydrolysis of nucleoside phosphate. V. Comparison of the metal-ion promoted dephosphorylation of the 5'-triphosphates of adenosine, inosine, guanosine and cytidine by manganese(+2), nickel(+2) and zinc(+2) in binary and ternary complexes, *Eur. J. Biochem.* 63:569–589.

Anderson, A. M., and Mertz, W., 1977. Glucose tolerance factor: An essential dietary agent, *Trends Biochem. Sci.* 2:277–279.

Anderson, O. R., 1981. Radiolarian fine structure and silica deposition, in *Silicon and Siliceous Structure in Biological Systems* (T. L. Simpson and B. E. Volcani, eds.), Springer-Verlag, Berlin, pp. 347–379.

Anderson, O. S., 1984. Gramicidin channels, *Annu. Rev. Physiol.* 46:531–548.

Anderson, W. P., 1975. Long distance transport in roots, in *Ion Transport in Plant Cells and Tissues* (D. A. Baker and J. L. Hall, eds.), North-Holland E.O., pp. 231–266.

Anderson, A. Forsen, S., Thulin, E., and Vogel, H. J., 1983. Cadmium-113 nuclear magnetic resonance studies of proteolytic fragments of calmodulin: Assignment of strong and weak cation binding site, *Biochemistry* 22:2309–2313.

Antonini, E., Brunori, M., Colosimo, A., Greenwood, C., and Wilson, M. T., 1977. Oxygen pulsed cytochrome oxidase: Functional properties and catalytic relevance, *Proc. Natl. Acad. Sci. USA* 74:3128–3132.

Antonio, M. R., Teo, B.-K., Orme-Johnson, W. H., Nelson, M. J., Groh, S. E., Lindahl, P. A., Kauzlarich, S. M., and Averill, B. A., 1982. Iron EXASFS of the iron-molybdenum cofactor of nitrogenase, *J. Am. Chem. Soc.* 104:4703–4705.

Antonovics, J., Bradshaw, A. O., and Turner, R. G., 1971. Heavy metal tolerance in plants, *Adv. Ecol. Res.* 7:1–85.

Appleton, D. W., and Sarker, B., 1971. The absence of specific copper(II)-binding site in dog albumin, a comparative study of human and dog albumins, *J. Biol. Chem.* 246: 5040–5046.

Arceneaux, J. E. L., Davis, W. B., Dower, D. N., Haydon, A. H., and Byers, B. H., 1973. Fate of labeled hydroxamates during iron transport from hydroxamate-iron chelates, *J. Bacteriol.* 115:919–927.

Arciero, D. M., Lipscomb, J. D., Huynh, D. H., Kent, T. A., and Munck, E., 1983. EPR and Mössbauer studies of protocatechuate 4,5-dioxygenase, *J. Biol. Chem.* 258: 14981–14991.

Armstrong, C. M., 1981. Sodium channels and gating currents, *Physiol. Rev.* 61:644–683.

Ashida, J., 1965. Adaptation of fungi to metal toxicants, *Annu. Rev. Phytopathol.* 3:153–174.

Atkin, C. L., Thelander, L., Reichard, P., and Lang, G., 1973. Iron and free radicals in ribonucleotide reductase, *J. Biol. Chem.* 248:7464–7472.

Averill, B. A., 1983. Fe-S and Mo-Fe-S clusters as models for the active site of nitrogenase, *Struct. Bonding* 53:59–103.

Babior, B. M., Kipness, R. S., and Curnutte, J. T., 1973. Biological defense mechanisms, production by leukocytes of superoxide, a potential bactericidal agent, *J. Clin. Invest.* 52:741–744.

Babu, Y. S., Sack, J. S., Greenhough, T. J., Bugg, C. E., Means, A. R., and Cook, W. J., 1985. Three-dimensional structure of calmodulin, *Nature* 315:37–40.

Baetjer, A. M. (ed.), 1974. *Chromium*, National Academy of Sciences, Washington, D.C.

Barber, D. A., and Shone, M. G. T., 1966. Absorption of silica from aqueous solutions by plants, *J. Exp. Bot.* 17:569–578.

Barber, J., 1984. Has the manganese protein of the water-splitting reaction of photosynthesis been isolated? *Trends Biochem. Sci.* 9:79–80.

Bassler, R., 1965. Effects of beryllium on synthesis of nuclear proteins and deoxyribonucleic acids in rat fibroblasts, *C. R. Soc. Biol.* 159:1620–1623.

Beinert, H., Emptage, M. H., Dreyer, J.-L., Scott, R. A., Hahn, J. E., Hodgson, K. O., and Thomson, A. J., 1983. Iron–sulfur stoichiometry and iron–sulfur clusters in three-iron proteins: Evidence for 3Fe–4S clusters, *Proc. Natl. Acad. Sci. USA* 80:393–396.

Bell, R. M., and Koshland, D. E., 1971. Covalent enzyme–substrate intermediate, *Science* 172:1253–1256.

Benemann, J. R., Smith, G. M., Kostel, P. J., and McKenna, C. E., 1973. Tungsten incorporation into *Azotobacter vinelandii* nitrogenase, *FEBS Lett.* 29:219–221.

Bereman, R. D., Ettinger, M., Kosman, D. J., and Kurland, R. J., 1977. Characterization of the copper(II) site in galactose oxidase, *Adv. Chem. Ser.* 162:263–280.

Berg, J. M., and Holm, R. H., 1982. Structures and reaction of iron–sulfur protein clusters and their synthetic analogs, in *Iron–Sulfur Proteins* (T. G. Spiro, ed.), Wiley, New York, pp. 1–66.

Berg, J. M., and Holm, R. H., 1984. Synthetic approach to the mononuclear active sites of molybdoenzymes: Catalytic oxygen atom transfer reactions by oxymolybdenum(IV, VI) complexes with saturation kinetics and without molybdenum(V) dimer formation, *J. Am. Chem. Soc.* 106:3035–3036.

Berridge, M. J., 1982. Phospholipids and cellular signaling, in *Calcium and Cell Function* (W. Y. Cheung, ed.), Vol. III, Academic Press, New York, pp. 1–36.

Berridge, M. J., 1985. Calcium mobilizing receptors: Membrane phosphoinositides and signal transduction, in *Calcium in Biological Systems* (R. P. Rubin, G. B. Weiss, and J. W. Putney, Jr., eds.), Plenum Press, New York, pp. 37–44.

Bertini, I., Luchinat, C., and Scozzafava, A., 1982. An insight into the zinc binding site and into the active cavity through metal substitution, *Struct. Bonding* 48:45–92.

Bielski, B. H., 1983. Evaluation of the reactivities of HO_2/O_2^- with compounds of biological interest, in *Oxy Radicals and Their Scavenger Systems* (G. Cohen and R. A. Greenwald, eds.), Vol. 1, Elsevier, Amsterdam, pp. 1–7.

Biggs, W. R., and Swinehart, J. H., 1976. Vanadium in selected biological systems, in *Metal Ions in Biological Systems* (H. Sigel, ed.), Vol. 6, Dekker, New York, pp. 141–196.

Blake, C. C. F., 1975. X-ray studies of glycolytic enzyme, *Essays Biochem.* 11:37–79.

Blaustein, M. P., 1985. Intracellular calcium as a second messenger: What's so special about calcium? in *Calcium in Biological Systems* (R. P. Rubin, G. B. Weiss, and J. W. Putney, Jr., eds.), Plenum Press, New York, pp. 23–33.

Boheim, G., 1974. Statistical analysis of alamethicin channels in black lipid membranes, *J. Membr. Biol.* 19:277–303.

Boheim, G., Kolb, H.-A., Bamberg, E., Appel, H.-J., Alpes, H., and Lauger, P., 1977. Gating processes in lipid membrane: Studies on the alamethicin and the gramicidin channel, in *Electrical Phenomena at the Biological Membrane Level* (E. Roux, ed.), Elsevier, Amsterdam, pp. 289–308.

Bolin, B., 1970. The carbon cycle, *Sci. Am.* 1970(9):125–132.

Borg, D. C., and Schaich, K. M., 1983. Reactions connecting autoxidation with oxy radical production, lipid peroxidation and cytotoxicity, in *Oxy Radicals and Their Scavenger Systems* (G. Cohen and R. A. Greenwald, eds.), Vol. 1, Elsevier, Amsterdam, pp. 122–129.

Bowen, J. E., 1969. Absorption of copper, zinc and manganese by sugarcane leaf tissue, *Plant Physiol.* 44:255–261.

Boyer, P. D., 1977. Coupling mechanisms in capture, transmission and use of energy, *Annu. Rev. Biochem.* 46:957–966.

Bradley, F. C., Lindstedt, S., Lipscomb, J. D., Que, L., Jr., Roe, A. L., and Rundgren, M., 1986. 4-Hydroxyphenylpyruvate dioxygenase is an iron-tyrosinate protein, *J. Biol. Chem.* 261:11693–11696.

Brady, F. O., 1976. The search for copper in L-tryptophan 2,3-dioxygenase, in *Iron and Copper Proteins* (K. T. Yasunobu, H. F. Mower, and O. Hayaishi, eds.), Plenum Press, New York, pp. 374–381.

Brady, F. O., and Udam, A., 1976. Is indoleamine 2,3-dioxygenase another heme and copper containing enzyme? in *Iron and Copper Proteins* (K. T. Yasunobu, H. F. Mower, and O. Hayaishi, eds.), Plenum Press, New York, pp. 343–353.

Bragg, P. D., 1974. Non-heme iron in respiratory chains, in *Microbial Iron Metabolism* (J. B. Neilands, ed.), Academic Press, New York, pp. 303–348.

Branden, C.-I., Jornvall, H., Eklund, H., and Furugren, B., 1975. Alcohol dehydrogenase, in *The Enzymes*, 3rd ed. (P. D. Boyer, ed.), Vol. XI, Part A, Academic Press, New York, pp. 103–109.

Bray, R. C., 1973. Problems in studying the role of molybdenum in xanthine oxidase, in *Chemistry and Uses of Molybdenum* (P. C. H. Mitchell, ed.), Climax Molybdenum Co., Ann Arbor E.O., pp. 216–223.

Breslow, R., Chin, J., Hilbert, D., and Trainor, G., 1983. Evidence for the general base mechanism in carboxypeptidase A catalyzed reactions: Partitioning studies on nucleophiles and $H_2^{18}O$ kinetic isotope effects, *Proc. Natl. Acad. Sci. USA* 80:4585–4589.

Briggs, R. W., and Armitage, I. M., 1982. Evidence for site selective metal binding calf liver metallothionein, *J. Biol. Chem.* 257:1259–1262.

Brown, D. G., Beckman, L., Asby, C. H., Vogel, G. C., and Reinprech, J. T., 1977. Oxygen-dependent ring cleavage in a copper coordinated catechol, *Tetrahedron Lett.* 1977(16): 1363–1364.

Brown, D. H., 1976. Mineral uptake by lichens, in *Lichenology: Progress and Problems* (D. H. Brown, D. L. Hawksworth, and R. H. Bailey, eds.), Academic Press, New York, pp. 419–439.

Brown, E. B., 1975. Transferrin–erythroblast interaction, in *Iron Metabolism and Its Disorders* (H. Kief, ed.), Excerpta Medica, Amsterdam, pp. 71–79.

Brown, J. M., Powers, L., Kincaid, B., Larrabee, J. A., and Spiro, T. G., 1980. Structural studies of the hemocyanin active site. I. Extended x-ray absorption fine structure analysis, *J. Am. Chem. Soc.* 102:4210–4216.

Bruening, R. C., Oltz, E. M., Furukawa, J., Nakanishi, K., and Kustin, K., 1985. Isolation and structure of tunichrome B-1, a reducing blood pigment from the tunicate *Ascidia nigra*, *J. Am. Chem. Soc.* 107:5298–5300.

Bryan, G. W., 1974. Adaptation of an estuarine polychaete to sediments containing high concentration of heavy metals, in *Pollution and Physiology of Marine Organisms* (F. J. Vernberg and W. B. Vernberg, eds.), Academic Press, New York, pp. 123–135.

Bryan, S. A., 1981. Heavy metals in cell's nucleus, in *Metal Ions in Genetic Information Transfer* (G. L. Eichhorn and L. G. Marzilli, eds.), Elsevier, Amsterdam, pp. 87–101.

Buckingham, D. A., 1977. Metal-OH and its ability to hydrolyze (or hydrate) substrates of biological interest, in *Biological Aspects of Inorganic Chemistry* (A. W. Addison, W. R. Cullen, D. Dolphin and B. R. James, eds.), Wiley–Interscience, New York, pp. 141–196.

Bull, C., Ballou, D. P., and Ohtsuka, S., 1981. The reaction of oxygen with protocatechuate 3,4-dioxygenase from *Pseudomonas putida*, *J. Biol. Chem.* 256:12681–12686.

Bunn, H. F., Forget, B. G., and Ranney, H. M., 1977. *Human Hemoglobin*, Saunders, Philadelphia.

Burgmayer, S. J. N., and Stiefel, E. I., 1985. Molybdenum enzymes, cofactors, and model systems, *J. Chem. Educ.* 62:943–953.

Burrows, W. D., 1977. Aquatic aluminum: Chemistry, toxicology and environmental prevalence, *Crit. Rev. Environ. Control* 7:167–216.

Burton, G. W., and Ingold, K. U., 1984. α-Carotene: An unusual type of lipid antioxidant, *Science* 224:569–573.

Burton, G. W., Cheeseman, K. H., Doba, T., Ingold, K. U., and Slater, T. F., 1983. Vitamin E as an antioxidant *in vitro* and *in vivo*, in *Biology of Vitamin E*. R. Porter and J. Whelan, eds.), Pitman, New York, pp. 4–14.

Byerrum, R. U. (ed.), 1974. *Vanadium*, National Academy of Sciences, Washington, D.C.

Byers, B. R., 1974. Iron transport in gram-positive and acid-fast bacilli, in *Microbial Iron Metabolism* (J. B. Neilands, ed.), Academic Press, New York, pp. 83–105.

Cairns-Smith, A. G., 1982. *Genetic Takeover and the Mineral Origins of Life*, Cambridge University Press, London.

Cairns-Smith, A. G., 1985. The first organisms, *Sci. Am.* 1985(6):90–100.

Campbell, A. K., 1983. *Intracellular Calcium, Its Universal Role as Regulator*, Wiley, New York.

Campbell, J. A., and Whiteker, R. A., 1969. A periodic table based on potential–pH diagrams, *J. Chem. Educ.* 46:90–92.

Cantley, L. C., 1981. Structure and mechanism of the (Na,K)-ATPase, *Curr. Top. Bioenerg.* 11:201–237.

Carroll, K. G., Spinelli, F. R., and Goyer, R. A., 1970. Electron probe microanalyser localization of lead in kidney tissues of poisoned rats, *Nature* 227:1056.

Carter, C. W., Jr., Kraut, J., Freer, S. T., Alfen, R. A., Sieker, L. G., Adman, E., and Jensen, L. H., 1972. A comparison of Fe₄S₄ clusters in high-potential protein and in ferredoxin, *Proc. Natl. Acad. Sci. USA* 69:3526–3529.

Challenger, F., 1945. Biological methylation, *Chem. Rev.* 36:315–361.

Chance, B., 1977. Electron transfer: Pathways, mechanisms and controls, *Annu. Rev. Biochem.* 46:967–980.

Chance, B., Saronio, C., and Leigh, J. S., Jr., 1975. The functional intermediates in the reaction of membrane-bound cytochrome oxidase with oxygen, *J. Biol. Chem.* 250: 9226–9237.

Chaney, M. O., Jones, N. D., and Debono, M., 1976. The structure of the calcium complex of A23187, a divalent cation ionophore antibiotic, *J. Antibiot.* 29:424–427.

Chasteen, N. D., 1983. The biochemistry of vanadium, *Struct. Bonding* 53:139–160.

Chatt, J., Perman, A. J., and Richards, H. L., 1975. The reduction of monocoordinated molecular nitrogen to ammonia in a protic environment, *Nature* 253:39–41.

Cheh, A. B., and Neilands, J. B., 1976. The α-aminolevulinate dehydratase: Molecular and environmental properties, *Struct. Bonding* 29:123–169.

Chen, G.J-J., McDonald, J. W., and Newton, W. E., 1976. Synthesis of Mo(IV) and Mo(V) complexes using oxo abstraction by phosphines, mechanistic implications, *Inorg. Chem.*, 15:2612–2615.

Chen, R. W., Wagner, P., Ganther, H. E., and Hoekstra, W. G., 1972. A low molecular weight cadmium-binding protein in testis of rats: Possible role in cadmium-induced testicular damage, *Fed. Proc.* 31:2725 (abstract).

Chen, R. W., Hoekstra, W. G., and Ganther, H. E., 1974. An unstable Cd-binding protein in the soluble fraction of rat testis, *Fed. Proc.* 32:3994 (abstract).

Chen, R. W., Whanger, P. D., and Weswig, P. H., 1975. Selenium-induced redistribution of cadmium binding to tissue proteins, possible mechanism of protection against cadmium toxicity, *Bioinorg. Chem.* 4:125–133.

Cheung, W. Y., 1980. Calmodulin plays a pivotal role in cellular regulation, *Science* 207: 19–27.

Cheung, W. Y. (ed.), 1982. *Calcium and Cell Function*, Vols. I–III, Academic Press, New York.

Chimiak, A., and Neilands, J. B., 1984. Lysine analogues of siderophores, *Struct. Bonding* 58:89–96.

Chio, K. S., Reiss, U., Fletcher, B., and Tappel, A. L., 1969. Peroxidation of subcellular organelles: Formation of lipofuscin like fluorescent pigments, *Science* 166:1533–1536.

Chiu, D., Lubin, B., and Shohet, S. B., 1982. Peroxidation reactions in red cell biology, in *Free Radicals in Biology* (W. A. Pryor, ed.), Vol. V, Academic Press, New York, pp. 112–160.

Chowdhury, P., and Louria, D. B., 1976. Influence of cadmium and other trace metals on human α_1-antitrypsin: An *in vitro* study, *Science* 191:480–481.

Christensen, H. N., 1975. *Biological Transport*, 2nd ed., Benjamin, New York, pp. 200–203.

Clare, D. A., Blum, J., and Fridovich, I., 1984. A hybrid superoxide dismutase containing both functional iron and manganese, *J. Biol. Chem.* 259:5932–5936.

Clarkson, D. T., 1974. *Ion Transport and Cell Structure in Plants*, Wiley, New York, pp. 218–284.

Clarkson, T. W., Nordberg, G. F., and Sayer, P. R. (eds.), 1983. *Reproductive and Developmental Toxicity of Metals*, Plenum Press, New York.

Cleland, W. W., 1977. Determining the chemical mechanisms of enzyme-catalyzed reactions by kinetic studies, *Adv. Enzymol.* 45:272–387.

Clementi, E., and Raimondi, D. L., 1963. Atomic screening constants from SCF functions, *J. Chem. Phys.*, 38:2686–2692.

Cloud, P., and Gibor, A., 1970. The oxygen cycle, *Sci. Am.* 1970(9):111–122.

Cohen, C., 1975. The protein switch of muscle contraction, *Sci. Am.* 1975(11):36–45.

Cohen, G., and Greenwald, R. A. (eds.), 1983. *Oxy Radicals and Their Scavenger Systems*, Vol. 1, Elsevier, Amsterdam.

Coleman, J. E., 1984. Carbonic anhydrase: Zinc and the mechanism of catalysis, *Ann. N.Y. Acad. Sci.* 429:26–48.

Coleman, J. E., and Vallee, B. L., 1966. Metallocarboxypeptidase: Stability constants and enzymatic characteristics, *J. Biol. Chem.* 236:2244–2249.

Collman, J. P., Gagne, R. R., Reed, C. A., Robinson, W. T., and Rodley, G. A., 1974. Structure of an iron(II) dioxygen complex: A model for oxygen carrying heme proteins, *Proc. Natl. Acad. Sci. USA* 71:1326–1329.

Collman, J. P., Brauman, J. I., Mennier, B., Raybuck, S. A., and Kodadek, T., 1984. Epoxidation of olefins by cytochrome P-450 model compounds: Mechanism of oxygen atom transfer, *Proc. Natl. Acad. Sci. USA* 81:3245–3248.

Collman, P. M., Freeman, H. C., Guss, J. M., Murata, M., Norris, V. A., Ramshwa, J. A. M., and Venkatapp, M. P., 1978. X-ray crystal structure analysis of plastocyanin at 2.7 A resolution, *Nature* 272:319–324.

Cone, J. E., Martin del Rio, R., Davis, J. N., and Stadtman, T. C., 1976. Chemical characterization of the selenoprotein component of *Clostridial* glycine reductase: Identification of selenocysteine as the organoselenium moiety, *Proc. Natl. Acad. Sci. USA* 73:2659–2663.

Cone, J. E., Martin del Rio, R., and Stadtman, T. C., 1977. *Clostridial* glycine reductase complex, *J. Biol. Chem.* 252:5337–5344.

Cook, J. S., DiLuzio, N. R., and Hoffman, E. O., 1975. Factors modifying susceptibility to bacterial endotoxin: The effect of Pb and Cd, *Crit. Rev. Microbiol.* 3:201–229.

Coon, M. J., Ballou, D. P., Haugen, D. A., Krezoski, S. O., Nordblom, G. D., and White, R. E., 1977. Purification of membrane-bound oxygenases: Isolation of two electrophoretically homogeneous forms of liver microsomal cytochrome P-450, in *Microsomes and Drug Oxidation* (V. Ullrich, I. Roots, A. Hildebrandt, R. Estabrook, and A. H. Conney, eds.), Pergamon Press, Elmsford, N. Y., pp. 82–94.

Corner, E. D., 1959. The poisoning of *Maia squinado* (Herbst) by certain compounds of mercury, *Biochem. Pharmacol.* 2:121–132.

Costerton, J. W., Geesey, G. G., and Chen, K.-J., 1978. How bacteria stick, *Sci. Am.* 1978(1):86–95.

Cotton, F. A., 1973. Structure and bonding in the molecular oxo-molybdenum compounds, in *Chemistry and Uses of Molybdenum* (P. C. H. Mitchell, ed.), Climax Molybdenum Co., Ann Arbor E.O., pp. 6–10.

Cramer, S. P., Gillum, W. O., Hodgson, K. O., Mortenson, L. E., Stiefel, E. I., Chisnell, J. R., Brill, W. J., and Shah, V. K., 1978a. The molybdenum site of nitrogenase.2. A Comparative study of Mo-Fe proteins and the iron-molybdenum cofactor by x-ray absorption spectroscopy, *J. Am. Chem. Soc.* 100:3814–3819.

Cramer, S. P., Hodgson, K. O., Gillum, W. O., and Mortenson, L. E., 1978b. The molybdenum site of nitorgenase: Preliminary structural evidence from x-ray absorption spectroscopy, *J. Am. Chem. Soc.* 100:3398–3407.

Cramer, S. P., Liu, C.-L., Mortenson, L. E., Spence, J. T., Liu, S.-M., Yamamoto, I., and Ljungdahl, L. G., 1985. Formate dehydrogenase molybdenum and tungsten sites— Observation by EXAFS of structural differences, *J. Inorg. Biochem.* 23:119–124.

Crichton, R. R., 1973. Structure and function of ferritin, *Angew. Chem. Int. Ed. Engl.* 12:57–65.

Crichton, R. R., 1984. Iron uptake and utilization by mammalian cells. II. Intracellular iron utilization, *Trends Biochem. Sci.* 9:283–285.

Cudd, A., and Fridovich, I., 1982. Electrostatic interactions in the reaction mechanism of bovine erythrocyte superoxide dismutase, *J. Biol. Chem.* 257:11443–11447.

Czech, M. P., 1977. Molecular basis of insulin action, *Annu. Rev. Biochem.* 46:359–384.

Dailey, H. A., and Fleming, J. E., 1983. Bovine ferrochelatase, *J. Biol. Chem.* 258:11453–11459.

David, C. N., 1974. Ferritin and iron metabolism in *Phycomyces,* in *Microbial Iron Metabolism* (J. B. Neilands, ed.), Academic Press, New York, pp. 149–158.

Davis, J. C., and Averill, B. A., 1982. Evidence for a spin-coupled binuclear unit at the active site of the purple acid phosphatase from beef spleen, *Proc. Natl. Acad. Sci. USA* 79:4623–4627.

Day, W., 1984. *Genesis on Planet Earth*, 2nd ed., Yale Univerrsity Press, New Haven, Conn.

DeLorenzo, R. J., 1982. Calmodulin in synaptic function and neurosecretion, in *Calcium and Cell Function* (W. Y. Cheung, ed.), Vol. III, Academic Press, New York, pp. 271–309.

Delwiche, C. C., 1970. The nitrogen cycle, *Sci. Am.* 1970(9):137–146.

Demaille, J. G., 1982. Calmodulin and other calcium-binding proteins, in *Calcium and Cell Function* (W. Y. Cheung, ed.), Vol. II, Academic Press, New York, pp. 111–144.

DePierre, J. W., and Ernster, L., 1977. Enzyme topology of intracellular membrane, *Annu. Rev. Biochem.* 46:201–262.

Dickerson, R. E., and Timkovich, R. 1975. Cytochromes C, in *The Enzymes*, 3rd ed. (P. D. Boyer, ed.), Vol. XI, Part A, Academic Press, New York, pp. 397–547.

Dingley, A. L., Kustin, K., Macara, I. G., McLeod, G. C., and Roberts, M. E., 1982. Vanadium-containing tunicate blood cells are not highly acidic, *Biochim. Biophys. Acta* 720:384–389.

Diplock, A. T., 1976. Metabolic aspects of selenium action and toxicity, *Crit. Rev. Toxicol.* 4:271–329.

Diplock, A. T., 1983. The role of vitamin E in biological membranes, in *Biology of Vitamin E* (R. Porter and J. Whelan, eds.), Pitman, New York, pp. 45–53.

Dixon, N. E., Gazzola, C., Blakely, R. L., and Zerner, B., 1976. Metal ions in enzymes using ammonia or amides, *Science* 191:1144–1150.

Dolphin, D. (ed.), 1982. B_{12}, Vols. 1 and 2, Wiley, New York.

Dolphin, D., and Felton, R. H., 1974. The biochemical significance of porphyrin pi cation radical, *Acc. Chem. Res.* 7:26–32.

Dresel, E. I. B., and Falk, J. E., 1956. Biosynthesis of blood pigments. III. Heme and porphyrin formation from α-aminolevulinic acid and from porphobilinogen in chicken erythrocyte, *Biochem. J.* 63:80–7.

Dykacz, G. R., Libby, R. D., and Hamilton, G. A., 1976. Trivalent copper as a probable intermediate in the reaction catalyzed by galactose oxidase, *J. Am. Chem. Soc.* 98: 626–628.

Eanes, E. D., and Termine, J. D., 1983. Calcium in mineralized tissues, in *Calcium in Biology* (Spiro, T.G., ed.), Wiley-Interscience, New York, pp. 201–233.

Eckhert, C. D., Sloan, M. V., Duncan, J. R., and Hurley, L. S., 1977. Zinc binding: A difference between human and bovine milk, *Science* 195:789–790. *Econ. Geol.*, 1973. 68:913–1179.

Egyed, A., 1973. Significance of transferrin bound bicarbonate in the uptake of iron by reticulocyte, *Biochim. Biophys. Acta* 304:805–813.

Eichhorn, G. L., 1981. The effect of metal ions on the structure and function of nucleic acids, in *Metal Ions in Genetic Information Transfer* (G. L. Eichhorn and L. G. Marzilli, eds.), Elsevier, Amsterdam.

Eichhorn, G. L., Berger, N. A., Butzow, C. P., Rifkind, M., Shin, Y. A., and Tarien, E., 1971. The effect of metal ions on the structure of nucleic acids, *Adv. Chem. Ser.* 100: 135–154.

Eigen, M., and Schuster, P., 1977. The hypercycle, a principle of natural self-organization: Part A, *Naturwissenschaften* 64:541–565.

Eigen, M., and Schuster, P., 1978. The hypercycle, a principle of natural self-organization: Parts B and C, *Naturwissenschaften* 65:7–41, 341–369.

Eigen, M., Gardiner, W., Schuster, P., and Winkler-Oswatitch, 1981. The origin of genetic information, *Sci. Am.* 1981(4):88–118.

Eklund, H., Nordstrom, B., Zeppezauer, M., Soderland, G., Ohlsson, I., Boiwe, T., and Branden, C.-I., 1974. Structure of horseliver alcohol dehydrogenase, *FEBS Lett.* 44: 200–204.

Ellefson, W. L., Whitman, W. B., and Wolfe, R. S., 1982. Nickel-containing factor F_{430}: Chromophore of the methyl reductase of Methanobacterium, *Proc. Natl. Acad. Sci. USA* 79:3707–3710.

Ellerton, H. D., Ellerton, N. F., and Robinson, H., 1983. Hemocyanin—A current perspective, *Prog. Biophys. Mol. Biol.* 41:143–248.

Emery, T., 1971. Hydroxamic acids of natural origins, *Adv. Enzymol.* 35:135–185.

Emery, T., 1974. Biosynthesis and mechanism of action of hydroxamate-type siderophores, in *Microbial Iron Metabolism* (J. B. Neilands, ed.), Academic Press, New York, pp. 107–123.

Emptage, M. H., Kent, T. A., Huynh, B. H., Rawlings, J., Orme-Johnson, W. H., and Munck, E., 1980. On the nature of the iron–sulfur centers in a ferredoxin from *Azotobacter vinelandii*: Mössbauer studies and cluster displacement experiments, *J. Biol. Chem.* 255:1793–1796.

Emptage, M. H., Kent, T. A., Kennedy, M. C., Beinert, H., and Munck, E., 1983. Mössbauer and EPR studies of activated aconitase: Development of a localized valence state at a subsite of the [4Fe-4S] cluster on binding citrate, *Proc. Natl. Acad. Sci. USA* 80: 4674–4678.

Engström, Y., Eriksson, S., Thelander, L., and Akerman, M., 1979. Ribonucleotide reductase from calf thymus, purification and properties, *Biochemistry* 18:2941–2948.

Ernster, L., 1977. Chemical and chemiosmotic aspects of electron transfer-linked phosphorylation, *Annu. Rev. Biochem.* 46:981–995.

Estabrook, R. W., and Werringloer, J., 1977. Active oxygen—Fact or fancy? in *Microsomes and Drug Metabolism* (V. Ullrich, I. Roots, A. Hildebrandt, R. Estabrook, and A. H. Conney, eds.), Pergamon Press, Elmsford, N.Y., pp. 748–757.

Ettinger, M. J., and Kosman, D. J., 1980. Chemical and catalytic properties of galactose oxidase, in *Copper Proteins* (T. G. Spiro, ed.), Wiley, New York, pp. 219–261.

Evans, G. W., 1973. Copper homeostasis, *Physiol. Rev.* 53:535–570.

Evans, G. W., 1976. Zinc absorption and transport, in *Trace Elements in Human Health and Diseases* (A. S. Prasad and D. Oberleas, eds.), Vol. 1, Academic Press, New York, pp. 181–187.

Faila, M. L., 1977. Zinc: Function and transport in microorganisms, in *Microorganisms and Minerals* (E. D. Weinberg, ed.), Dekker, New York, pp. 151–214.

Falke, J. J., Pace, R. J., and Chan, S. I., 1984a. Chloride binding to the anion transport binding sites of band 3, *J. Biol. Chem.* 259:6472–6480.

Falke, J. J., Pace, R. J., and Chan, S. I., 1984b. Direct observation of the transmembrane recruitment of band 3 transport sites by competitive inhibitors, *J. Biol. Chem.* 259: 6481–6491.

Fee, J. A., 1980. Superoxide, superoxide dismutase and oxygen toxicity, in *Metal Ion Activation of Oxygen* (T. G. Spiro, ed.), Wiley, New York, pp. 209–237.

Fee, J. A., and McClune, G. J., 1978. Mechanism of superoxide dismutase, in *Mechanism of Oxidizing Enzymes* (T. P. Singer, and Ondarza, R. N., eds.), Elsevier, Amsterdam, pp. 273–284.

Fee, J. A., and Palmer, G., 1971. Properties of parsley ferredoxin and its selenium-containing homolog, *Biochim. Biophys. Acta* 245:175–195.

Feeney, R. E., Osuga, D. T., Meares, C. F., Babin, D. R., and Penner, M. H., 1983. Studies on iron-binding sites of transferrin by chemical modification, in *Structure and Function of Iron Storage and Transport Proteins* (I. Urushizaki, P. Aisen, I. Listowsky, and J. W. Drysdale, eds.), Elsevier, Amsterdam, pp. 231–240.

Feigelson, P., 1976. Pseudomonad and hepatic L-tryptophan 2,3-dioxygenase, in *Iron and Copper Proteins* (K. T. Yasunobu, H. F. Mower, and O. Hayaishi, eds.), Plenum Press, New York, pp. 358–362.

Feigelson, P., and Brady, F. A., 1974. Heme-containing dioxygenases, in *Molecular Mechanisms of Oxygen Activation* (O. Hayaishi, ed.), Academic Press, New York, pp. 87–133.

Ferris, J. P., 1984. The chemistry of life's origin, *Chem. Eng. News* 62(Aug. 26):22–35.

Fielden, E. M., Roberts, P. B., Bray, R. C., Lowe, D. J., Mautner, G. N., Rotilio, G., and Calabrese, L., 1974. The mechanism of action of superoxide dismutase from pulse radiolysis and electron paramagnetic resonance, *Biochem. J.* 139:49–60.

Finch, C. A., and Loden, B., 1959. Body iron exchange in man, *J. Clin. Invest.* 38:392–396.

Finke, R. G., and Schiraldi, D. A., 1983. Model studies of coenzyme B_{12} dependent diol dehydratase. 2. A kinetic and mechanistic study focusing upon the cobalt participation or nonparticipation question, *J. Am. Chem. Soc.* 105:7605–7617.

Finke, R. G., McKenna, W. P., Schiraldi, D. A., Smith, B. L., and Pierpont, D., 1983. Model studies of coenzyme B_{12} dependent dehydratase. I. Synthetic, physical property and product studies of two key, cobalt-bound putative diol dehydratase intermediates, *J. Am. Chem. Soc.* 105:7592–7604.

Finke, R. G., Schiraldi, D. A., and Mayer, B. J., 1984. Towards the unification of coenzyme B_{12}-dependent diol dehydratase stereochemical and model studies: The bound radical mechanism, *Coord. Chem. Rev.* 54:1–22.

Fletcher, J., and Huehns, E. R., 1967. Significance of the binding of iron by transferrin, *Nature* 215:584–586.

Fletcher, J., and Huehns, E. R., 1968. Function of transferrin, *Nature* 218:1211–1214.

Fletterick, R. J., Bates, D. J., and Steitz, T. A., 1973. The structure of a yeast hexokinase monomer and its complexes with substrates at 2.7 A resolution, *Proc. Natl. Acad. Sci. USA* 70:38–42.

Flohe, L., 1982. Glutathione peroxidase brought into focus, in *Free Radicals in Biology* (W. A. Pryor, ed.), Vol. 5, Academic Press, New York, pp. 223–254.

Foote, C. S., 1977. Photosensitized oxidation and singlet oxygen: Consequence in biological systems, in *Free Radicals in Biology* (W. A. Pryor, ed.), Vol. 2, Academic Press, New York, pp. 85–133.

Forth, W., 1974. Iron absorption, a mediated transport across the mucosal epithelium, in *Trace Element Metabolism in Animals-2* (W. G. Hoekstra, J. M. Sutties, H. E. Ganther, and W. Mertz, eds.), University Park Press, Baltimore, pp. 199–215.

Forth, W., and Rummel, W., 1973. Iron absorption, *Physiol. Rev.* 53:724–792.

Foulkes, E. C. (ed.), 1982. *Biological Roles of Metallothionein*, Elsevier, Amsterdam.

Fowler, B. A., 1977. Toxicology of environmental arsenic, in *Toxicology of Trace Elements* (R. A. Goyer and M. A. Mehlman, eds.), Hemisphere, New York, pp. 79–122.

Fowler, B. A., Ishinishi, N., Tsuchiya, K., and Vahter, M., 1979. Arsenic, in *Handbook on the Toxicology of Metals* (L. Friberg, G. F. Nordberg, and V. B. Vouk, eds.), Elsevier, Amsterdam, pp. 293–319.

Fox, B., and Walsh, C. T., 1982. Mercuric reductase: Purification and characterization of a transposon-encoded flavoprotein containing an oxidation–reduction active disulfide, *J. Biol. Chem.* 257:2498–2503.

Fox, G. E., Stackebrandt, E., Hospell, R. B., Gibson, J., Maniloff, J., Dyer, T. A., Wolfe, R. S., Balch, W. E., Tanner, R. S., Magrum, L. J., Zablen, L. B., Blakemore, R., Gupta, R., Bonen, L., Lewis, B. J., Stahl, D. A., Luehrsen, K. R., Chen, K. N., and Woese, C. R., 1980. The phylogeny of prokaryotes, *Science* 209:457–463.

Fox, S. W., and Harada, K., 1958. Thermal copolymerization of amino acids to a product resembling protein, *Science* 128:1214.

Frausto da Silva, J. J. R., and Williams, R. J. P., 1976. The uptake of elements by biological systems, *Struct. Bonding* 29:67–121.

Frederick, J. F. (ed.), 1981. Origins and evolution of eukaryotic intracellular organelles, *Ann. N. Y. Acad. Sci.,* Vol. 361.

Friberg, L., Nordberg, G. F., and Vouk, V. B. (eds.), 1979. *Handbook on the Toxicology of Metals,* Elsevier, Amsterdam.

Fridovich, I., 1963. Inhibition of acetoacetic decarboxylase by anions, the Hofmeister lyotropic series, *J. Biol. Chem.* 238:592–598.

Fridovich, I., 1975. Superoxide dismutase, *Annu. Rev. Biochem.* 44:147–159.

Fridovich, I., 1976. Oxygen radicals, hydrogen peroxide and oxygen toxicity, in *Free Radicals in Biology* (W. A. Pryor, ed.), Academic Press, New York, pp. 239–277.

Frieden, E., 1971. Ceruloplasmin, a link between copper and iron metabolism, *Adv. Chem. Ser.* 100:292–321.

Frieden, E. (ed.), 1985a. *Biochemistry of the Essential Ultratrace Elements*, Plenum Press, New York.

Frieden, E., 1985b. New perspectives on the essential trace elements, *J. Chem. Educ.* 62: 917–923.

Frieden, E., and Osaki, S., 1974. Ferroxidase and ferrireductases: Their role in iron metabolism, in *Protein–Metal Interaction* (E. Frieden, ed.), Plenum Press, New York, pp. 235–265.

Frost, G., and Rosenberg, H., 1975. The relationship between the ton B locus and iron transport in *Escherichia coli*, *J. Bacteriol.* 124:704–712.

Fruton, J. S., 1976. Mechanism of the catalytic action of pepsin and related acid proteinases, *Adv. Enzymol.* 44:1–36.

Fujisawa, H., Hiromi, K., Uyeda, M., Okuno, S., Nozaki, M., and Hayaishi, O., 1972. Protocatechuate 3,4-dioxygenase. III. An oxygenated form of enzyme as reaction intermediate, *J. Biol. Chem.* 247:4422–4428.

Fujisawa, H., Yamaguchi, M., and Yamauchi, T., 1976. Partial purification and properties of 2,3-dihydroxybenzoate 2,3-dioxygenase, in *Iron and Copper Proteins* (K. T. Yasunobu, H. F. Mower, and O. Hayaishi, eds.), Plenum Press, New York, pp. 118–126.

Fung, C. H., Mildvan, A. S., Allerhand, A., Komonoski, R., and Scrutton, M. C., 1973. Interaction of pyruvate with pyruvate carboxylase and pyruvate kinase as studied by paramagnetic effects on ^{13}C relaxation rate, *Biochemistry* 12:620–629.

Fung, C. H., Mildvan, A. S., and Leigh, J. S., 1974. Electron and nuclear magnetic resonance studies of the interaction of pyruvate and transcarboxylase, *Biochemistry* 13:1160–1169.

Fung, C. H., Gupta, R. K., and Mildvan, A. S., 1976a. Magnetic resonance studies of the proximity and spatial arrangement of propionyl coenzyme A and pyruvate on a biotin-metalloenzyme, transcarboxylase, *Biochemistry* 15:85–92.

Fung, C. H., Feldman, R. J., and Mildvan, A. S., 1976b. ^{1}H and ^{31}P Fourier transform resonance studies of the conformation of enzyme-bound propionyl coenzyme A on transcarboxylase, *Biochemistry* 15:75–84.

Galbraith, R. A., Sassa, S., and Kappas, A., 1985. Heme binding to murine erythroleukemia cells, evidence for a heme receptor, *J. Biol. Chem.* 260:12198–12202.

Gale, P. H., and Egan, R. W., 1984. Prostaglandin endoperoxide synthase-catalyzed oxidation, in *Free Radicals in Biology* (W. A. Pryor, ed.), Vol. VI, Academic Press, New York, pp. 1–38.

Ganther, H. E., 1974. Biochemistry of selenium, in *Selenium* (R. A. Zingaro and W. C. Cooper, eds.), Van Nostrand–Reinhold, Princeton, N.J., pp. 546–614.

Ganther, H. E., Oh, S. H., Schaich, E., and Hoekstra, W. G., 1974. Studies on selenium in glutathione peroxidase, *Fed Proc.* 33:694.

Ganther, H. E., Gondie, C., Sunde, M. L., Kopecky, M. J., Wagner, P., Oh, S. H., and Hoekstra, W. G., 1972. Selenium: Relation to decreased toxicity of methylmercury added to diets containing tuna, *Science* 175:1122–1124.

Garrels, R. B., and Christ, C. L., 1965. *Solutions, Minerals and Equilibria*, Harper and Row, New York.

Garratt, R. C., Evans, R. W., Hasnain, S. S., and Lindley, P. F., 1986. An extended X-ray-absorption–fine-structure investigation of diferric transferrin and their ion-binding fragments, *Biochem. J.* 233:479–484.

Garrone, R., Simpson, T. L., and Pottu-Boumendil, J., 1981. Ultrastructure and deposition of silica in sponges, in *Silicon and Siliceous Structures* (T. L. Simpson and B. E. Volcani, eds.), Springer-Verlag, Berlin, pp. 495–550.

Gauch, H. G., 1972. *Inorganic Plant Nutrition,* Dowden, Hutchinson & Ross, Stroudsburg, Pennsylvania.

Gaykema, W. P. J., Hol, W. G. J., Vereijken, J. M., Soeter, N. M., Bak, H. J., and Beintema, J. J., 1984. 3.2 A structure of the copper-containing oxygen-carrying protein *Panulirus interruptus* hemocyanin, *Nature* 309:23–29.

Gelin, B. B., and Karplus, M., 1977. Mechanism of tertiary structural change in hemoglobin, *Proc. Natl. Acad. Sci. USA* 74:801–805.

George, S. G., 1982. Subcellular accumulation and detoxification of metals in aquatic animals, in *Physiological Mechanisms of Marine Pollutant Toxicity* (W. B. Vernberg, A. Calabrese, F. P. Thurberg, and E. T. Vernberg, eds.), Academic Press, New York pp. 23–52.

Gerencser, G. A. (ed.), 1984. *Chloride Transport Coupling in Biological Membranes and Epithelia,* Elsevier, Amsterdam.

Getzoff, E. C., Tainer, J. A., Weiner, P. K., Kollman, P. A., Richardson, J. S., and Richardson, D. C., 1983. Electrostatic recognition between superoxide and copper, zinc superoxide dismutase, *Nature* 306:287–290.

Glusker, J. P., 1971. Aconitase, in *The Enzymes,* 3rd ed. (P. D. Boyer, ed.), Vol. V, Academic Press, New York, pp. 413–439.

Glynn, I. M., 1985. The Na^+, K^+-transporting adenosine triphosphatase, in *The Enzymes of Biological Membranes,* 2nd ed. (A. Martonosi, ed.), Vol. 3, Plenum Press, New York, pp. 35–114.

Golding, B. T., 1982. Mechanism of action of the B_{12} coenzyme: Theory and models, in B_{12} (D. Dolphin, ed.), Vol. 1, Wiley, New York, pp. 543–582.

Gomez-Puyou, A., and Gomez-Lojero, C., 1977. The use of ionophores and channel formers in the study of the function of biological membrane, Curr. Top. Bioenerg. 6:221–257.

Gonzalez-Veraga, E., Hegenauer, J., and Saltman, P., 1982. Biological complexes of chromium: A second look at GTF, *Fed. Proc.* 41:286.

Gordon, L. G. M., and Haydon, D. A., 1976. Kinetics and stability of alamethicin conducting channels in lipid bilayers, *Biochim. Biophys. Acta* 436:541–556.

Goresky, C. A., Holmes, T. H., and Sass-Kortsalk, A., 1968. Initial uptake of copper in liver in the dog, *Can. J. Physiol. Pharmacol.* 46:771–784.

Govindjee, Wydrzunski, T., and Marks, S. B., 1977. The role of manganese in the oxygen evolving mechanism of photosynthesis, in *Bioenergetics of Biomembranes* (K. Packer, G. C. Papageogiou, and A. Trebst, eds.), Elsevier, Amsterdam, pp. 305–316.

Goyer, R. A., 1975. Comments on the paper by Miettinen, in *Heavy Metals in the Aquatic Environment* (P. A. Krenkel, ed.), Elsevier, Amsterdam, pp. 163–166.

Goyer, R. A., and Mehlman, M. A. (ed.), 1977. *Toxicology of Trace Elements,* Hemisphere, New York.

Goyer, R. A., and Moore, J. F., 1974. Cellular effects of lead, in *Protein–Metal Interaction* (M. Frieden, ed.), Plenum Press, New York, pp. 447–462.

Goyer, R. A., Leonard, D. L., Moore, J. F., Rhyne, B., and Krigman, M. R., 1970. Lead dosage and the role of the intranuclear inclusion body, *Arch. Environ. Health* 20: 705–711.

Gradman, D., 1984. Electronic Cl^- pump in the marine alga Acetabularia, in *Chloride Transport Coupling in Biological Membranes and Epithelia* (G. A. Gerencser, ed.), Elsevier, Amsterdam, pp. 13–61.

Granick, J. L., and Sassa, S., 1978. Hemin control of heme biosynthesis in mouse fried virus transformed erythroleukemia cells in culture, *J. Biol. Chem.* 253:5402–5406.

Graslund, A., Ehrenberg, A., and Thelander, L., 1982. Characterization of the free radical of mammalian ribonucleotide reductase, *J. Biol. Chem.* 257:5711–5715.

Gray, H. B., and Solomon, E. I., 1981. Electronic structures of the copper centers in proteins, in *Copper Proteins* (T. G. Spiro, ed.), Wiley, New York, pp. 1–39.

Green, N. M., Allen, G., Hebdon, G. M., and Thorley-Lawson, D. A., 1977, Structural studies on the Ca^{++} transporting ATPase of sarcoplasmic reticulum, in *Calcium-Binding Proteins and Calcium Function* (R. H. Wasserman, R. A. Corrandino, E. Carafoli, R. H. Kretsinger, D. H. McLennan, and F. L. Siegel, eds.), North-Holland, Amsterdam, pp. 164–172.

Gregory, E. M., Pennington, C. D., and Denz, D. L., 1983. Oxygen-dependent induction of superoxide dismutase in anaerobes, in *Oxy Radicals and Their Scavenger Systems* (G. Cohen and R. A. Greenwald, eds.), Vol. 1, Elsevier, Amsterdam, pp. 252–257.

Gresalfi, T. J., and Wallace, B. A., 1984. Secondary structural composition of the Na/K-ATPase E_1 and E_2 conformers, *J. Biol. Chem.* 259:2622–2628.

Griffin, B. W., and Peterson, J. A., 1975. *Pseudomonas putida* cytochrome P-450, the effect of complexes of the ferric hemeprotein on the relaxation of solvent water protons, *J. Biol. Chem.* 250:6445–6451.

Groth, D. H., Stettler, L., and Mackay, G. M., 1976. Interaction of Hg, Cd, Se, Te, As and Be, in *Effects and Dose-Relationships of Toxic Metals* (G. F. Nordberg, ed.), Elsevier, Amsterdam, pp. 527–543.

Groves, J. T., 1980. Mechanisms of metal-catalyzed oxygen insertion, in *Metal Ion Activation of Dioxygen* (T. G. Spiro, ed.), Wiley, New York, pp. 152–162.

Groves, J. T., and Chambers, R. R., 1984. Geometrical and stereochemical factors in metal-promoted amide hydrolysis, *J. Am. Chem. Soc.* 106:630–638.

Groves, J. T., and Dias, R. M., 1979. Rapid amide hydrolysis mediated by copper and zinc, *J. Am. Chem. Soc.* 101:1033–1035.

Groves, J. T., and Dias, R. M., 1983. Models of metalloenzymes: Carboxypeptidase A, in *The Coordination Chemistry of Metalloenzymes* (I. Bertini, R. S. Drago, and C. Luchinat, eds.), Reidel, Dordrecht, pp. 79–82.

Groves, J. T., and Nemo, T. E., 1982. Models of metalloenzymes: Peroxidase and cytochrome P-450, in *The Coordination Chemistry of Metalloenzymes* (I. Bertini, R. S. Drago, and C. Luchinat, eds.), Reidel, Dordrecht, pp. 328–341.

Groves, J. T., and Nemo, T. E., 1983. Epoxidation reactions catalyzed by iron porphyrins: Oxygen transfer from iodosylbenzene, *J. Am. Chem. Soc.* 105:5786–5791.

Groves, J. T., McClusky, G. A., White, R. E., and Coon, M. J., 1978. Aliphatic hydroxylation by highly purified liver microsomal cytochrome P-450, evidence for a carbon radical intermediate, *Biochem. Biophys. Res. Commun.* 81:154–160.

Guerrero, M. G., and Vega, J. M., 1975. Molybdenum and iron as functional constituents of the enzymes of nitrate reducing system in *Azotobacter chroococcum*, *Arch. Microbiol.* 102:91–94.

Gunn, S. A. (ed.), 1976. *Selenium*, National Academy of Sciences, Washington, D.C.

Hague, D. N., Martin, R. S., and Zetter, M. S., 1972. Role of metals in enzymic reactions. 5. Kinetics of ternary complex formation between magnesium and manganese (II) species and 8-hydroxyquinoline, *J. Chem. Soc. Faraday Trans.* 68:37–46.

Hahn, J. E., Hodgson, K. O., Anderson, L. A., and Dawson, J. H., 1982. Endogenous cysteine ligation in ferric and ferrous cytochrome P-450, *J. Biol. Chem.* 257:1793–1796.

Haiech, J., Derancourt, J., Pechere, J.-F., and Demaille, J. G., 1979. Magnesium and calcium binding to parvalbumins: Evidence for difference between parvalbumins and an explanation of their relaxing function, *Biochemistry* 18:2752–2758.

Haiech, J., Klee, C. B., and Demaille, J. G., 1981. Effects of cations on affinity of calmodulin for calcium: Ordered binding of calcium ion allows the specific activation of calmodulin-stimulated enzymes, *Biochemistry* 20:3890–3897.

Hall, D. D., 1977. Iron–sulfur proteins and superoxide dismutase in the biology and evolution of electron transport, *Adv. Chem. Ser.* 162:227–250.

Hall, D. D., Rao, K. K., and Mullinger, R. N., 1976. A role for Fe-S proteins in the origin and evolution of life, in *Protein Structure and Evolution* (J. L. Fox, Z. Deyl, and A. Balzy, eds.), Dekker, New York, pp. 233–256.

Halpern, J., 1985. Mechanisms of coenzyme B_{12}-dependent rearrangements, *Science* 227: 869–875.

Hamilton, G. A., 1980. Oxidases with monocopper active sites, in *Copper Proteins* (T. G. Spiro, ed.), Wiley, New York, pp. 193–218.

Hardman, K. D., and Ainsworth, C. F., 1972. Structure of concanavalin A at 2.4 A resolution, *Biochemistry* 11:4910–4919.

Harper, H. A., 1971. *Review of Physiological Chemistry*, 13th ed., Lange Medical Publ., Los Altos, Calif.

Harrison, R. M., Hoare, R. J., Hoyl, T. G., and Macara, I. G., 1974. Ferritin and hemosiderin, in *Iron in Biochemistry and Medicine* (A. Jacob and M. Worwood, eds.), Academic Press, New York, pp. 73–114.

Hartman, W. D., 1981. Form and distribution of silica in sponges, in *Silicon and Siliceous Structures in Biology* (T. L. Simpson, and B. E. Volcani, eds.), Springer-Verlag, Berlin, pp. 453–493.

Hartshorne, R. P., and Catterall, W. A., 1984. The sodium channel from rat brain, *J. Biol. Chem.* 259:1667–1675.

Hasselbach, W., 1978. The reversibility of the sarcoplasmic calcium pump, *Biochim. Biophys. Acta* 515:23–53.

Hasselbach, W., and Waas, W., 1982. Energy coupling in a sarcoplasmic reticulum Ca^{2+} transport: An overview, *Ann. N.Y. Acad. Sci.* 402:459–469.

Hatefi, Y., Ragan, C. I., and Galante, Y. M., 1985. The enzymes and the enzyme complexes of the mitochondrial oxidative phosphorylation system, in *The Enzymes of Biological Membranes*, 2nd ed. (A. N. Martonosi, ed.), Vol. 4, Plenum Press, New York, pp. 1–70.

Hauser, H., and Phillips, M. C., 1973. Structure of aqueous dispersions of phosphatidylserine, *J. Biol. Chem.* 248:8585–8591.

Hauser, H., Levine, B. A., and Williams, R. J. P., 1976. Interaction of ions with membranes, *Trends Biochem. Sci.* 1:278–281.

Hausinger, R. P., and Howard, J. B., 1982. The amino acid sequence of the nitrogenase iron protein from *Azotobacter vinelandii*, *J. Biol. Chem.* 257:2483–2490.

Hauska, G., and Trebst, A., 1977. Proton translocation in chloroplasts, *Curr. Top. Bioenerg.* 6:51–220.

Haylock, S. J., Buckley, P. D., and Blackwell, L. F., 1983. The relationship of chromium to glucose tolerance factor, II, *J. Inorg. Biochem.* 19:105–117.

Healy, F. P., 1973. Inorganic nutrient uptake and deficiency in algae, *Crit. Rev. Microbiol.* 3:69–113.

Hecky, R. D., Mopper, K., Kilham, P., and Degens, E. T., 1973. The amino acid and sugar composition of diatom cell walls, *Mar. Biol.* 19:323–331.

Hermann, T. R., and Shamoo, A. E., 1982. Isolation and use of ionophores for studying ion transport across natural membranes, in *Membranes and Transport* (A. N. Martonosi, ed.), Vol. 1, Plenum Press, New York, pp. 579–584.

Herzberg, G., and James, M. N. G., 1984. Structure of calcium regulatory muscle protein, troponin-C at 2.8 Å resolution, *Nature* 313:653–659.

Hess, B., Kunschmitz, D., and Engelhard, M., 1982. Bacteriorhopdopsin, in *Membranes and Transport* (A. N. Martonosi, ed.), Vol. 2, Plenum Press, New York, pp. 309–318.

Hider, R. C., 1984. Siderophore mediated absorption of iron, *Struct. Bonding* 58:25–87.

Hill, C. H., 1977. Toxicology of copper, in *Toxicology of Trace Elements* (R. A. Goyer and M. A. Mahlman, eds.), Hemisphere, New York, pp. 123–127.

Hille, B., 1971. The permeability of the sodium channel to organic cations in myelinated nerve, *J. Gen. Physiol.* 58:599–619.

Hille, B., 1975. Receptor for tetrodotoxin and saxitoxin, structural hypothesis, *Biophys. J.* 15:615–619.

Hinkle, P. C., and McCarty, R. E., 1978. How cells make ATP, *Sci. Am.* 1978(3):104–123.

Hirata, F., and Hayaishi, O., 1971. Possible participation of superoxide anion in the intestinal tryptophan 2,3-dioxygenase, *J. Biol. Chem.* 246:7825–7826.

Hirata, F., and Hayaishi, O., 1975. Studies on indoleamine 2,3-dioxygenase, I., Superoxide anion as substrate, *J. Biol. Chem.* 250:5960–5966.

Hirata, F., Shimidzu, T., Yoshida, R., Ohnishi, T., Fujiwara, M., and Hayaishi, O., 1976. Copper content of indoleamine 2,3-dioxygenase, in *Iron and Copper Proteins* (K. T. Yasunobu, H. F. Mower, and O. Hayaishi, eds.), Plenum Press, New York, pp. 354–357.

Hirata, F., Ohnishi, T., and Hayaishi, O., 1977. Indoleamine 2,3-dioxygenase, characterization and properties of enzyme–O_2^- complex, *J. Biol. Chem.* 252:4637–4642.

Holland, H. D., 1984. *The Chemical Evolution of the Atmosphere and Oceans*, Princeton University Press, Princeton, N.J.

Holm, H. V., and Cox, M. F., 1975. Transformation of elementary mercury by bacteria, *Appl. Microbiol.* 29:491–494.

Hopkins, L. L., Jr., 1974. Essentiality and function of vanadium, in *Trace Element Metabolism in Animals-2* (W. G. Hoekstra, J. M. Sutties, H. E. Ganther, and W. Mertz, ed.), University Park Press, Baltimore, pp. 397–405.

Hopkins, L. L., and Tilton, B. E., 1966.Metabolism of trace amount of vanadium 48 in rat organs and liver subcellular particles, *Am. J. Physiol.* 211:169–172.

Horvath, D. J., 1976. Trace elements and health, in *Trace Substances and Health* (P. M. Newberne, ed.), Part I, Dekker, New York, pp. 321–356.

Hozantko, R. B., Lauritzen, A. M., and Lipscomb, W. N., 1981. Metal cation influence on activity and regulation of aspartate carbamoyl transferase, *Proc. Natl. Acad. Sci. USA* 78:898–902.

Hsu, H. H. T., and Anderson, H. C., 1978. Calcification of isolated matrix vesicles and reconstituted vesicles from fetal bovine cartilage, *Proc. Natl. Acad. Sci. USA* 75: 3805–3808.

Huheey, J. E., 1983. *Inorganic Chemistry, Principles of Structure and Reactivity*, 3rd ed., Harper and Row, New York.

Hurley, L. C., Duncan, J. R., Sloan, M. V., and Eckhert, C. D., 1977. Zinc-binding ligands in milk and intestine: A role in neonatal nutrition, *Proc. Natl. Acad. Sci. USA* 74: 3547–3549.

Hutner, S. H., 1972. Inorganic nutrition, *Annu. Rev. Microbiol.* 26:313–346.

Hyde, B. E., Hodge, A. J., Kahn, A., and Birnstiel, M., 1963. Phytoferritin K: Identification and localization, *J. Ultrastr. Res.* 9:248–258.

Ikebuchi, H., Teshima, R., Suzuki, K., Terao, T., and Yamane, Y., 1986. Simultaneous induction of Pb-metallothionein-like protein and Zn-thionein in the liver of rats given lead acetate, *Biochem. J.* 233:541–546.

Inesi, G., Watanabe, T., Coan, C., and Murphy, A., 1982. The mechanism of sarcoplasmic reticulum ATPase, *Ann. N.Y. Acad. Sci.* 402:515–532.

Ishimura, Y., and Hayaishi, O., 1973. Noninvolvement of copper in the L-tryptophan 2,3-dioxygenase reaction, *J. Biol. Chem.* 248:8610–8612.

Ishimura, Y., and Hayaishi, O., 1976. On the prosthetic groups of L-tryptophan 2,3-dioxygenase from pseudomonad, in *Iron and Copper Proteins* (K. T. Yasunobu, H. F. Mower, and O. Hayaishi, eds.), Plenum Press, New York, pp. 363–373.

Iverson, W. P., 1968. Corrosion of iron and formation of iron phosphide, *Nature* 217: 1265–1267.

Jaspar, P., and Silver, S., 1977. Magnesium transport in microorganisms, in *Microorganisms and Minerals* (E. D. Weinberg, ed.), Dekker, New York, pp. 151–214.

Jencks, W. P., 1975. Binding energy, specificity and enzymic catalysis: The Circe effect, *Adv. Enzymol.* 43:219–410.

Johnson, L. J., and Rajagopalan, K. V., 1982. Structural and metabolic relationship between molybdenum cofactor and urothione, *Proc. Natl. Acad. Sci. USA* 79:6856–6860.

Johnson, L. J., Wand, W. R., Cohen, H. J., and Rajagopalan, K. V., 1974a. Molecular basis of the biological function of molybdenum: Molybdenum-free sulfite oxidase from livers from tungsten-treated rats, *J. Biol. Chem.* 249:5046–5055.

Johnson, L. J., Cohen, H. J., and Rajagopalan, K. V., 1974b. Molecular basis of the biological function of molybdenum: Molybdenum-free xanthine oxidase from livers of tungsten-treated rats, *J. Biol. Chem.* 249:5056–5061.

Johnson, L. J., Hainline, B. E., Rajagopalan, K. V., and Srison, B. H., 1984. The pterin component of the molybdenum cofactor, *J. Biol. Chem.* 259:5414–5422.

Johnson, M. K., Spiro, T. G., and Mortenson, L. E., 1982. Resonance Raman and electron paramagnetic studies on oxidized and ferricyanide-treated *Clostridium pasteurianum* ferredoxin, *J. Biol. Chem.* 257:2447–2452.

Jorgensen, P. L., Skriver, E., Herbert, H., and Maunsbach, A. B., 1982. Structure of the Na,K pump: Crystallization of pure membrane-bound Na,K-ATPase and identification of functional domains of the A-subunit, *Ann. N.Y. Acad. Sci.* 402:207–224.

Kambara, T., and Govindjee, 1985. Molecular mechanism of water oxidation in photosynthesis based on the functioning of manganese in two different environments, *Proc. Natl. Acad. Sci. USA* 82:6119–6123..

Kamp-Nielsen, L., 1971. The effect of deleterious concentrations of mercury on the photosynthesis and growth of *Chlorella pyrenoidosa*, *Physiol. Plant* 24:556–561.

Kappas, A., and Maines, M. D., 1976. Tin: A potent inducer of heme oxygenase in kidney, *Science* 192:60–62.

Kassner, R. J., 1973. A theoretical model for the effects of local nonpolar heme environment on the redox potentials in cytochromes, *J. Am. Chem. Soc.* 95:2674–2677.

Kasting, J. F., 1984. The evolution of the Precambrian atmosphere, in *Proc. 7th Internat. Conf. on Origins of Life* (K. Dose, A. Schwartz, and W. Thieman, eds.), Reidel, Dordrecht, pp. 75–82.

Kaufman, P. B., Dayanandan, P., Takeoka, Y., Bigelow, W. C., Jones, J. D., and Iler, D., 1981. Silica in shoots of higher plants, in *Silicon and Siliceous Structures in Biological Systems*, Springer-Verlag, pp. 408–449.

Kennedy, M. C., Kent, T. A., Emptage, M., Merkle, H., Beinert, H., and Munck, E., 1984. Evidence for the formation of a linear [3Fe-4S] cluster in partially unfolded aconitase, *J. Biol. Chem.* 259:14463–14471.

Kent, T. A., Dreyer, J.-L., Kennedy, M. C., Huynh, B. H., Emptage, M. H., Beinert, H., and Munck, E., 1982. Mössbauer studies of beef heart aconitase: Evidence for facile interconversions of iron–sulfur clusters, *Proc. Natl. Acad. Sci. USA* 79:1096–1100.

Kent, T. A., Emptage, M. H., Merkle, H., Kennedy, M. C., Beinert, H., and Munck, E., 1985. Mössbauer studies of aconitase, *J. Biol. Chem.* 260:6871–6881.

Kester, W. R., and Mathews, B. W., 1977. Crystallographic study of the binding of dipeptide inhibitors to thermolysin: Implications for the mechanism of catalysis, *Biochemistry* 16:2506–2510.

Kim, S. H., Quigley, G. J., Suddah, F. L., McPherson, A., Sneden, D., Kim, J.-J., Weinziel, J., and Rich, A., 1973. Three dimensional structure of yeast phenylalanine transfer RNA: Folding of the polynucleotide chain, *Science* 179:285–288.

Kirchgessner, M., Roth, H. P., and Weigand, E., 1976. Biochemical changes in zinc deficiency, in *Trace Elements in Human Health and Disease* (A. S. Prasad and D. Oberleas, eds.), Vol. 1, Academic Press, New York, pp. 189–225.

Klayman, D. L., and Gunther, W. H. H. (eds.), 1973. *Organic Selenium Compounds: Their Chemistry and Biology,* Wiley, New York.

Klee, C. B., Crouch, T. H., and Richman, P. G., 1980. Calmodulin, *Annu. Rev. Biochem.* 49:489–515.

Klevit, R. E., Dalgarno, D. C., Levine, B. A., and Williams, R. J. P., 1983. ¹H-NMR studies of calmodulin, the nature of the Ca^{2+}-dependent conformational change, *Eur. J. Biochem.* 139:109–114.

Klug, D., Babani, J., and Fridovich, I., 1972. A direct demonstration of the catalytic action of superoxide dismutase through the use of pulse radiolysis, *J. Biol. Chem.* 247:4839–4842.

Knowles, P. F., Lowe, D. J., Peters, J., Thorneley, R. N. F., and Yadov, K. D. S., 1982. Kinetic and magnetic resonance studies on amine oxidases, in *The Coordination Chemistry of Metalloenzymes* (I. Bertini, R. S. Drago, and C. Luchinat, eds.), Reidel, Dordrecht, pp. 159–176.

Koeman, J. H., Peeters, W. H. M., Koudstaas-Hol, C. H. M., Tjior, F. S., and de Goeij, J. J. M., 1973. Mercury–selenium correlation in marine mammals, *Nature* 245:385–386.

Kojima, N., Fox, J. A., Hausinger, R, P., Daniels, L., Orme-Johnson, W. H., and Walsh, C. T., 1983. Paramagnetic centers in the nickel-containing deazaflavin reducing hydrogenase from *Methanobacterium thermoautotrophicum, Proc. Natl. Acad. Sci. USA* 80: 378–382.

Kojima, Y., and Kagi, J. H. R., 1978. Metallothionein, *Trends Biochem. Sci.* 3:90–93.

Kojima, Y., Berger, C., Vallee, B. L., and Kagi, J. H. R., 1976. Amino acid sequence of equine renal metallothionein 1B, *Proc. Natl. Acad. Sci. USA* 73:3413–3417.

Kondo, I., Ishikawa, T., and Nakahara, H., 1974. Mercury and cadmium resistances mediated by penicillinase plasmid in *Staphylococcus aureus, J. Bacteriol.* 117:1–7.

Konetzka, W. A., 1977. Microbiology of metal transformation, in *Microorganisms and Minerals* (E. D. Weinberg, ed.), Dekker, New York, pp. 317–342.

Kono, Y., and Fridovich, I., 1983a. Isolation and characterization of the pseudocatalase of *Lactobacillus plantarum, J. Biol. Chem.* 258:6015–6019.

Kono, Y., and Fridovich, I., 1983b. Inhibition and reactivation of Mn-catalase, *J. Biol. Chem.* 258:13646–13648.

Koppenol, W. H., 1981. The physiological role of the charge distribution on superoxide dismutase, in *Oxygen and Oxy-Radicals in Chemistry and Biology* (M. A. J. Rodgers and E. L. Powers, eds.), Academic Press, New York, pp. 671–674.

Kosta, L., Byrne, A. R., and Zelenko, V., 1975. Correlation between selenium and mercury in man following exposure to inorganic mercury, *Nature* 254:238–239.

Kotyk, A., and Janacek, K., 1975. *Cell Membrane Transport: Principles and Techniques,* 2nd ed., Plenum Press, New York, pp. 199, 340.

Krampitz, G. P., 1982. Structure of the organic matrix in mollusc shells and avian egg shells, in *Biological Mineralization and Demineralization* (G. H. Nanchollas, ed.), Springer-Verlag, Berlin, pp. 219–232.

Kretsinger, R. H., and Nockolds, C. E., 1973. Carp muscle calcium-binding protein, *J. Biol. Chem.* 248:3313–3326.

Krinsky, N. I., 1977. Singlet oxygen in biological systems, *Trends Biochem. Sci.* 2:35–38.

Krüger, H. J., Huynh, B. H., Ljungdahl, P. O., Xavier, A. V., Der Vartanian, D. V., Moura, , Peck, H. D., Jr., Teixeira, M., Moura, J. J., and LeGall, J., 1982. Evidence for nickel and a three iron center in the hydrogenase of *Desulfovibrio desulfuricans, J. Biol. Chem.* 257:1462–1463.

Kulp, J. L., Schuber, A. R., and Hodges, E. J., 1960. Strontium-90 in man, IV, *Science* 132:448–454.

Kuo, L. C., and Makinen, M. W., 1982. Hydrolysis of esters by carboxypeptidase A requires a pentacoordinate metal ion, *J. Biol. Chem.* 257:24–27.

Kuo, L. C., Lipscomb, W. N., and Kantrowitz, E. R., 1982. Zn(II)-induced cooperativity of *E. coli* ornithine transcarbamoylase, *Proc. Natl. Acad. Sci. USA* 79:2250–2254.

Kuriki, Y., Halsey, J., Biltonen, R. B., and Racker, E., 1976. Calorimetric studies of the interaction of magnesium and phosphate with (Na^+,K^+)ATPase: Evidence for a ligand-induced conformational change in the enzyme, *Biochemistry* 15:4956–4961.

Kurtz, D. M., McMillan, R. S., Burgess, B. K., Mortenson, L. E., and Holm, R. H., 1979. Identification of iron–sulfur centers in the iron–molybdenum protein of nitrogenase, *Proc. Natl. Acad. Sci. USA* 76:4986–4989.

Kustin, K., McLeod, G. C., Gilbert, T. R., and Briggs, L. R., 4th, 1983. Vanadium and other metal ions in the physiological ecology of marine organisms, *Struct. Bonding* 53: 139–160.

Lai, C. Y., and Horecker, B. L., 1972. Aldolase: A model for enzyme structure–function relationships, *Essays Biochem.* 8:149–178.

Lammers, M., and Follman, H., 1983. The ribonucleotide reductase: A unique group of metalloenzymes essential for cell proliferation, *Struct. Bonding* 54:27–91.

Lamola, A. A., and Yamane, T., 1974. Zinc protoporphyrin in the erythrocytes of patients with lead intoxication and iron deficiency anemia, *Science* 186:936–938.

Lanyi, J. K., 1977. Light-induced transport in *Halobacterium* cell envelope vesicle, in *Bioenergetics of Membranes* (L. Packer, G. C. Papageogiou, and A. Trebst, eds.), Elsevier, Amsterdam, pp. 129–135.

Larrabee, J. A., and Spiro, T. G., 1980. Structural studies of the hemocyanin active site. 2. Resonance Raman spectroscopy, *J. Am Chem. Soc.* 102:4217–4223.

Latore, R., Coronado, R., and Verga, C., 1984. K^+ channels gated by voltage and ions, *Annu. Rev. Physiol.* 46:485–495.

Lawrence, G. D., and Sawyer, D. T., 1978. The chemistry of biological manganese, *Coord. Chem. Rev.* 27:173–193.

Leach, R. M., Jr., Muenster, A. M., and Wien, E. M., 1969. Role of manganese in bone formation. II. Effect upon chondroitin sulfate synthesis in chick epiphyseal cartilage, *Arch. Biochem. Biophys.* 133:22–28.

Lee, C. S., and O'Sullivan, W. J., 1985. The interaction of phosphorothioate analogues of ATP with phosphomevalonate kinase, *J. Biol. Chem.* 260:13909–13915.

Lee, I. P., and Dixon, R. L., 1973. Effects of cadmium on spermatogenesis studied by velocity sedimentation cell separation and serial mating, *J. Pharmacol. Exp. Ther.* 187: 641–652.

Lee, K. K., Erickson, R., Pan, S.-S., Jones, G., May, F., and Nason, A., 1974. Effect of tungsten and vanadium on the *in vitro* assembly of assimilatory nitrate reductase utilizing, *Neurospora crassa* mutant nit-1, *J. Biol. Chem.* 249:3953–3959.

Lerch, K., 1982. Primary structure of tyrosinase from *Neurospora crassa*, II, *J. Biol. Chem.* 257:6414–6419.

Levine, B. A., and Williams, R. J. P., 1982. Calcium binding to protein and anion centers, in *Calcium and Cell Function* (Cheung, W. Y., ed.), Academic Press, New York, Vol. II, pp. 1–38.

Levitt, J., 1972. *Responses of Plants to Environmental Stresses*, Academic Press, New York, pp. 524–530.

Levy, L., 1976. Antiinflammatory action of some compounds with antioxidant properties, *Inflammation* 1:333–345.

Lichtenberger, F., Nastainczyk, W., and Ullrich, V., 1976. Cytochrome P-450 as an oxene transferase, *Biochem. Biophys. Res. Commun.* 70:939–946.

Lieben, J. (ed.), 1973. *Manganese,* National Academy of Sciences, Washington, D.C.

Lindahl, P. A., Kojima, N., Hausinger, R. P., Fox, J. A., Boon, K. T., Walsh, C. T., and Orme-Johnson, W. H., 1984. Nickel and iron EXAFS of F_{420}-reducing hydrogenase from *Methanobacterium thermoautotrophicum, J. Am. Chem. Soc.* 106:3062–3064.

Lindskog, S., Henderson, L. E., Kannan, K. K., Liljas, A., Nyman, P. O., and Strandberg, B., 1971. Carbonic anhydrase, in *The Enzymes,* 3rd ed. (P. D. Boyer, ed.), Vol. V, Academic Press, New York, pp. 587–665.

Lindskog, S., Ibrahim, S. A., Jonsson, B.-H., and Simonsson, I., 1982. Carbonic anhydrase: Structure and mechanism, in *The Coordination Chemistry of Metalloenzymes* (I. Bertini, R. S. Drago, and C. Luchinat, eds.), Reidel, Dordrecht, pp. 49–64.

Lindstedt, S., and Rundgren, M., 1982. Blue color, metal content and the substrate binding in 4-hydroxyphenylpyruvate dioxygenase from *Pseudomonas* species strain P. J. 874, *J. Biol. Chem.* 257:11922–11931.

Lippard, S. J., 1982. New chemistry of an old molecule: cis-[Pt(NH₃)₂Cl₂], *Science* 218: 1075–1082.

Ljungdahl, L. G., 1976. Tungsten, a biologically active metal, *Trends Biochem. Sci.* 1:63–65.

LoBrutto, R., Scholes, C. P., Wagner, G. C., Gunsalus, I. C., and Debrunner, P. G., 1980. Electron nuclear double resonance of ferric cytochrome P-450$_{CAM}$, *J. Am Chem. Soc.* 102:1167–1170.

Loeb, L. A., 1974. Eukaryotic DNA polymerase, in *The Enzymes,* 3rd ed. (P. D. Boyer, ed.), Academic Press, New York, pp. 173–209.

Loeb, L. A., and Mildvan, A. S., 1981. The role of metal ions in the fidelity of DNA and RNA synthesis, in *Metal Ions in Genetic Information Transfer* (G. L. Eichhorn and L. G. Marzilli, eds.), Elsevier, Amsterdam, pp. 125–142.

Loeb, L. A., and Zakour, R. A., 1980. Metals and genetic miscoding, in *Nucleic Acid–Metal Ion Interaction* (T. G. Spiro, ed.), Wiley, New York, pp. 115–144.

Luckey, T. D., and Venugopal, B., 1977. *Metal Toxicity in Mammals,* Vol. 1, Plenum Press, New York.

Lynch, S. R., Lipschitz, D. A., Bothwell, T. H., and Charlton, R. W., 1974. Iron and the reticuloendothelial system, in *Iron in Biochemistry and Medicine* (A. Jacobs and M. Worwood, eds.), Academic Press, New York, pp. 563–587.

Macara, I. G., McLeod, G. C., and Kustin, K., 1979. Isolation, properties and structural studies on a compound from tunicate blood cells that may be involved in vanadium accumulation, *Biochem. J.* 181:457–465.

McBride, B. C., Merilee, H., Cullen, W. R., and Pickett, W., 1978. Anaerobic and aerobic alkylation of arsenic, *ACS Symp. Ser.* 82:94–115.

McBrien, D. C. H., and Hassal, K. A., 1967. The effect of toxic doses of copper upon respiration, photosynthesis and growth of *Chlorella vulgaris, Physiol. Plant* 20:113–117.

McCord, J., 1974. Free radicals and inflammation: Protection of synovial fluid by superoxide dismutase, *Science* 185:529–531.

MacManus, J. P., Watson, D. C., and Yaguchi, M., 1986. The purification and complete amino acid sequence of the 9000 Mr Ca-binding protein from rat placenta, *Biochem. J.* 235:585–595.

Makinen, M. W., and Yim, M. B., 1981. Coordination environment of the active site metal ion of liver alcohol dehydrogenase, *Proc. Natl. Acad. Sci. USA* 78:6221–6225.

Makinen, M. W., Yamamura, K., and Kaiser, E. T., 1976. Mechanism of action of carboxypeptidase A in ester hydrolysis, *Proc. Natl. Acad. Sci. USA* 73:3882–3886.

Makishima, S., 1960. Surface reaction zone of solid catalysts, *Catalyst (Tokyo)* 2:168–171.

Malacinski, G. M., and Konetzka, W. A., 1966. Bacterial oxidation of orthophosphite, *J. Bacteriol.* 91:578–582.

Malacinski, G. M., and Konetzka, W. A., 1967. Orthophosphite-nicotinamide adenine dinucleotide oxidoreductase from *Pseudomonas fluorescens*, *J. Bacteriol.* 93:1906–1910.

Margoshes, M., and Vallee, B. L., 1957. A cadmium protein from equine kidney cortex, *J. Am. Chem. Soc.* 79:4813–4814.

Margulis, L., 1981. *Symbiosis in Cell Evolution*, Freeman, San Francisco.

Martin, B., 1984. Bioinorganic chemistry of calcium, in *Metal Ions in Biological Systems* (H. Sigel, ed.), Vol. 17, Dekker, New York, pp. 1–49.

Martin, D. F., and Martin, B. B., 1976. Red tide, red terror, *J. Chem. Educ.* 53:614–616.

Martin, D. R., and Williams, R. J. P., 1976. Chemical nature and sequence of alamethicin, *Biochem. J.* 153:181–190.

Martin, D. W., Tanford, C., and Reynolds, J. A., 1984. Monomeric solubilized sarcoplasmic reticulum Ca pump protein: Demonstration of Ca binding and dissociation coupled to ATP hydrolysis, *Proc. Natl. Acad. Sci. USA* 81:6623–6626.

Martin, J.-F., 1977. Biosynthesis of polyene macrolide antibiotics, *Annu. Rev. Microbiol.* 31:13–38.

Marx, J. J. M., and Aisen, P., 1981. Iron uptake by rabbit intestinal mucosal membrane vesicles, *Biochim. Biphys. Acta* 649:296–304.

Martinez-Medellin, J., and Schulman, H. M., 1973. Preparation, properties, and metabolism of carbon-14-bicarbonate labeled transferrin, *Biochem. Biophys. Res. Commun.* 53:32–38.

Mason, H. S., 1957. Mechanism of oxygen metabolism, *Adv. Enzymol.* 19:79–233.

Mathews, B. W., Jansonius, J. N., Colman, P. M., Schoenborn, B. P., and Dupourque, D., 1972. Three dimensional structure of thermolysin, *Nature New Biol.* 238:37–41.

May, S. W., Philips, R. S., Mueller, P. W., and Herman, H. H., 1981. Dopamine-β-hydroxylase, comparative specificities and mechanisms of oxygenation reactions, *J. Biol. Chem.* 256:8470–8475.

Mayhew, S. G., and Ludwig, M. L., 1975. Flavodoxins and electron-transferring flavoproteins, in *The Enzymes*, 3rd ed. (P. D. Boyer, ed.), Vol. XII, Academic Press, New York, pp. 57–118.

Mayor, G. H., Keiser, J. A., Makdani, D., and Ku, P. K., 1977. Aluminum absorption and distribution: Effect of parathyroid hormone, *Science* 197:1187–1189.

Mazur, B. J., and Chui, C.-F., 1982. Sequence of the gene coding for beta-subunit of dinitrogenase from blue-green alga *Anabaena*, *Proc. Natl. Acad. Sci. USA* 79:6782–6786.

Mead, J. F., 1976. Free radical mechanisms of lipid damage and consequences for cellular membranes, in *Free Radicals in Biology* (W. A. Pryor, ed.), Vol. 1, Academic Press, New York, pp. 51–68.

Means, A. R., Tash, J. S., and Chaflouleas, J. G., 1982. Physiological implications of the presence, distribution and regulation of calmodulin in eukaryotic cells, *Physiol. Rev.* 62:1–39.

Mertz, W., 1969. Chromium occurrence and function in biological systems, *Physiol. Rev.* 49:163–239.

Meyer, J., Moulis, J.-M., and Lutz, M., 1984. Structural difference between [2Fe-2S] cluster in spinach ferredoxin and in the "red protein" from *Clostridium pasteurianum*, a resonance Raman study, *Biochem. Biophys. Res. Commun.* 119:828–835.

Meyer, O., and Rajagopalan, K. V., 1984. Selenite binding to carbon-monoxide oxidase from *Pseudomonas carboxydovorans*, *J. Biol. Chem.* 259:5612–5617.

Mietinen, J. K., 1975. The accumulation and excretion of heavy metals in organisms, in *Heavy Metals in the Aquatic Environment* (P. A. Krenkel, ed.), Elsevier, Amsterdam, pp. 155–162.

Mildvan, A. S., and Loeb, L. A., 1981. The role of metal ions in the mechanism of DNA and RNA polymerases, in *Metal Ions in Genetic Information Transfer* (G. L. Eichhorn and L. G. Marzilli, eds.), Elsevier, Amsterdam, pp. 103–123.

Mildvan, A. S., Kobes, R. D., and Rutter, J., 1971. Magnetic resonance studies of the role of divalent cation in the mechanism of yeast aldolase, *Biochemistry* 10:1191–1204.

Miller, R. K., and Shaikh, Z. A., 1983. Prenatal metabolism: Metals and metallothionein, in *Reproductive and Developmental Toxicity of Metals* (T. W. Clarkson, G. F. Nordberg, and P. R. Sayer, eds.), Plenum Press, New York, pp. 153–204.

Miller, R. S., Mildvan, A. S., Chang, H., Easterday, R., Maruyama, H., and Lane, M. D., 1968. The enzymatic carboxylation of phosphoenolpyruvate, *J. Biol. Chem.* 243:6030–6040.

Milne, D. B., 1984. Tin, in *Biochemistry of the Essential Ultratrace Elements* (E. Frieden, ed.), Plenum Press, New York, pp. 309–318.

Miquel, J., Oro, J., Bensch, K. G., and Johnson, J. E., 1977. Lipofuscin, fine structural and biochemical studies, in *Free Radicals in Biology* (W. A. Pryor, ed.), Vol. 3, Academic Press, New York, pp. 133–182.

Mitchell, P., 1977. Vectorial chemiosmotic processes, *Annu. Rev. Biochem.* 46:996–1005.

Mitchell, P., and Koppenol, W. H., 1982. Chemiosmotic ATPase mechanisms, *Ann. N.Y. Acad. Sci.* 402:584–601.

Moody, C. S., and Hassau, H. M., 1982. Mutagenicity of oxygen free radicals, *Proc. Natl. Acad. Sci. USA* 79:2855–2859.

Moog, R. S., McGuirl, M. A., Cote, C. E., and Dooley, D. M., 1986. Evidence for methoxatin (pyrroquinolinequinone) as the cofactor in bovine plasma amine oxidase from resonance Raman spectroscopy, *Proc. Natl. Acad. Sci. USA* 83:8435–8439.

Morgan, E. H., 1983. Cellular mechanism of iron uptake by developing erythroid cells, in *Structure and Function of Iron Storage and Transport Proteins* (I. Urushizaki, P. Aisen, I. Listowski, and J. W. Drysdale, eds.), Elsevier, Amsterdam, pp. 275–282.

Morningstar, J. E., Johnson, M. K., Cecchini, G., Ackrell, B. A. C., and Kearney, E. B., 1985. The high potential iron–sulfur cluster in *Escherichia coli* fumarate reductase is a three-iron cluster, *J. Biol. Chem.* 260:13631–13638.

Mortenson, L. E., and Chen, J.-S., 1974. Hydrogenase, in *Microbial Iron Metabolism* (J. B. Nielands, ed.), Academic Press, New York, pp. 231–282.

Mortenson, L. E., and Thorneley, R. N. F., 1979. Structure and function of nitrogenase, *Annu. Rev. Biochem.* 48:387–418.

Muir, W. A., Hopfer, U., and King, M., 1984. Iron transport across brush-border membranes from normal and iron-deficient mouse upper small intestine, *J. Biol. Chem.* 259:4896–4903.

Mukai, K., Huang, J. J., and Kimura, T., 1973. Electron paramagnetic resonance studies on the selenium-replaced derivatives of adrenodoxin, presence of the one selenium–one sulfur compound, *Biochem. Biophys. Res. Commun.* 50:105–110.

Munck, E., 1982. Mössbauer studies on [3Fe-3S] clusters and sulfite reductase, in *Iron–Sulfur Proteins* (T. G. Spiro, ed.), Wiley, New York, pp. 147–175.

Murphy, M. J., and Siegel, L. M., 1973. Siroheme and sirohydrochlorin, *J. Biol. Chem.* 248:6911–6919.

Myenell, G. G., 1972. *Bacterial Plasmids*, Macmillan Co., New York.

Nagatani, H. H., and Brill, W. J., 1974. Nitrogenase vanadium, effect of molybdenum, tungsten and vanadium on the synthesis of nitrogenase components in *Azotobacter vinelandii*, *Biochim. Biophys. Acta* 362:160–166.

Nakahara, H., Ishikawa, T., Sakai, Y., Kondo, I., and Mitsuhashi, S., 1977. Frequence of heavy-metal resistance in bacteria from patient in Japan, *Nature* 266:165–167.

Nakano, A.,Tasumi, M., Fujiwara, K., Fuwa, K., and Miyazawa, T., 1979. Nuclear magnetic relaxation study on the interaction of glycyl-L-tyrosine with manganese-carboxypeptidase A, *J. Biochem. (Tokyo)* 86:1001–1011.

Nakano, M., Noguchi, T., Sugioka, K., Fujiyama, H., Sato, M., Shimizu, Y., and Tsuji, Y., 1975. Spectroscopic evidence for the generation of singlet oxygen in the reduced nicotinamide adenine dinucleotide phosphate-dependent microsomal lipid peroxidation system, *J. Biol. Chem.* 250:2404–2406

Narins, D., 1980. Absorption of nonheme iron, in *Biochemistry of Nonheme Iron* (A. Bezkorovainy, ed.), Plenum Press, New York, pp. 47–126.

Nason, A., Lee, K.-Y., Pan, S.-S., Ketchum, P. A., Lambert, A., and Davies, J., 1971. *In vitro* formation of assimilatory reduced nicotinamide adenine dinucleotide from a *Neurospora* mutant and a component of molybdenum enzymes, *Proc. Natl. Acad. Sci. USA* 68:3242–3246.

Neilands, J. B., 1972. Evolution of biological iron binding center, *Struct. Bonding* 11: 145–170.

Neilands, J. B. (ed.), 1974a. *Microbial Iron Metabolism,* Academic Press, New York.

Neilands, J. B., 1974b. Iron and its role in microbial physiology, in *Microbial Iron Metabolism* (J. B. Neilands, ed.), Academic Press, New York, pp. 3–34.

Neilands, J. B., 1977. Siderophores: Biochemical ecology and mechanism of iron transport in eubacteria, *Adv. Chem. Ser.* 162:3–32.

Neilands, J. B., 1982. Microbial envelope proteins related to iron, *Annu. Rev. Microbiol.* 36:285–309.

Newton, W. H., and Otsuka, S. (eds.), 1980. *Molybdenum Chemistry of Biological Significance,* Plenum Press, New York.

Nexø, E., and Olesen, H., 1982. Intrinsic factor, transcobalamin and hepatocorrin, in B_{12} (D. Dolphin, ed.), Vol. 2, Wiley–Interscience, New York, pp. 57–85.

Nielsen, F. H., 1977. Nickel toxicology, in *Toxicity of Trace Elements* (R. A. Goyer and M. A. Mehlman, eds.) Hemisphere, New York, pp. 129–146.

Nielsen, F. H., Ollerich, D. A., Fosmire, G. J., and Sandstead, H. H., 1974. Nickel deficiency in chicks and rats, in *Protein–Metal Interaction* (M. Friedman, ed.) Plenum Press, New York, pp. 389–403.

Nishinaga, A., 1975. Oxygenation of 3-substituted indoles catalyzed by Co(II)–Schiff base complexes, *Chem. Lett.* 1975:273–276.

Nishinaga, A., Tojo, T., and Matsuura, T., 1974. A model catalytic oxygenation for reaction of quercetinase, *J. Chem. Soc. Chem. Commun.* 1974:896–897.

Nissen, P., 1974. Uptake mechanisms: Inorganic and organic, *Annu. Rev. Plant Physiol.* 25:53–79.

Noda, M., Shimidau, S., Tanaka, T., Takai, T., Kayano, T., Ikeda, T., Takahashi, H., Nakayama, H., Kanaoka, Y., Minamino, N., Kanagawa, K., Matsuo, H., Raftery, M. A., Hiroe, T., Inayama, S., Hayashida, H., Miyata, T., and Numa, S., 1984. Primary structure of *Electrophorus electricus* sodium channel from cDNA sequence, *Nature* 312:121–127.

Nordberg, G. F. (ed.), 1976a. *Effects and Dose–Response Relationships of Toxic Metals,* Sections 4.2 and 5.2, Elsevier, Amsterdam.

Nordberg, G. F. (ed.), 1976b. *Effects and Dose–Response Relationships of Toxic Metals,* Sections 4.3 and 5.3, Elsevier, Amsterdam.

Nordberg, G. F. (ed.), 1976c. *Effects and Dose–Response Relationships of Toxic Metals,* Sections 4.4 and 5.4, Elsevier, Amsterdam.

Nordblom, G. D., White, R. E., and Coon, M. J., 1976. Studies on hydroperoxide-dependent substrate hydroxylation by purified liver microsomal cytochrome P-450, *Arch. Biochem. Biophys.* 175:524–533.

Novick, R. P., and Roth, C., 1968. Plasmid-linked resistance to inorganic salt in *Staphylococcus aureus, J. Bacteriol.* 95:1335–1342.

Ochiai, E.-I., 1973a. Oxygenation of cobalt(II) complexes, *J. Inorg. Nucl. Chem.* 35:3375–3389.

Ochiai, E.-I., 1973b. Electronic structure and oxygenation of bis(salicylaldehyde) ethyl-enediamine cobalt(II), *J. Inorg. Nucl. Chem.* 35:1727–1734.

Ochiai, E.-I., 1974a. O–O bond cleavage in oxygenation, *Inorg. Nucl. Chem. Lett.* 10: 453–457.

Ochiai, E.-I., 1974b. Oxygen activation by heme, a theoretical and comparative study, *J. Inorg. Nucl. Chem.* 36:2129–2132.

Ochiai, E.-I., 1975. Bioinorganic chemistry of oxygen, *J. Inorg. Nucl. Chem.* 37:1503–1509.

Ochiai, E.-I., 1977. *Bioinorganic Chemistry: An Introduction,* Allyn & Bacon, Rockleigh, N.J.

Ochiai, E.-I., 1978a. Principles in the selection of inorganic elements by organisms, application to molybdenum enzymes, *BioSystems* 10:329–337.

Ochiai, E.-I., 1978b. Principles in bioinorganic chemistry, basic inorganic exercise, *J. Chem. Educ.* 55:631–633.

Ochiai, E.-I., 1978c. The evolution of the environment and its influence on the evolution of life, *Origins Life* 9:81–91.

Ochiai, E.-I., 1983a. Inorganic chemistry of ancient sediments: Bioinorganic aspects of origin and evolution of life, in *Cosmochemistry and Origin of Life* (C. Ponnamperuma, ed.), Reidel, Dordrecht, pp. 235–276.

Ochiai, E.-I., 1983b. Copper and the biological evolution, *BioSystems* 16:81–86.

Ochiai, E-I., and Morand, R., 1985. Formation constant of the calcium (ATP)$_2$ complex, *J. Coordin. Chem.* 14:83–86.

Octave, J.-N., Schneider, Y.-J., Sibille, J.-C., Crichton, R. R., and Trouet, A., 1983. Transferrin protein and iron uptake by cultured mammalian cells, in *Structure and Function of Iron Storage and Transport Proteins* (I. Urushizaki, P. Aisen, I. Listowski, and J. W. Drysdale, eds.), Elsevier, Amsterdam, pp. 321–325.

Oh, K. K., and Freese, E., 1976. Manganese requirement of phosphoglycerate phosphomutase and its consequences for growth and sporulation of *Bacillus subtilis, J. Bacteriol.* 127:739–746.

Ohnishi, K., Hirata, F., and Hayaishi, O., 1977. Indoleamine 2,3-dioxygenase: Potassium superoxide as substrate, *J. Biol. Chem.* 252:4643–4647.

Olson, J. A., 1965. The biosynthesis of cholesterol, *Ergeb. Physiol. Biol. Chem. Exp. Pharmakol.* 56:173–215.

Olson, J. S., Ballou, D. P., Palmer, G., and Massey, V., 1974. The mechanism of action of xanthine oxidase, *J. Biol. Chem.* 249:4363–4382.

Papa, S., 1982. Mechanism of active proton translocation by cytochrome systems, in *Membranes and Transport* (A. N. Martonosi, ed.), Vol. 1, Plenum Press, New York, pp. 363–368.

Parizek, J., 1976. Interrelationships among trace elements, in *Effects and Dose–Response Relationships of Toxic Metals* (G. F. Nordberg, ed.), Elsevier, Amsterdam, pp. 498–510.

Paschinger, H., 1974. Changed nitrogenase activity in *Rhodospirillum rubrum* after substitution of tungsten for molybdenum, *Arch. Microbiol.* 101:379–389.

Passow, H., Rothstein, A., and Clarkson, T., 1961. The general pharmacology of heavy metals, *Pharmacol. Rev.* 13:185–224.

Pateman, J. A., Core, D. J., Rever, B. M., and Roberts, D. B., 1964. A common cofactor for nitrate reductase and xanthine dehydrogenase which also regulates the synthesis of nitrate reductase, *Nature* 201:58–60.

Pedersen, P. L., 1982. H$^+$-ATPase in biological systems: An overview of their function, structure, mechanism and regulatory properties, *Ann. N.Y. Acad. Sci.* 402:1–20.

Perry, H. M., Jr., 1976. Review of hypertension induced in animals by chronic ingestion of cadmium, in *Trace Elements in Human Health and Disease* (A. S. Prasad and D. Oberleas, eds.), Vol. II, Academic Press, New York, pp. 417–430.

Perutz, M. F., 1978. Hemoglobin structure and respiratory transport, *Sci. Am.* 1978(12): 92–125.

Perutz, M. F., 1982. Nature of the iron–oxygen bond and control of oxygen affinity of the heme by the structure of the globin in hemoglobin, in *Structure and Function Relationship in Biochemical Systems* (F. Bossa, E. Chiancone, A. Finazzi-Agro, and R. Strom, eds.), Plenum Press, New York, pp. 31–48.

Pfaltz, A., Jaun, B., Eschenmoser, A., Jaenchen, R., Gilles, H. H., Diekert, G., and Thauer, R. K., 1982. Zur kentnis des faktors F430 aus methanogenen bakterien: struktur des porphinoiden ligandsystem, *Helv. Chim. Acta* 65:828–865.

Phillips, S. E. V., 1978. Structure of oxymyoglobin, *Nature* 273:247–248.

Philson, S. B., Debrunner, P. G., Schmidt, P. G., and Gunsalus, I. C., 1979. The effect of cytochrome P-450$_{CAM}$ on the NMR relaxation rate of water protons, *J. Biol. Chem.* 254:10173–10179.

Piras, R., and Vallee, B. L., 1967. Procarboxypeptidase A–carboxypeptidase A relationships: Metal and substrate binding, *Biochemistry* 6:348–357.

Piscator, M., 1976. The chronic toxicity of cadmium, in *Trace Elements in Human Health and Disease* (A. S. Prasad and D. Oberleas, eds.), Vol. II, Academic Press, New York, pp. 431–441.

Pitman, M. G., 1976. Ion uptake by plant roots, in *Transport in Plants II, Part B* (U. Luttuge and M. G. Pitman, eds.), Springer-Verlag, Berlin, pp. 95–128.

Pocker, Y., and Stone, J. T., 1967. The catalytic versatility of erythrocyte carbonic anhydrase. III. Kinetic studies of the enzyme-catalyzed hydrolysis of p-nitrophenyl acetate, *Biochemistry* 6:668–678.

Pocker, Y., and Stone, J. T., 1968. The catalytic versatility of erythrocyte carbonic anhydrase. IV. Kinetic studies of noncompetitive inhibition of enzyme-catalyzed hydrolysis of p-nitrophenyl acetate, *Biochemistry* 7:2936–2945.

Ponnamperuma, C. (ed.), 1983. *Cosmochemistry and Origin of Life,* Reidel, Dordrecht.

Postgate, J. R., 1974. Evolution within nitrogen-fixing systems, in *Evolution in the Microbial World* (M. J. Carlile and J. J. Skehel, eds.), Cambridge University Press, London, pp. 263–292.

Potter, J. D., and Johnson, J. D., 1982. Troponin, in *Calcium and Cell Function* (W. Y. Cheung, ed.), Vol. II, Academic Press, New York, pp. 145–173.

Poulos, T. L., Finzel, B. C., Gunsalus, I. C., Wagner, G. C., and Kraut, J., 1985. The 2.6 A crystal structure of *Pseudomonas putida* cytochrome P-450, *J. Biol. Chem.* 260: 16122–16130.

Powers, L., Chance, B., Ching, Y., and Angiolillo, P., 1981. Structural features and the reaction mechanism of cytochrome oxidase, *Biophys. J.* 34:465–498.

Prasad, A. S., and Oberleas, D. (eds.), 1976. *Trace Elements in Human Health and Disease,* Vol. 1, Academic Press, New York.

Pratt, J. M., 1972. *Inorganic Chemistry of Vitamin B$_{12}$,* Academic Press, New York.

Pratt, J. M., 1982. Coordination chemistry of the B$_{12}$ dependent reactions, in *B$_{12}$* (D. Dolphin, ed.), Vol. 1, Wiley, New York, pp. 325–392.

Premakumar, R., Winge, D. R., Wiley, R. D., and Rajagopalan, K. V., 1975. Copper-induced synthesis of copper-chelatin in rat liver, *Arch. Biochem. Biophys.* 170:267–277.

Prinz, R., and Weser, U., 1975. Naturally-occurring copper-thionein in *Saccharomyces cerevisiae, Hoppe-Seylers Z. Physiol. Chem.* 356:767–776.

Que, L., and Epstein, R. M., 1981. Resonance raman studies on protocatechuate 3,4-dioxygenase-inhibitor complexes, *Biochemistry* 20:2545–2549.

Que, L., Jr., Lipscomb, J. D., Zimmermann, R., Munck, E., Orme-Johnson, N. R., and Orme-Johnson, W. H., 1976. Mössbauer and EPR spectroscopy on protocatechuate 3,4-dioxygenase from *Pseudomonas aeruginosa, Biochim. Biophys. Acta* 452:320–334.

Que, L., Jr., Lipscomb, J. D., Munck, E., and Wood, J. M., 1977. Protocatechuate 3,4-dioxygenase, inhibitor studies and mechanistic implications, *Biochim. Biophys. Acta* 485:60–74.

Que, L., Heistand, R. H., Mayer, R., and Roe, A. L., 1980. Resonance raman studies of pyrocatechase-inhibitor complexes, *Biochemistry* 19:2588–2593.

Que, L., Jr., Widom, J., and Crawford, R. L., 1981. 3,4-Dihydroxyphenylacetate 2,3-dioxygenase: A Mn(II) dioxygenase from *Bacillus brevis, J. Biol. Chem.* 256:10941–10944.

Quigley, G. J., Teeter, M. M., and Rich, A., 1978. Structural analysis of spermine and magnesium ion binding to yeast phenylalanine t-RNA, *Proc. Natl. Acad. Sci. USA* 75: 64–68.

Quiocho, F. A., and Lipscomb, W. N., 1971. Carboxypeptidase A, *Adv. Protein Chem.* 25: 1–78.

Racker, E., 1977a. Proposal for a mechanism of Ca^{++} transport, in *Calcium-Binding Proteins and Calcium Function* (R. H. Wasserman, R. A. Corradino, E. Carafoli, R. H. Kretsinger, D. H. McLennan, and F. L. Siegel, eds.), North-Holland, Amsterdam, pp. 155–163.

Racker, E., 1977b. Mechanisms of energy transformation, *Annu. Rev. Biochem.* 46:1006–1014.

Ragsdale, S. W., Clark, J. E., Ljungdahl, L. G., Lundie, L. L., and Drake, H. L., 1983. Properties of purified carbon monoxide dehydrogenase from *Clostridium thermoaceticum*, a nickel, iron–sulfur protein, *J. Biol. Chem.* 258:2364–2369.

Ragsdale, S. W., Wood, H. G., and Antholine, W. E., 1985. Evidence that an iron–nickel–carbon complex is formed by reaction of CO with the CO dehydrogenase from *Clostridium thermoaceticum, Proc. Natl. Acad. Sci. USA* 82:6811–6814.

Ramachandran, G. N., and Sasisekharan, V., 1968. Conformation of polypeptides and proteins, *Adv. Protein Chem.* 23:283–438.

Recknagel, R. O., 1967. Carbon tetrachloride hepatotoxicity, *Pharmacol. Rev.* 19:145–208.

Reddy, K., Fletcher, B., Tappel, A., and Tappel, A. L., 1973. Measurement and spectral characteristics of fluorescent pigments in tissues of rats as a function of dietary polyunsaturated fats and vitamin E, *J. Nutr.* 103:908–915.

Reed, G. H., and Scrutton, M. C., 1974. Pyruvate carboxylase from chicken liver, *J. Biol. Chem.* 249:6156–6162.

Rees, D. C., Lewis, M., Hozantko, R. B., Lipscomb, W. N., and Hardman, K. D., 1981. Zinc environment and *cis*-peptide bonds in carboxypeptidase A at 1.75 Å resolution, *Proc. Natl. Acad. Sci. USA* 78:3408–3412.

Reeves, A. L., 1979. Beryllium, in *Handbook on the Toxicology of Metals* (L. Friberg, G. F. Nordberg, and B. Vouk, eds.), Elsevier, Amsterdam.

Reeves, A. L., and Vorwald, A. J., 1967. Beryllium carcinogenesis. II. Pulmonary deposition and clearance of inhaled beryllium sulfate in the rat, *Cancer Res.* 27:446–451.

Reptke, K. R. H., 1982. On the mechanism of energy release, transfer and utilization in Na,K-ATPase transport work: Old ideas and new findings, *Ann. N.Y. Acad. Sci.* 402: 272–285.

Rice, H. A., Leighty, D. A., and McLeod, G. C., 1973. The effects of some trace metals on marine phytoplankton, *Crit. Rev. Microbiol.* 3:27–49.

Rich, A., and Raj-Bhandary, U. L., 1976. Transfer RNA: Molecular structure, sequence and properties, *Annu. Rev. Biochem.* 45:805–860.

Richards, M. P., and Cousins, R. J., 1975. Mammalian zinc homeostasis: Requirement for RNA and metallothionein synthesis, *Biochem. Biophys. Res. Commun.* 64:1215–1223.

Richardson, J. R., Thomas, K. A., Rubin, B. H., and Richardson, D. C., 1975. Crystal structure of bovine Cu/Zn superoxide dismutase at 3 A resolution: Chain tracing and metal ligands, *Proc. Natl. Acad. Sci. USA* 72:1349–1353.

Richter, G. W., 1983. Cellular ferritin overload and formation of hemosiderin, in *Structure and Function of Iron Storage and Transport Proteins* (I. Urushizaki, P. Aisen, I. Listowski, and J. W. Drysdale, eds.), Elsevier, Amsterdam, pp. 155–162.

Ridley, W. P., Dizikes, L. J., and Wood, J. M., 1977. Biomethylation of toxic elements in the environment, *Science* 197:329–332.

Riedel, W. R., and Sanfilippo, A., 1981. Evolution and diversity of form in Radiolaria, in *Silicon and Siliceous Structure in Biological Systems* (T. L. Simpson and B. E. Volcani, eds.), Springer-Verlag, Berlin, pp. 322–346.

Ringe, D., Petsko, G. A., Yamamura, F., Suzuki, K., and Ohmori, D., 1983. Structure of iron superoxide dismutase from *Pseudomonas ovalis* at 2.9 A resolution, *Proc. Natl. Acad. Sci. USA* 80:3879–3883.

Ritchie, J. M., and Rogart, R. B., 1977. The binding of saxitoxin and tetrodotoxin to excitable tissue, *Rev. Physiol. Biochem. Pharmacol.* 79:1–50.

Roberts, J. J., 1981. The mechanism of action of antitumor platinum coordination compounds, in *Metal Ions in Genetic Information Transfer* (G. L. Eichhorn and L. G. Marzilli, eds.), Elsevier, Amsterdam, pp. 273–332.

Robson, R. L., and Postgate, J. R., 1980. Oxygen and hydrogen in biological nitrogen fixation, *Annu. Rev. Microbiol.* 34:183–207.

Robson, R. L., Eady, R. R., Richardson, T. H., Miller, R. W., Hawkins, M., and Postgate, J. R., 1986. The alternative nitrogenase of *Azotobacter chroococcum* is a vanadium enzyme, *Nature* 322:388–390.

Roe, A. L., Schneider, D. J., Mayer, R. J., Pyrz, J. W., Widom, J., and Que, L., 1984. X-ray absorption spectroscopy of iron-tyrosinate protein, *J. Am. Chem. Soc.* 106:1676–1681.

Rojas, E., 1977. Displacement currents and Na-channel gating in nerve membrane, in *Electrical Phenomena at the Biological Membrane Level* (E. Roux, ed.), Elsevier, Amsterdam, pp. 233–252.

Roodyn, D., and Wilkie, D., 1968. *The Biogenesis of Mitochondria*, Methuen, London.

Roos, D., 1977. Oxidative killing of microorganisms by phagocytic cells, *Trends Biochem. Sci.* 2:61–64.

Rosenberg, B., 1980. Platinum complexes for the treatment of cancer, in *Nucleic Acid–Metal Ion Interaction* (T. G. Spiro, ed.), Wiley, New York, pp. 1–29.

Rosenberg, H., and Young, I. G., 1974. Iron transport in the enteric bacteria, in *Microbial Iron Metabolism* (J. B. Neilands, ed.), Academic Press, New York, pp. 67–82.

Rosenfeld, I., and Beath, O. A., 1964. *Selenium*, Academic Press, New York.

Rotilio, R., Morpurgo, L., Calabrese, L., and Mondovi, B., 1973. Mechanism of superoxide dismutase reaction of the bovine enzyme with hydrogen peroxide and ferricyanide, *Biochim. Biophys. Acta* 302:229–235.

Rotilio, R., Calabrese, L., Rigo, A., and Fielden, E. M., 1982. The mechanism of action of Cu-Zn superoxide dismutase, in *Structure and Function Relationships in Biological Systems* (F. Bossa, E. Bossa, E. Chiancone, A. Finazzi-Agro and R. Strom, eds.), Plenum Press, New York, pp. 155–168.

Rubin, R. P., Weiss, G. B., and Putney, J. W., Jr. (eds.), 1985. *Calcium in Biological Systems*, Plenum Press, New York.

Sagers, R. D., 1974. Other iron-containing or iron-activated enzymes: Enzymes acting on certain amino acids, amines and acetyl phosphate, in *Microbial Iron Metabolism* (J. B. Neilands, ed.), Academic Press, New York, pp. 445–453.

Sanders, T. G., and Rutter, W. J., 1972. Molecular properties of rat pancreatic parotid alpha-amylase, *Biochemistry* 11:130–136.

Sawyer, D. T., and Bodini, M. E., 1975. Manganese(II) gluconate, redox model for photosynthetic oxygen evolution, *J. Am. Chem. Soc.* 97:6588–6590.

Schimpff-Weiland, G., Follman, H., and Auling, G., 1981. A new manganese-activated ribonucleotide reductase found in gram-positive bacteria, *Biochem. Biophys. Res. Commun.* 102:1276–1282.

Schloss, J. V., Porter, D. J. T., Bright, H. J., and Cleland, W. W., 1980. Nitro analogues of citrate and isocitrate as transition-state analogues for aconitase, *Biochemistry* 19: 2358–2362.

Schneider, D. J., Roe, L. A., Mayer, R. J., and Que, L., Jr., 1984. Evidence for synergistic anion binding to iron in ovatransferrin complexes from resonance Raman and extended X-ray absorption fine structure analysis, *J. Biol. Chem.* 259:9699–9703.

Schneider, E. G., Durham, J. C., and Sacktor, B., 1984. Sodium-dependent transport of inorganic sulfate by rabbit renal brush-border membrane vesicle, *J. Biol. Chem.* 259: 14591–14599.

Schneider, P. W., Bravard, D. C., McDonald, J. W., and Newton, W. E., 1972. Reactions of oxobis(N, N-dialkyldithiocarbamato)molybdenum(IV) with unsaturated organic molecules and their biochemical implications, *J. Am. Chem. Soc.* 94:8640–8641.

Schobert, B., and Lanyi, J. K., 1982. Halorhodopsin is a light-driven chloride pump, *J. Biol. Chem.* 257:10306–10313.

Schopf, J. W. (ed.), 1983. *Earth's Earliest Biosphere—Its Origin and Evolution,* Princeton University Press, Princeton, N.J.

Schopf, J. W., and Walter, M. R., 1983. Archean microfossils: New evidence of ancient microbes, in *Earth's Earliest Biosphere—Its Origin and Evolution* (J. W. Schopf, ed.), Princeton University Press, Princeton, N.J., pp. 214–234.

Schroepfer, G. J., Jr., 1982. Sterol biosynthesis, *Annu. Rev. Biochem.* 51:555–585.

Schutt, C., 1985. Hands on the calcium switch, *Nature* 315:15.

Schwartz, K., 1973. A band form of silicon in glycosaminoglycans and polyuroides, *Proc. Natl. Acad. Sci. USA* 70:1608–1612.

Schwartz, K., 1974. New essential trace elements (Sn, V, F, Si) progress report, in *Trace Element Metabolism in Animals-2* (W. G. Hoekstra, J. M. Suttie, H. E. Ganther, and W. Mertz, eds.), University Park Press, Baltimore, pp. 355–380.

Schwartz, K., 1977. Silicon, fibre and atherosclerosis, *Lancet* 1977:454–456.

Schwartz, W., and Passow, H., 1983. Ca^{2+}-activated K channels in erythrocytes and excitable cells, *Annu. Rev. Physiol.* 45:359–374.

Scopes, R. K., 1983. An iron-activated alcohol dehydrogenase, *FEBS Lett.* 156:303.

Shaanan, B., 1982. The iron–oxygen bond in human oxyhemoglobin, *Nature* 296:683–684.

Shah, V. K., and Brill, W. J., 1981. Isolation of molybdenum–iron cluster from nitrogenase, *Proc. Natl. Acad. Sci. USA* 78:3438–3440.

Shamberger, R. J., 1983. *Biochemistry of Selenium,* Plenum Press, New York.

Shamoo, A., and Goldstein, D. A., 1977. Isolation of ionophores from ion transport systems and their role in energy transduction, *Biochim. Biophys. Acta* 472:13–53.

Sharma, H. K., and Vaidyanathan, C. S., 1975. New mode of ring cleavage of 2,3-dihydroxybenzoic acid in *Tecoma stans,* partial purification and properties of 2,3-dihydroxybenzoate 2,3-dioxygenase, *Eur. J. Biochem.* 56:163–171.

Sheridan, R. P., Allen, L. C., and Carter, C. W., Jr., 1981. Coupling between oxidation state and hydrogen bond conformation in high potential iron sulfur protein, *J. Biol. Chem.* 256:5052–5057.

Shieh, J. J., and Yasunobu, K. T., 1976. Purification and properties of lung lysyl oxidase, in *Iron and Copper Proteins* (K. T. Yasunobu, H. F. Mower, and O. Hayaishi, eds.), Plenum Press, New York, pp. 447–462.

Sies, E., Wendel, A., and Burk, R. F., 1982. Se and non-Se glutathione peroxidase: Enzymology and physiology, in *Oxidases and Related Redox Systems* (T. E. King, ed.), Pergamon Press, Elmsford, N.Y., pp. 169–189.

Sigel, H., 1982. Metal ion promoted hydrolysis of nucleoside 5'-triphosphate, in *The Coordination Chemistry of Metalloenzymes* (I. Bertini, R. S. Drago, and C. Luchinat, eds.), Reidel, Dordrecht, pp. 65–78.

Sigel, H. (ed.), 1984. *Metal Ions in Biological Systems*, Vol. 17, Dekker, New York.

Sigel, H., and Amsler, P. E., 1976. Hydrolysis of nucleoside phosphates. 6. On the mechanism of the metal ion promoted dephosphorylation of purine nucleoside 5'-triphosphates, *J. Am. Chem. Soc.* 98:7390–7400.

Silbergeld, E. K., Fales, J. T., and Goldberg, A. M., 1974. Evidence for a junctional effect of lead on neuromuscular function, *Nature* 247:49–50.

Sillen, L. G., 1965. Oxidation states of the earth's ocean and atmosphere, a model calculation on earlier states and the myth of the "probiotic" soup, *Arkiv. Kemi* 24:431–456.

Sillen, L. G., 1966. Oxidation states of the earth's ocean and atmosphere, II, behavior of Fe, S and Mn in earlier states; regulation mechanisms for O and N_2, *Arkiv. Kemi* 25: 159–176.

Silver, S., 1977. Calcium transport in microorganisms, in *Microorganisms and Minerals* (E. D. Weinberg, ed.), Dekker, New York, pp. 49–103.

Silver, S., 1984. Bacterial transformation of and resistance to heavy metals, in *Changing Metal Cycles and Human Cycles* (J. D. Nriagu, ed.), Springer-Verlag, Berlin, pp. 199–223.

Silver, S., and Jaspar, P., 1977. Manganese transport in microorganisms, in *Microorganisms and Minerals* (E. D. Weinberg, ed.), Dekker, New York, pp. 105–149.

Silver, S., and Keach, D., 1982. Energy-dependent arsenate efflux: The mechanism of plasmid-mediated resistance, *Proc. Natl. Acad. Sci. USA* 79:6114–6118.

Simpson, T. L., and Volcani, B. E., (eds.), 1981. *Silicon and Siliceous Structures in Biological Systems*, Springer-Verlag, Berlin.

Singh, H. N., Vaishampayan, A., and Singh, R. K., 1978. Evidence for the involvement of a genetic determinant controlling functional specificity of group VI b elements in the metabolism of N_2 and NO_3^- in the blue-green alga *Nostoc muscorum, Biochem. Biophys. Res. Commun.* 81:67–74.

Sirover, M. A., and Loeb, L. A., 1976. Metal-induced infidelity during DNA synthesis, *Proc. Natl. Acad. Sci. USA* 73:2331–2335.

Skaar, H., Ophus, E., and Gullvag, B. M., 1973. Lead accumulation within moss leaf cells, *Nature* 241:215–216.

Slater, E. C., 1977. Mechanism of oxidative phosphorylation, *Annu. Rev. Biochem.* 46: 1015–1026.

Slater, J. P., Tarmir, I., Loeb, L. A., and Mildvan, A. S., 1972. The mechanism of *Escherichia coli* deoxyribonucleic acid polymerase. I. Magnetic resonance and kinetic studies of the role of metals, *J. Biol. Chem.* 247:6784–6794.

Solomon, E. I., 1981. Binuclear copper active site, in *Copper Proteins* (T. G. Spiro, ed.), Wiley, New York, pp. 41–108.

Spiro, T. G. (ed.), 1983. *Calcium in Biology*, Wiley, New York.

Spivack, B., and Dori, Z., 1975. Structural aspects of molybdenum(IV), molybdenum(V) and molybdenum(VI) complexes, *Coord. Chem. Rev.* 17:99–136.

Spudich, E. N., and Spudich, J. L., 1985. Biochemical characterization of halorhodopsin in native membranes, *J. Biol. Chem.* 260:1208–1212.

Stadtman, E. R., 1971. Role of multiple molecular forms of glutamine synthetase in the regulation of glutamine metabolism in *Escherichia coli, Harvey Lect.* 65:97–125.

Stadtman, T. C., 1974. Selenium biochemistry, *Science* 183:915–921.

Stadtman, T. C., 1980. Selenium-dependent enzymes, *Annu. Rev. Biochem.* 49:93–110.

Stallings, W. C., Powers, T. B., Pattridge, K. A., Fee, J. A., and Ludwig, M. L., 1983. Iron superoxide dismutase from *Escherichia coli* at 3.1 A resolution: A structure unlike that of copper/zinc protein at both monomer and dimer levels, *Proc. Natl. Acad. Sci. USA* 80:3884–3888.

Stallings, W. C., Pattridge, K. A., Strong, R. K., and Ludwig, M. L., 1985. The structure of manganese superoxide dismutase from *Thermus thermophilus* HB8 at 2.4 A resolution, *J. Biol. Chem.* 260:16424–16432.

Steiner, D. F., Kemmler, W., Tager, H. S., and Peterson, J. D., 1974. Proteolytic processing in the biosynthesis of insulin and other proteins, *Fed. Proc.* 33:2105–2115.

Stellwagen, E., 1978. Heme exposure as the determinate of oxidation–reduction potential of heme proteins, *Nature* 275:73–74.

Stenkampf, R. E., Sieker, L. C., and Jensen, L. H., 1976. Structure of the iron complex in methemerythrin, *Proc. Natl. Acad. Sci. USA* 73:349–351.

Stenkampf, R. E., Sieker, L. C., Jensen, L. H., McCallum, J. D., and Saunders-Loehr, J., 1985. Active site structure of deoxyhemerythrin and oxyhemerythrin, *Proc. Natl. Acad. Sci. USA* 82:713–716.

Sternweis, P. C., and Gilman, A. G., 1982. Aluminum: A requirement for activation of the regulatory component of adenylate cyclase by fluoride, *Proc. Natl. Acad. Sci. USA* 79:4888–4891.

Stiefel, E. I., and Gardner, J. K., 1973. Proton and electron transfer in molybdenum complexes: Implications for molybdenum enzymes, in *Chemistry and Uses of Molybdenum* (P. C. H. Mitchell, ed.), Climax Molybdenum Co., Ann Arbor, pp. 272–277.

Stiegman, W., and Weber, E., 1979. Structure of erythrocruorin in different ligand states refined at 1.4 A resolution, *J. Mol. Biol.* 127:309–338.

Stirling, C. E., 1975. Mercurial perturbation of brush border membrane permeability in rabbit ileum, *J. Membr. Biol.* 23:33–50.

Stoeckenius, W., 1976. The purple membrane of salt-loving bacteria, *Sci. Am.* 1976(6): 38–46.

Stoeckenius, W., and Bogomolni, R. A., 1982. Bacteriorhodopsin and related pigments of halobacteria, *Annu. Rev. Biochem.* 51:587–616.

Stoffyn, M., 1979. Biological control of dissolved aluminum in sea water, experimental evidence, *Science* 203:651–653.

Storm, D. R., and Gunsalus, R. P., 1974. Methyl mercury is a potent inhibitor of membrane adenyl cyclase, *Nature* 250:778–779.

Stout, C. D., 1982. Iron–sulfur protein crystallography, in *Iron–Sulfur Proteins* (T. G. Spiro, ed.), Wiley, New York, pp. 97–146.

Stout, C. D., Ghosh, D., Pattabhi, V., and Robbins, A. H., 1980. Iron–sulfur clusters in *Azotobacter* ferredoxin at 2.5 A resolution, *J. Biol. Chem.* 255:1797–1800.

Stroud, R. M., 1974. A family of protein-cutting proteins, *Sci. Am.* 1974(7):74–88.

Stynes, H. C., and Ibers, J. A., 1971. Effect of metal–ligand bond distances on rates of electron transfer reactions: The crystal structures of hexaamine ruthenium(II) iodide, $Ru(NH_3)_6I_2$ and hexaamine ruthenium(III) tetrafluoroborate, $Ru(NH_3)_6(BF_4)_3$, *Inorg. Chem.* 10:2304–2308.

Sugiura, Y., and Nomoto, K., 1984. Phytosiderophores: Structures and properties of mugineic acids and their metal complexes, *Struct. Bonding* 58:107–135.

Sugiura, Y., Kawabe, H., Tanaka, H., Fujimoto, S., and Ohara, A., 1981. Purification, enzymatic properties and active site environment of a novel Mn(III)-containing acid phosphatase, *J. Biol. Chem.* 256:10664–10670.

Summers, A. O., and Sugarman, L. I., 1974. Cell-free mercury(II)-reducing activity in a plasmid-bearing strain of *Escherichia coli*, *J. Bacteriol.* 119:242–249.

Sundaralingam, M., Bergstrom, R., Strasburg, G., Rao, S. T., Roychowdhury, P. Greser,

M., and Wang, B. C., 1985. Molecular structure of troponin-C from chicken skeletal muscle at 3 Å resolution, *Science* 227:945–948.

Sunderman, R. W. (ed.), 1975. *Nickel*, National Academy of Sciences, Washington, D.C.

Szebenyi, D. M. E., Obendorf, S. K., and Moffat, K., 1981. Structure of vitamin D-dependent calcium-binding protein from bovine intestine, *Nature* 294:327–332.

Tainer, J. A., Getzoff, E. D., Richardson, J. S., and Richardson, D. C., 1983. Structure and mechanism of copper, zinc superoxide dismutase, *Nature* 306:284–287.

Takai, Y., Kishimoto, A., and Nishizuka, Y., 1982. Calcium, phospholipids and protein kinase, in *Calcium and Cell Function* (W. Y. Cheung, ed.), Vol. II, Academic Press, New York, pp. 385–412.

Tan, S. L., Fox, J. A., Kojima, N., Walsh, C. T., and Orme-Johnson, W. H., 1984. Nickel coordination in deazaflavin and viologen-reducing hydrogenases from *Methanobacterium thermoautotrophicum*: Investigation by electron spin echo spectroscopy, *J. Am. Chem. Soc.* 106:3064–3066.

Tashian, R. E., and Hewett-Emmett, D. (eds.), 1984. *Biology and Chemistry of the Carbonic Anhydrase*, Ann. N.Y. Acad. Sci. 429.

Teeter, M. M., Quigley, G. J., and Rich, A., 1981. The binding of metals to t-RNA, in *Metal Ions in Genetic Information Transfer* (G. L. Eichhorn and L. G. Marzilli, eds.), Elsevier, Amsterdam, pp. 233–277.

Teixeira, M., Moura, I., Xavier, A. V., Der Vartanian, D. V., LeGall, J., Peck, H. D., Jr., Huynh, B. H., and Moura, J. J., 1983. *Desulfovibrio gigas* hydrogenase: Redox properties of the nickel and iron–sulfur centers, *Eur. J. Biochem.* 130:481–484.

Teixeira, M., Moura, I., Xavier, A. V., Huynh, B. H., Der Vartanian, D. V., Peck, H. D., Jr., LeGall, J., and Moura, J. J. G., 1985. Electron paramagnetic resonance studies on the mechanism of activation and the catalytic cycle of the nickel-containing hydrogenase from *Desulfovibrio gigas*, *J. Biol. Chem.* 260:8942–8950.

Tepper, L. B., 1972. Beryllium, *Critical Rev. Toxicol.*, 1:261–281.

Termine, J. D., Kleinman, H. K., Whitson, S. W., Conn, K. M., McGarvey, M. L., and Martin, G. R., 1981. Osteonectin, a bone-specific protein linking mineral to collagen, *Cell* 26:99–105.

Thauer, R. K., 1983. Die Neuen Nickelenzyme aus Anaeroben Bakterien, *Naturforscher* 70:60–64.

Thauer, R. K., Diekert, G., and Schonheit, P., 1980. Biological role of nickel, *Trends Biochem. Sci.* 5:304–306.

Thelander, L., and Reichard, P., 1979. Reduction of ribonucleotides, *Annu. Rev. Biochem.* 48:133–158.

Thomson, A. B. R., Olatunbosun, D. A., and Valberg, L. S., 1971a. Interrelationship of intestinal transport system for manganese and iron, *J. Lab. Clin. Invest.* 78:642–655.

Thomson, A. B. R., Valberg, L. S., and Sinclair, D. G., 1971b. Competitive nature of the intestinal transport mechanism for cobalt and iron in the rat, *J. Clin. Invest.* 50:2384–2394.

Tiffin, L. O., 1967. Translocation of manganese, iron, cobalt and zinc in tomato, *Plant Physiol.* 42:1427–1432.

Tornabene, T. G., and Edwards, H. W., 1972. Microbial uptake of lead, *Science* 176: 1334–1335.

Triggle, D. T., 1972. Calcium and neurotransmitter action, *Prog. Surf. Membr. Sci.* 5: 267–331.

Tsibris, J. C. M., Namtvedt, M. J., and Gunsalus, I. C., 1968. Selenium as an acid labile sulfur replacement in putidaredoxin, *Biochem. Biophys. Res. Commun.* 30:320–327.

Tsien, R. W., 1983. Calcium channels in excitable cell membrane, *Annu. Rev. Physiol.* 45: 341–358.

Tullius, T. D., Gillan, W. O., Carlson, R. M. K., and Hodgson, K. O., 1980. Structural study of the vanadium complex in living ascidian blood cells by x-ray absorption spectroscopy, *J. Am. Chem. Soc.* 102:5670–5676.

Tynecka, Z., Gos, Z., and Zajac, J., 1981. Energy-dependent efflux of cadmium coded by a plasmid resistance determinant in *Staphylococcus aureus*, *J. Bacteriol.* 147:313–319.

Udenfriend, S., and Cardinale, G., 1982. α-Ketoglutarate coupled dioxygenase, in *Oxygenases and Oxygen Metabolism* (M. Nozaki, M. Nozaki, S. Yamamoto, Y. Ishimura, M. Coon, L. Ernster and R. W. Estabrook, eds.), Academic Press, New York, pp. 99–107.

Ulbricht, W., 1977. Selective blocking of channels, in *Electrical Phenomena at the Biological Membrane Level* (E. Roux, ed.), Elsevier, Amsterdam, pp. 203–217.

Ullrich, V., and Duppel, W., 1975. Iron- and copper-containing monooxygenases, in *The Enzymes*, 3rd ed. (P. D. Boyer, ed.), Vol. XII, Academic Press, New York, pp. 253–297.

Ullrich, V., Roots, I., Hildebrandt, A., Estabrook, R., and Conney, A. H. (eds.), 1977. *Microsomes and Drug Oxidation*, Pergamon Press, Elmsford, N.Y.

Underwood, E. J., 1977. *Trace Elements in Human and Animal Nutrition*, 4th ed., Academic Press, New York.

Urushizaki, I., Aisen, P., Listowski, I., and Drysdale, J. W. (eds.), 1983. *Structure and Function of Iron Storage and Transport Proteins*, Elsevier, Amsterdam.

Vallee, B. L., 1977. Recent advances in zinc biochemistry, in *Biological Aspects of Inorganic Chemistry* (A. W. Addison, Cullen, R. W.,Dolphin, D. and James, B. R., eds.), Wiley–Interscience, New York, pp. 37–70.

Vallee, B. L., and Galdes, A., 1984. The metallobiochemistry of zinc enzymes, *Adv. Enzymol. Relat. Areas Mol. Biol.* 56:283–430.

Vallee, B. L., and Ullmer, D. D., 1972. Effects of mercury, cadmium and lead, *Annu. Rev. Biochem.* 41:61–128.

Vallee, B. L., and Williams, R. J. P., 1968. Metalloenzymes: The entatic nature of their active site, *Proc. Natl. Acad. Sci. USA* 59:498–505..

Valentine, J. S., 1973. The dioxygen ligand in mononuclear group VIII transition metal complexes, *Chem. Rev.* 73:235–245.

Valentine, J. S., and Pantoliano, M. W., 1981. Protein–metal ion interactions in cuprozinc protein (superoxide dismutase), in *Copper Protein* (T. G. Spiro, ed.), Wiley, New York, pp. 291–358.

Veatch, W. R., and Blout, E. R., 1974. The aggregation of gramicidin A in solution, *Biochemistry* 13:5257–5264.

Venugopal, B., and Luckey, T. D., 1977. *Metal Toxicity in Mammals*, Vol. 2, Plenum Press, New York, pp. 30–31.

Vidal, G., 1984. The oldest eukaryotic cells, *Sci. Am.* 1984(2):48–57.

Villafranca, J. J., 1981. Dopamine-β-hydroxylase, in *Copper Proteins* (T. G. Spiro, ed.), Wiley, New York, pp. 263–289.

Volcani, B. E., 1981. Cell wall formation in diatoms: Morphogenesis and biochemistry, in *Silicon and Siliceous Structures in Biological Systems* (T. L. Simpson and B. E. Volcani, eds.), Springer-Verlag, Berlin, pp. 157–200.

Volpe, P., Slaviati, G., Di Virgilio, F., and Pozzan, T., 1985. Inositol 1,4,5-triphosphate induces calcium release from sarcoplasmic reticulum of skeletal muscle, *Nature* 316: 347–349.

Wacker, W. E. C., and Vallee, B. L., 1959. Nucleic acids and metals, *J. Biol. Chem.* 234: 3257–3262.

Walsh, M. P., and Hartshore, D. J., 1982. Actomyosin of smooth muscle, in *Calcium and Cell Function* (W. Y. Cheung, ed.), Vol. III, Academic Press, New York, pp. 223–269.

Walsh, T. A., and Ballou, D. P., 1983. Halogenated protocatechuates as substrates for protocatechuate dioxygenase from *Pseudomonas cepacia, J. Biol. Chem.* 258:14413–14421.

Walsh, T. A., Ballou, D. P., Mayer, R., and Que, L., Jr., 1983. Rapid reaction studies on the oxygenation reactions of catechol dioxygenase, *J. Biol. Chem.* 258:14422–14427.

Walter, M. R., 1983. Archean stromatolites: Evidence of earth's earliest benthos, in *Earth's Earliest Biosphere—Its Origin and Evolution* (J. W. Schopf, ed.), Princeton University Press, Princeton, N.J., pp. 187–213.

Waters, A. H., and Mollin, D. L., 1971. Vitamin B_{12}, in *Hematopoietic Agents I, Hematic Agents* (J. C. Dreyfus, ed.), Pergamon Press, Elmsford, N.Y., pp. 1–70.

Weinberg, E. D. (ed.), 1977. *Microorganisms and Minerals,* Dekker, New York.

Weiner, S., 1979. Aspartic acid rich proteins: Major components of the soluble organic matrix of mollusc shells, *Calcif. Tissue Invest.* 29:163–167.

Weiner, S., and Hood, L., 1975. Soluble protein of the organic matrix of mollusc shells: A potential template for shell formation, *Science* 190:987–988.

Weller, M. G., and Weser, U., 1982. Ferric nitriloacetate: An active-center analogue of pyrocatechase, *J. Am. Chem. Soc.* 104:3752–3754.

Wetzel, R. G., 1975. *Limnology,* Saunders, Philadelphia, p. 265.

Wetherill, G. W., 1981. The formation of the earth from planetesimals, *Sci. Am.* 1981(6): 163–174.

White, J. M., and Harvey, D. R., 1972. Defective synthesis of globin chains in lead poisoning, *Nature* 236:71–73.

White, R. E., and Coon, M. J., 1982. Heme ligand replacement reactions of cytochrome P-450: Characterization of the bonding atom of the axial ligand trans to thiolate as oxygen, *J. Biol. Chem.* 257:3073–3083.

Whitton, B. A., and Say, P. J., 1975. Heavy metals, in *River Ecology* (B. A. Whitton, ed.), Blackwell, Oxford, pp. 286–311.

Wiklung, P. A., and Brown, D. C., 1976. Synthesis and characterization of some cobalt(III) catechol complexes, *Inorg. Chem.* 15:396–400.

Wiklund, P. A., Beckman, L. S., and Brown, D. G., 1976. Preparation and properties of a stable semiquinone complex, *Inorg. Chem.* 15:1996–1997.

Wilber, C. G., 1969. *The Biological Aspects of Water Pollution,* Thomas, Springfield, Ill., pp. 58–72.

Wilkström, M. K. F., 1977. Proton pump coupled to cytochrome c oxidase in mitochondria, *Nature* 266:271–273.

Wilkström, M., 1982. Proton translocation by cytochrome oxidase, in *Membranes and Transport* (A. N. Martonosi, ed.), Vol. 1, Plenum Press, New York, pp. 357–362.

Williams, R. J. P., and Wentworth, R. A. D., 1973. Molybdenum in enzymes, in *Chemistry and Uses of Molybdenum* (P. C. H. Mitchell, ed.), Climax Molybdenum Co., Ann Arbor, pp. 212–215.

Wilson, D. B., 1978. Cellular transport mechanisms, *Annu. Rev. Biochem.* 47:933–965.

Wilson, R. C. H., 1972. Prediction of copper toxicity in receiving waters, *J. Fish. Res. Board Can.* 29:1500–1502.

Winge, D. R., and Miklosky, K. A., 1982. Domain nature of metallothionein, *J. Biol. Chem.* 257:3471–3476.

Witshi, H. P., 1970. Effects of beryllium on deoxyribonucleic acid-synthesizing enzymes in regenerating rat liver, *Biochem. J.* 120:623–634.

Woese, C. R., 1981. Archeabacteria, *Sci. Am.* 1981(6):98–122.

Wolff, S. P., Garner, A., and Dean, R. T., 1986. Free radicals, lipids and protein degradation, *Trends in Biochem Sci.* 11:27–31.

Wolff, T. E., Berg, J. M., Hodgson, K. O. H., Holm, R. H., and Frankel, R. B., 1978. The molybdenum–iron–sulfur cluster complex $[Mo_2Fe_2S_6(SC_2H_5)_8]^{3-}$, a synthetic approach to the molybdenum site in nitrogenase, *J. Am. Chem. Soc.* 100:4630–4632.

Wood, H. G., and Barden, R. E., 1977. Biotin enzymes, *Annu. Rev. Biochem.* 46:385–413.

Wood, H. G., and Zwolinski, G. K., 1976. Transcarboxylase, *Crit. Rev. Biochem.* 4:47–122.

Woodwell, G. M., 1970. The energy cycle of the biosphere, *Sci. Am.* 1970(9):64–74.

Worwood, M., 1974. Iron and the trace metals, in *Iron in Biochemistry and Medicine* (A. Jacobs and M. Worwood, eds.), Academic Press, New York, pp. 335–367.

Wu, F. Y.-H., and Wu, C.-W., 1981. Intrinsic metals in RNA polymerases, in *Metal Ions in Genetic Information Transfer* (G. L. Eichhorn and L. G. Marzilli, eds.), Elsevier, Amsterdam, pp. 143–166.

Yadav, K. D. S., and Knowles, P. F., 1981. A catalytic mechanism for benzylamine oxidase from pig plasma-stopped flow kinetic studies, *Eur. J. Biochem.* 114:139–144.

Yamamoto, I., Saili, T., Liu, S.-M. and Ljungdahl, L. G., 1983. Purification of NADP-dependent formate dehydrogenase from *C. thermoaceticum*, a tungsten selenium-iron protein, *J. Biol. Chem.* 258:1826–1832.

Yamanaka, T., and Okunuki, K., 1974. Cytochromes, in *Microbial Iron Metabolism* (J. B. Neilands, ed.), Academic Press, New York, pp. 349–400.

Yamazaki, S., 1982. A selenium-containing hydrogenase from *Methanococcus vannielii*, *J. Biol. Chem.* 257:7926–7929.

Yariv, J., Smith, J. M. A., White J. L., Treffry, A., Rice, D. W., Ford, G. C., Harrison, P. M., Williams, J. M., Watt, G. D., and Stiefel, E. I., 1983. Bacterioferritin from *E. coli* and *A. vinelandii*, in *Structure and Function of Iron Storage and Transport Proteins* (I. Urushizaki, P. Aisen, I. Listowski, and J. W. Drysdale, eds.), Amsterdam, pp. 69–70.

Youngman, R. J., 1984. Oxygen activation: Is the hydroxyl radical always biologically relevant? *Trends Biochem. Sci.* 9:280–282.

Zeppezauer, M., 1982. Coordination properties and mechanistic aspects of liver alcohol dehydrogenase, in *The Coordination Chemistry of Metalloenzymes* (I. Bertini, R. S. Drago, and C. Luchinat, eds.), Reidel, Dordrecht, pp. 99–122.

Zumft, W. G., Mortenson, L. E., and Palmer, G., 1974. Electron paramagnetic studies on nitrogenase, *Eur. J. Biochem.* 46:525–535.

Zurer, P. S., 1983. The chemistry of vision, *Chem. Eng. News* 1983 (Nov. 28):24–35.

Index

Printed in the United States
By Bookmasters